P9-EFG-011

Lignin: Historical, Biological, and Materials Perspectives

ACS SYMPOSIUM SERIES **742**

Lignin: Historical, Biological, and Materials Perspectives

Wolfgang G. Glasser, EDITOR
Virginia Polytechnic Institute and State University

Robert A. Northey, EDITOR
University of Washington

Tor P. Schultz, EDITOR
Mississippi State University

American Chemical Society, Washington, DC

Library of Congress Cataloging-in-Publication Data

Lignin : historical, biological, and materials perspectives / Wolfgang G. Glasser, Robert A. Northey, Tor P. Schultz [editors]
p. cm.—(ACS symposium series : 742)

Includes bibliographical references and indexes.

ISBN 0-8412-3611-9

1. Lignin. I. Northey, Robert A., 1957– . II. Glasser, Wolfgang G., 1941– . III. Schultz, Tor P., 1953– . IV. Series.

QK898.L5L545 1999
572′.56682—dc21

99-16985
CIP

The paper used in this publication meets the minimum requirements of American National Standard for Information Sciences—Permanence of Paper for Printer Library Materials, ANSI Z39.48-94 1984.

Distributed by Oxford University Press

PRINTED IN THE UNITED STATES OF AMERICA

Advisory Board

ACS Symposium Series

Foreword

THE ACS SYMPOSIUM SERIES was first published in 1974 to provide a mechanism for publishing symposia quickly in book form. The purpose of the series is to publish timely, comprehensive books developed from ACS sponsored symposia based on current scientific research. Occasionally, books are developed from symposia sponsored by other organizations when the topic is of keen interest to the chemistry audience.

Before agreeing to publish a book, the proposed table of contents is reviewed for appropriate and comprehensive coverage and for interest to the audience. Some papers may be excluded in order to better focus the book; others may be added to provide comprehensiveness. When appropriate, overview or introductory chapters are added. Drafts of chapters are peer-reviewed prior to final acceptance or rejection, and manuscripts are prepared in camera-ready format.

As a rule, only original research papers and original review papers are included in the volumes. Verbatim reproductions of previously published papers are not accepted.

ACS BOOKS DEPARTMENT

Dedication

Joseph L. McCarthy
1997 Anselme Payne Award Recipient

Joseph McCarthy has made significant contributions on a wide range of topics dealing with the chemistry and processing of wood, with an emphasis on by-product utilization. Ever since his dissertation, "The Mechanism of Chlorination of Lignin", was patented and immediately put into commercial practice (at a Howard Smith Papermill near Montreal) as the basis for bleaching of kraft pulp, Joe's research contributions have met the requirements of practical relevance while providing excellent science. For example, during Joe's first days at the University of Washington, the Puget Sound Pulp and Timber Plant sulfite mill in Bellingham, Washington, came into conflict with effluent discharge regulations. McCarthy's research on both ethanol production from spent liquor and lignin sulfonates quickly became the basis for the Bellingham plant's (now owned by Georgia-Pacific) commercial leadership in the lignin-sulfonate markets. A "Pulp Mills Research" (PMR) program was started at the University of Washington with support from industry in response to the concerns of fishermen and oyster growers in the Puget Sound with the discharge of spent sulfite liquor from pulp mills. During a 13 year period, 1944–1957, this PMR program became the cornerstone of McCarthy's initial contribution to the science and engineering of chemical wood processes by focusing on kraft mill odor reduction and spent sulfite liquor utilization. Today, sulfite mills in the Pacific Northwest no longer discharge spent sulfite liquor. Other research on the industrial practice of pulp and papermaking has dealt with different areas including the thermodynamics of sulfur dioxide solutions, the composition and analysis of spent sulfite liquors, the recovery of heat and chemicals from spent pulping liquors, the thermodynamics of combustion of various spent sulfite liquor bases, sugar fermentation to ethanol, the structure and reaction of lignin sulfonates, the purification of kraft mill effluents by steam stripping, and the recovery of by-products of spent kraft black liquors by ultrafiltration.

Nearly 50 graduate students have graduated from McCarthy's laboratory and more than 25 postdoctoral research associates have been guided by him into fruitful careers. His work has been a living example of the pleasures of excelling at one's actions and his motto—"Don't do it unless it's fun"—serves as an inspiration for all scientists.

Contents

ANALYSIS

MODIFICATION AND UTILIZATION

PULPING AND BLEACHING

Preface

Lignin, the second most abundant organic substance on earth, is one of the three structural polymers present in all woody plants. Lignin influences our world in many different ways. If it had not been for lignin, plants would never have moved from the aquatic to the terrestrial environment during evolution. It is lignin that "stiffens" the plant stem to withstand the forces of gravity and wind. Lignin seals the water conducting system against the hydraulic pressure drop produced by the transport of water from the soil to the leaves and needles. Although lignin provides plants with a protective barrier against the attack by

multifaceted task of understanding lignin's formation, structure, and its reactions and behaviors during pulping, bleaching, and modification continues to attract the attention of scientists in all disciplines. Because these scientists have diverse scientific, engineering, and business interests, this research creates opportunities for broad discussions on many diverse issues dealing with lignin.

This book was developed from a symposium by the American Chemical Society's Cellulose, Paper, and Textile Division honoring the 1997 Anselme Payen award recipient, Joseph L. McCarthy, held at the 215th National ACS Spring meeting in Dallas, Texas. The Symposium provided a forum for organizing an integrated discussion of our current knowledge on lignin, and on its historical roots. It brought together international scientists from academia, government, and industry to discuss new insights and commercial developments on lignin. This book is divided into five sections: History and Structure, Biochemistry, Analysis, Modification and Utilization, and Pulping and Bleaching. Each section has at least one overview chapter, a comprehensive review written by one or several recognized experts that is followed by chapters dealing with current research findings.

The first section, History and Structure, provides an authoritative perspective on more than 50 years of scientific and engineering endeavors dealing with lignin structure, response to pulping and bleaching, and modification and utilization. The chapters dealing with biochemistry display the current understanding of biochemical pathways leading to precursor formation and the enzymes involved in the production of a cell-binding polymer of considerable nonuniformity. The chapters summarizing analytical advances address both chemical and molecular identities. The section on modification and utilization defines current practice and reviews recent advances in polymeric structure modification and utilization. The wide field of lignin reactions during pulping and bleaching are highlighted in the final part.

It is the hope of the editors that this book will provide practicing scientists with new ideas for studying or utilizing lignin, and also to furnish graduate students with a resource by which they can explore this yet under-understood biopolymer.

Acknowledgments

The editors express their appreciation to the ACS Cellulose, Paper, and Textile Division for sponsoring this symposium. Thanks also are given to the Georgia-Pacific Corporation, Westvaco, Weyerhaeuser, and the academic institutions of the editors for their financial and editorial support. The editors also wish to thank Anne Wilson and Kelly Dennis of the ACS Books Department for their competent and professional support and encouragement. Finally, the editors acknowledge the time and effort, which many of our professional friends contributed through the process of reviewing, critiquing, and otherwise improving this most recent contribution to the ACS Symposium Series.

WOLFGANG G. GLASSER
Department of Wood Science and Forest Products
Virginia Polytechnic Institute and State University
Blacksburg, VA 24061–0324

ROBERT A. NORTHEY
Paper Science and Engineering Division
College of Forest Resources
University of Washington
Seattle, WA 98195

TOR P. SCHULTZ
Forest Products Laboratory
Forest and Wildlife Research Center
Mississippi State University
Mississippi State, MS 39762–9820

HISTORY AND STRUCTURE

Chapter 1

Lignin Chemistry, Technology, and Utilization: A Brief History

Joseph L. McCarthy and Aminul Islam

Department of Chemical Engineering and College of Forest Resources, University of Washington, Seattle, WA 98195

In 1838, Anselme Payen discovered that treatment of wood with nitric acid and then an alkaline solution yielded a major insoluble residue that he called "cellulose," and dissolved incrustants which Schulze later designated "lignin." Following Payen's initial experiments were findings on how to selectively solubilize lignins with either sulfurous acid or alkaline solutions to yield useful cellulose fibers. Klason, Freudenberg, Nimz, Adler, Hibbert, Kratzl, Tischenko, Nakano and others conducted pioneering classical organic chemistry research that revealed much fundamental knowledge about the chemical structure of lignins. The emergence of polymer concepts provided an additional framework that advanced understanding of the physical behavior and chemical nature of lignins. Although much is known about the biosynthesis of lignin, new information is still emerging. Lignins in wood are now known to consist of a family of phenyl propane type polymers in a variety of structural units in combination with carbohydrates. Lignins are widely used as fuels in paper making processes and are sold commercially in expanding markets.

Anselme Payen (*1, 2*), a wealthy chemical manufacturer in France, first identified "cellulose" and *le materiel incrustant* or "lignin" in 1838 as separate components of wood "se compose de deux parties chimiquement tres distinguees." This discovery was made about half a century after the French Revolution (1789). Since that time many thousands of scientific papers, patents, and hundreds of books, have been published concerning these two most important natural polymers. To date, about 10,000-12,000 scientific papers have been published (estimate by authors) relating to lignins alone. Of these, over six hundred are incorporated in this compilation and some will be briefly discussed. The impossible task of selecting the "most important" papers and books has been avoided by choosing works that serve only as examples of lignin research, technological applications and utilization activities.

The history of contributions to the lignin field is presented in seven arbitrary time periods and each of these is divided into three main categories, Chemistry, Technology and Books. Contributions are not presented in rigorous chronological order, but are arranged to provide easy reading. To facilitate the use of this history, Appendix 1 presents all cited contributions chronologically although the main text of this paper contains only part of the entries in Appendix 1. In Appendix 2, a list is presented of many of the important research facilities that have made contributions to the understanding of lignin chemistry. The authors deeply regret, and apologize to investigators at centers not reported. Examples of proposed lignin structures are included in Figures 1-9. Table I includes information on the frequency of linkages in these structures. Examples of these linkages are presented in Figure 10.

To assist the inexpert reader, certain currently accepted characteristics of lignins are as follows. The term lignin refers to a family of heterogeneous biopolymers that contain limited branching and/or cross-linking. Lignin is assembled from coniferyl alcohol type monomers by enzymatic polymerization providing three-dimensional molecular architecture. It is amorphous, optically inactive and comprises usually about 20-40% of the mass of plant tissues in higher plants. Lignin functions in several ways. As a composite with cellulose fibers, it provides strength to the stem and the bole of higher plants and thus permits photosynthesis far above ground level. As a matrix of phenolic and aliphatic substances, it protects carbohydrates from attack by microorganisms and insects. It participates in a network of processes by which the organism conserves, recovers and reuses highly significant atoms and small molecules essential for cambial and cytoplasm functioning. For man, it is highly useful as a structural material, as a component of wood that must be removed to make paper, and as a potential phenolic raw material of increased importance as available oil and fuel resources are depleted. About 50 million tons of chemically separated lignins are produced annually throughout the world, of which about 1% is currently sold.

The two main motivations for lignin product research have always been the vast availability of this chemical around the world and that it is an inevitable by-product of the chemical manufacturing processes for cellulose and of pulp.

Life of Anselme Payen. Anselme Payen (*1,2*) was born in January 1795 to Marc and Jean Baptiste Piere Payen, a distinguished scholar in science and law as well as an owner of chemical factories. Young Payen studied chemistry, physics, and mathematics in the Polytechnique in Paris. After his father's death, Payen, at the age of 25, took over full responsibility for the chemical factories. In 1835, he left active participation in industry and became Professor of Industrial and Agricultural Chemistry at the Ecole Centrale des Arts et Manufactures. Four years later, he accepted the additional post of Professor of Applied Chemistry at the Conservatoires des Arts. Both posts he held until his death in 1871. He published over 200 papers in major French scientific journals and ten books on subjects ranging from dextrans, sugars, lignin, cellulose, and starch, to plant and animal biology. In 1838, he published his paper (*1*) describing cellulose and lignin isolation.

1838 – 1874

During this time, the existence of lignin was established. Nearly thirty years later the first industrial process for chemical pulping of wood by dissolving lignin and

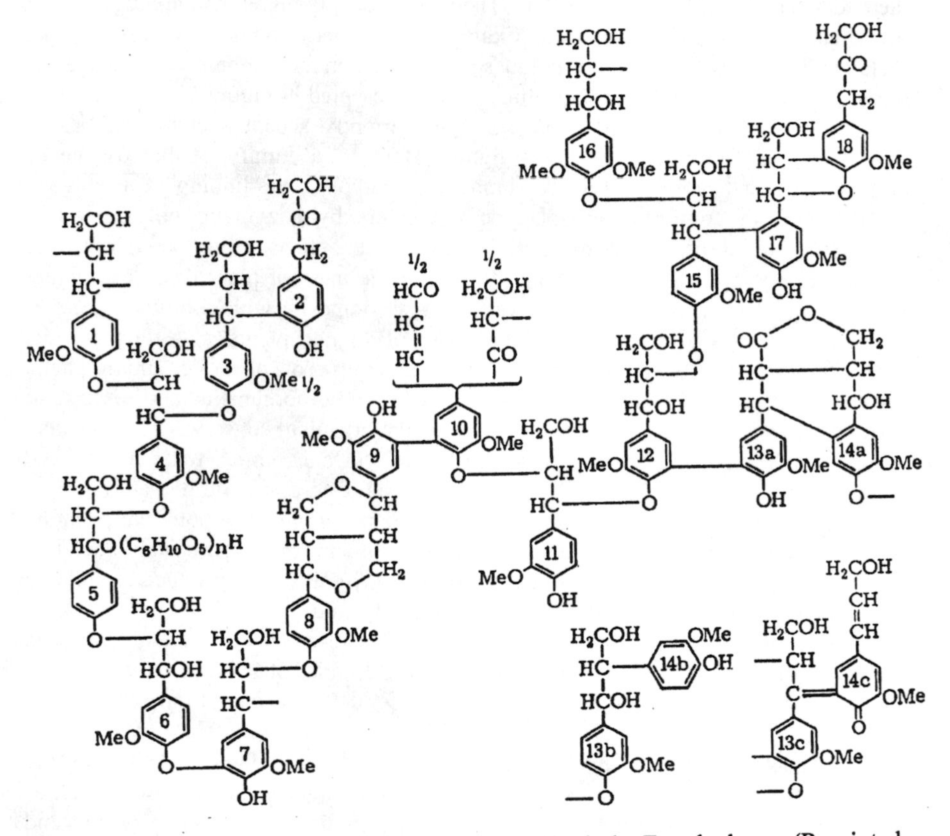

Figure 1. Proposed model structure of Spruce lignin by Freudenberg. (Reprinted with permission from ref. 362. Copyright 1968)

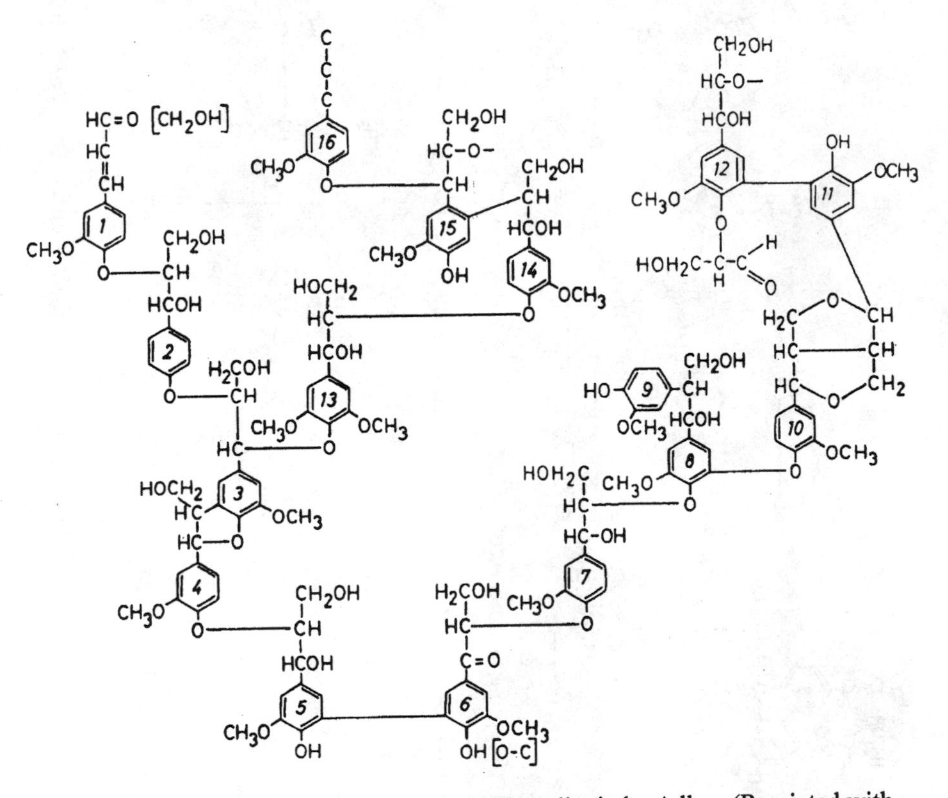

Figure 2. Proposed model structure of Spruce lignin by Adler. (Reprinted with permission from ref. 423. Copyright 1977)

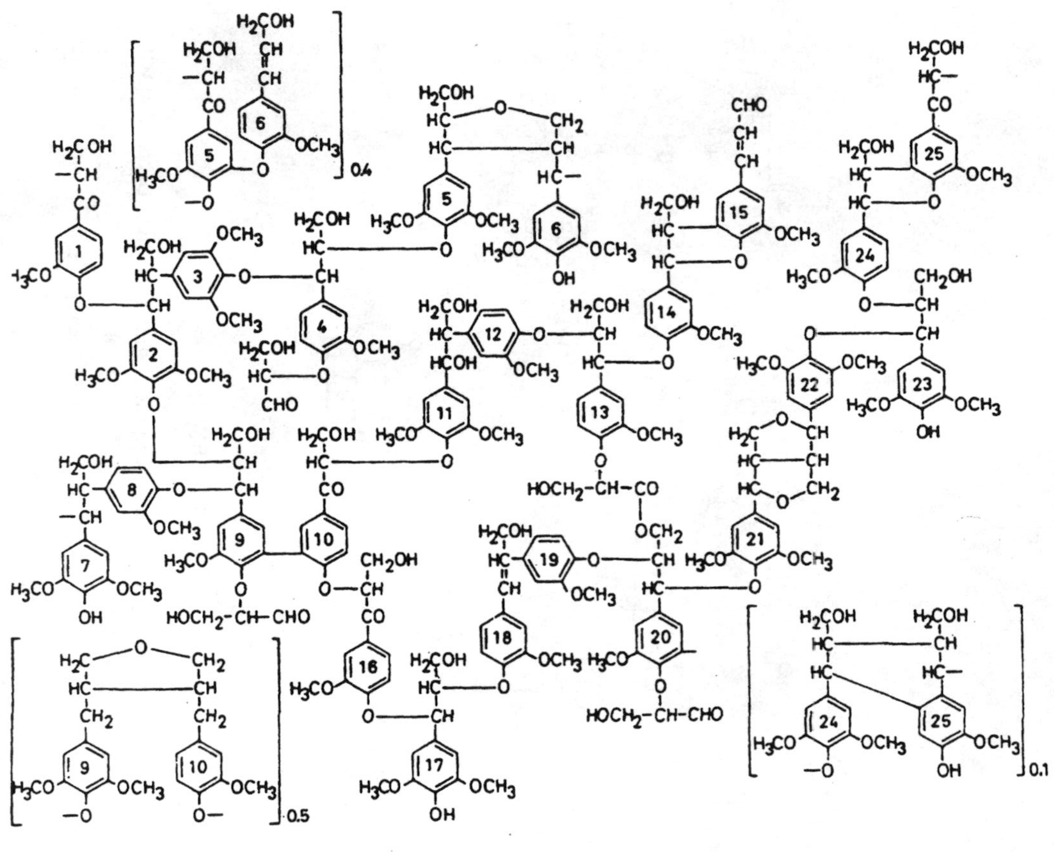

Figure 3. Proposed model structure of Beech lignin by Nimz. (Reprinted with permission from ref. 396. Copyright 1977)

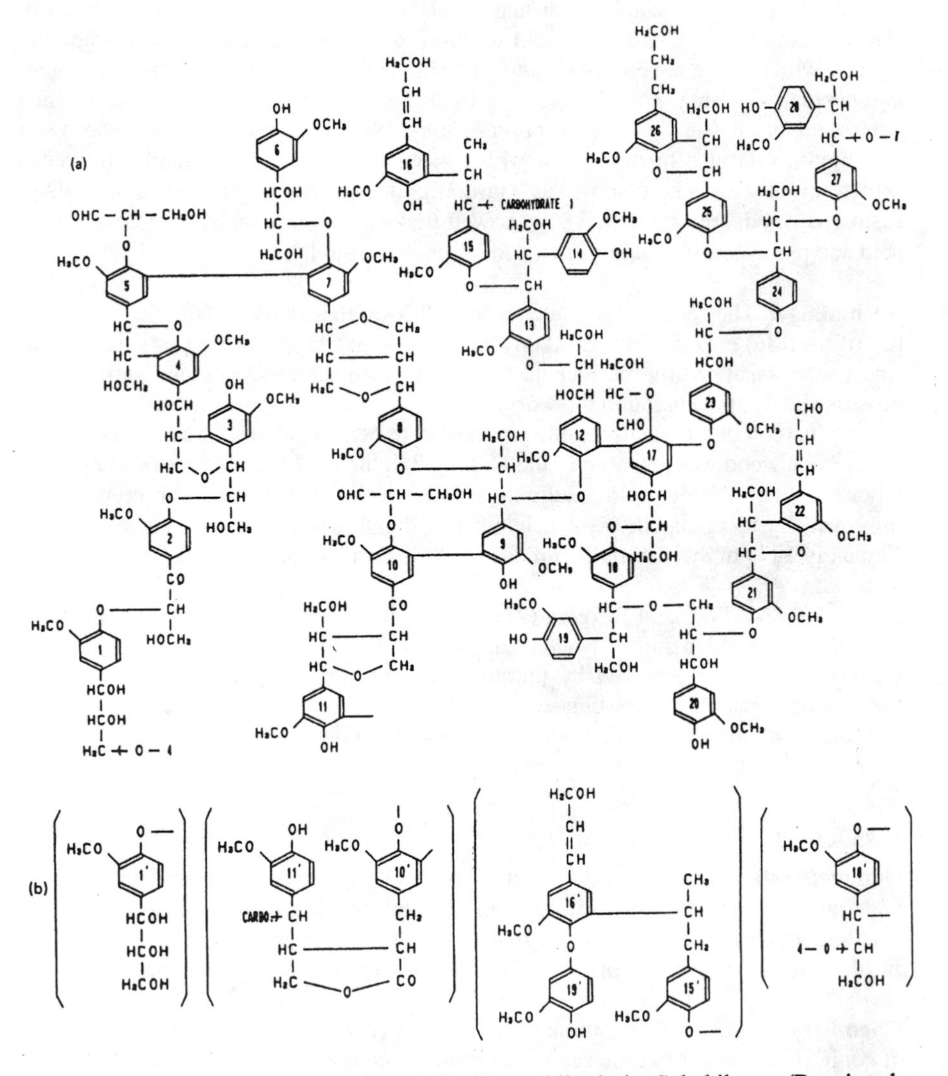

Figure 4. Proposed model structure of softwood lignin by Sakakibara. (Reprinted with permission from ref. 447. Copyright 1980)

producing useful cellulose fibers was patented. Research on lignin gave indications of its aromatic nature.

Chemistry. Payen's 1838 experiments (*1, 2*) revealed that wood contained a "cellulose" and an oxidizable incrustant which Schulze designated lignin in 1857 (*8*). Payen used a concentrated nitric acid solution to oxidize wood and then washed the residue with a dilute aqueous sodium hydroxide solution to dissolve lignin and other components. Using von Liebig's method for carbon, hydrogen, and oxygen determination, Payen was able to observe substantial differences between the wood and lignin. He also treated wood with concentrated sulfuric acid and observed a brown residue that is now known as Klason lignin. Possoz in 1858 (*9*) applied alkali fusion to wood. Erdmann in 1866 showed that this process yielded protocatechuic acid and pyrocatechol that he concluded originated from lignin.

Technology. The chemical production of cellulose fibers began 1600 years after the discovery (*4-6*) of paper in 105 AD by Ts' Ai Lun in China. During this time, plant fibers were separated by mechanical means and then screened to make paper. This process slowly spread around the world.

In 1866, the first successful process for the chemical production of cellulosic fibers from wood was patented in the U.S. by Tilghman (*11*). In this sulfite process, a hot aqueous sulfurous acid solution was used to dissolve lignin. The process was improved by using bisulfites of calcium or other bases to create a buffer system. Ekman in 1874 in Sweden built and operated the first sulfite pulp mill (*14*) using this technology.

In 1854, Watt and Burgess (*7*) proposed but did not commercialize a system to delignify wood using a NaOH solution at elevated temperatures. Chemical recovery was to be achieved by burning the spent liquor solids to obtain sodium carbonate that would be causticized with CaO. In 1870, Eaton (*13*) proposed the use of sodium sulfate instead of carbonate for makeup although this system was also not commercialized.

1875 – 1899

During this time, early studies on the nature of lignin and lignosulfonates were undertaken. Coniferin was isolated from cambial sap. Klason suggested that lignin was structurally related to coniferyl aldehyde based on elemental analysis. The sulfite pulping process was in operation and the kraft process had been patented.

Chemistry. In 1875, Tiemann and Mendelsohn (*15*) isolated coniferin, a glycoside of coniferyl alcohol, from the sap of the growing cambial layer. Their determination of the structure of this compound gave early insight into the composition of lignin. Wiesner in 1878 reported the classic color reaction of lignin with phloroglucinol (*18*).

At the Royal Institute of Technology in Stockholm (KTH), Peter Johan Klason (Gellerstedt, KTH, personal communication, 1998) pioneered Swedish lignin research. He taught at Lund from 1874 to 1887 as a Chemist and Docent (Assistant Professor) and at KTH as a "Laborator" from 1887 to 1890 and as a Professor of chemistry and chemical technology from 1890 to 1913. From 1915 to 1923 he

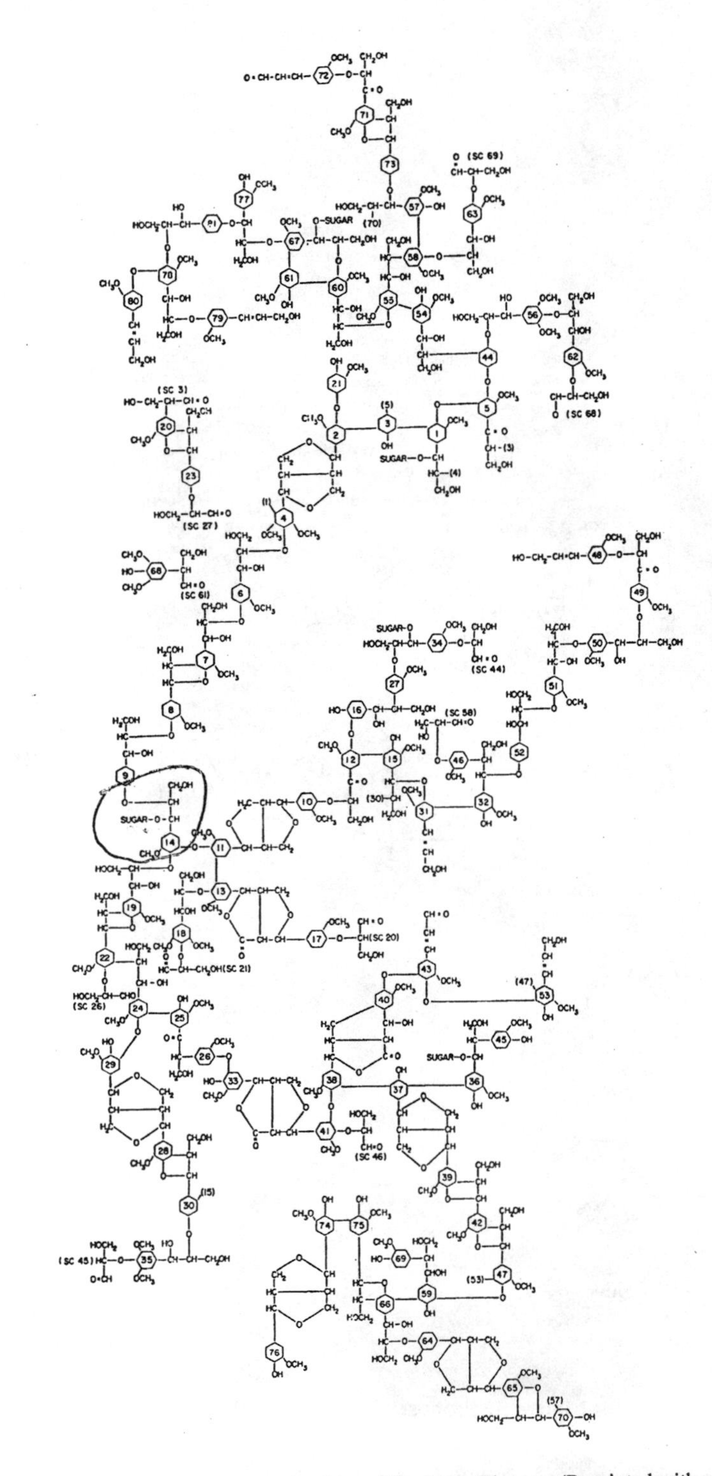

Figure 5. Proposed model structure of softwood lignin by Glasser. (Reprinted with permission from ref. 399. Copyright 1974) Larger copies of Figure 5 can be found on the next two pages.

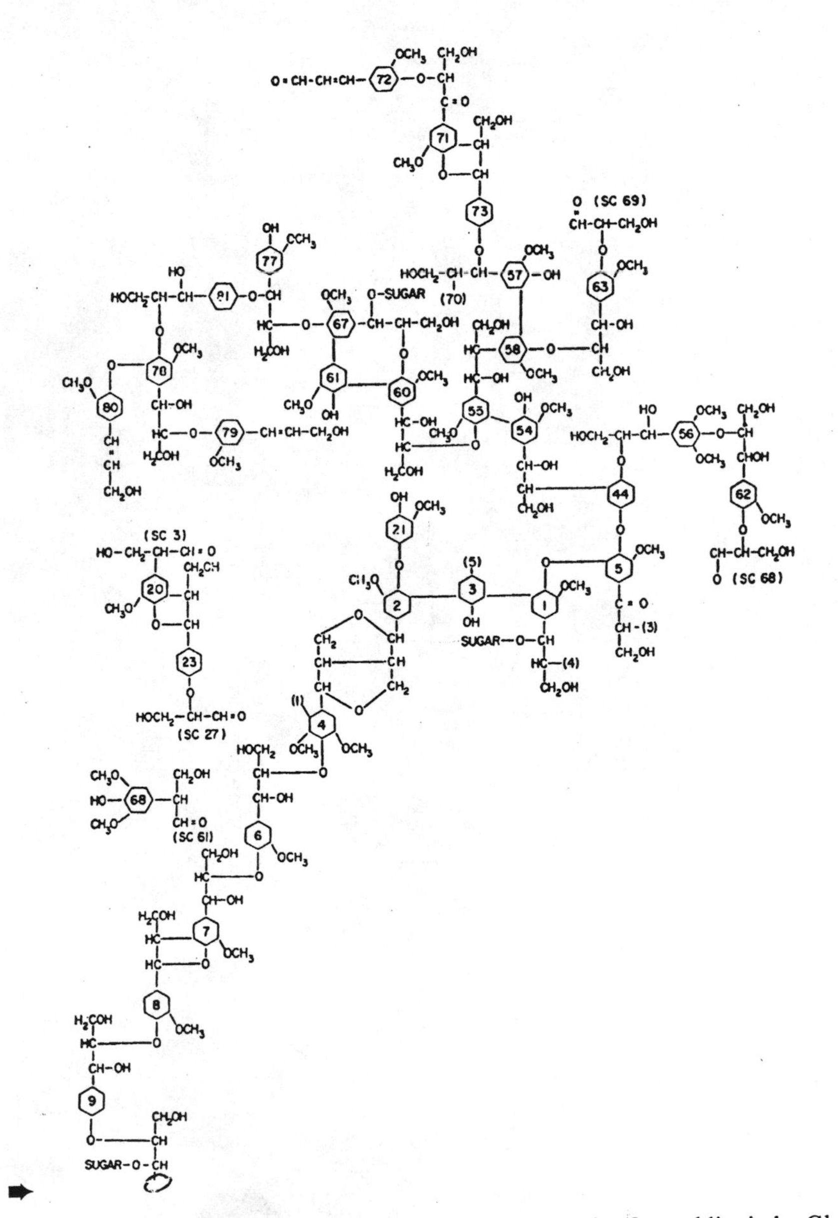

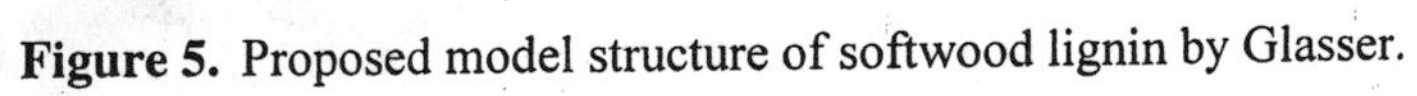

Figure 5. Proposed model structure of softwood lignin by Glasser.

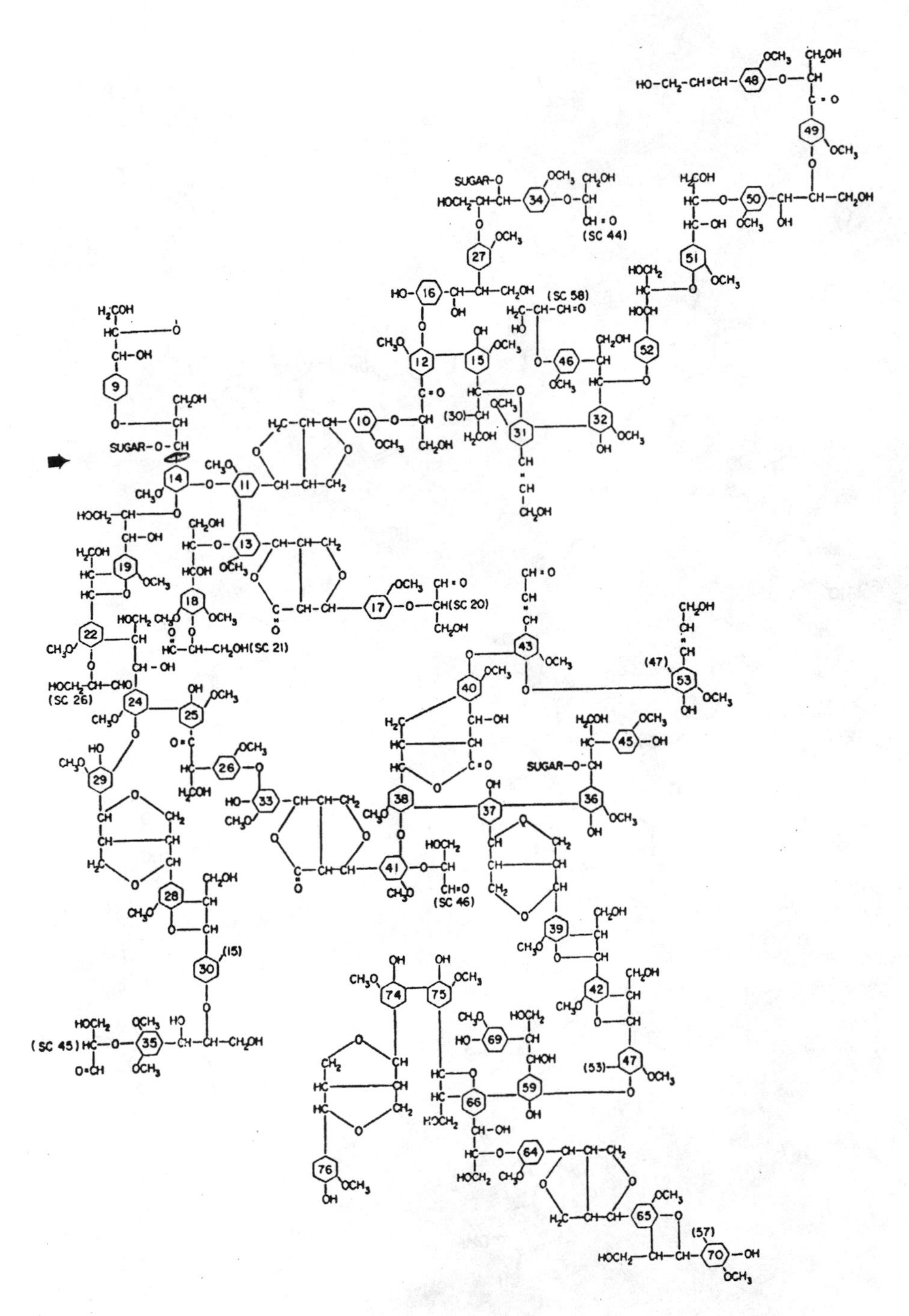

(Reprinted with permission from ref. 399. Copyright 1974)

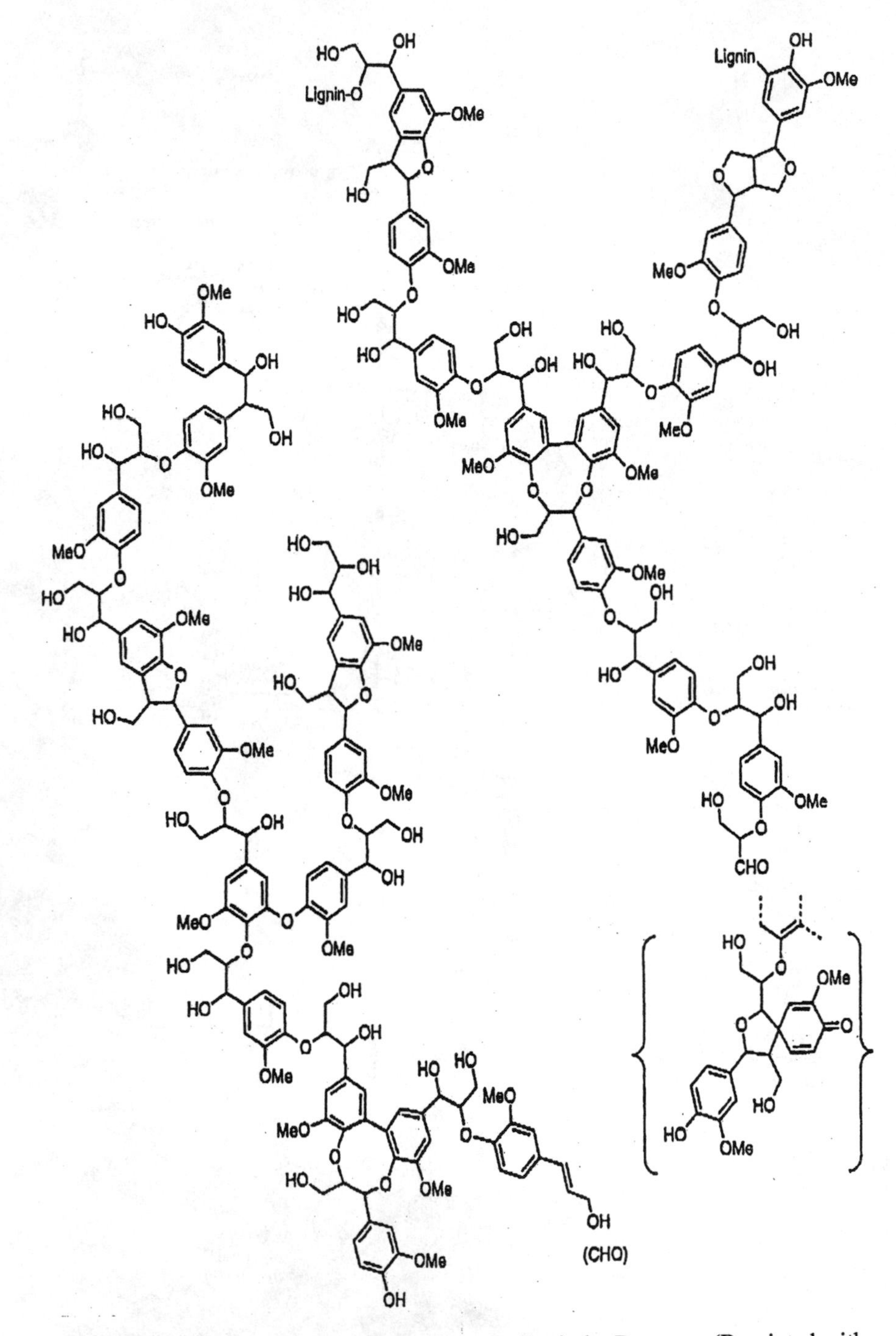

Figure 6. Proposed model structure of Spruce lignin by Brunow. (Reprinted with permission from ref. 687. Copyright 1998)

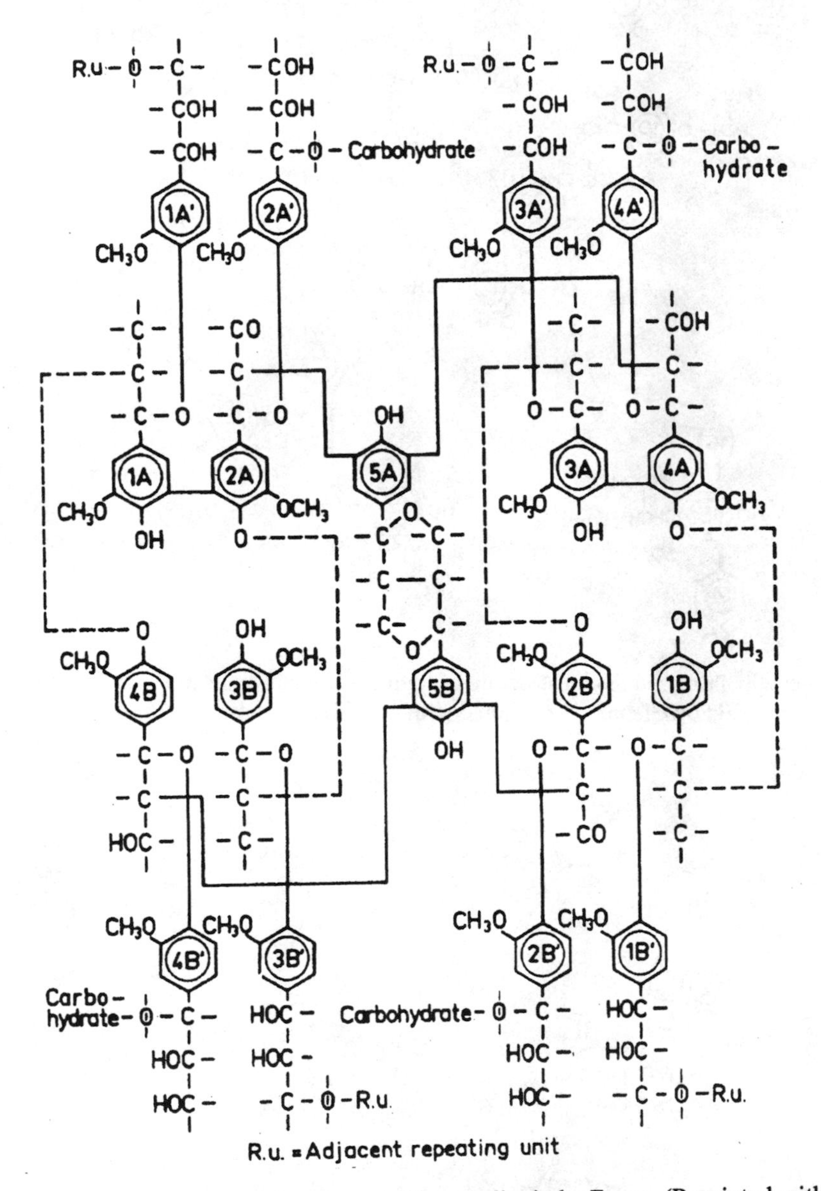

Figure 7. Proposed model structure of Spruce lignin by Forss. (Reprinted with permission from ref. 658. Copyright 1981)

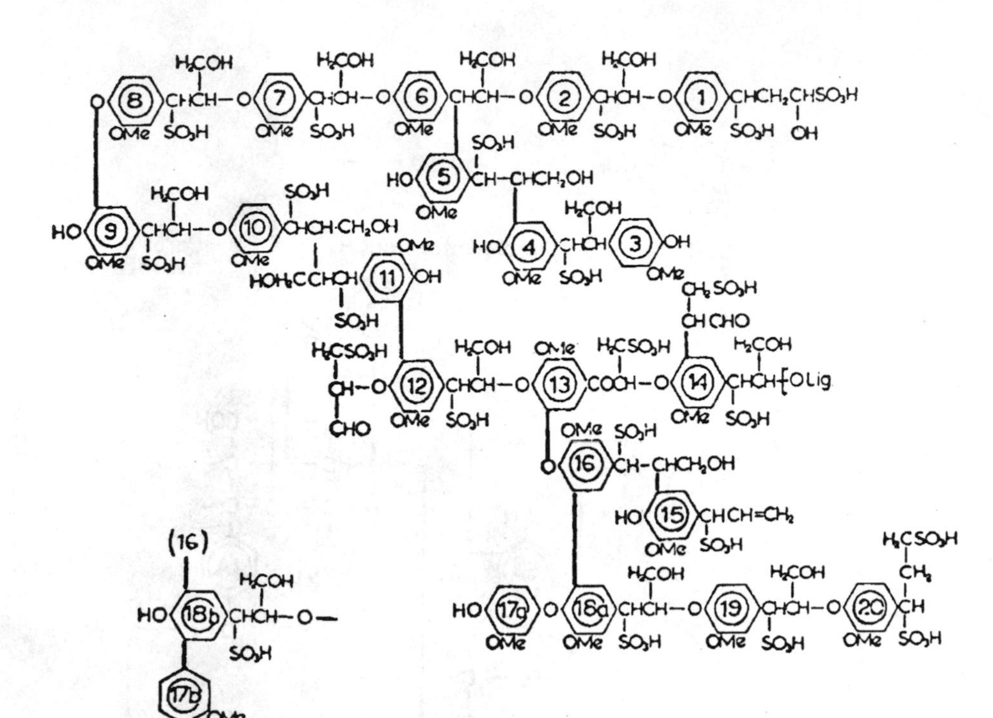

Figure 8. Proposed model structure of lignin sulfonate, extensively sulfonated lignin sulfonic acid by Glennie. (Reprinted with permission from ref. 378. Copyright 1977)

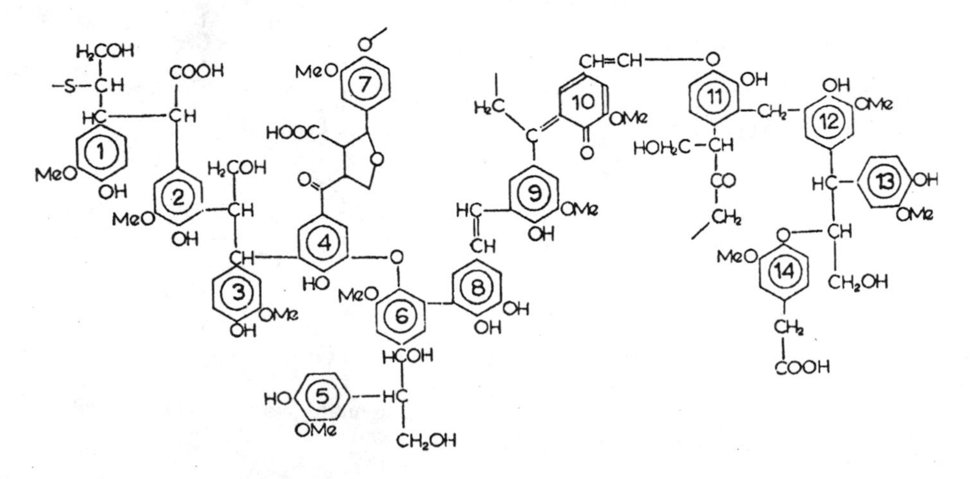

Figure 9. Proposed model structure of Pine Kraft lignin by Marton. (Reprinted with permission from ref. 385. Copyright 1971)

Table I. Frequency of Linkages and Functional Groups per 100 C_9 Units in Different Lignin Formulations

Author	Reference	Figure	Species[1]	C_6C_3 Units	[2]Linkages and Fuctional Groups/100 C_9 Units															
					β-O-4	α-O-4	β-5	β-1	5-5	β-β	4-O-5	α-6	γ-O-γ	γ-O-α	β-6/β-2	Aliphatic OH	Phenolic OH	Ketones	Aldehydes	Methoxyl
[2]Freudenberg	362	1	S	18	38.9	16.7	5.6	2.8	11.1	8.3	5.6	8.3	2.8	11.1	0.0	97.1	33.3	13.9	2.7	97.2
[2]Adler	423	2	S	16	43.8	18.8	6.3	6.3	12.5	6.3	6.3	0.0	0.0	12.5	6.3	19.5	4.5	1.0	1.5	16.0
[2]Nimz	396	3	H	25	48.0	32.0	8.0	16	2.0	6.4	1.6	0.4	2.0	8.0	0.0	81.6	14.4	15.6	16.0	124.0
[2]Sakakibara	447	4	S	28	42.9	21.4	14.3	11	10.7	7.1	3.6	0.0	0.0	17.8	3.6	100.0	32.1	7.1	14.3	89.3
[2]Glasser	399	5	S	81	39.5	16.0	12.3	7.4	16.0	11.1	11.1			19.8	1.2	104.0	29.0	7.5	16	91.4
[2]Brunow	687	6	S	25	64.0	20.0	12.0	4.0	8.0	4.0	4.0	0.0	0.0	8.0	0.0	158.0	12.0	0	6.0	100.0
Jurasek	669		S	2289	52.3	13.9	11.4	7.5	10.7	7.6	11.5	0.0	0.0	0.0	0.0					
Jurasek	669		H	2289	58.4	17.0	7.1	7.4	4.4	6.9	6.6	0.0	0.0	0.0	0.0					
Mikscke	394		S		49-51	6-8	9-12	2.0	9.5-11	2.0	3.5-4									
Mikscke	394		H		62.0	6-8	6.0	7.0	4.5	3.0	6.5									
[2]Glennie	378	8	LS	20	50.0	0.0	15.0	5.0	10.0	5.0	10.0	5.0	0.0	0.0	0.0	75.0	35.0	5.0	10.0	100.0
[2]Marton	385	9	KL	14	14.3	0.0	14.3	14.3	7.1	7.1	7.1	0.0	0.0	7.1	0.0	43.9	64.3	14.3	0	78.6

[1] S = Softwood, H = Hardwood, LS = Lignosulfonate, KL = Kraft Lignin

[2]Approximate values as calculated by authors from Figures 1-6, 8-9

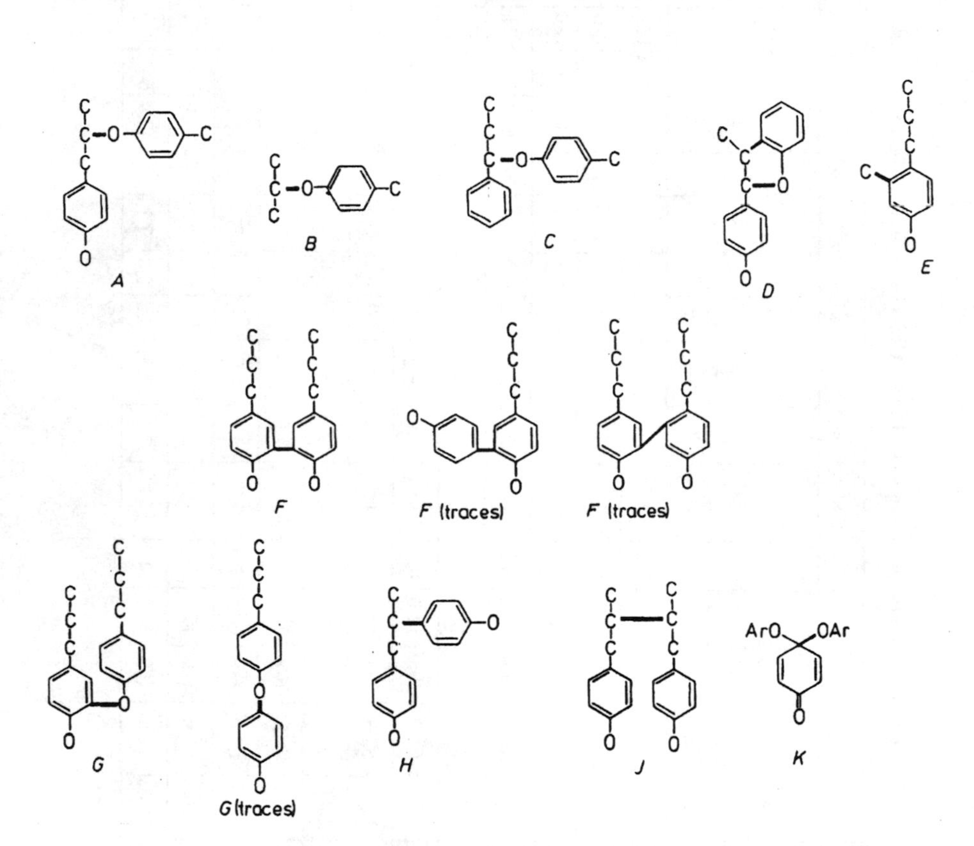

Figure 10. Typical linkage units in lignin by Sakakibara. (Reprinted with permission from ref. 447. Copyright 1980)

worked as a Teacher at the Forestry University in Stockholm. At the University of Lund, Klason studied organic sulfur compounds, thioglycolic acid, mercaptans, and others. At KTH, he started to work more in wood chemistry related research particularly the chemistry involved in kraft and sulfite pulping. He also made contributions in the fields of charcoal formation and lignin structure and chemistry (*27*). He proposed in 1897 that lignin was related structurally to coniferyl alcohol (*28*). For more than a century, research workers from KTH and the Swedish Forest Products Laboratory (STFI), in Stockholm, have contributed importantly to the knowledge of the structure and reactions of lignins.

Significant research was undertaken during this time period into the nature of spent sulfite liquor (SSL) because of the industrial discharge of this material into waterways. Lindsey and Tollens, in 1892, (*25*) proposed that sulfur was present in SSL as sulfonate groups on lignosulfonate molecules. Lange in 1890 precipitated lignosulfonates from alkaline solution (*24*). In 1898, Pollacsek (*29*) described the air-ferric chloride oxidation of lignosulfonates.

Technology. In 1880, Cross and Bevan (*19*) in England used aqueous chlorine and then sodium hydroxide solutions to yield purified cellulose from wood. In 1884, Dahl (*21*) in Danzig, Germany patented the use of a hot aqueous solution of sodium hydroxide and sodium sulfide to produce pulp. This process known as kraft (German word for strong) pulping is now the most important industrial process for delignification of wood. The spent liquor (KBL) is evaporated, burned and causticized with CaO using inexpensive sodium sulfate as the make-up chemical.

1900 - 1924.

Research on lignin had become extensive. Klason reported the first C, H, O and OCH_3 analysis for lignin and proposed that it consisted of coniferyl alcohol type units bound by ether linkages. Lignin had been isolated by several methods. Degradation experiments by pyrolysis and alkali fusion yielded low molecular weight compounds. Sulfite and kraft pulps were now produced in North America. Industrial ethanol was produced by wood saccharification.

Chemistry. Many of the lignin chemistry studies in this time involved the development of lignin isolation procedures and the characterization of isolated lignins through various degradation reactions. These reactions typically produced low molecular weight materials that provided some limited information on lignin structure.

Klason in 1908, suggested that lignin had a macromolecular structure (*33*) and by 1917 that it consisted of coniferyl alcohol type units bound together by ether linkages. In 1908, he inaugurated the analytical determination of Klason lignin (*33*) which is the insoluble residue remaining after treatment of woody tissues with 72% sulfuric acid to dissolve carbohydrates. In 1920, he published the first elemental composition of a lignin (Klason) as; C 66.67; H 5.49; O 27.84 and OCH3, 14.5% respectively (*34*). For many years following Klason, methods were proposed for the preparation of isolated lignins striving to obtain a lignin in a form unchanged from its structure in the plant tissue. These lignins were secured as either insoluble residues, as in the case of Klason lignin, or as soluble materials such as with sulfite, kraft, and alcohol lignins.

At the Technische Hochschule in Darmstadt, Germany, Heuser and Skioldebrand (*45*) in 1919 pyrolyzed Willstatter or HCl-lignin and obtained acetone, methanol, acetic acid, 13% tar, and 51% charcoal. Ungar (*41*) similarly investigated the properties of HCl lignin.

In 1921, Winsvold and Heurser (*50*) found that alkali fusion of lignin yielded substantial amounts of oxalic acid. With Samuelson (*51*) in 1922, Heurser found that oxalic acid in a yield of 25% could be secured by the oxidation of lignin in an alkaline solution after prehydrolysis and methylation. Reed (*42*) in 1917 also reported substantial yields of oxalic acid by lignin oxidation. Shrader (*49*) reported that the structure of lignin was modified such that it could be dissolved in alkaline solution by treatment with oxygen. In 1924, experimentation by Heuser and Hermann (*57*) with KOH fusion of lignin and lignosulfonates yielded pyrocatechol and protocatechuic acid in yields as high as 28%. This information led to the conclusion that lignin contained aromatic nuclei substituted at the 1-postion with carbon and at the 3 and 4 positions with hydroxyl groups.

Proctor and Hirst (*35*) in 1909 precipitated lignosulfonates with certain aromatic amines. Hagglund (*43*), at KTH in 1918, reported on the nitration of lignin. With C. B. Bjorkman (*56*) in 1924, he found that heating of lignin in 12% HCl yielded volatile aldehydes. Grafe (*30*) in 1904, heated SSL in alkaline solution and observed vanillin and acetaldehyde.

Technology. Wheelwright first produced sulfite pulp commercially in the United States in 1882 in Rhode Island (*20*). In 1907, Brompton operated the first kraft pulp mill in North America in Quebec (*31*). The first U. S. kraft paper was manufactured in 1909. Chlorine was first used to bleach pulp in 1913 by Heuser and Sieber (*40*).

G.H. Tomlinson I, a distinguished chemical engineering pioneer in the field of chemical processing of wood, designed and operated two ethanol plants that used wood saccharification during World War I. After pilot studies, a plant using sulfurous and then sulfuric acid was built at Georgetown, SC in collaboration with Dr. Arthur D. Little under an arrangement with the DuPont Co. It produced up to 2000 gal of ethanol per day, yielding up to 20 gallons of ethanol per ton of dry sawdust (*39*). In 1912, a second and much larger ethanol plant was designed and constructed at Fullerton, LA. The plant was closed in 1914 due to financial difficulties. This plant and the one at Georgetown together produced about 10,000,000 gallons of ethanol. Looking backwards, Tomlinson's overall conclusions were that the two plants were not successful financially because of the erroneous assumption that wood waste could be purchased "over an indefinite period at a nominal cost" (*44*).

1925 - 1949:

Expanded research activity during this time has provided many important results. Functional groups such as hydroxyls, benzyl alcohol and others were evaluated.

Degradation reactions yielded products in substantial quantities affirming the aromatic nature as well as the phenylpropane skeleton for lignin. Biosynthesis and biodegradation studies on lignins have been initiated. "Synthetic" lignins have been prepared from coniferyl alcohol by mushroom laccase or horseradish peroxidase through dehydrogenative polymerization. The kraft pulping recovery furnace was developed and widely adopted. Industrial plants were built to produce vanillin, ethanol, yeast, and lignin preparations from spent sulfite liquor.

Chemistry

Much fundamental work was undertaken during this time directed towards determining the structure of lignin. Many new techniques were developed to isolate lignin formulations to study. Generally, three approaches were taken towards investigating lignin structure: degradation reactions, biosynthetic work, and spectroscopy studies. In the most common approach, lignin formulations were subjected to various degradation reactions yielding identifiable products that gave useful structural information. Many new analytical techniques were developed during these times that were highly useful for the lignin chemists. In the final approach, model compounds were synthesized from postulated lignin precursors to produce new products for study.

Occurrence & Isolation. This was a time in which much work was applied to developing new methods to isolate lignin. In 1936, Bailey (*123*) at the University of Washington showed by microdissection that the middle lamella was composed mainly (72%) of lignin rather than pectin. In 1935, Van Beckum and Ritter (*121*) at the US Forest Products Laboratory (USFPL) removed lignin from plant tissue with hypochlorite followed by NaOH. The material that remained termed holocellulose consisted of the total carbohydrate mass present in the plant tissues.

In 1939, Brauns reported (*146,147*) that neutral solvent extraction of woody tissue and subsequent purification created a few percent of what he named native lignin or Braun's lignin. Presently, many investigators view this material as a mixture of lower molecular weight lignins, and/or lignans. This work reflected the continuing search by some classical organic chemists for a lignin that could be extracted simply by use of solvents without chemical reaction.

In 1947, Richie and Purves (*189*) at McGill University oxidized wood at pH 4 with aqueous 5% sodium periodate. The periodate lignin preparation thus obtained was 86-96% Klason lignin, insoluble in organic solvents even at boiling temperature, but completely soluble under conditions of sulfite pulping. A periodate lignin from spruce closely duplicated the behavior of spruce lignin in situ towards many degradative procedures.

Biosynthesis. In 1933, Professor H. Erdtman (*95-99*) at KTH, made the very important proposal based on phenol dimer model studies that lignin was formed in plants by dehydrogenative polymerization of p-hydroxycinnamyl alcohols. Nevertheless, Hilpert (*111,112*) believed that the precursors for carbohydrates and lignin were probably the same and that lignin was simply part of a carbohydrate complex.

Karl Freudenberg of the University of Heidelberg, was a major figure in early lignin biosynthesis studies. He was educated at the University of Berlin under the tutelage of Nobel Prize winner Emil Fischer (Nimz, B.F.H., personal communication, 1998). Freudenberg (*74*) apparently set forth the first formulation of the structure of cellulose in 1928. In 1933, Freudenberg (*100*) published a book with a chapter on lignin. Herein he suggested that lignin was a high polymer formed by dehydrogenation of coniferyl alcohol type monomers as suggested by Erdtman. He felt that lignin was a heteropolymer linked by ether and carbon-carbon bonds. Furthermore he felt that there were three or four types of structural units as happens with amino acid in proteins

Around the end of the 1940's coniferyl alcohol became easily available through the reduction of ethyl ferulate by $LiAlH_4$ (*271*). Freudenberg and coworkers carried out many experimental procedures to synthesize lignin *in vitro* using coniferyl alcohol with mushroom laccase and horseradish peroxidase (*227*). The result was a dehydrogenation polymer (DHP) and a series of lignols. From the average molecular weight of a repeating unit in these synthetic lignins, it was obvious that two hydrogens were lost from coniferyl alcohol during the polymerization. This result seemed to confirm Erdtman's 1933 hypothesis. The biochemistry of lignin formation was reviewed by Hibbert (*166*) in 1942.

Properties. The decades of 1940's and 1950's marked the introduction of much improved instrumentation. Specifically, spectroscopic techniques such as UV, IR, NMR, and MS as well as new molecular weight determination instruments such as the ultracentrifuge greatly assisted the lignin chemist. Additionally, chemists adopted multi stage separation equipment long used in chemical engineering. This happened first with the Craig Machine at Rockfeller University, and then with wonderfully powerful packed column chromatography apparatus and procedures.

The study of the physical and polymer chemistry of lignins was brought into prominence beginning in the 1920's by investigators influenced by the teachings of Mark, Flory, Staudinger, and others. In early theories, an association between lower molecular weight molecules explained the properties of cellulose, lignin, and other polymers. However, in the 1920's Staudinger proposed that cellulose and other polymers were chains held together by primary valence bonds (*234*).

Loughborough and Stamm (129,130) obtained early molecular weight determinations on lignins in 1936 and by Gralen (*162*) in 1946 with values of about M_n equal to 3900 and 3500, respectively.

The first comprehensive compilation of IR spectra of lignins was reported by Jones (*193*). Aulin-Erdtman in 1949 used UV spectra of lignins to demonstrate that isolated lignins were indeed aromatic. Studies of UV absorption of isolated lignins and lignin in woody tissue by Lange (*175*) in 1944 and by Aulin-Erdtman (*206*) in 1949, convinced most investigators that gymnosperm lignin was indeed a condensation polymer of coniferyl aldehyde as proposed by Klason some fifty years earlier.

It was determined that the lignin polymer, in contrast to other biopolymers, had no optical activity and no stereoregularity. This was presumed to be because the lignin polymer is formed by the random combination of dehydrogenation radicals.

Structural Studies. During these years, most researchers felt that lignin included aromatic nuclei. Beyond the presumed relationship with coniferyl alcohol and Freudenberg's carbon-hydrogen analytical results, no direct evidence was available to prove this. In 1939, Freudenberg (*149*) carried out the nitrobenzene oxidation of softwood and obtained vanillin. In 1940, Freudenberg, Lautsch and Engler (*152*) subjected spruce softwood lignin to alkaline nitrobenzene oxidation and obtained a 25% yield of vanillin and other aromatic fragments. Following this work, Creighton, McCarthy and Hibbert (*160*) at McGill University reported that from hardwood lignin, both vanillin and syringaldehyde were obtained in a total yield of about 43%. These studies confirmed that lignin contained aromatic nuclei.

Schultz (*156*) in 1940 found that nitrobenzene oxidation of SSL gave about 20% vanillin yield. Lautsch, Plankenhorn and Klink (*155*) in 1940 used metal oxide catalysts for air oxidation with wood and SSL and obtained about 10% vanillin yield. Pearl reported a careful study of the oxidation products of lignin using copper oxide catalyst in 1942 (*168*).

In 1938, another very important finding was made in the hydrogenolysis experiments of Harris and Adkins (*145*) at the University of Wisconsin, Madison and the USFPL. They conducted high-pressure hydrogenation of methanol lignin in dioxane solution with a Raney Ni catalyst to secure substantial yields of monomeric substituted cyclohexyl products. In 1941, Cooke, McCarthy, and Hibbert (*159*) found similar results with hardwood lignins. These findings pointed to the skeleton of the lignin monomers as guaiacyl propane type, as anticipated from coniferyl alcohol type monomers.

Freudenberg et al. (*125*) in 1936, showed that potassium permanganate oxidation of lignin yielded several aromatic acids affording significant structural information. In 1938, Freudenberg, et al.. (*143*) subjected the products of the reaction mixture generated from high temperature alkaline extraction of spruce wood to permanganate oxidation and then dimethyl sulfate methylation. They obtained a mixture of aromatic products including isohemipenic and dehydrodiveratric acid. These products indicated that it was possible that one of the linkages between the monomeric building blocks in lignin was through the 5 position on the aromatic ring. Pearl (*208*) in 1949 secured similar evidence from the products of metal oxide oxidation of lignosulfonates.

In 1935, King, Hibbert and Brauns (*119*) reported on organosolv lignins formed through methanolysis and ethanolysis. The soluble products of these reactions, Hibbert's ketones, were described in a series of papers with Cramer and Hunter (*148*), West (*171*), McGregor and Evans (*176*) and several others. These ketones were phenylpropane type molecules obtained in about 10% yield when wood was refluxed in acidic 3% methanol/ethanol solution. Dilution with water precipitated methanol/ethanol lignin leaving the Hibbert's ketones in solution. As a result of the studies, Hibbert proposed in 1942 apparently for the first time that lignin contained β-aryl ether type linkages.

In 1944, Hibbert and Gardner (*174*), as part of their studies on ethanolysis, synthesized several phenyl propane type model substances and subjected them to ethanolysis conditions. They found that acyloin side chain type ketols and dihydroxyacetone type substances apparently underwent rearrangement to ketols and diketones. They concluded from this that the entities present in lignin probably carried aryl glycerol-β-aryl ether or similar type structures that rearranged during ethanolysis to yield the observed ketols and diketones.

Pinoresinol, a β-β linked lignan, was first detected by Erdtman (*124*) in 1936 from spruce spent sulfite liquor. Isolation of this compound led to theories regarding the presence of this of type linkage in lignin.

Freudenberg, Meister and Flickinger in 1937 (*135*) proposed the presence of coumaran and related linkages in lignin. They sulfonated phenylcoumaran type model compounds generating a new phenolic hydroxyl group for each sulfonic acid group obtained. Erdtman suggested cleavage of cyclic ether linkages during the sulfonation process (*170*). Hagglund and Carlsson in 1933 (*102*) reported that sulfonation of lignin also resulted in the addition of one phenolic hydroxyl for each sulfonate group.

During the 1940's, Wacek and Kratzl (*205*) published a series of papers addressing the structure of the propyl side-chains of lignin. Phenyl propane structures with various ring and side-chain substituents were synthesized and subjected to well-known lignin reactions including sulfonation and nitrobenzene oxidation. They concluded from this work that lignin most likely contained side chain of the aldol type

Formulations. In the last few years of the 1940's based mainly on the work of Freudenberg and Adler, scientists began to set forth formulations for the structure of lignins. Rydholm in Sweden developed a very useful diagram illustrating the possible distribution of C_6C_3 units. The specific details of and the arguments supporting these early proposals have been very ably compiled in a book by Pearl (*357*) at the Institute of Paper Chemistry. However, even in 1948, a few investigators such as Schutz and associates (*196-198*) still doubted the aromatic nature of lignin.

In other significant work, Urban showed in 1926 that lignin contained hydroxyl groups that were methylated with dimethylsulfate (*65*). In 1927, Freudenberg and Harder (*66*) reported that the reaction of lignin with 12% HCl yielded HCHO. With Niemann and Belz (*78*) in 1929, Freudenberg found that HI split off methoxyl groups from lignin. Shorygina et al. in 1948 found that metallic Na in liquid NH_3 split lignin to yield 28% monomeric aromatic products (*199-202*). Sohn and Lenel (*209*) showed that heating softwoods in water at 1500°C for 2 hrs yielded a solution around pH 3-4, thus suggesting liberation of acetic acid and other acids from the woody tissue. Holmberg (*126,127*) studied benzyl alcohol groups and ether linkages in lignins at KTH in 1936.

In 1925, Hagglund (59) at KTH proposed that lignosulfonates consisted of a polydisperse system of sulfonated guaiacyl propane polymers. Holmberg (*118*) in 1935 suggested based on studies with model substances, that sulfonation of lignin takes place through reactions of benzyl alcohols.

Biodegradation: Phillips summarized the early work on the biodegradation of lignin, which began during the 1920's (*115*). Much of this work focused on the decomposition of lignin by white-rot fungi. In 1944, Waxman (*180*) reviewed the biodegradation studies ongoing at USFPL.

Technology

Pulping and Bleaching. Around 1930 came the enormously important development of the kraft recovery furnace by Tomlinson and Wilcoxson. Since the discovery of kraft pulping by Dahl in 1884, kraft black liquor had been burned with air in a rotary combustion furnace to yield carbonate which was then causticized with CaO. With this new furnace, operations were continuous and recoveries were high which dramatically improved mill economics. This development, along with the success of the kraft process in using all species of woods, reversed the economic balance that had previously favored the sulfite pulping process thus resulting in a tremendous expansion in the use of kraft process.

In 1936, Tomlinson generated the idea that MgSSL might be burned in a recovery furnace to yield MgO and SO_2 simplifying the recovery and recombination of heat, MgO and SO_2. This formed the basis for the Magnefite process (*141*). In cooperation with Babcock & Wilcox in 1937 and 1938, a pilot plant was operated at Cornwall, Canada. In 1948, the Weyerhaeuser Company under the leadership of Raymond Hatch, built a 300 ton/day pulp mill in Longview, WA

In 1941, Tomlinson, McCarthy, and Hibbert patented a process using first stage molecular chlorine in the multistage bleaching of kraft pulp to achieve a high degree of whiteness (*165*).

In 1946, McKee (*185*) described the delignification of wood and isolation of a lignin preparation by hydrotropic solvents such as Na xylene sulfonate.

Non-Lignin Chemicals from Wood. In the USSR around the beginning of World War II, many wood saccharification plants were built using the modified Scholler-Tornesch (*182*) type process in which wood was subjected to dilute sulfuric acid percolation. This resulted in the production of sugars and Klason type lignins. Much process and utilization research was done on these wood hydrolysis lignins in Leningrad, USSR, under the leadership of Sharkov (*183*).

During World War II, ethanol was in high demand leading to the construction of several plants to ferment SSL sugars to alcohol. This also provided a low sugar lignosulfonate product of higher quality. Two ethanol plants were built in Sweden (capacity of around 400,000 gal/yr.) and several in Germany as described by Kilp (*207*). In Canada, plants were built at Thorold, Ontario and Gatineau, Quebec as described by Nord and Sankey in 1944 (*178*). A plant was built in Bellingham, which produced about 9000 gal/day of 190 proof ethanol using *Saccharomyces cerevisiae* (*187*). Developments during this time period were summarized in 1954 (*245*).

Food and fodder yeast from SSL was produced in Germany during World War II by the Waldhof-Standard process. In the United States, a plant was built in Rhinelander, WI as described by Wiley (*247*) in 1954.

In 1937, Tomlinson and Hibbert (*142*) described the construction of a plant by the Howard Smith Paper Co. in Cornwall, Canada to produce vanillin from the alkaline oxidation of lignosulfonates. Vanillin was produced at a rate of 800 lb./day. Operations were continued until about 1970 when the plant was closed because SSL ceased to be available. Goheen reported that similar plants were built around 1936 by the Marathon Paper Co at Rothschild, WI using calcium lignosulfonates separated by the Howard process, and by the Borregaard Company in Sarpsborg, Norway. In 1933, Reidel (*110*) patented a process to synthesize vanillin from guaiacol and assigned it to E. de Hoen. Many years later this type of synthesis of vanillin from guaiacol made manufacture of vanillin from lignin unprofitable for most companies.

A laminated plastic product named Tomlinite that contained a lignin adhesive, was developed in about 1944 at the Howard Smith Co by G. H. Tomlinson. A slurry of precipitated lignin and pulp was screened to obtain a laminating paper, which was subjected to heat and compression yielding a useful board (*179*).

Industrial Lignins. In 1936, the Marathon Corporation constructed in Rothschild, WI, an industrial scale facility for the manufacture of lignosulfonate products including leather tanning agents and dispersants.

In 1942, the West Virginia Pulp and Paper Company (Westvaco) in Charleston, SC began semi-commercial production of lignin products from the kraft pulping of softwoods or hardwoods. Early products included INDULIN A and C. These products were applied in rubber, printing ink and ceramic industries.

Books. The 1925-1949 years marked the appearance of the first books on lignin: *Chemie der Ligninkorper* by Kurschner (*62*); *The Chemistry of Wood*, by Hawley and Wise (*181*), and *The Chemistry of Wood and Cellulose*, by Schorger (*64*). In 1947 came Hunter's informative book (*4*) on the History of Papermaking.

1950 - 1974:

This was a period of significant progress was made in the understanding of lignin structure spurned on by the development of new analytical tools and procedures. Investigators generally agreed that lignin is a high polymer constructed of coniferyl alcohol type monomers by enzymatic dehydrogenative polymerization. Based on quantification studies on linkages and functional groups, many lignin formulations were proposed. Industrially, plants were constructed to produce lignin products, ethanol, vanillin, dimethylsulfoxide, etc. The creation of the EPA in 1970 greatly influenced the activities of the pulp and paper industry. In 1971 Sarkanen and Ludwig (*388*) as editors, published their masterful treatise on lignins.

Occurrence and Isolation. Investigators found that wood contained phenolic substances such as lignans and tannins that were structurally similar to lignin and therefore interfered with isolation studies. Lewis (*675*) and Hemingway (*457*) have published excellent reviews on the chemistry of lignans and tannins respectively.

Many studies were undertaken to investigate lignin variations in different portions of the tree. In 1954, Lange (*244*) verified that the native lignin in cell walls was aromatic in nature by measuring the UV absorption of thin wood sections. Goring et al. (*370*), reported in 1969 differences in the content of guaiacyl units in primary and secondary walls and in the middle lamella of soft wood lignins. During the next year, similar differences were also found in hardwood lignin (*371*). At the NYSCFS, Timell (*344*) in 1965 and then Côté et al. (*346*) in 1966 published major information on compression wood.

During this time, researchers continued to work towards developing a lignin preparation that was substantially unchanged from its *in situ* structure in wood. The goal was approached in 1956 in A. Björkman's doctoral dissertation (*257*) from Chalmers University of Technology, in Göteborg, Sweden. He reported that after extensive milling in a non-swelling liquid such as toluene, up to about half of the lignin in softwood was extracted by dioxane to yield milled wood lignin (MWL) or Björkman lignin. In 1957, Björkman and Person (*261*) reported on the properties of both softwood and hardwood MWL.

In 1957, Björkman (*263*) extracted a material that he called a lignin-carbohydrate complex (LCC) from wood with solvents. In the same year, Merewether (*269*) reviewed early ideas about LCC's.

In 1950, Schubert and Nord (*213*) treated wood with brown rot enzymes releasing a small amount of enzyme lignin with properties similar to that of MWL.

Biosynthesis. During 1950-60's, important findings concerning the formation of coniferyl alcohol type substances in the cambial sap were made by using radioactive tracer techniques and enzyme studies. Although many other authors contributed heavily to this area, the work of Freudenberg and his many coworkers at the University of Heidelberg formed the foundation to our current knowledge of this field. The question of whether coniferyl alcohol was in fact the main precursor and intermediate in lignin formation was settled by tracer studies in 1955 by Freudenberg and Eisenhut (*251*) and by Kratzl et al. (*268*) in 1957. In Freudenberg's studies, radioactive labeled coniferyl alcohol and its precursors coniferin and phenylalanine were incorporated into lignin. Additionally, Freudenberg studied the products of the enzymatic dehydrogenation of coniferyl alcohol using both the *"zulauf"* and *"zutropf"* methods (*258*). In this work, Freudenberg discussed the principle of end-wise polymerization, a term apparently first used by Sarkanen in 1971 and also reviewed by Nimz, (404*)* and Adler (*423*). From dimeric and trimeric products obtained, they were able to identify ß-O-4, β-5 and β-β linkages. Additional work by Freudenberg included the detection of cell wall-bound β-glucosidase in lignifying tissue and the isolation of *p*-glucohydroxycinnamyl alcohols in the cambium of conifers. Freudenberg and Neish wrote an excellent review of the monomeric intermediates in lignin biosynthesis in 1968 (*362*).

In 1955, Brown and Neish (*250*) studied the involvement of the shikimic acid pathway in lignin biosynthesis. Higuchi (*267*) investigated the biochemistry of lignin formation in 1957. Kratzl (*288*) in 1960 studied the incorporation of ^{14}C labeled glucose into spruce phenylpropane units. In 1965, Hillis and Isoi (*336*) investigated the biosynthesis of polyphenols in eucalyptus.

Properties. Based on his studies of the Hildebrand solubility parameter of lignins in certain solvents, Schüerch in 1952 (*232*) proposed that the insolubility of lignin in neutral solvents was due to the fact that lignin was a network polymer. In 1955, Gardon and Mason (*252*) found that sulfonated lignin molecules behaved as flexible polyelectrolytes in measurements on conductivity, dyestuff adsorption, and viscosity. In addition, from studying the variation between intrinsic viscosity and molecular weight, they were able to determine that the degree of molecular chain branching was greater in higher molecular weight fractions.

In 1960, Rex (*290*) demonstrated the presence of free radicals in softwood dioxane acidolysis lignin via electron spin resonance (ESR). Comparison of ESR spectra by Kleinert (*355*) in 1967 for several lignins and DHP preparations showed that the amount of free radicals in each material was stable but varied between preparations. Evidence also indicated that free radical reactions took place during specimen preparation.

During this time, David Goring at McGill University was very involved with studies into the polymeric properties of lignin. With Rezanowich in 1960, he reported that lignin in solution assumed a nearly spherical shape (*291*). In 1963, he found that the glass transition temperature of lignin was markedly decreased by absorption of water (*318*). Goring proposed a degellation theory of delignification in 1968 (*365*) based on the assumption that wood contained two types of lignin gels. He proposed that degellation proceeded as the reverse of Flory's concept of the polymerization of trifunctional monomers. In 1971, Goring wrote an excellent review on the polymeric properties of lignin (*380*) that included information on molecular weight, molecular shape, electrolyte swelling, colloidal behavior, glass transition temperature, adhesive properties, electrophoresis fractionation, and associated solubility behaviors of lignins.

McCarthy and associates at the University of Washington were very active in studies on the polymer properties of lignin sulfonates. The results of early publications in this area were summarized in 1957 with Felicetta (*264*). Continuous counter current dialysis of spent sulfite liquor was used to prepare purified lignosulfonates fractions (*194*). Crystalline monomeric lignin sulfonates obtained from spent sulfite liquor were first reported (*361*). In 1957, McCarthy and Felicetta (*270*) used incremental sulfite delignification to yield five lignin sulfonate samples of *Mw* values between 3400 and 30,000. These fractions were hydrolyzed to yield "omega" preparations of *Mw* range 1900 to 13,000 through what the authors felt was the cleavage of α-O-4 ether linkages.

In 1962, Jensen, Fremer and Forss (*306*) delignified spruce wood with acidic Ca-bisulfite and fractionated the spent sulfite liquor using gelfiltration with Sephadex G-25 and G-75. The low molecular weight components had similar acidic and neutral UV spectra to the high molecular weight components but different alkaline spectra.

In 1970, Luner and Kempf (*373*) at NYSCFS estimated the size and shape of several lignins (MWL, dioxane and kraft lignins) by determining Langmuir force-area and potential-area isothermal diagrams of the lignins spread on water as monomolecular films.

Structural Studies

Functional Groups. Work towards the development of spectroscopy procedures for lignin analysis was very active during the 1950s. Analytical procedures for the determination of lignin content in solutions by UV absorbance was developed by Patterson (*218*). In 1950, Freudenberg (*211*) compared softwood lignin with DHP using infrared (IR) spectroscopy. Assignment of infrared bands in conifer lignin was made in 1960 by Hergert (*286*).

Aulin-Erdtman (*220*) and Goldschmid (*243*) showed how the $\Delta\varepsilon_i$ UV spectrophotometric differential method could be used to estimate the phenolic hydroxyl content of lignin solutions. In 1955, Adler and Hernestom (*249*) found though the sodium periodate oxidation of softwood lignin, that 30% of all guaiacyl units carried free phenolic hydroxyls and that 70% were etherified. In 1961, Adler and Lundquist (*293*) working with the oxidation of model compound orthoquinones and MWL by Fremy's salt estimated that the quantity of non-condensed phenolic units in softwood MWL was 0.15-0.18 units/OCH_3.

Hergert and Kurth (*239*) employed IR spectroscopy in 1953 to estimate the nature of carbonyl groups in lignins. In 1961, Adler and Marton (*294*) quantified carbonyl and ethylenic groups in MWL. Carbonyl groups were determined through oxime formation and borohydride reduction to be about 0.2 carbonyl groups per OCH_3, of which they felt half were conjugated. Through hydrogenation reactions, the number of ethylenic groups were reported at 0.03 per OCH_3 group.

In 1964, Ludwig, Nist and McCarthy (*326*) analyzed acetylated derivatives of MWL, dioxane, acetolysis, and Brauns "native", lignins using ^{1}H NMR spectroscopy apparently for the first time to estimate the quantities of several functional groups. Measurements of the chemical shifts of NMR signals from lignin-like model substances were used for calibration. Analysis of this data suggested that 50-60% of the monomeric structural units in MWL were condensed. It also indicated that hydrogen bonding existed in lignins. Lenz (*364*) confirmed and extended this work.

In 1975, Glasser et al. (*406*) at the University of Washington used ^{1}H-NMR spectroscopy to study the structure of acetyled lignosulfonates from Western Hemlock. Eight monomeric and seven dimeric lignin-like sulfonates were synthesized for use as standards.

In 1973, Lüdeman and Nimz (*402*) apparently were the first to use ^{13}C-NMR spectroscopy to characterize lignins. In 1974, Nimz and associates (*401-404*) made several significant contributions using ^{13}C-NMR that provided strong evidence that lignin is a branch-chain high polymer consisting of coniferyl alcohol type monomers and very probably combined chemically with carbohydrates.

At the University in Göteborg, Adler and his coworkers made many important contributions to the lignin field. In 1955, Adler and Gierer (*248*) determined by acidic methanolysis that the number of benzyl hydroxyl and ether sites in MWL was approximately 42 per 100 C9 units. Cyclic benzyl ethers in structures such as phenylcoumaran and pinoresinol were not included in this number. In further studies using oxidation and Δe_i procedures, the total number of non-cyclic benzyl ether groups was estimated at 24 per 100 C9 units (*423*).

Linkages. During this time, many procedures such as mild hydrolysis, acidolysis, thioacidolysis, and permanganate oxidation were developed or improved to break down lignin into fragments for analysis. The use of these procedures allowed for the determination and quantification of the linkages between lignin structural units.

Nimz, while in Freudenberg's group at Heidelberg, used mild hydrolysis (2% acetic acid, 100°C for several weeks) to degrade lignin with limited condensation. In 1965, Nimz and Gaber, (*338*), isolated syringaresinol from a hardwood thus indicating the presence of β-β links in lignin. In 1966 and 1967, many low molecular weight compounds (some eight dimers and one tetramer) assumed to have resulted from hydrolysis of benzyl ethers, were isolated and their structures determined (*350*). Nimz used ^{1}H-NMR (*356*) to obtain spectra of the important lignin degradation products from mild hydrolysis. Also with the objective of securing degradation of lignin without condensation, Sakakibara at Hokkaido University in Sapporo, Japan conducted studies using a relatively mild hydrolysis procedure. With Nakayama (*299*) in 1961, he found that heating softwood lignins in dioxane-water solution at 180°C dissolved 40-60% of the lignin and yielded several monomers and dilignols. Considerably more lignin was dissolved with hardwoods.

In 1964, Lundquist, found that acidolysis of softwood MWL gave rise to important monomeric and dimeric lignin types of molecules (*327*).

In 1969, Nimz et al. (*368*) described the very important thioacidolysis procedure for degradation whereby both α- and β-aryl ether bonds were cleaved. The mono through tetra lignols thus obtained were quantified and significant additional oligomers were identified.

Extensive research based on the mild hydrolysis in dioxane/water of lignin model compounds by Sakakibara et al. (*431*) yielded the isolation and identification of a variety of mono-, di-, and trilignols.

In 1966 Hrutfiord and McCarthy (*347*), used mild catalytic hydrogenolysis along with NMR and gas chromatography to estimate the number of condensed and uncondensed units in lignin. Chemical shift values for certain lignin signals were reported.

Miksche and associates (*384,394*) in the early 1970's took advantage of the powerful fractionation capabilities of gas chromatography to separate and identify the products of the permanganate oxidation of lignin. These analyses allowed for the quantification of linkage frequency in several lignin preparations from spruce and birch.

In 1957, Adler et al. (*262*) prepared model substances of the guaiacylglycerol β-aryl ether type and found that upon heating in acidic dioxane-water, about 0.3 phenolic hydroxyl groups per OCH_3 group were liberated. This was approximately the same proportion of terminal C-groups characteristic of Hibbert's type ketones. These results were taken to indicate that one fourth to one third of all guaiacyl propane units were of the aryl glycerol type joined by β-aryl ether linkages.

Following the suggestion by Freudenberg that phenylcoumaran structures (β-5) were significant lignin components, Adler et al. in 1959 (*278*) quantified these units in MWL at 0.08 phenylcoumaran entities per OCH_3 group through an acid dehydrogenation reaction yielding stilbene-type structures.

From studies on the catalytic air oxidation of model substances of the vanillin type, Pearl and Dickey (*231*) in 1952 proposed the presence of linkages of the α-α type. Later, Pearl (*357*) concluded that he had misinterpreted the data and that the correct linkage was β-1.

In 1955, Pew (*254*) proposed the presence of biphenyl linkages of the 5-5 type. Biphenyl linkages of the 5-1, 5-6 and 6-6 types were first detected by Chen (*301*) in 1962. In 1953, Freudenberg and Rasenak apparently first proposed the presence of biphenyl ether linkages (5-0-4 and 1-0-4) in lignin. Freudenberg and Chen (*354*) first discovered these linkages between 1962-67.

In 1959, to examine the aromatic character of lignin and its chemical binding with carbohydrates, Tischenko at the St. Petersburg Forestry Academy reported the ozonization of wood and lignin samples in non-aqueous solvents. Carbohydrate-free and carbohydrate-containing lignin isolated as triozonides were quantified through mild heating and a reaction with $KMnO_4$ producing oxalic acid. Lignin containing carbohydrates was found to consist of 60 to 68% lignin triozonide and 32 to 40% carbohydrates (Shevchenko, UBC, personal communication, 1998).

Reactions. In further work relating to vanillin, Kratzl and Silbernagel (*229*) showed that heating wood in water brought about a decrease in vanillin yield by nitrobenzene oxidation presumably by the condensation of lignin. The relationship between vanillin yield from nitrobenzene oxidation of lignin and the presence of phenolic hydroxyl groups was reported by Leopold (*230*).

In 1960, Sarkanen and Dence reviewed the cleavage of alkyl-aryl-ether linkages in lignin (*292*). In 1961, Gierer and Kunze at STFI (*298*) showed that the mechanism of cleavage of β-O-4 alkyl-aryl ethers involves a nucleophilic attack of neighboring hydroxyl groups on the β-carbon atom resulting in the formation of an epoxide and simultaneous removal of the aryl ether substituent as a phenoxide ion. In 1962, Gierer with Norén (*305*) clarified that the cleavage of α-aryl ether bonds during alkaline delignification proceeds via quinonemethide intermediates and thus requires a free phenolic hydroxyl group in the ρ-position to the propyl side-chain. In 1964 Gierer et al. (*324*) found with model substances that NaOH solution cleavage of β-0-4 linkages may occur if there is an adjacent hydroxyl or carbonyl group on the α- or γ-carbon of the propanoid side-chain.

In 1964, Gierer and Smedman (*325*) investigated the favorable effect of sulfide ions on the alkaline degradation of lignins during kraft delignification. It was found that the benzylthiol ion formed from the addition of a sulfide ion to a quinone methide intermediate, attacked the β-carbon atom forming a thiirane (episulfide) intermediate. The β-substituent was then eliminated through the sulfidolydic cleavage of the β-aryl ether bond. In 1962, Tischenko and Zhigalov (*311*) apparently first proposed sulfur dispositions in thiolignins.

In 1965, Sarkanen and Suzuki (*342*) showed that the peracetic acid oxidation of softwood lignin caused ring cleavage forming dicarboxylic acids.

Goring, Procter, and Yean (*358*) used UV microscopy in 1967 to find that in both kraft and acid sulfite pulping, lignin was removed preferentially from the secondary wall before the middle lamella. They observed no concentration gradients nor did they find that diffusion played a dominant role in delignification kinetics. In a later investigation, however, Goring (*397*) found that kraft lignin could be removed almost completely from fibers using a very long extraction time.

Gierer (*372*) summarized the chemistry of delignification in 1970 in a publication in which he discussed the mechanistic features of kraft, soda, neutral sulfite, and acidic sulfite pulping processes.

In 1957, Tischenko (*272*) proposed that since lignin consists of phenylalcohols or their ethers, lignin condensation should follow the pattern of formation of phenol-formaldehyde resins. In 1962, Ashorn and Enkvist showed that heating kraft black liquor with sulfur gave moderate yields of phenyl derivatives as well as dimethylsulfoxide (*300*).

Formulations. Freudenberg proposed schematic models of lignin in 1961, 1962, 1964, 1965 and 1968 (*362*). He started with 15 structural units in his formulation in 1961 and finished with eighteen structural units and three variants in 1968 (Figure 1).

In 1965, Forss and Fremer (*333*) in Helsinki at the Finish Pulp and Paper Research Institute proposed a structure (Figure 7) for coniferous lignin based primarily on the results of their own investigations. They felt that lignin was a heterogeneous material composed of a lignin polymer and a number of low molecular mass lignin-like hemilignins. The polymeric lignin was composed of identical repeating units. Each repeating unit consisted of 16 guaiacylpropane and 2 *p*-hydroxyphenylpropane units. Eight of the guaiacylpropane units and the two *p*-hydroxyphenylpropane units were interlinked to form an unhydrolyzable core. The remaining eight guaiacylpropane units were linked to the core by C4-O-C bonds. It must be noted that this structure is quite different from other proposed structures and that little account seems to have been taken of the evidence indicating the presence of several different configurations for structural units.

In 1973 and 1974, Nimz (*396, 403*) proposed a model structure for beech lignin (Figure 3) based in part on the results of his thioacetolysis and mild hydrolysis studies. His model consisted of twenty-five structural units of which six could be replaced by certain specific dilignols. The sequencing of the units is mostly arbitrary. His scheme can be rationalized in terms of the oxidative coupling of a mixture of 14, 10, and 1 molecules of coniferyl alcohol, sinapyl alcohol, and *p*-

coumaryl alcohol respectively. The proportions of the several types of bonds between the structural units are shown in Table I.

In 1971, Glennie (*378*) reviewed reactions in sulfite pulping and set forth the model structure of lignin sulfonates shown in Figure 8. The average molecular formula for a C9 unit in this structure is very similar to the formula obtained through ^{1}H-NMR studies by Glasser et al. (*406*) described earlier.

From 1964 to 1971, Marton (*328,385*) conducted extensive studies on kraft lignins and wrote an excellent review of alkaline pulping and lignins in which he set forth a tentative formulation of pine kraft lignin (Figure 9). In this formulation, about one half of the 14 structural units have lost side-chains, 30% are still non-phenolic; and 20% do not contain methoxyl groups. Portions of the phenylcoumaran structures from the original lignin have been converted to stilbenes and some condensation has occurred.

Glasser and Glasser (*398, 399*) developed several structural formulations for softwood and hardwood lignins using a computer-based simulation approach starting in 1974. Their structures were based upon certain analytical attributes along with the assumption that lignin formation occurs by random coupling of phenoxy radicals of coniferyl alcohol. To explain discrepancies between measured elemental compositions of lignin and their presumed structure (Figure 5) they incorporated certain undetected linkages and structural units.

Technology

Environmental. In 1970, the United States Congress approved the Environmental Protection Agency (EPA) Act that contained provisions of great importance to the pulp and paper industry. Many regulations were set down which stimulated significant research in environmental areas. Discharge of pulping effluents into watercourses was strictly controlled during this period, which forced calcium based sulfite mills to recover and sell the solids from their pulping effluents or convert to a different base that would allow a recovery system. The release of bleaching effluents also presented problems because of the presence of toxic entities such as chlorodioxins. The effluents also depleted oxygen in receiving waters. Discharge of volatile components into the atmosphere presented problems in both kraft and sulfite mills.

Pulping and Bleaching. In 1950, Sillen and Anderson (*233*) reported the equilibrium compositions of sulfite recovery furnace gases and solids that they calculated by thermodynamic methods. At economically feasible furnace temperatures, only the magnesium-based system gave MgO solids and SO_2 gas. For CaO and Na_2O recovery, substantially higher temperatures were needed. Ammonia was consumed but SO_2 recovery was feasible.

In 1956, Nikitin and Akim (*259*) described the use of an oxygen alkaline method for delignification and improved bleaching. In 1974, Kratzl et al. (*400*) studied the mechanisms of reaction of molecular oxygen with model substances in alkaline solutions. The formation of phenoxy radicals was discussed in terms of reactions such as side-chain elimination, ring opening, and demethoxylation. The effect of substituent groups on oxidizability correlated well with physico-chemical data. The calculation of critical oxidation potentials by quantum mechanical procedures was also reported.

Other Technologies. Gottlieb and Pelczar (*216*) reviewed the research undertaken through 1951 on the biodegradation of lignin. Kirk's summary (*382*) on the advances in lignin biodegradation research from 1951 until 1969, included information on the effect of white rot, brown rot, bacteria, and phenol oxidases on the degradation of lignins.

The "Rheinau" process for wood saccharification was described in 1958 by Hägglund and Riehm (*276*). Wood was reacted first with fuming and then dilute HCl producing nearly quantitative yields of sugars. In 1940, a full-scale plant of this type using wood waste was constructed at Regensburg, Germany which reportedly produced 500 tons/month of yeast.

Enkvist and Turunen were involved in early work on the cleavage of kraft black liquor in NaOH-NaSH melts yielding dimethyl sulfide and methyl mercaptan. The same products, according to Hägglund and Enkvist (*266*) in 1957, were obtained by pyrolysis of kraft black liquor with added sodium sulfide. In 1960, Goheen et al. (*281*) reported that the Crown Zellerback Co. built a ten million lbs/year plant at Bogalusa, LA to produce dimethyl sulfide from kraft black liquor. In a subsequent step, the dimethyl sulfide was oxidized to dimethyl sulfoxide.

In 1951, Fisher and Marshall (*215*) described the rebuilding of the vanillin plant at Thorold, Ontario to use CaO as a less expensive alkali. In 1953, the Monsanto Chemical Co. (*240*) established a vanillin plant in Seattle in which softwood spent sulfite liquor was alkaline air oxidized using a copper catalyst. Both of these plants were closed in the 1990's because of competition from synthetic vanillin from guaiacol.

Osahima, Maeda and Kashima (329) described phenol production from lignin using a high temperature process. Lignin sulfonates were desulfonated through heating in phenolic solvents at around 400°C for 4 hours yielding 20% monomeric phenols. It does not appear that this process was ever commercialized.

Industrial Lignins. In 1947, Puget Sound Pulp and Timber Co (now Georgia Pacific West, Inc.) began production of a fermented calcium lignosulfonate product (LIGNOSITE) in Bellingham, Washington. In 1960, King and Adolphson (*287*) patented the use of chrome and ferrochrome derivatives of LIGNOSITE as additives to oil well drilling muds.

In 1957, the lignin manufacturing facilities of the Marathon Corporation described earlier were acquired by the American Can Company. Production and sales of lignin chemicals from the kraft pulping of pines continued at Westvaco.

ITT Rayonier Inc began selling a calcium lignosulfonate product (Raylig) in 1937 from their Shelton, WA sulfite mill. Following the closure of this mill in 1956, the lignin business was transferred to Hoquiam, WA where it was subsequently expanded (Hergert, personal communication, 1998).

Books. During 1950-1975, several excellent books and chapters on lignin chemistry appeared: Included in these were:, Hägglund, *Chemistry of Wood*, 1951 (*217*); Brauns, *Chemistry of Lignin*, 1952 and 1960 (*221,285*); Nikitin, the *Chemistry of Cellulose and Wood*, 1962 and 1966 (*307,184*); Browning, the *Chemistry of Wood, and Methods of Wood Chemistry*, 1963 and 1967 (*312,353*); Marton *Lignin Structure and Reactions* in 1966 (*348*); and Sarkanen and Ludwig, *Lignins, Occurrence, Formation, Structure and Reactions*, 1971 (*388*); and Rydholm, *Pulping Processes*, in 1965 (*339*).

1975 - 1989.

During this time, studies continue to confirm the accepted concepts of lignin structures. Understanding of delignification and bleaching mechanisms is improved as a result of significant research. Anthraquinone is introduced as a pulping catalyst. EPA regulations have decreased the use of chlorine derivatives in favor of oxygen species. Use of industrial lignins has continued to increase as has research efforts directed towards finding new uses.

Chemistry

Occurrence and Isolation. Hergert at Oregon State University reported in 1977 (*427*) that there were differences between Braun's lignin and lignin in xylem tissue. He previously reported that bark lignins were very similar to xylem lignins when isolated by the same methods.

In 1979, Meshitsuka and Nakano (*167*) explained the mechanisms involved in the Mäule color reaction. In hardwoods, the purple color arises from the formation of chlorinated methoxy ortho-quinone structures from syringyl groups while in softwoods the brown color arises from the formation of chlorinated o-quinones from guaiacyl groups.

In 1981, Lee, Meshitsuka, Cho and Nakano (*458*) concluded from their earlier studies of stepwise preparation of MWL, that MWL from hardwood mainly originates from the compound middle lamella.

In 1985, Meshitsuka and Nakano (*491*) at the University of Tokyo developed a technique for collecting and differentiating xylem fractions from hardwoods. They were able to collect middle lamella and primary wall sections with only minor contamination from the secondary wall. From the analysis of these samples, they concluded that the middle lamella lignin in hardwood fibers is rich in guaiacyl units and highly condensed.

In 1987, Atalla (*498*) at the United States Forest Products Laboratory observed that aromatic rings in lignin seemed to be aligned preferentially in a direction tangential to the secondary wall in softwood xylem tissue.

Biosynthesis. In 1982, Westermark (*466*) at STFI reported that the reaction of an enzyme generated superoxide radical with coniferyl alcohol in the presence of calcium at pH 8, yielded dehydrodi conferyl alcohol and pinoresinol.

Properties. In 1976, Zarubin (*422*) proposed, apparently for the first time, that lignin is as an acid-base system and that the functional groups in lignin can be classified based on the principles of modern acid base theory. In 1979, Lindström (*439*) at STFI reported in some detail on the colloidal behavior, association and gelation properties of kraft lignin.

Connors, Sarkanen and McCarthy (*440*) in 1980 found that a bimodal molecular weight distribution for lignins seemed to arise from molecular association during chromatographic processes. In 1984, Pla and Robert (*478*) described the conformation of lignin in solution.

The viscoelastic properties of the amorphous components of wood were studied by Salmén (*480*) and by Kelley et al. (*523*) with the softening behavior modeled following the WLF equation. Distinct and separate glass-to-rubber transitions for lignin and non-crystalline carbohydrates were identified that revealed T_g values for lignin between 60 and 100°C.

In 1989, Elder (*515*) characterized certain coniferyl alcohol type monomers using computational chemistry methods.

Structural Studies. In 1976, phenylcoumaran structures with cinnamaldehyde and α-hydroxypropane side-chains were isolated through dioxane-water hydrolysis of lignin by Aoyama and Sakakibara (*412*). In the same year, Yasuda and Sakakibara (*420*) isolated β-6 phenyl-isochroman from the hydrogenolysis of compression wood.

During this time period, a significant amount of effort was applied towards investigating the structure of lignin using NMR. In 1977, Lundquist and Olsson (*430*) used ^{1}H-NMR to characterize lignin. At McGill, Argyropoulos (*578*) used ^{19}F-NMR and ^{31}P-NMR spectroscopy. At the EFP in Grenoble, Robert and Gagnaire (*460*) apparently reported the first quantitative structural analysis of lignin using ^{13}C-NMR. Robert and Brunow in 1984 found via ^{13}NMR (*479*), that MWL contained about 0.9 aliphatic OH/OCH_3. In 1989, Robert, et al. used ^{13}NMR to evaluate the distribution of *erythro* and *threo* forms of various types of β-0-4 structures. Surprisingly, in spite of its large number of asymmetric carbon atoms, lignin seems to be the only natural macromolecule that has no optical activity (*503*).

In 1986, Ishizu and Nakano (*677*) developed procedures for the characterization of lignin by ozone degradation and discussed *erythro* and *threo* ratio of the α-carbon of lignin side-chain based upon erythronic and threonic acid yields.

Formulations. In 1977, Adler proposed (*423*) a lignin structure (Figure 2) with sixteen different prominent structures in arbitrary sequence. For spruce lignin, he estimated the frequency of occurrence of certain linkages as shown in Table I. Adler also included the same linkage types and frequency data for birch.

Sakakibara (*447*) generated a lignin structure in 1980 (Figure 4) based on available analytical data. His model contains 28 structural units and several alternates as shown in Table I. Certain doubtful structures were omitted.

Glasser and Glasser continued their work with their computer-model (*453*) that provided a comparison between experimental and predicted analytical results for the structure of lignin. The results of many independent analytical methods such as elemental and functional group analyses, ^{1}H-NMR spectroscopy, permanganate oxidation degradation, and gel-permeation chromatography were incorporated in the program. A structural scheme for softwood lignin was generated (Figure 5). The linkage frequencies for this structure are given in Table I.

Reactions. In 1989, Gellerstedt and Northey (*517*) at STFI studied the structure of birch MWL from the quantification of over 30 different acids from permanganate oxidation. Their results suggested that α aryl-ether linkages were cleaved during milling.

Technology

Delignification and Bleaching. In 1977, Holton and Chapman (*429*) of Canadian Industries, Ltd. reported yield increases as a result of the addition of anthraquinone to alkaline pulping. Nomura (Meshitsuka, G., University of Tokyo, personal communication, 1998) independently developed a similar method using tetrahydro anthraquinone in Japan in 1974. This method is widely used in the world particularly in Japan, and other Asian countries. In 1977 Gratzl et al. (*425*) studied the effect of quinone additives on the soda pulping of hardwoods.

In 1976, Akim et al. (*411*) reported on changes in the molecular weight distribution of cellulose, which occur during the oxygen-alkaline purification of pulps. Gierer and Insgard in 1977 formulated mechanisms for reactions of lignin with oxygen and hydrogen peroxide in alkaline media (*426*). In 1980, Sjöström (*448*) summarized the knowledge of chemistry of oxygen delignification as well as the chemistry of sulfite delignification.

In 1980, kraft pulping was interpreted in chemical terms by Gierer (*452*) as a competition between alkali-promoted degradation and condensation reactions. Cleavage of the various types of lignin bonds was attributed to three phases of kraft pulping. Similarities in the degradation mechanisms of lignins and carbohydrates were noted. Based on the results from extensive model studies, Gierer (*486,487*) in 1985 divided the chemical reactions involved in pulping and bleaching into nucleophilic and electrophilic categories. This concept greatly facilitated the understanding of cooking and bleaching reactions and their complicated interplay in various sequences.

Saka, et al. (*465*) in 1982 investigated the topochemistry of delignification by soda, soda-AQ and kraft of thin sections of Douglas Fir (*Pseudotsuga menziesiii*) using a bromination technique monitored by SEM-EDXA. Kinetics were established for the removal of lignin from morphological regions. The data suggested structural differences between middle lamella and secondary wall lignins.

In 1981, Glasser (*434*) reported that based on the results of an extensive study on the structure of residual lignins in pulps, the bond between lignin and carbohydrates is a dynamic one and is regenerated following cleavage under pulping conditions. This dynamic behavior manifests itself in dramatic changes in the association of lignin with carbohydrates while having little input to linkage types and distributions. The regeneration of the lignin-carbohydrate bond was attributed to the formation of quinone-methide intermediates during delignification.

In 1986, Dolk, et al. (*495*) undertook delignification studies using a flow-through reactor to reduce the residence time of solutes to a few minutes to minimize solution changes in dissolved lignins. Softwood (Western hemlock) was soda pulped while a hardwood (Black cottonwood) was subjected to soda and organosolv pulping. During the course of the delignification of hemlock with alkali, the M_w values rose from about 3000 to 26,000 (M_w/M_n= 2.7 to 11.5). With Cottonwood, the M_w values rose from about 4700 to 55,000 (M_w/M_n = 2.1 to 8.5) during alkaline pulping and rose from 800 to 7400 (M_w/M_n = 1.8 to 10) in the organosolv system.

Kinetics and reaction mechanisms for the alkaline delignification of hemlock were discussed by Yan et al. (*482*) in terms of the gel delignification theory.

Biodegradation. In 1980, Hall (*444*) implicated reduced and otherwise activated oxygen-species such as superoxide radical anion in the degradation of lignin when efforts to identify extracellular lignin-degrading enzymes initially failed. This non-selective oxidative lignin degradation mechanism was subsequently shown to be supplemented by distinctive enzymes. Kuwahara, et al. (476), isolated the Mn-peroxidase enzyme in 1984 from ligninolytic cultures of *Phanerochaete chrysosporium.* Later three forms of Mn-peroxidases were isolated by Johansson and Nyman (*500*). Forrester et al. (*507*) reported in 1988 that Mn-peroxidase is unique because of its apparent ability to penetrate lignocellulosic complexes.

In 1976, Forss and Passiner (*414*) described the Pekilo process for production of food and feed protein through the continuous fermentation of SSL carbohydrates and acetic acid by microfungus *Paecïlomyces varioti.* Ovchinnikova et al.. (*470*) in Russia in 1983 studied the possibilities of using oxygen-alkaline bleaching effluents for production of yeast for animal feed.

In 1971, Goheen (*379*) reported that kraft black liquor has an energy content of 2948 Kcal/Kg in spite of containing more than 40% inorganics. Lignin alone has thermal content of about 61,000 Kcal/Kg which is about 1.5 times more than cellulose. Burning kraft black liquor solids, according to Goheen, was economically a marginal use and perhaps the best use of lignin would be to leave it in paper and thus improve the yield.

Lignin Products. During this time period, there were several publications on studies into development of lignin-based resins. Much of this work was concentrated on thermosetting polyurethanes, phenolics, epoxies and acrylates; and on thermoplastic esters, ethers and star-like block copolymers.

Forss and Fuhrmann reported in 1976 the production of Karatex which was a kraft black liquor based adhesive for plywood (*413*). In 1979, Gratzl (*438*) discussed the potential of using technical lignins as extenders in phenolic resins. He stated that for softwood kraft lignin; base-catalyzed hydroxymethylation occurs at the 5-postion for about every third structural unit. For acid catalyzed ketone-kraft lignin condensation, reaction is estimated to occur in 40 to 70 sites per 100 structural units.

In 1982, Mathur studied the characterization of SSL-phenol condensation product using several analytical techniques (*464*).

Lindberg, Kuuasla and Levon (*527*) of the University of Helsinki discussed the chemical and technological aspects of lignin modification and utilization paying particular attention to the ability of lignin to compete with oil-based products.

Glasser with Saraf (*408*) in 1975 at VPI reported on engineering plastics containing lignins. In 1985 with Saraf and others (*475, 477, 481*), Glasser discussed the effect of carboxylation and on hydroxyalkylation on lignin containing plastics. Glasser's work (*518*), provided the first successful approach to the development of thermosetting and thermoplastic structural polymers using lignin derivatives based upon kraft and organosolv lignins.

Industrial Lignins. Georgia Pacific continued to expand production and sales of its LIGNOSITE product line of lignosulfonates.

In 1982, the lignin manufacturing facilities of the American Can Corporation in Rothschild and Green Bay were sold to Reed's North American Paper Group out of Canada (Rasmussen, G., Lignotech-USA, personal communication, 1998). In 1988, Daishowa Chemicals Inc acquired the Division. In 1991, Borregaard LignoTech of Norway acquired the business (Aarsrud, W., Lignotech-Sweden, personal communication, 1998).

From the middle of 1970's, the Hoquiam, WA sulfite mill of ITT Rayonier sold all the sodium lignosulfonates produced at the mill until the mill was closed in 1992. During some years the profitability of the lignin business was greater than that of the pulp business. The Marathon vanillin plant was sold to ITT Rayonier in 1975 and transported to Hoquiam where it operated until the shutdown of the sulfite mill (Hergert, H., consultant, personal communication, 1998).

Tembec Inc, at Temiscarning, Quebec began selling evaporated ammonium lignosulfonates in 1983 for feed pelleting. In 1984, production increased to include spray-dried ammonium and sodium lignosulfonates. (Benson, R., Tembec Paper, Quebec, personal communication, 1998).

Alcell lignin was produced on a pilot plant scale in Valley Forge, PA from the delignification of hardwoods by acidic ethanol-water. Much research was undertaken to understand the characteristics of this product as well as possible applications. (Hergert, H., consultant, personal communication, 1998.)

In about 1990, the Aldrich Chemical Company began to list for sale a series of certain lignin preparations including organosolv and autohydrolysis lignins. (Glasser, VPI, personal communication, 1998.)

Books. In the 1975-1989 period, additional important books and major reviews appeared including: Adler, *Lignin Chemistry, Past, Present and Future*, 1977 (*413*); Nakano, *Chemistry of Lignin*, 1978 (*432*); Glasser, *Lignin*, 1980 (*442*); Fengel and Wegener, *Wood Chemistry, Ultrastructure and Reactions*, 1984 (*473*); Boye, *Utilization of Lignin and Lignin Derivatives*, 1984 (*472*); Glasser and Sarkanen, *Lignin, Properties and Uses*, 1989 (*518*), Kirk, Higuchi and Chang, *Lignin Biodegradation*, 1980 (*445,446*); Crawford, *Lignin Biodegradation and*

Transformation, 1981 (*450*); Higuchi, Chang, and Kirk, *Recent Advances in Lignin Biodegradation Research*, 1983 (*468*); Higuchi, *Biosynthesis and Biodegradation of Wood Component*, and *Bioprocess Engineering*, 1985 (*489*)and 1989 (*522*).

An excellent small but encompassing textbook was written by Sjöström entitled, *Wood Chemistry; Fundamentals and Applications*, 1981 and 1993 (*461,13*).

International Synposium on Wood and Pulping Chemistry. After World War II, several international wood chemistry related meetings were held in such locations as Seattle, New York, and Grenoble. At these meetings, discussions developed on the conflict of these international meetings with regional meetings. At a TAPPI wood chemistry meeting held during a conference in 1980 in Helsinki, Knut Kringstad of Norway and Ingemar Falkehag of the United States led a discussion where it was decided to assemble one truly international meeting on alternate years. Involved in this discussion were J. McCarthy (USA), H. Nimz (Germany), C. Heitner (Canada), J. Jensen (Finland), and others (K. Kringtad, private communication, 1999)

The Swedish researchers took the initiative of organizing the first symposium in 1981 in Stockholm as the Ekman Days, in honor of the man who established the first chemical pulp mill in 1874. The Symposium has met biennially in many cities around the world including Tokyo, Japan (1983), Vancouver, Canada (1985); Paris, France (1987), Raleigh, USA (1989); Melbourne, Australia (1991); Beijing, China (1993), Helsinki, Finland (1995) and Montreal, Canada (1997).

1990 - 1998.

Significant advances have been made in the study of the biosynthesis of lignin and lignans. The stereoselective coupling of coniferyl alcohol was observed and this finding suggested that lignin formation may be not random, but an ordered process. Gene manipulation of a *Populus* gave rise to a substantial increase in cellulose content, and a significant decrease in lignin content. An eight membered ring was found in lignin. A useful procedure for description of lignin structure using fractal geometry was suggested. Industrial lignin technology and opportunity continued to advance.

Chemistry

Occurrence. In 1997, Hansen and Björkman (*633*) at the Danish Technical University examined the ultrastructure of the wood fiber wall by using the Hansen solubility parameter concept. They found that while lignin and cellulose had no affinity, hemicelluloses had affinity for both and therefore acted as a type of surfactant binding cellulose and lignin together.

Biosynthesis. During this time, knowledge increased greatly concerning many aspects of the biochemistry and genetics of plant growth and metabolism (*652*). In 1998, Lewis and Davin (*672*) reviewed what was known about lignin biosynthesis in three phases: pathways to monolignols, coupling of phenylpropanoids to lignin or lignan dimers, and conversion of dimers to polymers, i.e. lignins.

Several investigators described the manipulation of genes in trees hoping that bioengineering would produce more easily pulped trees. Chang et al.. (*668*), reported that the cellulose and lignin content of *Populus tremuloides* could be changed substantially through this process. In 1998, Lapierre, Petit-Conil, and Leplé (*670*) studied the effect on lignin structure and kraft pulping of genetic engineering of *Populus* lignin through downgrading certain key enzymes of the lignin biosynthetic pathway.

Lewis, Sarkanan and associates (*673,640*) reported that when an oxidase such as lacase is incubated alone with E coniferyl alcohol, three racemic dimers are formed. However, when a particular 78 kva dirigent protein (isolated from a *Forcythi*a) is added to the enzyme mixture, the coupling is redirected to afford essentially only (+)- pinoresinol. This was the first example of *in vitro* steroselective bi-molecular phenoxy radical coupling; an essential step in the overall biosynthesis process. The subsequent polymerization of these dimers by a template mechanism (*683*) to lignins and lignans is postulated by S. Sarkanan to occur.

Higuchi (*635*) recently suggested that all enzymes involved in lignin formation have now been recognized. Landucci (*638*) prepared and characterized a DHP through a dehydropolymerization of ρ-hydroxycinnamyl alcohol with metal oxidants.

Properties and Charactrization. Evstighneyav, et al. (*553*) at the St. Petersburg Forest Academy described the electrochemistry of quinonemethides. In 1989, Lundquist, Von Unge, Bardet and Robert reported the distribution of *erythro* and *threo* types of β-O-4 structures in aspen lignin as determined by ^{13}C-NMR using the INADEQUATE pulse sequence (*531*). In 1997, Li, Lundquist and Stomberg (*645*) studied the stereochemistry of lignin structures and products of alkaline degradation.

Many procedures have been developed for the characterization of lignins by physical methods. Examinations of MWL using the new MALDI mass spectrographic procedure were described by Metzger et al. (*567*). It was determined that individual lignin species could be identified and quantified up to a molecular weight of as high as perhaps 50,000. Using the same procedure, Angelis et al. (*613*), evaluated the molecular weight distribution of a DHP and found reversibility in the formation.

In 1993, Buchholz et al. (*552*) characterized certain purified *pauci* disperse lignosulfonate fractions of differing M_w ranging from 2400 to 140,000. Hydrogen, oxygen, and methoxyl contents per C9 were found to be similar while hydroxyl and sulfonate increased with decreasing molecular weight. Constants for the Mark-Houwink equation were determined. As a function of M_w, values were obtained for diffusivity, UV absorptivity, intrinsic viscosity, hydraulic radius, and zeta potential.

Tsutsumi, et al. (*543*), described the relative utility of permanganate oxidation versus ozonation procedures for characterizing structures of lignins and model compounds. *Erythro* and *threo* ratios for α-carbon configurations in softwoods, hardwoods and certain grasses determined by ozonation were estimated by Sarkanen, Islam, and Anderson (*569*). Many characteristics of jute lignin were reported by Sarkanen and Islam (*585*).

For rapid characterization of wood and lignins, Meier and Faix (*566*) showed the value of analytical pyrolysis. In 1992, Faix (*655*) summarized the utility of FTIR spectroscopy in rapid lignin characterization.

In 1998, Akim, Cordeire and Neto (*653*) characterized cork lignin by dry HI splitting with ^{1}H-NMR and GPC analysis. Their study indicated that cork lignin is a three-dimensional cross-linked network polymer rather than a branched type polymer.

Chemical degradation of lignins by the use of thioacidolysis was again employed by Monties, Lapierre et al. (*631*) in France to establish linkages in certain dimers. Applications were described (*587*) for this procedure to provide molecular profiles of lignins.

Stepwise cleavage of ethers in a hardwood lignin by the use of trimethylsilyl iodide by Hori and Meshitsuka (*636*) revealed heterogeneity in structural units. Certain β-5 types of lignin model compounds were synthesized by Li and Lundquist (*641, 642*). Alkaline hydrolysis of non-phenolic β-O-4 model compounds was again reported by Schultz and Fisher (*648*). Kadala, et al. (632), investigated high temperature reactions of model compounds and lignins with stabilized hydrogen peroxide.

In 1998, Omori, Aoyama, and Sakakibara (*681*) at Hokkaido University studied the reactions of β-0-4 lignin model compounds in the presence of carbohydrates. Recombination of cinnamyl alcohol radicals produced during heating of dimeric model compounds was not observed if polysaccharides were present. Carbohydrates were considered to act as hydrogen donor to stabilize the formed radical species.

Isolated lignins and model substances were studied at McGill University by Argyropoulos et al. (*627*) using ^{19}F -NMR & ^{31}P -NMR procedures.

Formulations. Forss and Fremer (*658*) in 1998 again pointed to the concept that coniferous lignin is a heterogeneous material composed of polymeric glycolignin and a substantial proportion of generically different low molecular mass hemilignins. According to this view, it is not possible to describe adequately polymeric lignin and the different hemilignin molecular species by specific structural formulas.

Karhunen, Rummakko, Sipilia and Brunow (*606*) discovered the presence of a new type of intermonomer linkage in softwood lignin representing an 8-membered ring (dibenzodioxocin). The structure was synthesized by Karhunen, Rummakko, Pajunen and Brunow (*620*) in 1996. A structure containing this unit is presented in Figure 6 (*687*). A similar 8-membered ring configuration also appears in Taxol, a medically significant component from the bark and wood of certain trees of the *Taxus* genus (*578*).

As a new approach to describe lignin structures, fractal geometry was applied by Gravitis (*558*) in 1992. Jurasek (*619*) also used fractal dimensions to describe middle lamella, secondary wall and DHP lignins.

Three-dimensional computer modeling of middle lamella, secondary wall, and DHP lignins was continued by Jurasek in 1998 (*669*) using seven bond types with specified frequencies reported in Table I. Lignin-hemicellulose cross-linkages were used in the secondary wall modeling. Lignin structures have been created for

DHP entities using both *zutropf* and *zulauf* procedures. Consideration was given to both hydrogen bonding and primary valence cross-linking. Secondary wall structural units and their aromatic rings were preferentially oriented in parallel with the main fiber axis with a fractal geometry order parameter of 0.19 and 0.18 for softwood and hardwoods, respectively. Atalla (*498*) has reported observing similar orientation from Raman microprobe experimentation.

Technology

Reactions. In 1992, Hrutfiord and Negri (*559*) studied the role of residual lignin in the chlorine bleaching process as the possible source of PCDD and PCDF. Working with DHP lignin, it was found that phenols substituted in the 4-position with a chlorine-displacement side chain were good dioxin precursors. Compression wood possessed much more of these structures than normal woods. The mechanisms involved in this process were discussed in a separate paper (*560*).

In 1993, Cole studied brightness stability in high yield pulps (*579*). Heitner studied the generation of free radical species during the light-induced yellowing of pulp (*665*). Liu, Gustafson, Callis and McKean (*643*) described a method to determine the residual lignin content of individual pulp fibers through measurement of fluorescence after staining using a microphotometric method.

The increase of the content of carboxyl groups in industrial lignins by alkali fusion was reported by Nakano and Yamada (*609*).

Pulping. In 1991 SSA-Billerud in Sweden commercialized the "shock" pyrolysis method to convert sodium based spent sulfite liquor to sulfur-free sodium carbonate (Björkman, DTU, personal communication, 1996). Recently this method has been modified to supplement or to replace Tomlinson kraft recovery furnaces.

Leclerc and Olsson (*563*) reported in 1992 that lignin removal during pulping can be pictured by de-gelation models assuming the presence in wood of two distinct cross linked lignins; one in the middle lamella and the other in the secondary wall. Delignification mechanisms for hardwoods in a soda-AQ system were proposed by Venica, Chen and Gratzl (*651*). Gellerstedt (*614*) described reactions during CMP pulping.

Literature relating to delignification using flow-through reactors has been reviewed by Islam et al. (*618*) in 1996. In this procedure, plant tissue as thin platelets remains stationary in the reactor while the reaction liquid flows through at a nearly constant composition.

Compositional changes in structural units during kraft and Alcell (acidic aqueous ethanol) delignification were studied in 1998 by Lin, Pye and Argyropoulos (*676*) using ^{31}P- NMR and other procedures. In the Alcell process, significant lignin condensation at C-5 was found to occur but at different degrees for softwood and hardwood lignins. Aspen MWL was found to contain about 4-ρ-hydroxyphenyl propane structures per 100 structural units which were easily dissolved during pulping.

Lonneberg (*644*) evaluated the reactivity of residual lignins in pulps. Treimanis (*573*) reported on the influence of fiber structure on pulp properties.

Bleaching. The mechanisms of kraft pulp bleaching were reviewed by Gellerstedt (*614*) in 1996. The shift away from chlorine bleaching has been reported by Moran (*680*). 1998 estimates show the dominant use of ClO_2, with major decreases in Cl_2 and rising use of H_2O_2 and O_2, and ozone (O_3). The trend toward chlorine free and use of oxygen derivatives is strong and doubtless will continue.

During the 1988-1995 period, the reactions between oxygen radicals and prominent lignin structures were extensively studied by Gierer and coworkers (*594*). Oxygen radicals were generated by γ-rays using determination methods based on chemiluminescence. In 1990, Gierer (*534*) rationalized the complex course of bleaching of pulps by dividing the reactions involved into three categories: cationic, radical, and anionic reactions. Gierer in 1997 (*632*) paying particular to the role played by superoxide and hydroxyl radicals reviewed the chemistry of totally chlorine free (TCF) bleaching

In 1994, Akim (*592*) compared bleaching of sulfite pulp using oxygen, peroxide and ozone technologies to older methods using chlorine and hypochlorite. Akim reported that oxygen and alkaline bleaching allows for the complete elimination of chlorination, the direct discharge of bleaching effluents into the kraft black liquor recovery system, increased yield and quality of bleached pulp, and significantly bleaching costs. The decrease in volume and contamination of bleaching effluents in this system assists the goal of closed cycle water system. . The kinetics of oxygen bleaching of hardwood sulfite pulp were described by Koksharov and Akim (*562*).

Application of ligninolytic enzymes to bleaching of kraft pulp with manganese peroxidase was investigated by Harazono, Kondo and Sakai (*616*) in 1996.

Biodegradation. In 1993, Gierer (*582*) proposed that the microbial degradation of lignin is brought about by superoxide and hydroxyl radicals rather than by enzymes such as ligninases, phenol oxidases and peroxidases. These enzymes are considered to caytalyse the formation of the degrading radicals rather than the degradation proper as suggested earlier.

Wariishi, Valli and Gold (*550*) reported that a manganese peroxidase enzyme plays an important role in fungal degradation of lignin. Lignin peroxidases were used by Kurek and Monties (*586*) to oxidize a spruce lignin and by Hattori and Shimada (*625,634*) to degrade a β-O-4 type lignin-carbohydrate model compound. Jonsson, Karlsson, Lundquist and Nyman in Sweden (*538*) in 1990 described the stereospecificity in enzymic and non-enzymic oxidation of β-O-4 lignin model compounds.

Degradation of lignin and phenolic compounds by basidial fungi was studied by Medvedeva et al. (*588*).

Utilization of Lignin

Lignins in Medicine. Some lignin formulations properties make them well-suited for certain medicinal purposes. Kraft lignins and lignosulfonates posses certain antibiotic activity due to the presence of phenolic and carboxyl groups Nakano et al. (*599*). The sorption properties of lignins open up new avenues for their utilization in medicinal products (*589*).

In human nutrition, the term "dietary fiber" includes the plant polysaccharides and lignin, which is resistant to hydrolysis by the digestive enzymes of humans (*419, 611*). It is known that lignified cell wall carbohydrates are well digested but the limits of this digestion are set by lignin. (*393*). Trowell et al. (*419*) tested the effect of lignin on digestibility in animals by regenerating blends of lignin and cellulose and found increased digestibility in the presence of lignin.

Lignin Polymer Products. The cure behavior of lignin-containing resins was described by Glasser and Hofmann (*584*) in 1993 by application of the Gillham - TTT modeling approach. With this information it was possible to assess lignin with regard to its ability to become an integral part of infinite networks as opposed to being a glass-like filler. Phase-compatabilizing lignin modifications are revealed for the incorporation of lignin into such thermoplastic polymers as polyolefins, polyacrylics, PVC and others by Oliveira and Glasser in 1994 (*599-600*). The first prototype printed circuit board (PCB) for the electronics industry was formulated by IBM with a 50% lignin-containing epoxy resin (*623*). The synthesis, properties and degradation of organo-silicon derivatives of lignin were reported by Telycheva (571).

Oliveira and Glasser in 1994 (*597*) described a series of star-shaped lignin block copolymers showing extraordinary compatibility characteristics with synthetic polymers. Thermoplastic blends of lignin and lignin-derivatives in synthetic polymers were shown to be subject to property engineering via lignin content and lignin modification (*597-600*). Monties and Chabbert (*697*) correlated structural variations with physicochemical properties of certain synthetic materials containing industrial lignins. Thermol and mechanical properties of kraft lignin urethanes were determined by Nakano et al. (*646,647*)

Industrial Lignins. In 1998, about 1% of all lignin generated in paper production worldwide was isolated and sold. Most of this material was lignosulfonates, which are produced by Fraser Paper Inc., Georgia-Pacific Inc., Nippon Paper Industries Ltd, LignoTech-USA, Borregard Lignotech-Norway and Tembec, Inc. Westvaco and Lignotech-Norway produce kraft lignins. In 1990, Lin and Zhong (*540*) listed that the world production of lignin products at 138.5 million Kg/year. In 1998, Gargulak and Lebo (*661*) reviewed the current uses of industrial lignins and reported on new and unique possible uses. Total sales in 1980 and 1996 were reported to be about $180 and $600 million, respectively.

There are numerous uses for lignin products including binders for animal feed pellets, bricks, ceramics, and road dust, dispersants for oil well drilling products, dyestuffs, pesticides, carbon black, concrete, gypsum, etc., water treatment additives, and battery expanders. Utilization and markets for lignins have often been discussed. Terent'ev (*549*) reviewed the utilization of lignins in Russia.

In 1998 Fraser Paper Inc (formerly Flambeau Paper Co) in Park Falls, WI, USA produced 170 tons/day of lignosulfonate solids from birch wood. Sales of calcium lignosulfonates from this site began 1965. Xylose is recovered from the lignosulfonates for the production of xylitol. (McColloch, J., Frazer Paper, personal communication, 1998).

Georgia-Pacific Inc. sold over 220,000 tons of lignosulfonates and 5,000,000 gallons of ethanol in 1998. The majority of products were fermented calcium and sodium lignosulfonates along with a large variety of modified specialty products.

LignoTech USA, with its manufacturing facilities in Rothschild, WI continued to produce and sell a variety of lignosulfonate products. Its parent company, Borregaard LignoTech of Norway which belongs to the publicly traded company Orkla group, has eight production plants in Europe and in the USA with customers in more than 70 countries. Vanillin from lignosulfonate is produced in Norway and this is apparently the only lignin-vanillin plant still operation in the world. SAPPI of South Africa and Borregaard of Norway have agreed to produce large amounts of hardwood lignosulfonate products in South Africa (Aarusrud, W., Lignotech-Sweden and (Rasmussen, G., Lignotech-USA personal communication, 1998).

Established in 1993 by merger of Jujo Paper Co (founded 1949) and Sanyo-Kokasaku Pulp, Nippon Paper Industries Ltd. markets wood chemicals through its Chemical Products Division. Several products are marketed including lignosulfonates (San X, Vanilex, Pearlex, Sanflo) and also ribonucleic acid (Sanyo RNA) and yeast (Sanyo Yeast) from wood sugars in the lignosulfonates. Vanillin was also produced from lignosulfonates but production terminated in 1994. (Machihara, A., Nippon Paper, Tokyo, personal communication, 1998.)

Tembec of Canada sells about 75,000 tons/yearr (50% basis) of liquid and powdered NH_3SSL as well as some NaSSL products. Ethanol (18 million liters/year) is produced from SSL sugars. (Benson, R., Tembec Paper, Quebec, personal communication, 1998).

In 1998, the polychemicals division of Westvaco marketed a variety of specialty lignin chemicals produced from kraft black liquor. These products find uses in the following fields: dyes and pigment chemicals, mineral technology, asphalt, agricultural, lead storage batteries and phenolic resins (Hintz, H., Westvaco, personal communication, 1998).

The Repap Company in Valley Forge, PA operated an ethanol pulping plant that produced Alcell ethanol lignins until operations were terminated in 1998 because of financial difficulties. These lignins competed in the same markets as lignosulfonates and kraft lignins. (Hergert, H., consultant, personal communication, 1998).

Books. Since 1990, important books have continued to appear including Nakano, *Chemistry of Lignin*, 1990 (*541*); Faix and Meier, *Proceedings of the 1st EWLP Utilization and Ayalyses of Lignin*, 1990 (*533*); Lewis and Goldstein, *Wood Structure* *and Composition*, 1991 (*545*); Hon and Shirashi, *Wood and Cellulose Chemistry*, 1991 (*548*); Lin and Dence, *Methods in Lignin Chemistry*, 1992 (*564*); Kennedy, Phillips and Williams, *Lignocellulosics, Science, Technology, Development and Uses* 1992 (*561*); Heitner and Scaiano, *Photochemistry of Lignocellulocis Materials*, 1993 (*583*); Zakis. *Functional Analysis of Lignins and their Derivatives*, 1994 (*601*); Hon, *Chemical Modification of Lignocellulosic Materials*, 1996 (*617*); Higuchi, *Biochemistry and Molecular Biology of Wood*, 1997 (*635*); Lewis and Sarkanen, *Lignin and Lignan Biosynthesis*; 1998 (ACS Symposia Ser. 697) (*675*) and Young and Akhtar, *Environmentally Friendly Pulping Processes*, 1998 (*686*).

Concluding Comments

Chemistry. After more than a century and a half of research, we should think of lignins as a family of molecules for which the chemical structure is generally known. However, many immediate problems remain to be resolved, such as the nature and extent of chain branching and/or crosslinking, and the sequence of structural units, etc. Differences in lignins in individual plant tissues, and certainly among the varieties of higher plants need to be elucidated.

Biochemistry and genetic studies of biosynthesis have opened up new opportunities for understanding of how lignins and lignans are formed, dimerize and polymerize. The new book on Lignin and Lignan Biosynthesis, edited by N. Lewis and S. Sarkanan, reveals many exciting advances and challenges. The several chapters, including some by the editors themselves, contain many papers (too numerous and extensive to review here) by such leading researchers as Atalla, Brunow, Ericsson, Ralph Shimada, Tershima and others.

It has long been recognized that lignin cannot be isolated without change. Thus each family of lignin calls for characterization, which uniquely depends on the procedures, and chemical reactions used in isolation, and this diversity must be superimposed upon that inherent in the plant and its growing environs. Much information of this sort has already been collected and doubtless will be carefully compiled but this task does not seem an appropriate part of this paper.

Current models and formulations of lignins are probably close to correct but many details need to be added. The three-dimensional configuration of lignin is particularly challenging with respect to its biosynthesis and its relation to the architecture of cellulose and hemicellulose structures.

All these matters must be taken into account in descriptions of *in vivo* lignins constructed by Nature.

Technology and Utilization. Available industrial lignins apparently now are always byproducts, and thus the main product substantially controls their properties and costs. The largest use of lignins is probably wood and reformed products. This is because in construction materials, it is inexpensive, easily worked and available throughout the world. Better technology for preserving wood via biodegradation studies will enhance this type of utilization.

The next largest use of wood is for production of cellulose products where the byproduct lignins are mainly employed as a fuel to produce process steam. The heat generated by burning effluent solids is approximately equal to that required for evaporating and processing. Delignification, however, is not complete, and costly and potentially polluting bleaching steps are required. What is needed is technology for "brightening" residual lignins to produce white paper so the lignin would then be sold at the price of paper itself.

The cost of an isolated kraft lignin, for example, might be estimated mainly as the cost of a replacement fuel such as natural gas, oil or coal. Thus the market for isolated lignins, such as kraft lignin, must be found in relatively high priced products, and this seems the pathway followed by the Westvaco Company.

When the effluent solids are not burned, environmentally friendly disposal presents a substantial problem and, indeed, significant cost. Effluent sugars may be permeated to produce industrial ethanol. Some byproduct lignin sulfonates, usually after modification, are sold as concrete oil well drilling mud additives and other products by such companies as Lignotech and Georgia Pacific.

Alcohol delignification produces a cellulose product and sulfur free lignin. However the promise of this process had been darkened by the recent closure of the ethanol pulping trial, apparently mainly because of shortcomings in paper quality and uncertainty in profits from the lignin which is said to have many useful and reliable properties.

Thus, the large-scale challenge which lignins brings to technology, at least on a short term basis, is to find methods for utilization of lignins which will bring profits sufficiently high to make advantageous use of alternate fuels such as exhaustible natural gas, oil and coal, now so liberally available.

Acknowledgements

The authors greatly appreciate the good advice, valuable information and most helpful review of drafts of this manuscript by the following scientists and their colleagues and coworkers: Anders Bjorkman of DTU in Copenhagen Denamrk; Oscar Faix of UH in Hamburg Germany; Kaj Forss of CL in Helsinki, Finland; Herbert Hergert as consultant and recently associated with Repap Inc. in Valley Forge, USA; Bjorn Hrutfiord of UW in Seattle, USA, Josef Gierer of KTH in Stockholm, Sweden; Wolfgang Glasser; of VPI in Blacksburg, USA; David Goring of UT in Toronto, Canada; Josef Gratzl of NCSU in Raleigh, USA; Norman Lewis of WSU in Pullman, USA; Gyosuka Meshitsuka of UT in Tokyo; Mikio Hatakayama in Fukui Institute of Technology in Fukui, Japan; Horst Nimz of *Holzforschung* in Germany; Robert Northey of UW, Seattle, and Richard Buchholz until recently associated with Georgia Pacific West Inc, in Bellingham, USA; Danielle Robert of CERMAV in Grenoble, France; Yoshhiro Sano of UH in Kita, Japan; Mikio Shimada of KU in Kyoto, Japan; Adrian Wallis of CSIRO in Melbourne, Australia; Terery Fullerton of PAPRO in Rotorua, New Zealand; and also Michael Zarubin and Sergi Shevshenko of SPFTA and Eduard Akim of VNIB both in St Petersburg, Russia.

Thanks is especially expressed to Editors R. Northey, R. Schulz and W. Glasser for their guidance and to Professor J.W. Rogers, Jr., and Ms. La Donna Kennedy of the University of Washington Department of Chemical Engineering for their support and help in completing this manuscript.

Finally, one of the authors (JLM) records his most sincere appreciation to the members and officers of the Cellulose Division of the ACS for honoring him with the 1997 Anselme Payen Award.

LITERATURE CITED

1. Payen, A. *Compt. Rend.* **1838**, *7*, 1052.
2. Payen, A. *Compt. Rend.* **1839**, *8*, 52.
3. Reid, J.D.; Dryde, E.C. *Textile Colororist* **1940**, *62*, 43.
4. Hunter, D. *Papermaking: The History and Technique of an Ancient Craft*; Dover Publications, Inc.: N.Y., 1947; 483.
5. *In* Hunter, Ref. 4, p. 466.
6. Plumlee, C. *Paper Age* **1998**, *114*, 4, 24-25.
7. Watt, C.; Burgess, H., U. S. Patent No. 1,448 and 1,449, 1854.
8. Schulze, F. *Chem. Zentr.* **1857**, 321.
9. Possoz, L. *Dingler Polytech. J.* **1858**, 127, 150, 382.
10. Erdmann, J. *Annal.* **1866**, *138*, 1.
11. Tilghman, B.C. British Patent No. 2,924; 1866.
12. Payen, A. *Dinglers Polytech. J.* **1867**, *185*, 308.
13. Eaton, *In* Sjöström, E. *Wood Chemistry: Fundamentals and Applications*; Second Edition; Academic Press, Inc.: N.Y., 1993; 116.
14. Ekman, in Sjöström, Ref. 13, p. 104.
15. Tiemann, F.; Mindelsohn, B. *Ber.* **1875**, *8*, 1139.
16. Nikintin, N.I. *The Chemistry of Cellulose*; 1966; 478.
17. Tiemann, F.; Haarman, W. *Ber.* **1874**, *7*, 606.
18. Wiesner, J. *Sitzber Akad. Wiss. Wien, Math.-naturw. Klasse* **1878**, *17*, 511.
19. Bevan, E.J.; Cross, C.F. *J. Chem. Soc.* **1880**, *38*, 666.
20. Wheelwright, in Hunter, Ref. 4, p. 575.
21. Dahl, in Hunter, Ref. 4, p. 575.
22. Tiemann, F.; Haarman, W. *Ber.* **1875**, *8*, 509.
23. Riordin, in Hunter, Ref. 4, p. 576.
24. Lange, G. *Z. Physiol. Chem.* **1890**, *14*, 15, 217.
25. Lindsey, J.B.; Tollens, B. *Annal.* **1892**, *267*, 341, 353.
26. Klason, P. *Tek. Tid. Uppl. C. Kemi.* **1893**, 23, 33, 53, 55.
27. Klason, P. *Beritäge zur Kenntnis der chemischen Zusammensetzung des Fichenholzes*, Berlin, 1911.
28. Klason, P. *Svensk. Kem. Tidskr.* **1897**, *9*, 133.
29. Pollacsek, E. *Austrial Prov.* **1898**, 1524.
30. Grafe, V. *Monatsh. Chem.* **1904**, *25*, 987.
31. Brompton, in Hunter, Ref. 4, p. 581.

32. Klason, P. *Hauptversammlungsber. Ver. Zellstoff- u. Papier-Chemiker u. Ingenieure* **1908**, 52.
33. Klason, P. *Ark. Kem. Mineral. Geol.* **1908**, *3*, 5, 9.
34. Klason, P. *Ber.* **1920**, *53*, 705, 1862, 1864.
35. Procter, H.R.; Hirst, S. *J. Soc. Chem. Ind.* **1909**, *28*, 293.
36. Ruttan, R.F. *J. Soc. Chem. Ind.* **1909**, *28*, 1290.
37. Klason, P. *Arkiv Kemi, Mineral Geol.* **1910**, *4*, 1.
38. Klason, P.; Segerfelt, B. *Arkiv Kemi, Mineral Geol.* **1911**, *4*, 6.
39. Tomlinson, G.H. *Ind. Eng. Chem.* **1918**, *10*, 859.
40. Heuser, E.; Sieber, R.Z. *Angew. Chem.* **1913**, *26*, 801.
41. Ungar, E., *Dissertation*, Zürich, 1914.
42. Reed, H.C. U. S. Patent No. 1,217,218; Feb. 27, 1917.
43. Hägglund, E. *Arkiv. Kemi.* **1918**, *7*, 20.
44. Tomlinson, G.H. *Chem. Ind.* **1948**, *24*, 371.
45. Heuser, E.; Skiöldebrand, C. *Z. Angew. Chem.* **1919**, *32*, 41.
46. Tomlinson, G.H. *National Research Council of Canada* **1919**, *No. 7*.
47. Dore, W.H. *Ind. Eng. Chem.* **1920**, *12*, 475.
48. Klason, P., in Sjöström, Ref. 13, p. 71.
49. Schrader, H. *Ges. Abhand. Kenntnis Kohle* **1920**, *5*, 276.
50. Heuser, E.; Winsvold, A. *Cellulosechem* **1921**, *2*, 113.
51. Heuser, E.; Samuelsen, S. *Cellulosechem.* **1922**, *3*, 78.
52. Klason, P. *Chem. Ber.* **1922**, *55*, 448.
53. Ritter, G.J.; Fleck, L.C. *Ind. Eng. Chem.* **1922**, *14*, 1050.
54. Ritter, G.J.; Fleck, L.C. *Ind. Eng. Chem.* **1923**, *15*, 1055.
55. Ritter, G.J.; Fleck, L. C. *Ind. Eng. Chem.* **1926**, *18*, 608.
56. Hägglund, E.; Björkman, C.B. *Biochem. Z.* **1924**, *147*, 74.
57. Heuser, E.; Herrman, F. *Cellulose Chem.* **1924**, *5*, 1.
58. Fierz-David, H.E.; Hannig, M. *Helv. Chim. Acta.* **1925**, *8*, 900.
59. Hägglund, E. *Svensk Kem. Tid.* **1925**, *37*, 116, 120.
60. Krohn, V. *Ann. Acad. Sci. Fennicae* **1924**.
61. Hawley, L.F.; Wise, E.L. *The Chemistry of Wood: Lignin; ACS Monograph Ser.*; The Chemical Catalog Company, Inc.: NY, 1926; 43-96.
62. Kürschner, K. *Zur Chemie der Ligninkorpe* **1926**, 116.
63. Kürschner, K. *Wochbl. Papierfabr.* **1926**, *57*, 26.
64. Schorger, A.W. *The Chemistry of Cellulose and Wood: Lignins*; First Edition; McGraw-Hill: NY., London, 1926; 70-121.
65. Urban, H. *Cellulose Chemie* **1926**, *7*, 73.
66. Freudenberg, K.; Harder, M. *Chem. Ber.* **1927**, *60*, 581.
67. Hägglund, E.; Johnson, T. *Biochem. Z.* **1927**, *187*, 98.
68. Herzog, R.O.; Hillmer, A. *Chem. Ber.* **1927**, *60*, 365.
69. Mason, W.H. *Paper Trade J.* **1927**, *55*, 131.
70. Mason, W.H. U. S. Patent No. 1, 655, 618; 1928.
71. Scholler, H. *Zellstoffaser* **1933**, *32*, 64.
72. Scholler, H. German Patent No. 676,967; 1939.
73. Scholler, H. U. S. Patent No. 1,641,771; 1927.

74. Freudenberg, K. *Liebigs Ann, Chem.* **1928**, *461*, 130.
75. Freudenberg, K.; Zocher, H.; Dürr, W. *Ber. Deutsch. Chem. Ges.* **1929**, *62*, 1814.
76. Freudenberg, K.; Zocher, H.; Dürr, W. *Ber.* **1929**, *62B*, 1814.
77. Freudenberg, K.; Zocher, H.; Dürr, W. *Ber.* **1938**, *71*, 1810.
78. Freudenberg, K.; Belz, W.; Niemann, C. *Chem. Ber.* **1929**, *62*, 1554, 1561.
79. Hägglund, E.; Klingstedt, F.W. *Svensk Kem. Tid.* **1929**, *41*, 185.
80. Hägglund, E.; Klingstedt, F.W. *Z. physik. Chem.* **1931**, *A 152*, 295.
81. Rassow, B.; Zickmann, P. *J. Prakt. Chem.* **1929**, *123*, 189.
82. Schaarschmidt, A.; Nowak, P.; Zetzsche, W. *Z. Angew. Chem.* **1929**, *42*, 618.
83. Campbell, W.B.; Maass, O. *Can. J. Research.* **1930**, *2*, 42.
84. Freudenberg, K.; Dürr, W. *Ber.* **1930**, *63*, 2713.
85. Holmberg, B. *Svensk Papperstidn* **1930**, *33*, 679.
86. Freudenberg, K. *Cellulosechemie* **1931**, *12*, 267.
87. Hägglund, E. *Svensk Papperstidn.* **1931**, *34*, 160.
88. Nihlén, H. *Svensk Papperstidn.* **1934**, *37*, 754.
89. Howard, G.C. U. S. Patent No. 1,699,845; 1929.
90. Howard, G.C. *Ind. Eng. Chem.* **1930**, *22*, 1184.
91. Howard, G.C. *Ind. Eng. Chem.* **1934**, *26*, 614.
92. Kullgren, C. *Svensk Kem. Tid.* **1932**, *44*, 15.
93. Hägglund, E. *Finish Paper Timber J.* **1934**, *16*, 383.
94. Stamm, A.J.; Semb, J.; Harris, E.E. *J. Phys. Chem.* **1932**, *36*, 1574.
95. Erdtman, H. *Biochem. Z.* **1933**, *258*, 172, 288.
96. Erdtman, H. *Ann. Chem.* **1933**, *503*, 283.
97. Erdtman, H. *Proc. Roy. Soc.* **1933**, *143A*, 177, 191, 223, 228.
98. Erdtman, H. *Svensk Kem. Tidskr.* **1935**, *40*, 243.
99. Erdtman, H. *Svensk Paperstidn.* **1941**, *44*, 243.
100. Freudenberg, K. *Tannin, Cellulose, Lignin*; Julius Springer: Berlin, 1933; 133,165.
101. Kullgren, C.; Du Rietz, C. *Svensk Kem. Tid.* **1930**, *42*, 179.
102. Hägglund, E.; Carlsson, G.E. *Biochem. Z.* **1933**, *257*, 467.
103. Kullgren, C. *Svensk Papperstidn* **1933**, *36*, 499.
104. Kullgren, C.; Du Rietz, C. *Svensk Kem. Tid.* **1933**, *45*, 185.
105. Kullgren, C.; Du Rietz, C. *Svensk Kem. Tid.* **1931**, *43*, 99, 161.
106. Kullgren, C.; Du Rietz, C. *Svensk Kem. Tid.* **1932**, *44*, 15.
107. Kullgren, C.; Du Rietz, C. *Svensk Kem. Tid.* **1933**, *45*, 185.
108. Kullgren, C.; Du Rietz, C. *Svensk Kem. Tid.* **1934**, *46*, 136.
109. Kullgren, C.; Du Rietz, C. *Svensk Kem. Tid.* **1937**, *49*, 52.
110. Reidel, J., British Patent No. 401,562, Nov. 16, 1933.
111. Hilpert, R.S.; Littmann, E. *Ber.* **1934**, *67*, 1551.
112. Hilpert, R.S.; Littmann, E. *Ber.* **1935**, *68*, 16.
113. Pauly, H. Austrian Patent No. 83, 306; 1917.
114. Pauly, H. *Ber.* **1934**, *67*, 1188.
115. Phillips, M. *Chem Review* **1934**, *14*, 103.
116. Asplund, A.J.A. U. S. Patent No. 2,008,892; 1935.
117. Brauns, F.E.; Hibbert, H. *Can. J. Research* **1935**, *13B*, 28.

118. Holmberg, B. *Papir, J.* **1935**, *23*, 81, 92.
119. King, E.G.; Brauns, F.E.; Hibbert, H. *Can. J. Research* **1935**, *13B*, 35.
120. Thomas, E.N.M.; Hewitt, J. *Nature* **1935**, *136*, 3428, 69-70.
121. Van Beckum, W.G.; Ritter, G.J. *Paper Trade* **1935**, *105*, 18, 127.
122. Wilcoxon, L.S. *Paper Trade J.* **1935**, *100*, 24, 72.
123. Bailey, A.J. *Industrial Eng. Chem. Analytical Edition* **1936**, *8*, 1.
124. Erdtman, H. *Svensk Kem. Tid.* **1936**, *48*, 230, 236.
125. Freudenberg, K.; Janson, A.; Knopf, E.; A., H. *Chem. Ber.* **1936**, *69*, 1415.
126. Holmberg, B. *Svensk Papperstidn* **1936**, Special no. 113, 39, 117.
127. Holmberg, B. *Chem. Ber.* **1936**, *69*, 115.
128. Anon. *Paper Mill Wood Pulp News* **1936**, *59*, 17.
129. Loughborough, D.L.; Stamm, A.J. *J. Phys. Chem.* **1936**, *40*, 1113.
130. Loughborough, D.L.; Stamm, A.J. *J. Phys. Chem.* **1941**, *46*, 1137.
131. Nikitin, N.I.; Orolova, L.M. *Zh. P. Kh.* **1936**, *9*, 2210.
132. Phillips, M. *Science Progress* **1936**, *30*, 442.
133. Sandborn, L.T.; Salvesen, J.R.; Howard, G.C. U.S. Patent No. 2,057,117; Oct. 13, 1936.
134. Tomlinson II, G.H.; Hibbert, H. *J. Am. Chem. Soc.* **1936**, *58*, 345.
135. Freudenberg, K.; Meister, M.; Flickinger, E. *Ber.* **1937**, *70*, 500.
136. Haworth, R.D. *Ann. Repts. Prog. Chem. Soc.* **1937**, *33*, 267.
137. Maass, O.; Calhoun, J.M.; Yorston, F.H. *Can. J. Research.* **1937**, *5B*, 457.
138. Maass, O.; Calhoun, J.M.; Yorston, F.H.; Cannon, J.J.R. *Can. J. Research.* **1938**, *16B*, 242.
139. Maass, O.; Calhoun, J.M.; Yorston, F.H.; Cannon, J.J.R. *Can. J. Research.* **1939**, *17B*, 121.
140. Olson, F.; Peterson, W.; Scherard, E. *Ind. Eng. Chem.* **1937**, *29*, 1026.
141. Tomlinson, G.H. U.S. Patent No. 2,070,632; Feb 16, 1937.
142. Tomlinson II, G.H.; Hibbert, H. U.S. Patent No. 2,069,185; Jan. 26, 1937.
143. Freudenberg, K.; Engler, K.; Flickinger, E.; Sobek, A.; Klink, F. *Ber.* **1938**, *71*, 1817.
144. Freudenberg, K.; Müller, H.F. *Chem. Ber.* **1938**, *71*, 1281.
145. Harris, E.E.; Adkins, H. *Paper Trade J.* **1938**, *107*, 20, 38.
146. Brauns, E.F. *J. Am. Chem. Soc.* **1939**, *61*, 2120.
147. Brauns, E.F. *Paper Trade J.* **1940**, *111*, 14, 35.
148. Cramer, A.B.; Hunter, M.J.; Hibbert, H. *J. Am. Chem. Soc.* **1939**, *61*, 509.
149. Freudenberg, K. *Z. Angew. Chem.* **1939**, *52*, 362.
150. Tomlinson, G.H.; Wilcoxson, L.S. U.S. Patent No. 2,151,110; June 8, 1939.
151. Anon. *Sulfite Waste Liquors, An Annotated Bibliography*; Institute of Paper Chemistry: Appleton, WI, 1940.
152. Freudenberg, K.; Lautsch, W.; Engler, K. *Chem. Ber.* **1940**, *73*, 167.
153. Freudenberg, K.; Neish, A.C. *Constitution and Biosynthesis of Lignin*; Springer: Berlin, Heidelberg, 1968.Anon. *Sulfite Waste Liquors, An Annotated Bibliography*; Institute of Paper Chemistry: Appleton, WI, 1940.
154. Glading, R.E. *Paper Trade J.* **1940**, *111*, 23, 32.
155. Lautsch, W.; Plankenhorn, E., P.; Klink, F. *Z. Angew. Chem.* **1940**, *53*, 450.

156. Schultz, L. U. S. Patent No. 2,187,366; 1940.
157. Tomlinson, G.H.; Wilcoxson, L.S. *Paper Trade J.* **1940**, *110*, 209, 31.
158. Tomlinson, G.H. *Pulp Paper Mag. Can.* **1944**, *45*, 1.
159. Cooke, L.M.; McCarthy, J.L.; Hibbert, H. *J. Am. Chem. Soc.* **1941**, *63*, 3052.
160. Creighton, R.H.J.; McCarthy, J.L.; Hibbert, H. *J. Am. Chem. Soc.* **1941**, *63*, 312, 3049.
161. Erdtman, H. *Svensk Papperstidn.* **1941**, *44*, 243.
162. Gralen, N. *J. Colloid Sci.* **1946**, *1*, 453.
163. Hägglund, E. *Svensk Papperstidn.* **1941**, *44*, 183.
164. Pearl, I.A.; Bailey, A.J.; Benson, H.K. *Paper Trade J.* **1941**, *113*, 17, 47.
165. Tomlinson, G.H.; McCarthy, J.L.; Hibbert, H. U.S. Patent No. 2,226,356; Dec. 24, 1941.
166. Hibbert, H. *Ann. Rev. Biochem.* **1942**, *11*, 183.
167. Meshitsuka, G.; Nakano, J. *Mokuzai Gakkaish* **1979**, *25*, 588.
168. Pearl, I.A. *J. Am. Chem. Soc.* **1942**, *64*, 1429.
169. Aulin-Erdtman, G. *Svensk Papperstidn.* **1944**, *47*, 91.
170. Erdtman, H. *Svensk Papperstidn.* **1943**, *46*, 226.
171. West, E.; S., M.A.; Hibbert, H. *J. Am. Chem. Soc.* **1943**, *65*, 1187.
172. Creighton, R.H.J.; Gibbs, R.D.; Hibbert, H. *J. Am. Chem. Soc.* **1944**, *66*, 32, 37.
173. Hachihama, Y.; Nira, K.; Kyogoku, T. *Kogyo Kagaku Zasshi* **1944**, *47*, 209.
174. Gardner, J.A.F.; Hibbert, H. *J. Am. Chem. Soc.* **1944**, *66*, 610.
175. Lange, P.W. *Svensk Papperstidn.* **1944**, *47*, 262.
176. McGregor, W.S.; Evans, T.H.; Hibbert, H. *J. Am. Chem. Soc.* **1944**, *66*, 41.
177. Nord, F.F.; Sankey, C.A. *Can. Chem. Process Ind.* **1944**, *28*, 464.
178. Nord, F.F.; Sankey, C.A. *Pulp and Paper Mag. Can.* **1949**, *45*, 171.
179. Tomlinson II, G.H. *Pulp and Paper Mag. Can.* **1944**, *45*, 817.
180. Waxman, S.A. In *Wood Chemistry: Lignin;* Wise, E.L., ed. ACS Monograph Ser.; Reinhold Publishing Corporation: N.Y., 1944, 853-860.
181. *Wood Chemistry: Lignin*; Wise, E.L. ed. ACS Monograph Ser.; Reinhold Publishing Corporation: N.Y., 1944; 272-368.
182. Faith, W.L. *Ind. Eng. Chem.* **1945**, *37*, 9.
183. Sharkov, V.I. *Gidroliznoe Proizvodstvo (The Hydrolysis Industry)* **1945**, *1*, 94.
184. Nikitin, N.I. *The Chemistry of Cellulose and Wood; Israel Program for Scientific Translation Ltd.*; Jerusalem, 1966; 548-569.
185. McKee, R.H. *Ind. Eng. Chem.* **1946**, *38*, 382.
186. Anon *Pulp & Paper Mag. of Can.* **1948**, *49*, 2, 57.
187. Ericsson, E.O. *Chem Eng. Progress* **1947**, *43*, 165.
188. McCarthy, J. U.S. Patent 2,430,355, Nov. 4, 1947.
189. Ritchie, P.E.; Purves, C.B. *Pulp Paper Mag. Can.* **1947**, *48*, 12, 74.
190. Samuelson, O.; Öhgren, T. *Svensk Papperstidn.* **1947**, *49*, 499.
191. Adler, E.; Ellmer, L.R. *Acta. Chem. Scand.* **1948**, *2*, 839.
192. Brewer, C.P.; Cooke, L.M.; Hibbert, H. *J. Am. Chem. Soc.* **1948**, *70*, 57.
193. Jones, E.J. *J. Am. Chem. Soc.* **1948**.
194. Peniston, Q.P.; McCarthy, J.L. *J. Am. Chem. Soc.* **1948**, *70*, 1329.
195. Poljak, A. *Angew. Chem.* **1948**, *60*, 45.

196. Schütz, F.; Sarten , P.; Meyer, H. *Angew.* **1948**, *60*, 115, 56, 59, 96.
197. Schütz, F.; Sarten, P.; Meyer, H. *Holtzforschung* **1947**, *1*, 2.
198. Schütz, F. *Holtzforschung* **1948**, *2*, 33.
199. Shorygina, N.N.; Kefeli, T.Y. *Zhurn. Obsch. Khimii.* **1947**, *17*, 2058.
200. Shorygina, N.N.; Kefeli, T.Y. *Zhurn. Obsch. Khimii.* **1948**, *18*, 528.
201. Shorygina, N.N.; Kefeli, T.Y. *Dokl. Akad. Nauk.* **1949** *14,* 5.
202. Shorygina, N.N.; Kefeli, T.Y.; Semechkina, A.F. *Gidrolizanya Promyshennost,* **1949**, 2, 6.
203. Tomlinson, G.H. *Pulp Paper Mag. Can.* **1948**, *49*, 7, 63.
204. Tomlinson II, G.H. *Northeastern Wood Utilization Council* **1948**, *No. 19*, 71-7.
205. Wacek, A.; Kratzl, K. *J. Polymer. Sci.* **1948**, *3*, 539.
206. Aulin-Erdtman, G. *Tappi J.* **1949**, *32*, 160.
207. Kilp, W. *Wochbl. Papierfabr.* **1949**, *77*, 9.
208. Pearl, I.A. *J. Am. Chem. Soc.* **1949**, *71*, 2196.
209. Sohn, A.W.; Lenel, P.O. *Das Papier* **1949**, *3*, 109.
210. Cho; Lee; Meshitsuka, G.; Nakano, J. *Mokuzai Gakkaishi*, **1980**, *25*, 527.
211. Freudenberg, K.; Siebert, W.; Heirnberger, W.; Kraft, R. *Chem. Ber.* **1950**, *83*, 533.
212. Richtzenhain, H. *Svensk Papperstidn.* **1950**, *53*, 644.
213. Schubert, W.J.; Nord, F.F. *J. Am. Chem. Soc.* **1950**, *72*, 977, 3835.
214. Schubert, W.J. *Holzforschung* **1951**, *5*, 1.
215. Fisher, J.H.; Marshall, H.B. U. S. Patent No. 2,576,752-3; 1951.
216. Gottlieb, S.; Pelczar Jr., M.J. *Bacteriol. Revs.* **1951**, *15*, 55.
217. Hagglund, E. *Chemistry of Wood: Lignins*; Academic Press: NY., 1951; 181-389.
218. Patterson, R.F.; Keays, J.L.; Hart, J.S.; Strapp, R.K.; Luner, P. *Pulp Paper Mag. Can.* **1951**, *52*, 12, 105.
219. Adler, E.; Lindgren, B.O.; Saeden, U. *Svensk Papperstidn.* **1952**, *55*, 245.
220. Aulin-Erdtman, G. *Svensk Paperstidn.* **1952**, *55*, 745.
221. Brauns, F.E. *The Chemistry of Lignin*; Academic Press: NY., 1952; 808.
222. Edling, G. *Svensk Papperstidn.* **1952**, *55*, 863.
223. Edling, G. *Papier* **1953**, *7*, 159.
224. Edling, G. *Paper Trade J.* **1953**, *153*, 15, 14.
225. Freudenberg, K., et. al. *Chem. Ber.* **1952**, *85*, 641-647.
226. Freudenberg, K. *Holzforschung* **1952**, *6*, 37-42.
227. Freudenberg, K.; Hübner, H. *Chem. Ber.* **1952**, *85*, 1181-1191.
228. Kawamura, J.; Higuchi, T.J. *Soc. Tex. Ind.* **1952**, 335.
229. Kratzl, K.; Silbernagal, H. *Monatsh. Chem* **1952**, *83*, 1022.
230. Leopold, B. *Acta. Chem. Scand.* **1952**, *6*, 38.
231. Pearl, I.A.; Dickey, E.E. *J. Am. Chem. Soc.* **1952**, *74*, 614.
232. Schuerch, C. *J. Am. Chem. Soc.* **1952**, *74*, 5061.
233. Sillén, L.G.; Anderson, T. *Svensk Papperstidn.* **1952**, *55*, 622.
234. Flory, P.J. *Principles of Polymer Chemistry*; Cornell University Press: Ithaca, NY, 1953.
235. Freudenberg, K.; Rasenak, D. *Chem. Ber.* **1953**, *86*, 755-758.

236. Freudenberg, K.; Bitter, F. *Chem. Ber.* **1953**, *86*, 155.
237. Freudenberg, K.; Dietrich, H. *Chem. Ber.*, **1953**, *86*, 4-10.
238. Hägglund, E.; Liungren, S. *Papierfabrikant,* **1953**, *31*, 35.
239. Hergert, H.L.; Kurth, E.F. *J. Am. Chem. Soc.* **1953**, *75*, 1622.
240. Monsanto Chemical Co. Britin Patent No. 695,301; Aug. 5, 1953.
241. Felicetta, V.F.; Markham, A.E.; McCarthy, J.L. *Tappi J.* **1954**, *37*, 431.
242. Gierer, J. *Acta Chem. Scand.* **1954**, *8*, 1319.
243. Goldschmid, O. *Anal. Chem.* **1954**, *26*, 1421.
244. Lange, P.W. *Svensk Papperstidn.* **1954**, *57*, 525, 533, 235, 501.
245. McCarthy, J.L. in Unterkofler, Ref. 246, pp. 95-134.
246. *Industrial Fermentations*; Unterkofler, A.E. ed. Chemical Publishing Co.: NY, 1954; Vol 1.
247. Wiley, A.J. in Unterkofler, Ref. 246, pp. 307-347.
248. Adler, E.; Gierer, J. *Acta. Chem. Scand.* **1955**, *9*, 84.
249. Adler, E.; Hernestam, S. *Acta. Chem. Scand.* **1955**, *9*, 319.
250. Brown, S.A.; Neish, A.C. *Nature* **1955**, *175*, 688.
251. Freudenberg, K.; Eisenhut, W. *Chem. Ber.* **1955**, *88*, 626.
252. Gardon, L.J.; Mason, G.S. *Canadian J. Chem.* **1955**, *33*, 1491.
253. Keilen, J.J. *Chemurgic Digest* **1955**, *14*, 3, 13.
254. Pew, J.C. *J. Am. Chem. Soc.* **1955**, *77*, 2831.
255. Smith, D.C. *Nature,* **1955**, *176*, 267, 927 .
256. Williams, M.L.W.; Laudel, R.F.; Ferry, J.D. *J. Am. Chem. Soc.* **1955**, *77*, 3701.
257. Björkman, A. *Svensk Papperstidn.* **1956**, *59*, 477.
258. Freudenberg, K. *Angew. Chem.* **1956**, *68*, 508.
259. Nikitin, V.M.; Akim, G.L. *Trudy Leningrad. Esotekhn. Acad.* **1956**, 75, 145-156
260. Siegel, S.M. *J. Am. Chem. Soc.* **1956**, *78*, 1753.
261. Björkman, A.; Person, B. *Svensk Papperstidn.* **1957**, *60*, 158.
262. Adler, E.; Pepper, J.M.; Erikson, E. *Ind. Eng. Chem.* **1957**, *49*, 1391.
263. Björkman, A. *Svensk Papperstidn.* **1957**, *60*, 243.
264. Felicetta, V.F.; McCarthy, J.L. *Tappi J.* **1957**, *40*, 851-866.
265. Felicetta, V.F.; McCarthy, J.L. *J. Am. Chem. Soc.* **1957**, *79*, 4499.
266. Hägglund, E.; Enkvist, T. U. S. Patent No. 2,711,430; June 21, 1955, Reissue 24,293, Mar. 19, 1957.
267. Higuchi, T. *Physiol. Plantarum* **1957**, *10*, 356, 621.
268. Kratzl, K.; Billek, G.; Klein, E.; Buchtela, K. *Monatsh Chem.* **1957**, *88*, 721.
269. Merewether, J.W.T. *Holzforschung* **1957**, *35*, 247.
270. Nokihara, E.; Tuttle, M.J.; Felicetta, V.F.; McCarthy, J.L. *J. Am. Chem. Soc.* **1957**, *79*, 4495.
271. Schlesinger, H. U.S. Patent No. 2,576,311, Nov. 27, 1957.
272. Tischenko, D.V. *Bumazhn. Prom.* **1957**, N12, 5.
273. Tischenko, D.V. *Zhurn. Prikl. Khimii.* **1959**, *32*, 157.
274. Vroom, K.E. *Pulp Pap. Mag. Can.* **1957**, *58*, 3, 228.
275. Aulin-Erdtman, G.; Haghom, L. *Svensk Papperstidn.* **1958**, *61*, 187-210.
276. Hägglund, V.E.; Riehm, T. *Svensk Paperstidn.* **1958**, *61*, 18B, 665-668.
277. Papadakis, M. *La Revue Des Materiax de Construction,* **1958**, 519.

278. Adler, E.; Delin, S.; Lundquist, K. *Acta. Chem. Scand.* **1959**, *13*, 2149.
279. Felicetta, V.F.; Lung, M.; McCarthy, J.L. *Tappi J.* **1959**, *42*, 496.
280. Freudenberg, K. *Biochemistry of Wood; The 4th International Congress of Biochemistry*; Pergamon Press: London, 1959; 121.
281. Goheen, D.W.; Hearon, W.M.; Cisney, M.E.; Wethern, J.D. U. S. Patent No. 2,914,568; Nov. 24, 1959.
282. Tischenko, D.V. *Zhurnal Prikladnoi Khimii.* **1959**, *32*, 686.
283. Watson, C.A. *Forest Prod. J.* **1959**, *9*, 3, 25.
284. Acerbo, S.N.; Schubert, W.J.; Nord, F.F. *J. Am. Chem. Soc.* **1960**, *82*, 735.
285. Brauns, F.E.; Brauns, D.A. *The Chemistry of Lignin*; Academic Press: NY, London, 1960; 804.
286. Hergert, H.L. *J. Org. Chem.* **1960**, *25*, 405.
287. King, E.G.; Adolphson, C. U. S. Patent Nos. 2,935,473 and 2,935,504; 1960.
288. Kratzl, K.; Billek, K. *Monatsch Chem.* **1959**, *90*, 536.
289. Kratzl, K.; Faigle, H. *Z. Naturf.* **1960**, *15b*, 4.
290. Rex, R.W. *Nature* **1960**, *188*, 1185.
291. Rezanowich, A.; Goring, D.A.I. *J. Colloid Sci.* **1960**, *15*, 452.
292. Sarkanen, K.V.; Dence, C.W. *J. Org. Chem.* **1960**, *25*, 715.
293. Adler, E.; Lundquist, K. *Acta Chem. Scand.* **1961**, *15*, 223.
294. Adler, E.; Marton *J. Acta Chem. Scand.* **1961**, *15*, 357, 370.
295. Enkvist, T. *Paperi Puu* **1961**, *43*, 657.
296. Freudenberg, K.; Jovanovic, V.; Topfmeier, F. *Chem. Ber.* **1961**, *94*, 3227.
297. Freudenberg, K.; Sidhu, G. *Holzforschung* **1961**, *15*, 33-39.
298. Gierer, J.; Kunze, I. *Acta Chem. Scand.* **1961**, *15*, 803.
299. Sakakibara, A.; Nakayama, N. *Mokuzai Gakkaishi* **1961**, *7*, 13.
300. Ashorn, T.; T., E. *Acta. Chem. Scand.* **1962**, *16*, 548.
301. Chen, C.L. *Inaugural Dissertation*; Universite Heidelberg: Heidelberg, 1962; 1-75.
302. Chirkin, G.; Tishchenko, D. *Zhur. Priklad. Khim.* **1962**, *35*, 153.
303. Chudakov *Industrial Use of Lignin*; 1962, 2nd ed.1972, 3rd ed. 1983.
304. Freudenberg, K.; Nimz, H. *Chem. Ber.* **1962**, *95*, 2057-2062.
305. Gierer, J.; Norén, I. *Acta Chem. Scand.* **1962**, *16*, 1713, 1976.
306. Jensen, W.; Fremer, K.E.; Forss, K. *Tappi J.* **1962**, *45*, 122.
307. Nikitin, N.I. *Chemistry of Wood and Cellulose; Lignin.*; USSR Academy of Science: Moscow, Leningrad, 1962; 426-489, 711.
308. Pearson, D.A.; Ericsson, E.O.; McCarthy, J.L. In *Sulfite Pulping in Pulp and Paper Science and Technology, Vol. 1;* Libby, C.E., ed. McGraw-Hill: NY, 1962.
309. Pew, J.C. *Nature* **1962**, *193*, 250.
310. Sakakibara, A.; Nakayama, N. *J. Japan Wood Res. Soc.* **1962**, *8*, 157.
311. Tischenko, D.V.; Zhigalov, Y.V. *Zhurn. Prikl. Khimii.* **1962**, *35*, 147.
312. Browning, B.L. *The Chemistry of Wood*; Interscience Publishers: N.Y., 1963; 249-311.
313. Freudenberg, K.; Harkin, J.M. *Phytochemistry* **1963**, *2*, 189.
314. Freudenberg, K.; Lehmann, B. *Chem. Ber.* **1963**, *96*, 1860.

315. Freudenberg, K.; Lehmann, B. *Chem. Ber.* **1963**, *96*, 1050.
316. Freudenberg, K.; Jovanovic, V.; Topfmeier, F. *Chem. Ber.* **1963**, *96*, 2178.
317. Freudenberg, K.; Jones, K.; Renner, H. *Chem. Ber.* **1963**, *96*, 1846.
318. Goring, D.A.I. *Pulp Paper Mag. Can.* **1963**, *64*, 12, T517.
319. Nimz, H. *Chem. Ber.* **1963, *96*,** 478-485.
320. Yashchanko, A.V. *Lesnoi Zh.* **1963**, *6*, 1, 62.
321. Enkvist, T.; Turunen, J. *Chemie et Biochimie de la Lignine, de la Cellulose et des Hemicelluloses, Actes de Symposium International de Grenoble*; Universite de Grenoble: 1964.
322. Freudenberg, K. *Holzforschung* **1964**, *18*, 3.
323. Freudenberg, K.; Sidhu, G. *Holzforschung* **1964**, *18*, 3-9.
324. Gierer, J.; Lenz, B.; Norén, I.; Soderberg, S. *Tappi J.* **1964**, *47*, 233.
325. Gierer, J.; Smedman, L.A. *Acta Chem. Scand.* **1964**, *18*, 1244.
326. Ludwig, C.H.; Nist, B.J.; McCarthy, J.L. *J. Am. Chem. Soc.* **1964**, *86*, 1196.
327. Lundquist, K. *Acta Chem. Scand.* **1964**, *18*, 1316.
328. Marton, J. *Tappi J.* **1964**, *47*, 713.
329. Oshima, M.; Maeda, Y.; Kashima, K. Canadian Patent No. 700,209; Dec. 22, 1964.
330. Bergins, V. *Tappi J.* **1965**, *48*, 15-20.
331. Berlyn, G.P.; Mark, R.E. *Forest Products J.* **1965**, *15*, 140.
332. Dubsy, G.A.; McElhinney, T.R.; Wiley, A.J. *Tappi J.* **1965**, *48*, 95.
333. Forss, K.; Fremer, K.E. *Paperi Puu,* **1965**, *47*, 443.
334. Freudenberg, K. *Science* **1965**, *14*, 595.
335. Freudenberg, K.; Chen, C.L.; Harkin, J.M.; Nimz, H.; Renner, H. *Chem. Comm.* **1965**, 224.
336. Hillis, W.E.; Isoi, K. *Phytochemistry* **1965**, *4*, 905.
337. Nimz, H. *Chem. Ber.* **1965**, *98*, 533, 3153, 3160.
338. Nimz, H.; Gaber, H. *Chem. Ber.* **1965**, *98*, 538.
339. Rydholm, S.A. *Chemical Pulping*; Interscience: NY., London, Sydney, 1965; 439-714, 1231.
340. Rydholm, S.A. *Pulping Processes, Wood Chemistry*; 1965; 90-254.
341. Sapotnitsky *The use of sulfite liquors*; 3rd; 1981.
342. Sarkanen, K.V.; Suzuki, J. *Tappi J.* **1965**, *48*, 459. Schubert, W.J. *Lignin Biochemistry*; Academic Press: NY, London, 1965; 131.
343. Schubert, W.J. *Lignin Biochemistry*; Academic Press: NY, London, 1965; 131.
344. Timell, T.E. In *Cellular Ultrastructure of Woody Plants;* Côté, W.A.J., ed. Syracuse Univ. Press: 1965, 127.
345. Adler, E.; Lundquist, K.; Miksche, G.E. *Adv. Chem. Ser.* **1966**, *59*, 22.
346. Côté, W.A., Jr.; Simon, B.W.; Timell, T.E. *Svensk Paperstidn.* **1966**, *69*, 547-558.
347. Hrutfiord, B.F.; McCarthy, J.L. In *Lignin, Structure and Reaction;* ACS Symposium Series: Washington.D. C., 1966.
348. Marton, J. In *Lignin, Structure and Reactions;* ACS Symposium Ser.: Washington, DC, 1966, 267.
349. Nimz, H. *Chem. Ber.* **1966**, *99*, 469, 2638.

350. Nimz, H. *Holzforschung* **1966**, *20*, 105.
351. Nimz, H. *Angew. Chem. Int. Ed. Engl.* **1966**, *5*, 843.
352. Sewalt, V.; Oliveira, W.d.; Glasser, W.; Fontenot, J. *J. Sci. Food Agric.* **1996**, *71*, 204.
353. Browning, B.L. *Methods of Wood Chemistry*; Interscience Publishers: N.Y., 1967; 1-2.
354. Freudenberg, K.; Chen, C.L. *Chem. Ber.* **1967**, *100*, 3683-3688.
355. Kleinert, T.N. *Tappi J.* **1967**, *50*, 3, 120.
356. Nimz, H. *Chem. Ber.* **1967**, *100*, 181.
357. Pearl, I.A. *The Chemistry of Lignin*; Marcel Dekker Ltd.: N.Y., 1967; 339.
358. Procter, A.R.; Yean, A.Q.; Goring, D.A.I. *Pulp and Paper Mag. Can.* **1967**, *68*, 9, T445.
359. Rinaudo, P.M.M.; Pla, F. *Chimie Analytique,* **1967**, *49*, 6, 320.
360. Sarkanen, K.V.; Chang, H.M.; Allan, G.G. *Tappi J.* **1967**, *50*, 587.
361. Schubert, S.W.; Andrus, M.G.; Ludwig, C.; Glennie, D.; McCarthy, J.L. *Tappi J.* **1967**, *50*, 186.
362. *Constitution and Biosynthesis of Lignin;* Freudenberg, K.; Neish, A.C., eds. Springer-Verlag: Berlin, 1968, 129.
363. Kiryushina, M.F.; Tischenko, D.V. *Proc. Sov. -Fin. Symposium* **1968**.
364. Lenz, B.L. *Tappi J.* **1968**, *51*, 511.
365. Szabo, A.; Goring, D.A.I. *Tappi J.* **1968**, *51*, 440.
366. Kringstad, K.P.; Chang, C.W. *Tappi J.* **1969**, *52*, 2382-2385.
367. Matsukura, M.; Sakakibara, A. *Mokuzai Gakkaishi* **1969**, *15*, 35.
368. Nimz, H. *Chem. Ber.* **1969**, *102*, 799.
369. Bolker, H.I.; Brenner, H.S. *Science* **1970**, *170*, 173.
370. Fergus, B.J.; Procter, A.R.; Scott, J.A.N.; Goring, D.A.I. *Wood Sci. Technol.* **1969**, *3*, 117-118.
371. Fergus, B.J.; Goring, D.A.I. *Holzforschung* **1970**, *24*, 113.
372. Gierer, J. *Svensk Papperstidn.* **1970**, *73*, 571.
373. Luner, P.; Kempf, U. *Tappi J.* **1970**, *53*, 11.
374. Pearl, I.A. *Annual Review of Lignin Chemistry, 1970-1975*; Institute of Paper Chemistry: Appleton, WI.
375. Anon *United States Government Manual, Edition 1997-98*; 508.
376. Dence, C.W. in Sarkanen and Ludwig, Ref. 389, pgs. 373-422.
377. Gellerstedt, G.; Gierer, J. *Svensk Papperstidn.* **1971**, *74*, 117.
378. Glennie, D.W. in Sarkanen and Ludwig, Ref. 389, pgs. 597-631.
379. Goheen, D.W. in Sarkanen and Ludwig, Ref. 389, pgs. 797-824.
380. Goring, D.A.I. in Sarkanen and Ludwig, Ref. 389, pgs. 695-761.
381. Hergert, H.L. *Infrared Spectra.* **1971**, 267.
382. Kirk, T.K. *Ann. Rev. Phytopath* **1971**, *9*, 185.
383. Kiryushina, M.F.; Tischenko, D.V. *Zh. Prikl. Khim.* **1971**, *44*, 1, 459-466.
384. Larsson, S.L.; Miksche, G.E. *Acta. Chem. Scand.* **1971**, *25*, 647.
385. Marton, J. in Sarkanen and Ludwig, Ref. 389, p. 679.
386. Nimz, H.; Das, K. *Chem. Ber.* **1971**, *104*, 2359.
387. Sarkanen, K.V.; Hergert, H.L. in Sarkanen and Ludwig, Ref. 389, pgs. 43-89.

388. Sarkanen, K.V.; Ludwig, C.H. in Sarkanen and Ludwig, Ref. 389, p. 867.
389. Wallis, A.F.A. In *Lignins, Occurrence, Formation, Structure and Reactions;* Sarkanen, K.V.; Ludwig, C.H., eds. Wiley-Interscience: N.Y., 1971, 351.
390. Forss, K.G.; Gadd, G.O.; Lampila, M.; Lundell, R.O.; Williamson, H.W. In *International Sulfite Pulping and Recovery Conference* TAPPI/CPPA: Boston, MA, 1972, 165-170.
391. Obiaga, T.I., *Ph.D. Dissertation*, Univ. of Toronto, Toronto, 1972.
392. Samuelson, O.; Sjöberg, L.A. *Svensk Papperstidn.* **1972**, *75*, 583.
393. Smith, L.W.; Goering, H.K.; Gordon, C.H. *J. Dairy Sci.* **1972**, *55*, 1140.
394. Erikson, M.; Larson, S.; Miksche, G.E. *Acta Chem. Scand.* **1973**, *27*, 903.
395. Grushnikov. In *Elkin Progress and Problems in Lignin Chemist,* Leningrad, 1991
396. Nimz, H.H. *Tappi J.* **1973**, *56*, 124.
397. Wood, J.R.; Goring, D.A.I. *Pulp Pap. Mag. Can.* **1973**, *74*, T309.
398. Glasser, W.G.; Glasser, H.R. *Macromol* **1974**, *7*, 17.
399. Glasser, W.G.; Glasser, H.R. *Holzfoschung* **1974**, *28*, 5.
400. Kratzl, K.; Claus, P.; Lonsky, W.; Gratzl, J.S. *Wood Sci. and Technol.* **1974**, *8*, 35-49.
401. Ludemann, H.D.; Nimz, H. *Biochem. Biophys. Res. Commun.* **1973**, *52*, 1162-1169.
402. Lüdemann, H.D.; Nimz, H. *Makromol. Chem.* **1974**, *175*, 2409.
403. Nimz, H. *Angew. Chem. Int. Ed. Engl.* **1974**, *13*, 313.
404. Nimz, H.; Mogharab, I.; Lüdemann, H.D. *Makromol. Chem.* **1974**, *175*, 2563.
405. Sudo, K.; Sakakibara, A. *Mokuzai Gakkaishi* **1974**, *20*, 396.
406. Glasser, G.W.; Gratzl, S.J.; Collins, J.J.; Forss, K.; McCarthy, J.L. *Macromol.* **1975**, *8*, 565.
407. Goldstein, I.S. *J. Appl. Polymer Sci.* **1975**, *28*, 259-267.
408. Hsu, O.; Glasser, W.G. *Appl. Polym. Symp.* **1975**, *28*, 297.
409. Malinen, R.; Sjöström, E. *Pap. Puu* **1975**, *57*, 728.
410. McGinnis, G.D.; Parikh, S. *Wood Sci.* **1975**, *7*, 4, 295.
411. Akim, G.L.; Richter, N.E.; Stromsky, S.V.; Chupka, E.I.; Nikitin, V.M. *Wood Chemistry,* **1976**, 4, 6-9.
412. Aoyama, M.; Sakakibara, A. *Mokuzai Gakkaishi* **1976**, *22*, 591-592.
413. Forss, K.; Fuhrmann, A. *Pap. Puu.* **1976**, *58*, 817.
414. Forss, K.; Passiner, K. *Pap. Puu.* **1976**, *58*, 608.
415. Glasser, W.G.; Glasser, H.R. *Cell. Chem. Technol.* **1976**, *10*, 1, 23, 39.
416. Glasser, W.G.; Glasser, H.R.; Nimz, H.H. *Macromol.* **1976**, *9*, 866.
417. Shafizadeh, F. In *Thermal Uses and Properties of Carbohydrates and Lignins;* Shafizadeh, F.; Sarkanen, K.V.; Tillman, D.A., eds. Academic Press: NY, San Francisco, London, 1976, 1-17.
418. Shorygina; Reznikov *Elkin The Chemical Reactivity of Lignin*; 1976.
419. Trowell, H.C.; Southgate, D.A.; Wolever, T.M.S.; Leeds, A.R.; Gassull, M.A.; Jenkins, D. *J. Lancet* **1976**, 967.
420. Yasuda, S.; Sakakibara, A. *Mokuzai Gakkaishi* **1976**, *22*, 606.
421. Zarubin, M.Y. *Proc. 4th ISWPC* **1987**, *1*, 407-413.

422. Zarubin, M.Y., *D. Sc. Thesis*, Leningrad Forest Technical Academy, 1976.
423. Adler, E. *Wood Sci. and Tech.* **1977**, *11*, 169-218.
424. Eriksson, Ö.; Lindgren, B.O. *Svensk Paperstidn.* **1977**, *80*, 59-63.
425. Ghosh, K.L.; Venkatesh, V.; Chin, W.J.; Gratzl, J.S. *Tappi J.* **1977**, *60*, 11, 127.
426. Gierer, J.; Insgard, F. *Svensk Paperstidn.* **1977**, *80*, S510.
427. Hergert, H.L. In *Cellulose Chem. Technol. Symp.;* Arthur, J.C., ed. ACS Symp. Ser, Washington D.C: 1977, 48; 227-248.
428. Higuchi, T.; Shimada, M.; Nakatsubo, F.; Tanahashi, M. *Wood Sci. Technol.* **1977**, *11*, 153-167.
429. Holton, H.H.; Chapman, F.L. *Tappi J.* **1977**, *60*, 11, 121.
430. Lundquist, K.; Olsson, T. *Acta. Chem. Scand.* **1977**, *B31*, 788.
431. Sakakibara, A.; Sanyo, A.; Endo, S. *Mukozai Gakkaishi* **1977**, *23*, 193.
432. *Chemistry of Lignin*; Nakano, J. ed. Yuni Publishers: Tokyo, 1978.
433. Anon. *Weyerhaeuser Science Symp. on Biological Deliginification: Present status and future directions*; Tacoma, WA, 1978.
434. Glasser, W.G.; Barnett, C.A. *Tappi J.* **1979**, *62*, 8, 101.
435. Glasser, W.G.; Honeycutt, S.; Barnett, C.A.; Moroshi, N. *Tappi J.* **1979**, *62*, 11, 111.
436. Glasser, W.G.; Moroshi, N. *Tappi J.* **1979**, *62*, 12, 101.
437. Glasser, W.G. *Sv. Papperstidning* **1981**, *84*, 6.
438. Gratzl, J. *Weyerhaeuser Science Symposium on Phenolic Resins, Chemistry and Applications*; Tacoma, WA, 1979.
439. Lindström, T. *Colloid & Polymer Sci.* **1979**, *257*, 277.
440. Connors, W.J.; Sarkanen, S.; McCarthy, J.L. *Holzforschung* **1980**, *34*, 80.
441. Eriksson, O.; Goring, D.A.I.; Lindgreen *Wood Sci. Technol.* **1980**, *14*, 267.
442. Glasser, W. In *Lignin, in Pulp and Paper;* Casey, J.P., ed. Wiley: NY., 1980, 39-111.
443. *Chemistry of Delignification with Oxygen, Ozone, and Peroxides*; Gratzl, J.S.; Nakano, J.; Singh, R.P. eds. Uni Publishers Co. Ltd.: Tokyo, Japan, 1980; 280.
444. Hall, P.L. *Enzyme and Microbial Technol.* **1980**, *2*, 3, 170. Kirk, T.K.; Higuchi, T.; Chang, H.M. *Lignin Biodegradation: Microbiology, Chemistry, and Potential Applications*; CRC Press Boca Raton: FL, 1980; 1, 241.
445. Kirk, T.K.; Higuchi, T.; Chang, H.M. *Lignin Biodegradation: Microbiology, Chemistry, and Potential Applications*; CRC Press Boca Raton: FL, 1980; 1, 241.
446. Kirk, T.K.; Higuchi, T.; Chang, H.M. *Lignin Biodegradation: Microbiology, Chemistry, and Potential Applications*; CRC Press Boca Raton: FL, 1980; 2, 255.
447. Sakakibara, A. *Wood Sci. Technol.* **1980**, *14*, 89.
448. Sjöström *E. EUCEPA Symp. Helsinki* **1980**, *1*, 4.
449. *Eckman Days Proc. Intern. Symp. on Wood and Pulping Chem.*, 1981, 86-89.
450. Crawford, R.L. *Lignin Biodegradation and Transformation*; Wiley: NY., 1981.
451. Detroit, J.; Lin, S.Y. *Proc. 1st ISWPC* **1981**, *4*, 44-52.
452. Gierer, J. *Wood Sci. Technol.* **1980**, *14*, 241.
453. Glasser, W.G.; Glasser, H.R. *Pap. Puu.* **1981**, *63*, 71.

454. Glasser, W.G.; Glasser, H.R. *Sv. Papperstidning* **1981**, *84*, 6, R25.
455. Glasser, W.G.; Glasser, H.R.; Moroshi, N. *Macromol.* **1981**, *14*, 253.
456. Glasser, W.G.; Saraf, V.; Newman, W. *J. Adhesion* **1982**, *14*, 233.
457. Hemingway, R.W. In *Organic Chemicals from Biomass;* Goldstein, I.S., ed. CRC Press: Boca Ratan, FL, 1981, 189-248.
458. Lee; Meshitsuka, G.; Cho; Nakano, J. *Mokuzai Gakkaishi* **1981**, *27*, 678.
459. Nikitin *Theoretical Principles of Delignification.*
460. Robert, D.; Gagnaire, D. *Eckman Days Proc. ISWPC* **1981**, *1*, 86-89.
461. Sjöström, E. *Wood Chemistry: Fundamentals, and Applications*; Academic Press: 1981; 68-82, 223.
462. Deschamps, A. *Informations Chimie* **1982**, *223*, 187-190.
463. Ertel, J.R.; Hedges, J.I. *Anal. Chem.* **1982**, *54*, 174-178.
464. Mathur, V.K., *Ph.D. Dissertation*, University of Washington, Seattle, WA, 1982.
465. Saka, S.; Thomas, R.J.; Gratzl, J.S.; Abson, D. *Wood Sci. and Technol.* **1982**, *16*, 138-153.
466. Westermark, U. *Wood Sci. Technol.* **1982**, *16*, 71.
467. Anon. *Proc. 2nd ISWPC*, 1983.
468. *Recent Advances in Lignin Biodegradation Research;* Higuchi, T.; Chang, H.M.; Kirk, T.K., eds.Uni. Publishers Co. Ltd.: Tokyo, Japan, 1983; 279.
469. Kirk, T.K. in Higuchi, et al., Ref. 468, pp. 1-11.
470. Ovachinnikova, V.V.; Soboleva, L.E.; Meliachenko, S.I.; Bystrova, T.N.; Akim, G.L. *Wood Chemistry* **1983**, 5, 72-75.
471. Tanahashi, M. *Wood Res. Techn. Notes.* **1983**, 18, 34.
472. Boye, F. *Utilization of Lignins and Lignin Derivatives; Bib. Ser. No. 292, Part 1 & 2*; Inst. of Paper Chem.: Appleton, USA, 1984.
473. Fengel, D.; Wegener, G. *Wood: Chemistry, Ultra Structure and Reactions*; Walter de Gruyer: NY., Berlin, 1984; 132-181, 613.
474. Gellerstedt, G.; Lindfors, E. *Holzforschung* **1984**, *38*, 151-158.
475. Glasser, W.G.; Leitheiser, R. *Polymer Bulletin* **1984**, *12*, 1.
476. Kuwahara, M.; Glenn, J.K.; Morgan, M.A.; Gold, M.H. *FEBS Lett.* **1984**, *169*, 247-250.
477. Muller, D.; Kelly, S.; Glasser, W.G. *J. Adhesion.* **1984**, *17*, 3, 185.
478. Pla, F.; Robert, A. *Holzforschung* **1984**, *38*, 37.
479. Robert, D.R.; Brunow, G. *Holzforschung* **1984**, *38*, 85.
480. Salmén, L. *J. Material Sci.* **1984**, *19*, 3090.
481. Saraf, V.; Glasser, W.G. *J. Appl. Polym. Sci.* **1984**, *29*, 5, 1831.
482. Yan, J.F.; Pla, F.; Kondo, R.; Dolk, M.; McCarthy, J.L. *Macromol.* **1984**, *17*, 2137-2142.
483. Allen, N.S. In *New Trends in the Photochemistry of Polymers;* Allen, N.S.R., J. F., ed. Elsevier Appl. Sci. Publ.: London, 209-246.
484. Anon. *Proc. 3rd ISWPC*, 1985.
485. Atalla, R.H.; Agarwal, U.P. *Science* **1985**, *227*, 636.
486. Gierer, J. *Wood Sci. Technol.* **1985**, *19*, 289.
487. Gierer, J. *Wood Sci. Technol.* **1985**, *20*, 1-33.
488. Gratzl, J.S. *Das Papier* **1985**, *39*, 10A, v17-v23.

489. *Biosynthesis and Biodegradation of Wood Components*; Higuchi, T. ed. Academic Press, Inc.: London, 1985.
490. Kirk, T.K.; Shimada, M. in Higuchi, Ref. 468, pp. 579-605.
491. Meshitsuka, G.; Nakano, J. *J. Wood Chem. Tech.* **1985**, *5*, 391.
492. Saraf, V.; Glasser, W.G.; Wilkes, G.; Grath, J.M. *J. App. Polym. Sci.* **1985**, *30*, 2207.
493. Gierer, J. *Wood Sci. Technol.* **1986**, *20*, 1.
494. Lewis, N.G.; Luthe, C.E. *Holzforschung* **1986**, *40*, 153-157.
495. Pla, F.; Dolk, M.; Yan, F.J.; McCarthy, J.L. *Macromol.* **1986**, *19*, 1471.
496. Rials, T.G.; Glasser, W.G. *Holzforschung* **1986**, *40*, 6, 353.
497. Anon. *Proc. 4rd ISWPC*, 1987.
498. Atalla, R.H. *J. Wood Chem. Tech.* **1987**, *7*, 115.
499. Gratzl, J.S. *Das Papier* **1987**, *41*, 3, 120-149.
500. Johanson, T.; Nyman, P.O. *Acta. Chem. Scan.* **1987**, *B 41*, 762-765.
501. Kelly, S.S.; Rials, T.G.; Glasser, W.G. *J. Material Sci.* **1987**, *22*, 617.
502. Lewis, N.G.; Dubelsten, P.; Eberhardt, T.L.; Yamamoto, E.; Towers, G.H.N. *Phytochem.* **1987**, *26*, 2729-2734.
503. Lewis, N.G.; Yamamoto, E.; Hooten, J.B.; Just, G.; Ohashi, H.; Towers, G.H. *Science* **1987**, 1344-1346.
504. Ohashi, E.; Yamamoto, H.; Lewis, N.G.; Towers, G.H. *Phytochemistry* **1987**, *26*, 1915-1916.
505. Terashima, N.; Fukushima, K.; Takabe, K. *Proc. 4th Internat. Symp. Wood Pulp. Chem.* **1987**, *1*, 267.
506. Ciemniesti, S.; Glasser, W. *Polymer* **1988**, *29*, 1021.
507. Forrester, I.T.; Grabski, A.C.; Burgess, R.R.; Leatham, G.F. *Biochem. Biophys. Res. Commun.* **1988**, *157*, 992-999.
508. Glasser, W.; Knudsen, J.; Chang, C. *J. Wood Chem.Tech.* **1988**, *8*, 2, 221.
509. Smith, D.C.; Glasser, W.G.; Glasser, H.R.; Ward, T.C. *Cellulose Chem. Technol.* **1988**, *22*, 171.
510. Alén, R.; Hartus, T.; Sjöström, E. *Proc. 5th ISWPC* **1989**, 159.
511. Anon. *Proc. 5th ISWPC*, 1989.
512. Barnett, C.A.; Glasser, W.G. in Glasser and Sarkanen, Ref. 518, pgs. 436-451.
513. Dolk, M.; Yan, F.J.; McCarthy, J.L. *Holtzforschung,* **1989**, *43*, 2, 91-98.
514. Dutta, S.; Garver, T.M.; Sarkanen, S. in Glasser and Sarkanen, Ref. 518, pgs. 155-176.
515. Elder, T. in Glasser and Sarkanen, Ref. 518, pgs. 262-271.
516. Forss, K.; Kokkonen, R.; Sågfors, P.E. in Glasser and Sarkanen, Ref. 518, pgs. 29-41.
517. Gellerstedt, G.; Northey, R.A. *Wood Sci. Technol.* **1989**, *23*, 1, 75.
518. *Lignins: Properties and Materials*; Glasser, W.G.; Sarkanen, S. eds. ACS Symp. Ser. No 397: Washington, DC, 1989.
519. Glasser, W.G.; Sarkanen, S. in Glasser and Sarkanen, Ref. 518, pgs. 524-528.
520. Goring, D.A.I. in Glasser and Sarkanan, Ref. 518, pgs. 2-12.
521. Hatakeyama, H.; Hirose, S.; Hatakeyama, T. in Glasser and Sarkanen, Ref. 518, pgs. 205-218.

522. Higuchi, T. In *Bioprocess Engineering;* Ghose, T.K., ed. Ellis Horwood: Chichester, 1989, 39-58.
523. Kelley, S.S.; Glasser, W.G.; Ward, T.C. in Glasser and Sarkanen, Ref. 518, pgs. 402-413.
524. Koshijima, T.; Watanabe, T.; Yaku, F. in Glasser and Sarkanen, Ref. 518, pgs. 11-28.
525. Lange, H.P.; Wagner, B.; Yan, J.F.; Kaler, E.W.; McCarthy, J.L. *5th ISWPC* **1989**, 577.
526. Lewis, N.G.; Razul, R.A.; Yamamoto, E.; Hooten, J.B.; Bokelman, G.H. In *Biosynthesis and Biodegradation of Plant Cell Wall Polymers;* Lewis, N.G.; Pai; eds. ACS Symp. Ser. 389: 1989, 169-181.
527. Lindberg, J.J.; Kuusela, T.A.; Levon, K. in Glasser and Sarkanen, Ref. 518, pgs. 190-204.
528. Loran, J.H.; Wu, C.F.; Pye, E.K.; Balatinecz, J.J. in Glasser and Sarkanen, Ref. 518, pgs. 312- 323.
529. Mörck, R.; Reimann, A.; Kringstad, K.P. in Glasser and Sarkanen, Ref. 518, pgs. 390-401.
530. Narayan, R.; Stacy, N.; Ratcliff, M.; Chum, H.L. in Glasser and Sarkanen, Ref. 518, pgs. 476-485.
531. Robert, D.B., M.; Lundquist, K.; Von Unge, S.; *Proc. 5th ISWPC*, 1989, 21-25.
532. Detroit, W.J. U.S. Patent No. 2,446,133; August 29, 1995.
533. Faix, O.; Meier, D. In *Proc. of 1st European Workshop on Lignocellulosics and Pulp (EWLP);* Faix, O.; Meier, D., eds. Hamburg-Bergedort: Komissionsverlag Buchhandlung Max Wiedebusch: 1990, 400.
534. Gierer, J. *Holzforschung.* **1990**, *44*, 397.
535. Hofmann, K.; Glasser, W.G. *Thermochimica Acta.* **1990**, *166*, 169.
536. Higuchi, T. *Wood Sci. Technol.* **1990**, *24*, 23.
537. Johansson; Nyman; Stoffer; Weliender. In *Biotechnology in Pulp & Paper Manufacture. Applications and Fundamental Investigations;* 1990, 429-438.
538. Jonsson, L.; Karlsson, O.; Lundquist, K.; Nyman, P.O. *FEBS Lett.* **1990**, *276*, 45-48.
539. Lewis, N.G.; Yamamoto, E. *Annu. Rev. Plant Phys. Mol. Biol.* **1990**, *41*, 455-497.
540. Lin, S.Y.; Zhong, X.J. *China Pulp Pap.* **1990**, *9*, 2, 45-53.
541. *Chemistry of Lignin; (in Japanese)*; Nakano, J. ed. Uni Publisher Co. Ltd.: Tokyo, Japan, 1990.
542. Shoemaker, H.E. *Recl. Trav. Chim. Pays-Bas.* **1990**, *109*, 255-272.
543. Tsutsumi, Y.; Islam, A.; Anderson, C.D.; Sarkanen, K.V. *Holzforschung* **1990**, *44*, 1, 59-66.
544. Anon. *Proc. 6th ISWPC*, 1991.
545. Chen, C.L. In *Lignins: Occurrence in Woody Tissues, Isolation, Reactions, and Structure in Wood Structure and Composition;* Lewin, M.; Goldstein, I.S., eds. Marcel Dekker: NY, 1991, 183-261.

546. Davin, L.B.; Umezawa, T.; Lewis, N.G. In *Enantioselective Separations in Phytochemistry: Modern Phytochemical Mehtods;* Stafford, H.A.; Fischer, N.H., eds. Plenum: Paris, 1991, 25; 75-112.
547. Dilling, P. Japan Patent No. Kokai 80,684/94; March 22, 1994.
548. *Wood and Cellulosic Chemistry*; Hon, D.N.S.; Shiraishi, N. eds. M. Dekker Inc.: NY., Basel, 1990.
549. Terent'ev, O.A. *Mitt. Bundesforschungsanst. Forst-Holzwirtsch* **1991**, *168*, 44-49.
550. Wariishi, H.; Valli, K.; Gold, M.H. *Biochem. Biophys. Res. Commun.* **1991**, *176*, 1, 269-75.
551. Alén, R., in Kennedy et al., Ref. 561, pgs. 803-808.
552. Buchholz, F.R.; Neal, A.J.; McCarthy, J.L. *J. Wood Chem. and Technol.* **1992**, *12*, 447-469.
553. Evstigneyav, E.I.; Shevchenko, S.M.; Apushkinsky, A.G.; Semenov, S.G., in Kennedy et al., Ref. 561, pg. 657.
554. Faix, O. *Das Papier* **1992**, *46*, 12, 733-740.
555. Faix, O. In. *Methods in Lignin Chemistry,* Linn, S.; Dence, C.W. eds. Springer-Verlag, 1992, p. 686, 687.
556. Gellerstedt, G., in Kennedy et al., Ref. 561, pgs. 219-304.
557. Gierer,J.; et. al. *Holzforschung* **1992**, *46*, 495.
558. Gravitis, J., in Kennedy et al., Ref. 561, pgs. 613-627.
559. Hrutfiord, B.F.; Negri, A.R. *Tappi J.* **1992**, *75*, 129.
560. Hrutfiord, B.F.; Negri, A.R. *Chemosphere.* **1992**, *25*, 1-2, 53.
561. *Lignocellulosics1, Science, Technology, Development and Use*; Kennedy, J.F.; Phillips, G.O.; Williams, P.A. eds. Ellis Horwood Ltd.: Chichester, England, 1992.
562. Koksharov, A.; Akim, G., in Kennedy et al., Ref. 561, pgs. 247-251.
563. Leclerc, D.F.; Olson, J.A. *Macromol.* **1992**, *25*, 1667.
564. *Methods in Lignin Chemistry*; Lin, S. Y.; Dence, C. W., Eds.;. Springer Series in Wood Science; Springer-Verlag, Berlin, Heidelberg, 1992; 578.
565. Makarova, O.V.; Delneko, I.P., in Kennedy et al., Ref. 561, pgs. 327-330.
566. *Methods in Lignin Chemistry*; Meier, D.; Faix, O.; Linn, S.; Dence, C.W.; Eds.; Springer Verlag: 1992; 83-109, 178-199, 233-241 (In Japanese).
567. Metzger, J.O.; Bicke, C.; Faix, O.; Tuszynski, W.; Angermann, R.; Karas, M.; Strupat, K. *Angew. Chem. Int. Ed.* **1992**, *31*, 6, 762-764.
568. Qian, P.; Islam, A.; Sarkanen, K.V.; McCarthy, J.L. *Holzforschung* **1992**, *46*, 4, 321-324.
569. Sarkanen, K.; Islam, A.; Anderson, C., in Ref. 566, pp. 387-406.
570. Struszczyk, H., in Kennedy et al., Ref. 561, pgs. 791-801.
571. Telysheva, G.M., in Kennedy et al., Ref. 561, pgs. 643-655.
572. Wrzesniewska-Tosik, K.; Struszczyk, H., in Kennedy et al., Ref. 561, pgs. 629-633.
573. Treimanis, A.P., in Kennedy et al., Ref. 561, pgs. 419-428.
574. Tuor, U.; Wariishi, H.; Shoemaker, H.E.; Gold, M.G. *Biochem.* **1992**, *31*, 4986-4995.

575. Afanasiev, N. In *7th ISWPC,* Ref. 577, pp. 869-880.
576. Akim, G.L. *Pulp Paper Board.* **1993**, N5, 25-27.
577. Anon. *Proc. 7th ISWPC*, 1993.
578. Argyropoulos, D.S. In *7th ISWPC,* Ref. 577, pp. 776-787.
579. Cole, B.J.W.; Huth, S.P.; Runnels, P.S. In *Photochemistry of Lignocellulosic Materials;* Heitner, C.; Scaiano, J.C., eds. ACS Symp. Series: Washington DC., 1993, 205-213.
580. Faix, O.; Böttcher, J.H.; Bremer, J.P. In *7th ISWPC,* Ref. 577, pp. 829-836.
581. Faix, O.; Böttcher, J.H. *Holzforschung.* **1993**, *47*, 1, 45-49.
582. Gierer, J. *Holzforschung.* **1993**, *47*, 181.
583. Heitner, C. In *Photochemistry of Lignocellulosic Materials;* Heitner, C.; Scaiano, J.C., eds. ACS Symp. Series: Washington DC., 1993, 192-204.
584. Hofmann, K.; Glasser, W.G. *J. Adhesion.* **1993**, *40*, 229.
585. Islam, A.; Sarkanen, K.V. *Holzforschung* **1993**, *47*, 123.
586. Kurek, B.; Monties, B. In *7th ISWPC,* Ref. 577, p. 668.
587. Lapierre, C.; Pollet, B.; Tollier, M.T.; Chabbert, B.; Monties, B. In *7th ISWPC,* Ref. 577, p. 818.
588. Medvedeva, S.A.; Volchatova, I.V.; Seredkina, S.G.; Belousova, I.A.; Antipova, I.A.; Babkin, V.A. , 1993.
589. Sjöström, E. in Sjöström, Ref. 13, p. 293.
590. Wang, R.; Chen, C.L.; Gratzl, J.S. In *7th ISWPC,* Ref. 577, pp. 942-950.
591. Wong, A. In *7th ISWPC,* Ref. 577, p. 547.
592. Akim, G.L. *Pulp Paper Board* **1994**, 1-2, 22-23.
593. Englewood; *Ind. Bioprocessing* 1994, 16, no. 1, 4-5.
594. Gierer, J.; Yang, E.; Reitberger, T. *Holzforschung* **1994**, *48*, 405.
595. Hofmann, K.; Glasser, W.G. *Macromol. Chem. Phys.* **1994**, *195*, 65.
596. Lewis, N.G.; Davin, L.B. In *Evolution of Biochemical Pathway to Lignins and Lignans;* Nes, W.D., ed. ACS Symp. Ser.: 1996, 563, 202-246.
597. Oliviera, W.D.; Glasser, W. *Macromol.* **1994**, *27*, 5.
598. Oliviera, W.D.; Glasser, W. *J. Appl. Polym. Sci.* **1994**, *51*, 563.
599. Oliviera, W.D.; Glasser, W. *Polymer* **1984**, *35*, 9, 1977.
600. Oliviera, W.D.; Glasser, W. *J. Wood Chem. Tech.* **1994**, *14*, 1, 119.
601. Zakis, G.F. *Functional Analysis of Lignins and Their Derivatives*; Tappi Press: Atlanta, GA., 1994; 94.
602. Anon. *Proc. 8th ISWPC*, 1995.
603. Bernards, M.A.; Lopes, M.L.; Zajicek, J.; Lewis, N.G. *J. Biol. Chem.* **1995**, *270*, 7382-7386.
604. Blanchette, R.A.; Goni, M.A.; Hedges, J.I.; Nelson, B.C. *Holzfoschung* **1995**, *49*, 1-10.
605. Higuchi, T. *Biosynthesis and Biodegradation of Wood Components*; Academic Press, Inc.: Orlando, FL, 1985; 679.
606. Karhunen, P.; Rummakko, P.; Sipilia, J.; Brunow, G. *Tetrahedron Lett.* **1995**, *36*, 1, 169-170.
607. Faix, O.; Meier, D.; Berns, J. *Wochenbl. Papierfabr.* **1995**, *123*, 14/15, 656-660.

608. Proc. EC-Contractor's Meeting 7th October, 1992, ed. Meier, D., *et al.* eds. 240. EU-Comission, Directorate-General XIII, L-2920 Luxemburg.: Florence, Italy.
609. Nakano, J.; Yamada, M. *Proc. Pulp and Pap. Research Conference* **1995**, *no. 6*, 24-27.
610. Orth, A.B.; Tien, M. In *Genetics and Biotechnology A Chapter on Biotechnology of Lignin Degradation. Mycota II.;* Kuck, U., ed. Springer-Verlag: Berlin, Heidelberg, 1995, 287-302.
611. Theander, O.; Aman, P.; Westerlund, E.; Anderson, R.; Petterson, D. *J. AOAC Internatl.* **1995**, *78*, 4, 1030-1044.
612. Yakubova, M.R.; Ryabchenko, V.P.; Stancheva, A.A.; Pulatov, B.K.; Abduazimov, K.A.; Dzhakhangirov. F.N. *Khimiya Prirodnykh Soedinenii,* **1995**, *4*, 630-631.
613. Angelis, F.D.; Fregonese, P.; Verì, F. *Rapid Comm. in Mass Spectr.* **1996**, *10*, 1504-1508.
614. Gellerstedt, G. In *Pulp Bleaching: Principles and Practice;* Dence, C.; Reeve, D., eds. TAPPI Press: 1996, 91-111.
615. Gerischer, G. *Pap. S. Afr.* **1996**, *16*, 2, 18, 20.
616. Harazono, K.; Kondo, R.; Sakai, K. *Tappi J.* **1996**, *4*, 497-706.
617. *Chemical Modification of Lignocellulosic Materials*; Hon, D.N.S. ed. M. Dekker: NY., 1996; 20-31; 129-157.
618. Islam, A.; Wang, S.H.; McCarthy, J.L. *J. Pulp and Pap. Sci.* **1996**, *22*, 3, J97-J103.
619. Jurasek, L. *J. Pulp. Pap. Sci.,* **1996**, *22*, 10, J376.
620. Karhunen, P.; Rummakko, P.; Pajunen, A.; Brunow, G. *J. Chem. Soc., Perkin Trans.* **1996**, *1*, No. 18, Organic and Biochemistry, 2303.
621. Richards, L.D. *Chemical Marketing Reporter* **1996**, Oct. 21, 828.
622. Sewalt, V.; Glasser, W.; Fontenot, J.; Allen, V. *J. Sci. Food Agric.* **1996**, *71*, 195.
623. Shaw, J.; Buchwalter, S.; Hedrick, J.; Kang, S.; Kosban, L.; Gelorme, J.; Lewis, D.; Pu'uishothaman, S.; Saraf, R.; Vrehbeck, A. *Printed Circuit Fabrication* **1996**, *19*, 11, 38.
624. Stránel, O.; Sebök, T. *Pap. Celul.* **1996**, *51*, 12, 275.
625. Tokimatsu, T.; Miyata, S.; Ahn, S.; Umezawa, T.; Hittori, T.; Shimada, M. *J. Jap. Wood Res. Soc.* **1996**, *42*, 173-179.
626. Van Haerden, P.; Towers, G.H.N.; Lewis, N.G. *J. Biol. Chem.* **1996**, *271*, 12350-12355.
627. Ahvazi, B.C.; Argyropoulos, D.S. In *9th ISWPC*, Ref. 628, pp. 7-1-7-5.
628. Anon. *Proc. 9th ISWPC*, Montreal, Canada, 1997.
629. Africa, P.S. *Pap. S. Afr.* **1997**, *17*, 4, 12.
630. Davin, L.B.; Wang, H.B.; Crowell, A.L.; Bedgan, D.L.; Marlin, D.M.; Sarkanen, S.; Lewis, N.G. *Science* **1997**, *275*, 362-366.
631. Fukushima, K.; Pollet, B.; Monties, B.; Lapierre, C. In *9th ISWPC*, Ref. 628, pp. 29-1-29-4.
632. Gierer, J. *Holzforschung* **1997**, *51*, 34.
633. Hansen, C.M.; Björkman, A. In *9th ISWPC*, Ref. 628, pp. 35-1-35-5.

634. Miyata, S.; Tokimatsu, T.; Umezawa, T.; Hattori, T.; Shimada, M.; In *9th ISWPC*, Ref. 628, pp. 72-1-72-3.
635. Higuchi, T. *Biochemistry and Molecular Biology of Wood*; Springer-Verlag: Berlin, Heidelberg, 1997; 362.
636. Hori, K.; Meshitsuka, G., Ref. 652, Abstract No. 12.
637. Lewis, N.G.; Crowell, A.L.; Costa, M.A.; Davin, L.B. In *9th ISWPC*, Ref. 628, pp. H6-1-H6-4.
638. Landucci, L. In *9th ISWPC*, Ref. 628, pp. 54-1 -- 54-4.
639. Lewis, N.G.; Crowell, A.L.; Costa, M.A.; Davin, L.B. In *9th ISWPC*, Ref. 498, pp. H6-1-H6-4.
640. Lewis, N.G.; Sarkanen, S.; Davin, L.B.; Wang, H.B.; Crowell, A.L.; Bedgar, D.L.; Martin, D.M. *Science* **1997**, 275, 362.
641. Li, S.; Lundquist, K. *J. Wood Chem. Technol.* **1997**, *17*, 4, 391-397.
642. Li, S.; Lundquist, K. *Acta Chem. Scand.* **1997**, *51*, 12, 1224-1228.
643. Liu, Y.; Gustafson, R.; Callis, J.; McKean, B. In *9th ISWPC*, Ref. 628, pp. T2-1-T2-5.
644. Lönnberg, B.; Lindström, S. In *9th ISWPC*, Ref. 628, pp. 62-1-62-4.
645. Li, S.; Lundquist, K.; Stomberg, R., In *9th ISWPC*, Ref. 628, pp. B4-1-B4-5.
646. Nakano, J.; Izuta, Y.; Orita, T.; Hatakeyama, H.; Kobashigawa, K.; Teruya, K.; Hirose, S. *J. Soc. Fiber Sci. Technol.* **1997**, *53*, 10, 416.
647. Nonaka, Y.; Tomita, B.; Hatano, Y. *Holzforschung* **1997**, *51*, 183-187.
648. Schultz, T.P.; Fisher, T.H.; *Proc* In *9th ISWPC*, Ref. 628, pp. B5-1-B5-4.
649. Sewalt, V.; Glasser, W.; Beauchemiu, K. *J. Agric. Food Chem.* **1997**, *14*, 1823.
650. Toffey, A.; Glasser, W.G. *Holzforschung* **1997**, *51*, 71.
651. Venica, A.; Chen, C.; Gratzl, J. In *Anthraquinone Pulping;* Goyal, G., ed. TAPPI Press: Anthology of Published Papers 1977-1996, 1997; Chapter 4: 511-517.
652. *215th ACS National Meeting, Cellulose Division* 1998, Dallas, TX.
653. Akim, G.L.; Cordeiro, N.; Neto, C.P.; Gandini, A., Ref. 652, No. 35.
654. Berthold, F.; Lindfors, E.; Gellerstedt, G. *Holzforschung* **1998**, *52*, 481-489.
655. Camarero, S.; Barrasa, J.M.; Pelayo, M.; Martinez, A.T., *J. Pulp and Pap. Sic.* **1998,** 24, No. 7, 197-203.
656. Cathala, B., et al., Ref. 652, Abstract No. 33.
657. Faix, O.; Meier, D., Ref. 652, Abstract No. 29.
658. Forss, K.G.; Fremer, K.E., *The Ekman-Days 1981 International Symposium on Wood and Pulping Chemistry, Stockholm,* June 9-12, 1981, Vol. 4, 29-38
659. Fujita, M.; Gang, D.R.; Davin, L.B.; Lewis, N.G. *J. Biol. Chem. (In Press).*
660. Gandini,. A.; Guo, Z.; Montanari, S.; Belgecem, N., Ref. 652, Abstract No. 17.
661. Gargulak, J.D.; Lebo, S.E., Ref. 652, Abstract No.7.
662. Gellerstedt, G. see Ref. 652, Abstract No. 3.
663. Glasser, W.G., Ref. 652, Abstract No. 5.
664. Gratzl, J., Ref. 652, Abstract No. 2.
665. Heitner, C., et al., Ref. 652, Abstract No. 22.
666. Herbert, L., Ref. 652, Abstract No. 14.
667. Hori, K.; Meshitsuka, G., Ref. 652, Abstract No. 12.

668. Hu, W.J.; L., P.J.; Lung, J.R.; Kawaoka, A.; Kao, Y.Y.; Hideki, S.; Stokke, S.S.; J., T.C.; Chiang, V.L., Ref. 652, Abstract No. 11.
669. Jurasek, L. *J. Pulp and Pap. Sci.* **1998**, *24*, 7, 209-212.
670. Lapierre, C.; Pollet, B., Ref. 652, Abstract No. 10.
671. Lewis, N., Ref. 652, Abstract No.1.
672. Lewis, N.; Davin, L.B. in Lewis and Sarkanan, Ref. 675, p. 343.
673. Lewis, N.; Davin, L.B.; Sarkanan, J. in Lewis and Sarkanan, Ref. 675, pp. 1-29.
674. Lewis, N., et al., see Ref. 652, Abstract No. 9.
675. *Lignin and Lignan Biosynthesis*; Lewis, N.S.; Sarkanen, S. eds. ACS: 1998; 426.
676. Liu, Y.; Pye, E.K.; Argyropoulos, D.S., Ref. 652, Abstract No. 25.
677. Matsumoto, Y.; Ishizu, A.; Nakano, J., *Holzfarschung* **1998** *40,* Suppl. 81-85.
678. *International Fact & Price Book; Pulp and Paper International*; Matussek, H.; Stefan, V. eds. Jaakko Poyry Consulting: Finland, 1998; 429.
679. Monties, B.; Chabbert, B.; Baumberger, S., Ref. 652, Abstract No. 13.
680. Moran, R.A. *Paper Age* **1998**, *114*, 4, 10-12.
681. Omori, S.; Aoyama, M.; Sakakibara, A. *Holzforschung* **1998**, *52*, 391-397.
682. Robert, D., Ref. 652, Abstract No. 37.
683. Sarkanan, S. in Lewis and Sarkanan, Ref. 675, p. 194-209.
684. Sarkanen, S.; Li, Y., Ref. 652, Abstract No. 15.
685. Schultz, T.; Nicholas, D., Ref. 652, Abstract No. 38.
686. Young, R.A.; Akhtar, M., eds. In *Environmentally Friendly Technologies for the Pulp and Paper Industry,* John Wiley & Sons Inc.: New York, 1998, 5-67.
687. Brunow, G., et al. In Lewis and Sarkanen, ref. 675, p 131-147.

Appendix 1. Cited Contributions in a Chronological Order

Year	Investigators	Contributions	Ref
1838	Payen	Discovery of lignin	1
1839	Payen	Mechanical incrustation theory of lignin deposition	2
1839	Reid	Discovery of lignin	3
1840	Hunter	History of paper and lignin	4
1840	Hunter	History of paper and lignin	5
1840	Plumlee	History of paper and lignin	6
1853	Watt/Burgess	Proposed heating of wood with alkali under pressure to accomplish delignification	7
1857	Schulze	Named lignin from lignum Latin word for wood	8
1858	Possoz	Alkali fusion: first applied to lignin in wood	9
1866	Erdmann	Alkali fusion on isolated lignin yielded protocatechuic acid and pyrocatechol	10
1866	Tilghman	Delignified wood with sulfurous acid solutions. First successful chemical pulp process	11
1867	Payen	Treated wood with hot 10% HCl yielding fermentable sugars with pulp & lignin	12
1870	Eaton	First proposed NaOH process with sodium sulfate make up	13
1874	Ekman	First pulp mill constructed, CaMg base sulfite pulp mill in Sweden.	14
1874	Tiemann/ Mendelsohn	Isolation and structure of coniferin studied	15
1875	Nikintin	Conferin studies	16
1875	Tiemann	Conferin studies	17
1878	Wiesner	Classic color reaction: lignins /phloroglucinol	18
1880	Cross/Bevan	Used aqueous Cl_2 & NaOH to remove lignin	19
1882	Wheelwright	First sulfite process in US was operated	20
1884	Dahl	Invented sulfate (kraft) process	21
1885	Tiemann/Haarman	Conferin studies	22
1888	Riordon	Sulfite pulp first produced in Canada	23
1890	Lange	First lignin precipitated from alkaline solution	24
1892	Lindsey/Tollens	Sulfur present as sulfonate in SSL lignin; first found sulfite liquor lactone (conidendrin)	25
1893	Klason	Studied sulfurous acid bound to lignin; first attempted to dissolve lignin with alcohol and mineral acids; first analyzed soda black liquor; mainly found hydroxy acids and lactones	26
1897	Klason	Lignin studies	27

Year	Investigators	Contributions	Ref
1898	Klason	Lignin properties similar to coniferyl alcohol polymer	28
1898	Pollacsek	Formed vanillin through oxidation of SSL with air and ferric chloride	29
1904	Grafe	Alkaline hydrolysis of SSL produces vanillin	30
1907	Brompton	Sulfate pulp first produced in North America	31
1908	Klason	Klason lignin determined quantitatively as residue from 72% H_2SO_4 treatment	32
1908	Klason	Macromolecular structure for lignin suggested.	33
1908	Klason	First elemental analysis of lignins	34
1909	Procter/Hirst	Aromatic amines precipitates lignosulfonic acids	35
1909	Ruttan	Wood scarification for EtOH production	36
1910	Klason	Polymeric nature of kraft lignin	37
1911	Klason/Segerfelt	Kraft black liquor lignin contains organically bound sulfur	38
1911	Tomlinson	Two wood scarification plants using sulfurous and sulfuric acids were established in USA	39
1913	Heuser/Sieber	Pulp purification using aqueous chlorine	40
1914	Ungar	Investigated HCl lignin in Willstätter's lab	41
1917	Reed	Oxidation of lignins yielded oxalic acid	42
1918	Hägglund	Nitrated lignin	43
1918	Tomlinson	Career review	44
1919	Heuser/ Skioldebrand	Pyrolysis of lignin yielded acetone, methanol, acetic acid with 13% tar and 51% charcoal	45
1919	Tomlinson	Ethanol from scarification of wood	46
1920	Dore	First established that acetyl group carried by xylan, not by lignin	47
1920	Klason	Coniferyl alcohol similarity to lignin	48
1920	Schrader	Lignin was dissolved in alkali solution with O_2	49
1921	Heuser/Winsvold	Oxalic acid formed from alkali fusion of lignin	50
1922	Heuser/Samuelsen	Lignin in alkaline oxidation gave 25% oxalic acid after prehydrolysis and methylation	51
1922	Klason	Lignin methylene groups converted to CO and COOH by H_2O_2 oxidation in neutral solution	52
1922	Ritter/Fleck	First found highest lignin content for early wood **See also** Refs. 54, 55	53
1924	Hägglund/ Björkman, C.B.	Lignin +12% HCl yielded volatile aldehyde on heating	56
1924	Heuser/Hermann	KOH fusion of lignin yielded aromatics	57

Year	Investigators	Contributions	Ref
1925	Fierz-David/ Hannig	First carried out pressure hydrogenation of lignin, wood, and cellulose	58
1925	Hägglund	Lignosulfonates are a polydisperse system of guaiacylpropane polymers; higher acidity equates to faster sulfonation	59
1925	Krohn	Investigation of baker's, fodder and nutritional yeast from sugars of sulfite waste liquor	60
1926	Hawley/Wise	*The Chemistry of Wood. Lignin*, 43-96.	61
1926	Kürschner	*Zur Chemie der Ligninkörper*, 116. **See also** Ref. 63	62
1926	Schorger	*The Chemistry of Cellulose and Wood; Lignins*, 70-121.	64
1926	Urban	Methylated hydroxyl with dimethylsulfate	65
1927	Freudenberg/ Harder	Lignin + 12%HCl yielded HCHO	66
1927	Hägglund/Johnson	Phlorglucinol trimethylether can participate in lignin modification	67
1927	Herzog/Hillmer	Lignin UV spectra was first reported	68
1927	Mason	Explosion process is developed for production of fiber board (Masonite) **See also** Ref. 70	69
1927	Scholler	Dilute acid saccharification of wood: Scholler-Tornesch process and lignin. **See also** Refs. 72, 73	71
1928	Freudenberg	First chemical formulation of cellulose	74
1929	Freudenberg/ Zocher/Dürr	Isolated lignin by boiling with dilute H_2SO_4 and cuprammonium hydroxide extraction	75
1929	Freudenberg/ Zocher/Dürr	Refractive index of Lignin of 1.61; typical of aromatic compounds. **See also** Ref. 77	76
1929	Freudenberg/Belz/ Niemann	Split off methoxyl groups from lignin by HI in Zeisel type method	78
1929	Hägglund/ Klingstedt	Studied spectrochemical properties of isolated lignin from beech wood. **See also** Ref. 80	79
1929	Rassow/Zickmann	Lignin alkali fusion gave acetic, butyric acids	81
1929	Schaarschmidt/ Nowak/Zetzsche	Dicarboxylic acids formed from HCl lignin and NO_2 suggested presence of lignin double bonds	82
1930	Campbell/Maass	In sulfite pulping, hydrogen ion concentration decreased rapidly with rising temperature	83
1930	Freudenberg/Dürr	Oxidation by ozone yielded 1-2% acetic acid	84
1930	Holmberg	Mercaptolysis; Thioglycolic acid with α–alkoxy function of lignin to give thioglycolic acid lignin	85

Year	Investigators	Contributions	Ref
1931	Freudenberg	Lignin appeared to be weakly oriented or an amorphous substance, no melting point, decomposed & carbonized on heating	86
1931	Hägglund/Nihlén	Pulping with high bisulfite ions gives highly sulfonated lignin. **See also** Ref. 88	87
1931	Howard	Howard process: sugar free calcium lignosulfonates precipitated from SSL by CaO. **See also** Refs. 90, 91	89
1932	Kullgren/Hägglund	Presence of acetal groups was detected in lignin. **See also** Ref. 93	92
1932	Stamm/Semb/ Harris	Lignin may be of alicyclic nature	94
1933	Erdtman	Lignin formed by enzyme dehydrogenation of phenolic α, β-unsaturated C_6C_3 progenitors of coniferyl alcohol type. **See also** Refs. 96, 97, 98, 99	95
1933	Freudenberg	*Tannin, Cellulose, Lignin* (in German); first postulated the existence of γ-0-4 linkage without any experimental support	100
1933	Freudenberg/Sohns Duerr/Niemann	Nitrated lignin in alkaline solution was optically inactive	101
1933	Hägglund/Carlson	Sulfonation of lignin yielded one new sulfonate & one new phenolic hydroxyl group	102
1933	Kullgren	High temperature behavior of lignosulfonic acid in water and alcohol. **See also** Refs. 104, 105, 106, 107, 108 and 109	103
1933	Reidel	Synthesis of vanillin from guaiacol	110
1934	Hilpert/Littmann	No fundamental differences between lignins & carbohydrates in straw & beech. **See also** Ref. 112	111
1934	Pauly	First dissolved lignin of wood or straw with boiling 85% acetic or formic acid with small addition of sulfuric or hydrochloric acid. **See also** Ref. 114	113
1934	Phillips	Summarized early lignin biodegradation studies	115
1935	Asplund	Asplund process: wood chips were disintegrated to fiber in the presence of steam at 160 to 180°C	116
1935	Brauns/Hibbert	Introduced alkoxyl groups into lignin	117
1935	Holmberg	First sulfonation of carbinol groups by sulfite and thioglycolic acid; lignin sulfonation proceeds via benzyl alcohol or ether positions	118

Year	Investigators	Contributions	Ref
1935	King/Brauns/ Hibbert	Organosolv lignins prepared by methanolysis and ethanolysis	119
1935	Thomas/Hewitt	X-ray patterns indicate lignin is amorphous	120
1935	VanBeckum/Ritter	Holocellulose purified by aqueous Cl_2/NaOH	121
1935	Wilcoxon	Tomlinson alkaline pulping chemical and heat recovery furnace using water cooled walls	122
1936	Bailey	Isolated middle lamella by microdissection, possessed high concentration of lignin	123
1936	Erdtman	First isolated pinoresinol β-β from SSL	124
1936	Freudenberg/ Janson/Knopf/Haag	$KMnO_4$ oxidation, following alkaline hydrolysis or directly, & methylation, yielded aromatic acids	125
1936	Holmberg	Lignin benzyl groups sensitive to solvolysis. Lignin thioglycolic and lignosulfonic acids are similar. **See also** Ref. 127	126
1936	Howard	Howard process: sugar free calcium lignosulfonates commercialized	128
1936	Loughborough/ Stamm	Early determination of M_n. **See also** Ref. 130	129
1936	Nikitin/Orlova	Extracted lignin with dilute acidic dioxane	131
1936	Phillips	Biodegradation of lignins was summarized	132
1936	Sanborn/Salverson/ Howard	Vanillin produced from calcium lignosulfonates precipitated from SSL by CaO	133
1936	Tomlinson II/ Hibbert	Alkaline oxidation of SSL yielded 6-9% vanillin	134
1937	Freudenberg/ Meister/Flickinger	First detected phenyl cowman β-5 linkages in lignin	135
1937	Haworth	Diarylbutanes or phenylnaphthalenes designated lignans	136
1937	Maass/Calhoun/ Yorston	In sulfite pulping, delignification rate is approximately proportional to the amounts of undissolved lignin. **See also** Ref. 138, 139	137
1937	Olson/Peterson/ Scherard	Lignin protects wood from the destructive action of microorganisms	140
1937	Tomlinson I	Mg sulfite pulping process with chemical recovery system was described	141
1937	Tomlinson II/ Hibbert	Commercial oxidation of SSL in NaOH to yield vanillin at 800 lbs/day	142
1938	Freudenberg/Engler Flickinger/Sobek/ Klink	Permanganate oxidation of lignin indicated the presence of 5-5 and β-5 linkages	143

Year	Investigators	Contributions	Ref
1938	Freudenberg/ Muller	Ether linkages cleaved in alkali	144
1938	Harris/Adkins	Hydrogenolysis of methanol lignin in dioxane/Cu- chromite yielded mono cyclohexyl alcohols	145
1939	Brauns	Isolated, methylated native lignin/Brauns lignin. **See also** Ref. 147	146
1939	Cramer/Hunter/ Hibbert	Ethanolysis of spruce wood	148
1939	Freudenberg	First reported 20 +% vanillin by alkaline nitrobenzene oxidation of spruce lignin	149
1939	Tomlinson/ Wilcoxson	Patented a chemical recovery furnace for kraft pulp mills	150
1940	Anon	*Sulfite Waste Liquors. An Annotated Bibliography*	151
1940	Freudenber/Lautsch Engler	Alkali nitrobenzene oxidation of softwood lignin gave 25% vanillin and other aromatic fragments	152
1940	Freudenberg/Neish	Supported Erdmann's polymerization mechanism	153
1940	Glading	Attributed characteristic absorption spectrum of lignin to the presence of flavon and flavanone rings	154
1940	Lautsch/ Plankenhorn/Klink	Air oxidation of SSL with metal oxide gave about 10% vanillin	155
1940	Schultz	Nitrobenzene oxidation of lignins or SSL gave about 20% of aromatic aldehydes	156
1940	Tomlinson	Developed magnifite process **See also** Ref. 158	157
1941	Cooke/McCarthy/ Hibbert	Hydrogenolysis of wood in dioxane /Cu – chromite yielded mono-cyclohexyl alcohols	159
1941	Creighton/ Hibbert McCarthy	High yields of vanillin & syringaldehyde from nitrobenzene oxidation of hardwood lignin	160
1941	Erdtman	Naturally occurring dimeric phenylpropanes contain, C-C linkages between β-carbon atoms in lignans such as pinoresinol, lariciresinol, & conidendrin	161
1941	Gralén	Determination of *MW* by ultracenterfuge	162
1941	Hägglund	Sulfidation of thiolignin: scission took place only in small number of oxygenated lignin heterocycles, S adds to benzyl groups of lignin	163
1941	Pearl/Bailey/ Benson	Desulfonation of lignosulfonic acid with lime	164

Year	Investigators	Contributions	Ref
1941	Tomlinson/ McCarthy/Hibbert	Molecular Cl_2 used in multistage bleaching of kraft pulp achieves high whiteness	165
1942	Hibbert	Presence of β-O-4 linkages first postulated on the basis of ethanolysis experiments; biochemistry of lignin formation in plants	166
1942	Meshitsuka/ Nakano	Maule color reaction mechanism was concluded for both hardwood and softwood	167
1942	Pearl	First CuO oxidation of lignin yielded vanillin	168
1943	Aulin-Erdtman	Lignin spectral studies	169
1943	Erdtman	Proposed sulfonation cleaves cyclic ether linkages	170
1943	West/Hibbert	Refluxing softwood wood flour in 2% HCl/ethanol yields four guaiacyl ketones @ 10% yield	171
1944	Creighton/Hibbert/ Gibbs	First applied nitrobenzene oxidation for characterization of lignins and wood identification. Nitrobenzene oxidation of grass lignins yielded *p*-hydroxy benzaldehyde, vanillin & syringaldehyde	172
1944	Hachihama/Nira/ Kyokogu	Free phenolic hydroxyl groups reported in Lignosulfonates from UV spectra	173
1944	Hibbert/Gardner	Affirmed β-O-4 linkages in lignin	174
1944	Lange	UV proof that lignin in wood is aromatic	175
1944	McGregir/Evans/ Hibbert	Hardwood ethanolysis yielded lignin oligomers; Hibbert's ketones	176
1944	Nord/Sankey	Studied enzymatic isolated lignin. **See also** Ref. 178	177
1944	Tomlinson II	Laminated fiberboard, Tomlinite or Arborite was made from precipitated kraft lignin	179
1944	Waxman	Biodegradation literature is reviewed	180
1944	Wise	*Wood Chemistry*; Lignins; Ch. X; 272-368.	181
1945	Faith	Scholler-T process in US	182
1945	Sharkov/Nikitin	Wood saccarification in USSR. **See also** Ref. 184	183
1946	McKee	Delignification of lignin by hydrotropic solvents; Na-xylenesulfonate	185
1946	Tomlinson	Tomlinite fiberboard using kraft lignin	186
1947	Ericsson	Described Bellingham, USA plant which produced 9000 gal/ day of ethanol from SSL	187
1947	McCarthy	Steam stripping of SO_2 from SSL prior to fermentation	188

Year	Investigators	Contributions	Ref
1947	Ritchie/Purves	Prepared insoluble lignin by periodate oxidation of wood to dissolve carbohydrates	189
1947	Samuelson/Öhgren	Lignin studies	190
1948	Adler/Ellmer	Softwood contained 2-3% coniferyl groups by phloroglucinol-HCl	191
1948	Brewer/Cooke/ Hibbert	First time C9 units with intact guaiacyl & syringyl groups obtained by wood hydrogenolysis	192
1948	Jones	First study of lignin IR spectra	193
1948	Peniston/McCarthy	Purified lignosulfonates from SSL by counter-current dialysis	194
1948	Poljak	Peracetic acid oxidation of lignin in pulp	195
1948	Schütz/Sarten/ Meyer	Maintained that lignin is an aliphatic structure. **See also** Ref. 197, 198	196
1948	Shorygina/Kefeli	Na in liquid NH_3 yielded 28% mono-aromatics. **See also** Ref. 200, 201, 202	199
1948	Tomlinson	Vanillin from lignosulfonates	203
1948	Tomlinson II	Tomlinite fiberboard	204
1948	Wacek/Kratzl	Aldol type structures are present in lignins	205
1949	Aulin-Erdtman	Isolated lignins are aromatic; proved by UV	206
1949	Kilp	German SSL ethanol plants and lignin uses	207
1949	Pearl	Oxidation of lignin suggested the presence of 5-5 & β-5 linkages	208
1949	Sohn/Lenel	Heating of Pine wood in water @ 150°C for 2 hours gave pH 3.6-4 solution	209
1950	Cho/Lee/Nakano/ Meshitsuka	Confirmed structural heterogenetity of lignin	210
1950	Freudenberg/ Siebert/Heirnberger /Kraft	Compared softwood lignin with DHP by IR	211
1950	Richtzenhain	Estimated the amount of lignan or isolignan groups to be about 5% of all lignin units	212
1950	Schubert/Nord	Prepared enzyme isolated lignins. **See also** Ref. 214	213
1951	Fisher/Marshall	Ontario Paper Co. (Thorold) rebuilt vanillin plant using CaO as alkaline source	215
1951	Gottlieb/Pelczar	Extensive review of lignin biodegradation	216
1951	Hägglund	*Chemistry of Wood; Lignins*; Chapt III	217
1951	Patterson/Keays/ Hart/Strapp/Luner	Lignin content of solution estimated by UV absorption	218
1952	Adler/Lindgren/ Saeden	Mechanism of formation of Hibbert's ketones & guaiacol from ethanolysis of β-ether	219

Year	Investigators	Contributions	Ref
1952	Aulin-Erdtmann	Lignin UV $\Delta\varepsilon_i$ method showed phenolic hydroxyl and other characteristics	220
1952	Brauns	*The Chemistry of Lignin*, 808.	221
1952	Edling	Calculated sulfite and kraft heat with chemical recovery. Heat generated approx. equals process heat required for producing unbleached pulp. **See also** refs. 223, 224	222
1952	Freudenberg, et. al.	Occurrence of a cell wall-bound β-glucosidase in the lignifying tissue. **See also** Ref. 226	225
1952	Freudenberg/ Hübner	Isolation of dilignols from dehydrogenation of coniferyl alcohol in vitro: β-5	227
1952	Kawamura/Higuchi	LCC found in dioxane lignin from bark	228
1952	Kratzl/Silbernagel	Heating softwood in water decreased vanillin yield	229
1952	Leopold	Reported *p*-hydroxyl phenyl structural units by nitrobenzene oxidation of softwoods	230
1952	Pearl/Dickey	Found evidence for α- α linkages in lignins	231
1952	Scheurch	Solvent solubility of lignins; Hildebrand parameter	232
1952	Sillen / Anderson	Calculated sulfite recovery furnace gases & solids composition by thermodynamic methods	233
1953	Flory	*Principles of Polymer Chemistry*	234
1953	Freudenberg/ Rasenak	First identified β-β linkage in pinoresinol by enzymatic dehydrogenation of coniferyl alcohol	236
1953	Freudenberg/ Bitters	Proof of incorporation of coniferyl alcohol into lignin with radioactive precursors, D-coniferin	236
1953	Freudenberg/ Dietrich	Isolation of dilignols: β-β (syringaresinol)	237
1953	Hägglund/Liungren	First measured exceptionally low methoxyl content of spruce compression wood lignins	238
1953	Hergert / Kurth	Carbonyl groups in lignins estimated by IR	239
1953	Monsanto Chemical Inc.	Established vanillin plant (Seattle) using SSL in alkaline solution with Cu-catalyst	240
1954	Felicetta/McCarthy	Cations in SSL can be base exchanged using ion exchange resins (Dowex)	241
1954	Gierer	Total benzyl alcohol & noncyclic benzyl ethers in softwood MWL is about 0.43 unit/OCH_3 by acidic MeOH methylation	242
1954	Goldschmid	Measured lignin phenolic OH by UV $\Delta\varepsilon_i$ method	243

Year	Investigators	Contributions	Ref
1954	Lange	Cell walls lignins are aromatic, by UV	244
1954	McCarthy	Industrial fermentation of SSL to EtOH	245
1954	Unterkofler	*Industrial Fermentations*	246
1954	Wiley	Food and fodder yeast produced from SSL in Germany by Waldhof-Standard process and in US by SPM process in Rhinelander, WI	247
1955	Adler/Gierer	In MWL, α-O-H plus α-O-4 groups are about $0.43/OCH_3$ not including cyclic ethers; noncyclic benzyl ethers are about $0.24/OCH_3$	248
1955	Adler/Hernestam	30% free phenolic groups in gymnosperms	249
1955	Brown/Neish	^{14}C labeled tracer & enzyme studies revealed formation of coniferyl alcohol type substances in cambial sap. Shikimic acid pathway in lignin biosynthesis	250
1955	Freudenberg/ Eisenhut	Coniferyl alcohol is the precursor of lignin in biosynthesis	251
1955	Gardon/Mason	Lignosulfonates behave as polyelectrolytes	252
1955	Keilen	Kraft lignins modified to change properties: solubility, solution viscosity, polyelectrolyte,	253
1955	Pew	Proposed the 5-5 linkage in lignin	254
1955	Smith	Brauns bagasse lignin contained 11% *p*-coumaric, small amounts of ferulic, *p*-hydroxy benzoic, vanillic, & syringic acids	255
1955	Williams/Laudel/ Ferry	Mechanical deformation of polymers	256
1956	Björkman, A.	Developed method for preparation of MWL. After extensive milling in a non swelling solvent, half or more of softwood lignin is extracted by dioxane	257
1956	Freudenberg	Zulauf and Zutropf DHP synthetic lignin Freudenberg DHP	258
1956	Nikitin/Akim	Pioneered use of oxygen-alkali for delignification and improved bleaching	259
1956	Siegel	Cellulose surface enhance the oxidative polymerization of eugenol	260
1957	A. Björkman/ Person	Properties of MWL from softwoods & hardwoods	261
1957	Adler/Pepper/ Erikson	Acid hydrolysis of MWL liberates about 0.3 phenolic OH/OCH_3 which was nearly equal to number of terminal C -methyl groups characteristic of side-chains in Hibbert's ketones; 1/3-1/4 all guaiacyl groups in lignin joined by β-O-4 linkages.	262

Year	Investigators	Contributions	Ref
1957	Björkman	Lignin carbohydrate complex from MWL	263
1957	Felicetta/ McCarthy	PMR program at the University of Washington	264
1957	Felicetta/ McCarthy	SSL M_w in acidic hydrolysis	265
1957	Hägglund/Enkvist	Addition of sodium sulfide to kraft black liquor, pyrolysis yielded dimethylsulfoxide & dimethylmercaptan	266
1957	Higuchi	Biochemical studies of lignin formation	267
1957	Kratzl/Billek/ Klein/Buchtela	Demonstrated the formation of radioactive Hibbert's ketones from lignin synthesized in spruce fed with ^{14}C labeled coniferin	268
1957	Merewether	Reviewed early ideas about LCC	269
1957	Nokihara/Tuttle/ Felicetta/ McCarthy	Incremental acid sulfite delignification showed for hemlock, M_w initially 2,000, progressively to 12,000. For maple, M_w initially 1500, to 2000	270
1957	Schlesinger	Reduction of ethylferulate by $LiAlH_4$ to coniferyl alcohol	271
1957	Tischenko	Condensation of lignin similar to phenol-formaldehyde resins. **See also** Ref. 273	272
1957	Vroom	H factor to express delignification time-temperature effects	274
1958	Aulin-Erdtman/ Haghom	Detected 0.05/OCH_3 biphenyl structures in lignin by UV $\Delta\varepsilon_I$ studies on Brauns lignin.	275
1958	Hägglund/Riehm	Wood saccharification by Rheinau process	276
1958	Papadakis	lignosulfonates & sulfonated kraft lignin are useful as grinding aids in cements & dispersing aids in concrete	277
1959	Adler/Delin/ Lundquist	MWL contains about 0.08 phenylcoumaran structures / OCH_3	278
1959	Felicetta/Lung/ McCarthy	Reported substantial separation of SSL sugars from lignosulfonates by ion exchange chromatography	279
1959	Freudenberg	*Biochemistry of Wood*	280
1959	Goheen/Hearon/ Cisney/Wethern	Crown-Zellerbach Co built a ten million lbs/yr plant to make dimethylsulfide from kraft lignin at Bogalusa, LA	281
1959	Tischenko	Reported evidence favoring the aromatic structure of lignin by ozonation studies	282
1959	Watson	SSL hexose sugars fermented by Georgia-Pacific West at Bellingham WA, to yield industrial ethanol and purified lignosulfonates	283

Year	Investigators	Contributions	Ref
1960	Acerbo/Schubert/ Nord	Introduced [1-^{14}C] glucose and [6-^{14}C] glucose into spruce phenyl propane units	284
1960	Brauns/Brauns	*The Chemistry of Lignin*, 804.	285
1960	Hergert	IR spectra of lignin and related compounds	286
1960	King/ Adolphason	Chrome and ferrous derivatives of lignosulfonates work well as additives to oil well drilling muds	287
1960	Kratzl/Faigle	Introduced [1-^{14}C] glucose into spruce phenyl propane units. **See also** Ref. 289	288
1960	Rex	Free radicals in softwood lignin via EPR; paramagnetism in dioxane acidolysis lignins	290
1960	Rezanowich/ Goring	Lignin macromolecule in solution behaves like a compact spherical microgel	291
1960	Sarkanen/Dence	Reviewed cleavage of alkyl aryl ether linkages in lignins	292
1961	Adler/Lundquist	6-6 & β-aromatic bonds are present in 45% of condensed lignins; uncondensed phenyl unit in softwood MWL contained 0.15-0.18 unit/OCH_3	293
1961	Adler/Marton	Proposed total allocation of CO in lignins; Studied groups in lignins	294
1961	Enkvist	Kraft black liquor contains small quantities of vanillin, guaiacol, acetoguaiacone, vanillic acid, *p*-hydroxybenzoic acid	295
1961	Freudenberg/ Jovanovic/ Topfmeier	5-, 6- or 2-deuterium labeled coniferyl alcohol	296
1961	Freudenberg/Sidhu	Structural scheme of spruce lignin	297
1961	Gierer / Kunze	Formation of epoxide involved in the cleavage of β-O-4 alkyl-aryl ethers	298
1961	Sakakibara/ Nakayama	40-60% of softwood lignin was dissolved by heating wood powder at 180°C in dioxane-water to yield mono and dilignols	299
1962	Ashorn/Enkvist	Heating kraft black liquor with sulfur gave moderate yields of phenyl derivatives and dimethylsulfide etc.	300
1962	Chen	First detected biphenyl 5-1, 5-6, 6-6 linkages in permanganate oxidation products. The structure of tetralin (cyclolignan) bearing β-β and α-6 linkages was postulated as a precursor	301
1962	Chirkin/ Tishchenko	Detected α-1, α-5, α-6 linkages in alkaline pulping	302

Year	Investigators	Contributions	Ref
1962	Chudakov	*Industrial Use of Lignin*	303
1962	Freudenberg/Nimz	Isolation of trilignols from dehydrogenation of coniferyl alcohol in vitro: β-0-4/β-β	304
1962	Gierer/Norén	NaOH reactions: benzyl ether & β-O-4 cleavage Simultaneous removal of aryl -ether substituent as phenoxide ion	305
1962	Jensen/Fremer/ Forss	Characterization of high & low molecular weight lignosulfonates fractions	306
1962	Nikitin	*Chemistry of Wood and Cellulose*	307
1962	Pearson/Ericsson McCarthy	Characterization of lignosulfonates	308
1962	Pew	Biphenyl group in lignin	309
1962	Sakakibara/ Nakayama	Heating wood powder in dioxane/water gave more soluble lignols in hardwoods than softwoods	310
1962	Tischenko/ Zhigalov	Elementary S is predominant in thiolignins	311
1963	Browning	*The Chemistry of Wood*; Chapt. **6**; 249-311.	312
1963	Freudenberg/ Harkin	Evidence for the occurrence of p-glucohydroxycinnamyl alcohols in the cambium of spruce and other conifers	313
1963	Freudenberg/ Lehman	Proof of incorporation of phenylalanine into lignin	314
1963	Freudenberg/ Lehman	Coniferyl alcohol labeled at carbon atoms α, β or γ	315
1963	Freudenberg/ Jovanovic/ Topfmeier	6- or 2-deuterium labeled coniferyl alcohol	316
1963	Freudenberg/Jones/ Renner	Coniferyl alcohol	317
1963	Goring	Glass transition temperature in lignin markedly changed by sorption of water	318
1963	Nimz	Isolation of β-5/β-5	319
1963	Yashchenko/ Sharkov	Suggested lignin's use for soil consolidation	320
1964	Enkvist/Turunen	Lignin cleavage in NaOH-NaSH melts	321
1964	Freudenberg	Worked on structural model for spruce lignin	322
1964	Freudenberg/Shidu	Spruce lignin formulation	323
1964	Gierer/Lenz/ Norén Soderberg	NaOH solution cleavage of β-O-4 linkages may occur in model substances if there is adjacent hydroxyl or carbonyl group in α- or γ-carbon of the propanoid side-chain	324

Year	Investigators	Contributions	Ref
1964	Gierer/Smedman	Sulfidolytic cleavage of β-aryl ether bonds take place during kraft delignification	325
1964	Ludwig/Nist/ McCarthy	High resolution ^{1}H-NMR of model compounds. Chemical shifts reported. Functional groups evaluated. Suggested the presence of hydrogen bonding in lignins	326
1964	Lundquist	Mono and dilignols isolated by acidolysis of softwood MWL	327
1964	Marton	Kraft lignin	328
1964	Oshima/Maeda/ Kashima	Naguchi process: SSL lignins heated to 400°C in phenolic solvent yielding 20% monomeric phenols	329
1965	Bergins	Kappa number of pulp	330
1965	Berlyn/Mark	More than half the lignin in wood is located in the secondary wall	331
1965	Dubsy/ McElhinney/ Wiley	SSL sugars separated by preparation of an acetonal derivative soluble in acetone leaving partially purified insoluble lignosulfonates	332
1965	Forss/Fremer	Proposed lignin structure with a repeating unit of 16 guaiacylpropane & 2 *p*-hydroxy phenyl propane units. Proposed existence of hemilignins which are different in origin than polymeric lignin	333
1965	Freudenberg	The constitution and formation of lignin from *p*- hydroxycinnamyl alcohols	334
1965	Freudenberg/Chen/ Harkin/Nimz/ Renner	First isolated β-5 linkage dehydrodiconiferyl alcohol by 0.5% HCl-methanolic treatment of spruce wood	335
1965	Hillis/Isoi	Biosynthesis of polyphenols in eucalyptus	336
1965	Nimz	Several oligolignols were isolated by water hydrolysis of softwood powder at 100°C	337
1965	Nimz/Gaber	Isolated syringaresinol from mild hydrolysis of hardwood lignin indicating β-β linkage	338
1965	Rydholm	*Pulping Processes, Wood Chemistry, Chemical Pulping*. **See also** Ref. 340	339
1965	Sapotnitsky	*The use of sulfite liquors, 3rd ed.1981*	341
1965	Sarkanen/Suzuki	Peracetic acid oxidation of softwood lignin cleaves rings yielding dicarboxyl functions	342
1965	Schubert	*Lignin Biochemistry, 1965*; 131.	343
1965	Timell	Lignin in compression wood may be 34-41% higher than normal wood	344

Year	Investigators	Contributions	Ref
1966	Adler/Lundquist/ Miksche	Identified stilbene (α, β-diphenylethylene skeleton) in the acidolysis of spruce MWL, & traced its genesis to be β-1 lignol	345
1966	Côté/Simon/ Timell	Chemical composition of wood and bark from normal and compression region of fifteen species of softwoods	346
1966	Hrutfiord/ McCarthy	Estimated condensed/ uncondensed units in lignin by hydrogenolysis	347
1966	Marton	*Lignin, Structure and Reactions*	348
1966	Nimz	First experimentally detected α-O-4 type linkage showing that neutral warm water solutions alone hydrolyzed certain sensitive lignin bonds; possibly α-O-4	349
1966	Nimz	First isolated β-1 lignol structures, diarylpropanediols from spruce & beech protolignins by mild hydrolysis	350
1966	Nimz	β-0-4 and β-1 indicated by degradation of spruce lignin to oligomers	351
1966	Sewalt/Glasser et al	Lignin impact on fiber degradation: model study	352
1967	Browning	*Methods of Wood Chemistry*; **1 & 2**	353
1967	Freudenberg/ Chen	Permanganate oxidation of lignin yields evidence for β-6, β-2, 4-O-5, and 1-O-4 linkages.	354
1967	Kleinert	Stable free radicals demonstrated in lignins	355
1967	Nimz	Glycerol side–chains exist in hydrolysis of both softwood and hardwood lignins. NMR study of lignols; structure determined for several similar lignols and other oligolignols.	356
1967	Pearl	*The Chemistry of Lignin*; 339.	357
1967	Procter/Yean/ Goring	Topochemistry of delignification in pulping: secondary wall delignified faster than middle lamella	358
1967	Rinaudo/Pla	Studied fractionation of lignin polymers by gel filtration	359
1967	Sarkanen/Chang/ Allan	Hardwood lignin UV absorbtivity vs OCH_3 gave linear relationship at 280 nm	360
1967	Schubert/Andrus Ludwig/Glennie/ McCarthy	Crystalline lignosulfonates monomers isolated and characterized	361

Year	Investigators	Contributions	Ref
1968	Freudenberg/Neish	*Constitution and Biosynthesis of Lignin*; 129. Proposed model structure for spruce lignin; biosynthesis of lignins from coniferyl alcohol type monomers; bonding between lignin and wood carbohydrates	362
1968	Kiryushina/ Tischenko	Lignin and wood carbohydrate bonds	363
1968	Lenz	^{1}H NMR characterization of lignins	364
1968	Szabo/Goring	Reversing Flory's theory of trifunctional polymerization and applying to a random three dimensional gel such as lignin, showed that degradation should yield small molecules initially and then progressively larger ones	365
1969	Kringstad/Chang	Lignin-hemicellulose complexes	366
1969	Matsukura/ Sakakibara	Found heterogeneous distribution of structural units in lignins	367
1969	Nimz	Thioacetolysis degradation of spruce and beech lignin cleaved both α- and β- aryl ether bonds yielding 71 and 94% mono-tetrameric lignols	368
1970	Bolker/Brenner	Lignin is a three dimensional gel structure	369
1969	Fergus/Goring Proctor/Scott	Measured distribution of lignin in different parts of cell wall and middle lamella of softwoods	370
1970	Fergus/Goring	Measured guaiacyl and syringyl propane groups in different parts of cell wall and middle lamellae of hardwoods	371
1970	Gierer	Chemical reactions during kraft pulping	372
1970	Luner/Kempf	Lignins form gel like monolayers at air-water interfaces	373
1970	Pearl	*Annual Review of Lignin Chemistry, 1970-1975*	374
1970	US Govt. Manual (1997-98)	The EPA was established in the Executive Branch as an independent agency to permit coordinated and effective Governmental action on behalf of the environment	375
1971	Dence	Halogenation and nitration reactions of lignin model compounds studied	376
1971	Gellerstedt/ Gierer	In acid sulfite pulping, α -ether bonds cleaved to yield additional sulfonate and free phenolic hydroxyl groups	377

Year	Investigators	Contributions	Ref
1971	Glennie	Proposed structural formulation sketches for (1)insoluble lignin sulfonate by sulfonation at pH 6 and 135°C; and (2) extensively sulfonated lignin sulfonic acid	378
1971	Goheen	Sulfonated lignins are good dye dispersants, emulsifiers, binders, adhesives, resin additives	379
1971	Goring	Reviewed polymeric properties of lignins and lignin derivatives	380
1971	Hergert	IR absorption bands in lignins	381
1971	Kirk	Review of lignin biodegradation	382
1971	Kiryushina/ Tischenko	Lignin-carbohydrate chemical bonding by ozonation	383
1971	Larsson/Miksche	Determination of structural units bond frequency in lignin	384
1971	Marton	Proposed tentative structural formulation sketch for segment of pine Kraft lignin	385
1971	Nimz/Das	Isolated α-β type compounds and tetrahydrofuran dilignols involving γ-O-γ and β- β linkages	386
1971	Sarkanen/ Hergert	Uncondensed units in lignin quantified through determination of total yield of aromatic aldehydes from nitrobenzene oxidation of wood	387
1971	Sarkanen/Ludwig	*Lignins; Occurrence, Formation, Structure and Reactions*	388
1971	Wallis	Diversity of Lignins	389
1972	Forss/Gadd/ Lundell/ Williamson	Pekilo process produces food and feed protein through continuous fermentation of SSL carbohydrates and acetic acid by microfungus, *Paecïlomyces varioti*	390
1972	Obiaga	Softwood MWL showed M_w at 22 K with M_w/M_n = 2.6; Softwood DHP showed M_w at 3.7, 11.1, 31.4 K M_w with M_w/M_n = 1.5, 2.2 and 3.7 respectively	391
1972	Samuelson/Sjoberg	Oxygen-alkali delignification of wood	392
1972	Smith/Goering/ Gordon	Lignified cell wall carbohydrates are well digested; limit of this digestion set by lignin	393
1973	Erickson/Larsson Miksche	Types and frequency of structural units links in lignins	394
1973	Grushnikov/Elkin	*Progress and Problems in Lignin Chemistry*	395
1973	Nimz	First proposed structural scheme of beech lignin	396

Year	Investigators	Contributions	Ref
1973	Wood/Goring	Distribution of lignin in spruce kraft and acid sulfite fibers	397
1974	Glasser/Glasser	First computer simulation (SIMREL) of lignin **See also** Ref. 399	398
1974	Kratzl/Claus/ Lonsky/Gratzl	Reported mechanisms of molecular oxygen on a model substance in alkaline solution	400
1974	Lüdemann/Nimz	^{13}C-NMR studies of lignins. **See also** Ref. 402	401
1974	Nimz	Model structure of beech lignin; types and frequency of linkages	403
1974	Nimz/Mogharab/ Lüdemann	DHP structure is different by *Zulauf* versus *Zutropf* procedures	404
1974	Sudo/ Sakakibara	Hydrogenolysis of lignin yielded dimer with one C_β -C_6 and trimer with two C_β -C_5 links	405
1975	Glasser/McCarthy/ Gratzl/Collins/ Forss	Studies of NMR of ligninsulfonate and model substances	406
1975	Goldstein	Lignin studies	407
1975	Hsu/Glasser	Demonstrated technical feasibility for structural polymers from carboxyl derivatives of non-sulfonated lignins	408
1975	Malinen/Sjöström	Formation of carboxylic acids: kraft pulping	409
1975	McGinnis/Parikh	Alkali extracted softwood bark yielded about 20% lignin	410
1976	Akim/Nikitin/et al.	Molecular weight distribution changes in pulps by oxygen-alkaline purification	411
1976	Aoyama/ Sakakibara	Isolation of phenylcoumaran compound from hydrolysis products of hardwood lignin	412
1976	Forss/Fuhrmann Passiner	Karatex lignin based adhesive for plywood See also Ref. 414	413
1976	Glasser/Glasser	Found discrepancies in properties proposed for computer-simulated lignins reported in 1974	415
1976	Glasser/Glasser/ Nimz	Incorporation of several hypothetical dialkyl ether bonds in proposed 1974 computer-simulated lignin resolved reported discrepancies	416
1976	Shafizadesh/ Sarkanen/Tillman	*Thermal Uses and Properties of Carbohydrates and Lignins*	417
1976	Shorygina/ Reznikov/Elkin	*The Chemical Reactivity of Lignin*	418

Year	Investigators	Contributions	Ref
1976	Trowell/Southgate/ Wolever/Leeds/ Gassull/Jenkins	The nutritional term dietary fiber includes plant polysaccharides and lignin which are resistant to hydrolysis by the digestive enzymes of humans	419
1976	Yasuda/ Sakakibara	β-6 phenylisochroman isolated from hydrogenolysis of compression wood	420
1976	Zarubin	Lignin groups are classified based on modern acid-base theory. **See also** Ref. 422	421
1977	Adler	*Lignin Chemistry, Past, Present and Future.* Structural model of spruce lignin proposed	423
1977	Eriksson/Lindgren	Linkage between lignin and hemicelluloses	424
1977	Ghosh/Gratzl	Quinone additives in soda pulping of hardwoods	425
1977	Gierer/Insgard	Lignin reactions with oxygen & hydrogen peroxide in alkaline media	426
1977	Hergert	Differences between bark and xylem lignins	427
1977	Higuchi/Shimada/ Tanahashi	Showed differences in biosynthesis of softwoods and hardwoods	428
1977	Holton/Chapman	Kraft pulping with anthraquinone	429
1977	Lundquist/Olsson	^{1}H NMR studies of lignins	430
1977	Sakakibara/Sanyo/ Endo	Hydrolysis of lignins with dioxane and water	431
1978	Nakano	*Chemistry of Lignin*; (in Japanese)	432
1978	Weyerhaeuser Co.	*Biological Deliginification: Present status and future directions*	433
1979	Glasser	The structure of lignins in pulps. **See also** Ref. 435, 436, 437	434
1979	Gratzl	*Potential of Technical Lignins as Extenders in Phenolic Resins. Assessment of Reactive Sites*	438
1979	Lindström	Colloidal behavior of kraft lignin in aqueous solution, association and gelation	439
1980	Connors/Sarkanen/ McCarthy	Bimodal molecular weight distribution for lignins from molecular association	440
1980	Eriksson/Goring/ Lindgreen	Evidence for lignin carbohydrate linkages	441
1980	Glasser	*Lignin. In Pulp & Paper: Chemistry & Technology*	442
1980	Gratzl/Nakano/ Singh	*Chemistry of Delignification with Oxygen, Ozone and Peroxides*	443
1980	Hall	First proposed that lignin biodegradation proceeds via the generation of super oxide radical anions or similarly activated oxygen species; independent of specific enzymes;	444

Year	Investigators	Contributions	Ref
1980	Kirk/Higuchi/ Chang	*Lignin Biodegradation: microbiology, chemistry, and potential applications*, 1, 242.	445
1980	Kirk/Higuchi/ Chang	*Lignin Biodegradation: microbiology, chemistry, and potential applications*, 2, 255.	446
1980	Sakakibara	Structural model for softwood lignins	447
1980	Sjöstrom	Chemistry of oxygen delignification	448
1981	Anon.	*Ekman-Days, 1st ISWPC*	449
1981	Crawford	*Lignin Biodegradation and Transformation*	450
1981	Detroit/Lin	Structural heterogeneity of technical lignins; effect on utilization	451
1981	Gierer	Attributed cleavage in kraft pulping to three phases, reported degradation mechanisms for lignin and carbohydrates	452
1981	Glasser/Glasser	Designed a computer model for softwood lignin structure, **See also** Ref. 454	453
1981	Glasser/Glasser/ Moroshi	Computer modeling: based on analytical results obtained from isolated lignins	455
1981	Glasser/Saraf/ Newman	Lignin-carbohydrate bond regenerated via quinone methide intermediates during pulping	456
1981	Hemingway	Condensed tannins: base catalyzed reactions	457
1981	Lee/Meshitsuka Cho/Nakano	Hardwood MWL mainly originates from the compound middle lamella layer. Determined by milling, sieving and sedimentation methods	458
1981	Nikitin	*Theoretical Principles of Delignification*	459
1981	Robert/ Gagnaire	First quantitative structural analysis of lignin made by ^{13}C NMR	460
1981	Sjöström	*Wood Chemistry: Fundamentals, and Applications*	461
1982	Deschamps	Chemical and biotechnological lignin uses	462
1982	Ertel/Hedges	CuO oxidation product of lignin identified by GC	463
1982	Mathur/Sarkanen	Characterization of SSL-phenol reactions	464
1982	Saka/Thomas/ Gratzl/Abson	Compared soda, soda-AQ and kraft delignification of middle lamella vs. secondary wall	465
1982	Westermark	Phenolic coupling by Ca superoxide as a lignification reaction in wood	466
1983	Anon.	*2nd ISWPC.*	467
1983	Higuchi/Chang/ Kirk	*Recent Advances in Lignin Biodegradation Research*	468
1983	Kirk	Research perspective on lignin biodegradation	469

Year	Investigators	Contributions	Ref
1983	Ovichinnikova/ Akim et al.	Animal feed yeast production from oxygen-alkali bleaching effluents	470
1983	Tanahashi	Defibration behaviors of softwoods and hardwoods quite different in the explosion process.	471
1984	Boye	*Utilization of Lignins and Lignin Derivatives*	472
1984	Fengel/Wegener	*Wood: Chemistry, Ultra structure and Reactions*, Part V, Lignin; 132-181, 613.	473
1984	Gellerstedt/ Lindfors	Studied structural changes in lignin during kraft pulping	474
1984	Glasser/Leitheiser	Hydroxypropyl lignins as fire resistant foam compounds	475
1984	Kuwahara/Glenn/ Morgan/Gold	First discovered Mn-peroxidase enzyme from ligninolytic cultures of *Phanerochaete chrysosporium*	476
1984	Muller/Kelly/ Glasser	Lignin phenolic resin characterization and performance	477
1984	Pla/Robert	Conformation of lignin in solution	478
1984	Robert/Brunow	Softwood MWL contains about 0.9 unit aliphatic and phenolic OH/ OCH by ^{13}C NMR	479
1984	Salmén	Viscoelastic properties of lignin in wet wood; related to glass transition of the *in situ* lignin	480
1984	Saraf/Glasser	Structural properties in solution of cast lignin/polyurethane films	481
1984	Yan/Pla/Kondo/ Dolk/McCarthy	Proposed delignification stages for lignin	482
1985	Allen	Photochemistry of polymers	483
1985	Anon	*3rd ISWPC*	484
1985	Atalla/Agarwal	Raman microprobe determines lignin orientation in cell walls of woody tissue	485
1985	Gierer	Reactions of lignins during pulping and bleaching. **See also** Ref. 487	486
1985	Gratzl	Brightness stabilization in pulps	488
1985	Higuchi	*Biosynthesis and Biodegradation of Wood Components*, 679.	489
1985	Kirk/Shimada	Lignin biodegradation	490
1985	Meshitsuka/ Nakano	Structural characteristics of compound middle lamella hardwood lignin	491
1985	Saraf/Glasser/ Wilkes/McGrath	Structures-property relationships in PEG containing polyethylene networks	492
1986	Gierer	Chemistry of delignification, Part II. Bleaching reactions	493

Year	Investigators	Contributions	Ref
1986	Lewis/Luthe	Identification of Forss hemilignins as mono, di and tri lignosulphonic acids	494
1986	Pla/Dolk/Yan/ McCarthy	Characterization of alkaline and organosolv lignins from flow-through delignification of Black Cottonwood	495
1986	Rials/Glasser	Effect of lignin structure on polyurethane network formation	496
1987	Anon.	*4th ISWPC*	497
1987	Atalla	Aromatic rings in lignins may be aligned preferentially tangential to secondary wall	498
1987	Gratzl	Pulps stabilization by non-chlorine bleaching agents	499
1987	Johnsson/Nyman	Isolation of Mn(II) dependent extracellular peroxidase from white-rot fungus	500
1987	Kelley/Rials/ Glasser	Viscoelastic properties of amorphous components (lignin and non-crystalline carbohydrates) of wet wood studied. Softening behavior modeled following the WLF equation. For lignin, Tg is 60-100°C	501
1987	Lewis/Dubelsten/ Eberhardt/ Yamamoto/Towers	Discovery of enzymatic process in beech bark which converts trans-coniferyl alcohol into cis-coniferyl alcohol	502
1987	Lewis/Yamamoto/ Hooten/Just/ Ohaski/Towers	First demonstration of ^{13}C specific labeling of lignin in situ by administration of ^{13}C precursors	503
1987	Ohashi/Yamamoto/ Lewis/Towers	Discovery of the final step in monolignal formation	504
1987	Terashima/ Fukushima et al.	Lignification is controlled by the individual cell	505
1988	Ciemmiesti/Glasser	Polyvinyl alcohol hydroxypropyl lignin blend	506
1988	Forrester/Grabski/ Burgess /Leatham	Mn-dependent peroxidases, and the biodegradation of lignin	507
1988	Glasser/Knudsen/ Chang	Polyblends of ethylene-vinyl acetate copolymers with lignin	508
1988	Smith/ Glasser/ Glasser/Ward	Simulated hydrolytic depolymerization of a lignin computer model of 10,000 structural units: branched lignin with limited cross-linking	509
1989	Alén/Hartus/ Sjöström	Chemistry of bleaching	510
1989	Anon.	*5th ISWPC*	511
1989	Barnett/Glasser	Bleaching hyroxypropyl lignin: with H_2O_2	512

Year	Investigators	Contributions	Ref
1989	Dolk/Yan/ McCarthy	Kinetics of delignification in flow-through reactor	513
1989	Dutta/Sarkanen	Kraft lignin association model	514
1989	Elder	Computational methods: chemistry of lignin	515
1989	Forss/Kokkonen/ Sagfors	Low molecular weight glycolignin & hemilignins: extracted by heating spruce wood in neutral buffered dioxane H_2O_2	516
1989	Gellerstedt/ Northey	Permanganate oxidation study of birch suggested aryl ether linkages cleaved during milling. Analysis yielded 30 different acids identified by GC-MS	517
1989	Glasser/Sarkanen	*Lignin: Properties and Materials*	518
1989	Glasser/Sarkanen	Industrial lignin & derivatives:	519
1989	Goring	Proposed that native lignin in middle lamella is a random, three dimension polymer, in the secondary wall a non-random two dimensional network polymer	520
1989	Hatakewyama/ Hatakeyama	High performance lignin polymers	521
1989	Higuchi	*Bioprocess Engineering.*	522
1989	Kelley/Glasser/ Ward	Effect of soft-segment content on properties of lignin based polyurethanes	523
1989	Koshijima/Yaku	Lignin-carbohydrate complex is amphoteric	524
1989	Lange/Wagner/ Yan/Kaler/ McCarthy	Lignin network generated by computer	525
1989	Lewis/Razul Yamamoto/Hooten/ Bokelman	*Biosynthesis and Biodegradation of Plant Cell Wall Polymers;* Lewis, N.G.; Pai; eds. ACS Symp. Ser. 389: 1989, 169-181.	526
1989	Lindberg/Levon/ Kuusela	Described specialty polymers from lignin	527
1989	Loran/Wu/Pye/ Balatinecz	Production, properties of Alcell lignin from ethanol pulping described	528
1989	Mork/Kringstad	Kraft lignin elastomeric polyurethanes	529
1989	Narayan/Chum/ Stacy/Ratcliff	Engineering lignin polystyrene	530
1989	Robert/Bardet/ Von Unge/ Lundquist	First NMR study of a large number of diastereo-isometic forms of β-O-4 dimers	531
1990	Detroit	Nitric acid oxidized lignosulfonates with enhanced solubility and dispersant/surfactant properties	532

Year	Investigators	Contributions	Ref
1990	Faix/Meier	*Proceedings of 1st (EWLP). Utilization and Analysis of Lignins;* 400	533
1990	Gierer	Bleaching reactions divided into cationic, anionic and radical reactions	534
1990	Glasser/Hoffman	Application of Gillham's modeling technique to lignin containing resins	535
1990	Higuchi	Lignin biochemistry: biosynthesis and biodegradation	536
1990	Johansson/ Nyman/ Stoffer/Weliender	*In Biotechnology in Pulp & Paper Manufacture. Applications and Fundamental Investigations.* 429-438	537
1990	Jonsson/Lundquist Nyman/Karlsson	Stereospecificity in enzymic & non-enzymic oxidation of β-O-4 lignin model compounds	538
1990	Lewis/Yamamoto	Critical review of the lignification process	539
1990	Lin/Zhong	Physical & chemical properties of industrial lignins and MWL. Uses & applications for lignins in China and in the West	540
1990	Nakano	*Chemistry of Lignin;* (in Japanese)	541
1990	Schoemaker	Chemistry of lignin biodegradation	542
1990	Tsutsumi/Islam/ Anderson/ Sarkanen	Characterization of the products of acidic permanganate oxidation, vs. ozonolysis of lignins and model compounds	543
1991	Anon	*6th ISWPC*	544
1991	Chen	*Lignins in Wood Structure and Composition. pp183-261*	545
1991	Davin/Urmezawa/ Lewis	Resolution of optical isomers of lignan model compounds by critical HPLC	546
1991	Dilling	Preparation of thiolignin	547
1991	Hon/Shiraishi	*Wood and Cellulosic Chemistry*	548
1991	Terent'ev	Lignin utilization in the USSR	549
1991	Wariishi/Valli/ Gold	First report of fungal degradation of lignin by Mn-peroxidase enzyme	550
1992	Alen	Thermochemical conversion of black liquor organics into heavy fuel oil	551
1992	Buchholz/Neal/ McCarthy	Certain physical properties of paucidisperse lignosulfonates were determined as a function of MW	552
1992	Evstigneyav/ Shevchenko/ Apushkinsky/ Semenov	Electrochemistry of lignin model quinonemethides	553
1992	Faix	Lignin utilization	554
1992	Faix	Pyrolysis for characterization of lignins	555

Year	Investigators	Contributions	Ref
1992	Gellerstedt	Chemical aspects of kraft pulp bleaching	556
1992	Gierer, et al.	Reactions between oxygen derived radials and prominent lignin structures	557
1992	Gravitis	Fractal geometry of lignin: structure and properties	558
1992	Hrutfiord/Negri	Toxic chlorinated dibenzofurans (DBF) and dibenzodioxins (DBD), more prevalent in DHP's from coumaryl alcohol than from coniferyl alcohol	559
1992	Hrutfiord/Negri	Proposed mechanism for formation of chlorinated DBF and DBD	560
1992	Kennedy/Phillips/ Williams	*Lignocellulosics Science, Technology, Development and Use*, 865.	561
1992	Koksharov/Akim	Bleaching kinetics: O_2; hardwood sulfite pulp	562
1992	Leclerc/Olson	Delignification a result of chemical degradation of two distinct cross-linked lignin gel fractions (middle lamella and secondary wall)	563
1992	Lin/Dence	*Methods in Lignin Chemistry*; 578. Also in Japanese	564
1992	Makarova/Deineko	Oxidation of wood by O_2 in alcohols	565
1992	Meier/Faix	Pyrolysis gas chromatography- Mass spectrometry	566
1992	Metzger/Bicke/ Faix/Tuszynski/ Angerman/ Karas/Strupat	First applied MALDI-MS to lignin; MWL yielded precise distribution of individual molecular weights up to 50,000	567
1992	Qian/Islam/ Sarkanen/ McCarthy	*Erythro*/ *threo* ratios and molecular weights of lignins from three hardwood plants species	568
1992	Sarkanen/Islam/ Anderson	Determination of *erythro*/ *threo* ratios of softwoods, hardwoods and grasses *in situ* lignins by ozonation.	569
1992	Struszczyk	Plastics and high performance lignin based polymers	570
1992	Telysheva	Lignin modified with organosilicon compounds	571
1992	Tosik/ Struszczyk	Increased thermal stability obtained with modified steam explosion-lignin	572
1992	Treimanis	Influence of fiber structure on pulp properties	573

Year	Investigators	Contributions	Ref
1992	Tuor/Wariishi/ Schoemaker/ Gold	Oxidation of phenolic aryl glycerol-β-aryl ether and α- carbonyl lignin model compounds by Mn-peroxidase from *Phanerochaete chrysosporium.*	574
1993	Afanasiev	Confirmed the polydisperse and polyelectrolyte characteristics of lignosulfonates	575
1993	Akim	Comparison of modern methods of sulfite pulp bleaching to Cl_2 and HOCL bleaching	576
1993	Anon	*7th ISWPC.*	577
1993	Argyropoulos	^{31}P NMR in wood chemistry	578
1993	Cole/Huth/Runnels	Brightness stability on high-yield pulps	579
1993	Faix/Böttcher/ Bremer	Characterization of lignocellulosics and lignins using multi- variate calibration technique for analytical pyrolysis and FTIR. **See also** Ref. 581	580
1993	Gierer	Hydroxy radical activity in growth of white-rot fungi	582
1993	Heitner/Scaiano	*Photochemistry of Lignocellulosic materials*	583
1993	Hofmann/Glasser	Cure behavior of lignin resin	584
1993	Islam/Sarkanen	Characterization of jute lignin by FTIR, UV, ^{1}H-NMR and ^{13}C-NMR	585
1993	Kurek/Monties	Oxidation of spruce lignin in colloidal state by horse-radish peroxidase and lignin peroxidase from *Phanerochacte chrysosporium*	586
1993	Lapierre/Pollet/ Tollièr/Chabbert/ Monties	Lignin molecular profiling: thioacidolysis, advantages & limitations	587
1993	Medvedeva/ Volchatova/ et al.	Oxidative destruction of lignin and phenol compounds by basidial fungi *Phanerochaete Sanguinca & coriolus villosus*	588
1993	Sjöström	*Wood Chemistry: Fundamentals, and Applications*; 293.	589
1993	Wang/Chen/ Gratzl	Decolorization & dechlorination of organics in pulp bleach plant effluents by photo-oxidation	590
1993	Wong	8-membered ring configuration in Taxol	591
1994	Akim	Review of sulfite pulp bleaching methods	592
1994	Englewood	Markets for lignosulfonates and lignins	593
1994	Gierer/Young/ Reitberger	Lignin oxidation by superoxide radical	594
1994	Hofmann/Glasser	Network formation in lignin based epoxy resins	595

Year	Investigators	Contributions	Ref
1994	Lewis/Davin	Proposed evolutionary pathway to lignan and lignins in vascular plants	596
1994	Oliviera/Glasser	Described star-shaped lignin block copolymers of high compatibility with synthetic resins. **See also** Ref. 598	597
1994	Oliviera/Glasser	Multiphase materials of lignin copolymers with cellulose components and with styrene. **See also** Ref. 600	599
1994	Zakis	*Functional Analysis of Lignins and Their Derivatives*, 94.	601
1995	Anon.	*8th ISWPC.*	602
1995	Bernards/Lopez/ Zajicek/Lewis	Discovers that principle phenolic component of suberin is hydroxycinnamic acid.	603
1995	Blanchette/Goni/ Hedges/Nelson	Soft-rot fungal degradation of lignin in 2700 year old archeological woods	604
1995	Higuchi	Status of lignin biodegradation research	605
1995	Karhunen/Brunow/ Rummakko/Sipilia	Dibenzodioxocin, a new type of linkage in softwood lignins	606
1995	Meier/Bernd/Faix	Lignins converted thermochemically to higher value compounds. **See also** Ref. 608	607
1995	Nakano/Yamada	COOH & PhOH groups in industrial lignins	609
1995	Orth/Tien	*A Chapter on Biotechnology of Lignin Degradation. Mycota II. In Genetics and Biotechnology*, 287-302	610
1995	Theander/Aman/ Westerlund/ Anderson/ Petterson	Lignin is designated as a constituent of dietary fiber with neutral sugars and uronic acid residues	611
1995	Yakubova et al	Sorption Properties of lignins and perspectives for their utilization in medical applications	612
1996	Angelis/Fregonese/ Veri	DHP's investigation by MALDI Time-of-flight MS. DHP formation is reversible	613
1996	Gellerstedt	Chemical characteristics of lignin in CMP and chemical pulps; chemical modifications during kraft and sulfite pulping	614
1996	Gerischer	3rd International Forum on markets for sulfur free lignins	615
1996	Harazono/Kondo/ Sakai	Application of ligninolytic enzymes (Mn-peroxidase) to bleaching of kraft pulp	616
1996	Hon	*Chemical Modification of Lignocellulosic Materials*; Chapts. 2 & 5; 20-31; 129-157	617
1996	Islam/Wang/ McCarthy	Reviewed delignification studies in flow-through reactors	618

Year	Investigators	Contributions	Ref
1996	Jurasek	Fractal dimensions from computer modeled lignin structures in three dimensions	619
1996	Karhunen/Brunow	Synthesized dibenzodioxocin lignin model structure	620
1996	Richards	Petrochemical industry feed stock	621
1996	Sewalt, et al	Digestability fermented corn stover	622
1996	Shaw/Lewis/et al	IBM made first printed circuit board and printed wire-board containing 50% lignin containing epoxy resin	623
1996	Stránel/Sebök	Review of structure, molecular weight range, polyelectrolyte properties, and utilization of lignosulfonates	624
1996	Tokimatsu/Miyata/ Ahn/Umezawa/ Hittori/Shimada	The non-phenolic β-O-4 type LCC in *erythro* & *threo* forms oxidized by lignin peroxidase	625
1996	VanHearden/ Towers/Lewis	Discovery of nitrogen cyclic mechanism in sustaining lignification	626
1997	Ahvazi/ Agyropoulos	Quantitative detection of carbonyl groups in technical and native lignins by using ^{19}F-NMR	627
1997	Anon.	*9th ISWPC*	628
1997	Anon.	S. Africa makes lignosulfonate products	629
1997	Davin/Wang/ Crowell/Bedgar/ Marlin/ Sarkanen/Lewis	Discovery of first protein controlling phenolic coupling	630
1997	Fukushima/Pollet/ Monties/Lapierre	Thioacidolysis dimer analyses of lignols	631
1997	Gierer	Chemical aspects of kraft pulping	632
1997	Hansen/Björkman	Hansen solubility parameter concept was examined for cell wall ultrastructure	633
1997	Hattori/Shimada Miyata/Tokimatsu Umezawa	Degradation of β-0-4 lignin carbohydrate model compounds by lignin peroxidase	634
1997	Higuchi	*Biochem. and Molecular Biol. of Wood*; 362.	635
1997	Hori/Meshitsuka	Stepwise cleavage of ether linkages by trimethylsilyliodide, heterogeneity of hardwood lignin	636
1997	Kadla/Chang/ Jameel/Gratzl	High temperature reactions of lignin and lignin model compounds with stabilized H_2O_2	637
1997	Landucci	DHP from *p*-hydroxycinnamyl alcohol and metal salt oxidants	638
1997	Lewis/Crowell/ Costa/Davin	Stereoselective bimolecular phenoxy radical coupling by α dirigent protein	639

Year	Investigators	Contributions	Ref
1997	Lewis/Sarkanen/ Martin/Bedgar/ Crowell/Wang/ Davin	A 78-kilodalton protein has been isolated that, in the presence of an oxidase or one electron oxidant, effects stereoselective bimocular phenoxy radical coupling *in vitro*.	640
1997	Li/Lundquist	Synthesis of lignin model compounds β-5	641
1997	Li/Lundquist	NMR of propionate lignin derivatives	642
1997	Liu/Gustafson/ Callis/McKean	Fluorescence microphotometry measures the lignin content of single pulp fibers	643
1997	Lönnberg/ Lindström	Reactivity of residual lignin in pulps	644
1997	Lundquist/ Stomberg	Stereochemistry of lignin structures and lignin degradation products	645
1997	Nakano/Izuta/ Hatakeyama et al.	Thermal and mechanical properties of kraft black liquor polyurethanes	646
1997	Nonaka/Tomita/ Hatano	Synthesis of lignin/epoxy resins in aqueous systems and their properties	647
1997	Schultz/Fisher	Alkaline hydrolysis of nonphenolic β-0-4 lignin model compounds	648
1997	Sewalt/Glasser et al	Removal of enzyme hydrolysis inhibition by chemical modification of lignin and adhesives	649
1997	Toffey/Glasser	Cure characterization polyurethanes and lignins	650
1997	Venica/Chen/ Gratzl	Mechanism of soda-AQ delignification of hardwoods	651
1998		215 ACS National Meeting	652
1998	Akim/Cordeiro/ Neto	Cork lignins appear to be relatively cross-linked vs. chain branching	653
1998	Berthold/Lindfors/ Gellerstedt	Degradation of guaiacyl-glycerol β-guaiacol	654
1998	Camarero/Barrasa/ Pelayo/Martinez	Biopulping of wheat straw using Pleurotus fungi and 60 day incubation time	655
1998	Criss/Schultz/ Ingram	*Investigation of low temperature internal SnAR reaction on nonphenolic β-O-4 lignin model dimers*	656
1998	Faix/Meier	*Comparative investigations of milled wood lignins*	657
1998	Forss/Fremer	Overall lignin structures include lower molecular weight hemilignins	658
1998	Fujita/Gang/Davin/ Lewis	Isolation and characterization of first genes in heartwood formation	659
1998	Gandini/Guo/ Montanari/ Belgecem	*Lignin-based polyesters and polyurethanes*	660

Year	Investigators	Contributions	Ref
1998	Gargulak/Lebo	Commercial use of lignin based materials	661
1998	Gellerstedt	*Current and future bleaching technology*	662
1998	Glasser	Computer program for generating lignin structures	663
1998	Gratzl	*Principal delignification mechanisms*	664
1998	Heitner/Manley/ Wang/Ahvazi	*Light-induced free radicals in lignin*	665
1998	Hergert	*Level-off D.P. of lignin derived from pulping*	666
1998	Hori/Meshitsuka	*Structural heterogeneity of hardwood lignin*	667
1998	Hu Popko/Lung/ Kawaoka/Kao/ Hideki/Stokke/ Tsai/Chiang	Transgenic aspen (*Populus tremuloides*) trees with severely suppressed expression of a lignin pathway gene had a ~45% reduction in lignin quantity and a 15% increase in cellulose content	668
1998	Jurasek	Three dimensional computer model for softwoods & hardwood DHP, middle lamella, secondary wall, and kraft pulping lignins	669
1998	Lapierre/Pollet	Genetic engineering of poplar lignins initiated	670
1998	Lewis	*Lignin biosynthesis*	671
1998	Lewis/Davin/ Sarkanan, S.	Lignin and lignan biosynthesis	672
1998	Lewis/Davin/ Sarkanan, S.	Lignin and lignan biosynthesis	673
1998	Lewis et al	*Lignans, lignins and abnormal lignins: a biochemical clarification*	674
1998	Lewis/Sarkanen	*Lignin, Lignan: Biosynthesis ACS Symp. Ser. 697, in Press*	675
1998	Liu/Pye/ Argyropoulos	Softwood and hardwood kraft and Alcell dissolved and residual lignins from different stages of the cook analyzed by ^{31}P NMR	676
1998	Matsumoto/Ishizu Nakano	Lignin studies	677
1998	Matussek/Stefan	*Intl. Fact & Price Book; Pulp and Paper Int.*	678
1998	Monties/ Chabbert/ Baumberger	Correlations between structural variations and physico-chemical properties of synthetic materials containing industrial lignins	679
1998	Moran	Biotechnology applications in the pulp and paper industry	680
1998	Omori/Aoyama/ Sakakibara	Reactions of β-0-4 lignin model compounds in the presence of carbohydrates	681
1998	Robert	*Lignin fractions extracted from thermochemical pulps*	682
1998	Sarkanan, S.	Template polymerization	683

Year	Investigators	Contributions	Ref
1998	Sarkanen/Li	*Thermoplastics with very high lignin contents*	684
1998	Schultz/Nicholas	*Lignin structure susceptibility to white-rot fungus*	685
1998	Young/Akhtar	*Environmentally Friendly Technologies for the Pulp and Paper Industry*, pp. 5-67	686

Appendix 2 –Research Centers and Locations

Research Center	Location
Auburn University (UA)	Auburn, USA
Bundesforschungsansalt fur Forst und Holtzwirtschaft (BFH)	Hamburg, Germany
Centre Etude Recherche Macro. Vegetale (CERMAV)	Grenoble, France
Centre Technique du Papier (CTP)	Grenoble, France
Commonwealth Scientific and Industrial Research Organisation (CSIRO)	Melbourne, Australia
Ecole Française de Papeterie et des Industries Graphiques (EFPG)	Grenoble, France
Empire State Paper Research Institute (ESPRI)	Syracuse, USA
The Finnish Pulp and Paper Research Institute (KCL)	Helsinki, Finland
Georgia Pacific West Inc.	Bellingham, USA
Institute of Paper Science and Technology (IPST)	Atlanta, USA
Institute National de la Recherche Agronomique (INRA)	Thiverval-Gringon, France
Kyoto University	Kyoto, Japan
Latvia Academy of Science (LAS)	Riga, Latvia
Ligno-Tech Inc.	Rothschild, USA
McGill University	Montreal, Canada
Michigan Technological University (MT)	Houghton, USA
Mississippi State University (MSU)	Mississippi State, USA
Moscow State University (MSU)	Moscow, Russia
Nanjing Forestry University	Nanjing, China
National Renewable Energy Laboratory (NREL)	Golden, USA
New Mexico Tech (NMT).	Socorro, USA
Oregon State University (OSU)	Corvallis, USA

Research Center	Location
Pulp and Paper Research Institute of Canada (Paprican)	Montreal, Canada
Royal Institute of Technology (KTH)	Stockholm, Sweden
Repap Technologies	Valley Forge, USA
North Carolina State University (NCSU)	Raleigh, USA
Saint Petersburg Forestry Academy	Petersburg, Russia
State University of New York, College of Environmental Science & Forestry (ESF)	Syracuse, USA
Swedish Forest Products Res. Lab. (STFI)	Stockholm, Sweden
University of Aveiro	Aveiro, Portugal
University of British Columbia (UBC)	Vancouver, Canada
University of California-Berkeley (UCB)	Berkeley, USA
University of Hamburg	Hamburg, Germany
University of Minnesota (U of M)	Minnesota, USA
University of Maine	Orono, USA
University of Mississippi (UM)	Oxford, USA
University of Tokyo, (UT)	Tokyo, Japan
University of Toronto (U of T)	Toronto, Canada
University of Vienna	Vienna, Austria
University of Washington (UW)	Seattle, USA
University of Wisconsin (UW	Madison, USA
US Department of Agriculture (USDA)	Madison, USA
US Forest Products Laboratory (USFPL)	Madison, USA
Virginia Tech (VPI)	Blacksburg, USA
Washington State University (WSU)	Pullman, USA

Chapter 2

The Nature of Lignin: A Different View

K. Forss[1] and K-E. Fremer[2]

[1]Ståhlbergsvägen 6 D 34, 00570 Helsinki, Finland
[2]Tölögatan 19A, 00260 Helsinki, Finland

The aim of this paper is to acquaint the reader with our understanding of the nature of coniferous lignin. However, as our view differs radically from the generally accepted view, a complete account would require more pages than are here available. We shall therefore limit ourselves to a few examples, which may be of interest. A complete version including all experimentally obtained data will be published elsewhere.

Glycolignin and Hemilignins

Back in 1962 we reported (*1*) that, on elution with water through Sephadex G-25, the lignin sulfonates produced on heating Norway spruce wood in an acid bisulfite solution are fractionated over a broad retention volume range. In order to avoid confusion, we shall use the designation *glycolignin* for the polymeric lignin, which comprises 80-85 percent of the lignins and which gives rise to both *high* and *low molecular mass glycolignin sulfonates*. We use the collective designation *hemilignins* to refer to the low molecular mass lignins, which form 15-20 percent of total lignins in spruce wood and give rise to a multitude of low molecular mass *hemilignin sulfonates*.

The division of the lignins into glycolignin and hemilignins, which, like cellulose and the hemicelluloses, are presumably of different generic origin, is a cornerstone in our comprehension of the nature of coniferous lignin.

We regard the observation that the hemilignins participate in neither the polymerization nor depolymerization reactions of glycolignin as support for our belief that the hemilignins and the glycolignin are generically different lignins. However, among the hemilignin sulfonates is a low molecular mass fraction formed through sulfitolytic liberation of sulfonated guaiacylglycerol units.

By taking into account in our experimental work the difference between the hemilignins and the glycolignin, we have reached the conclusion that glycolignin is not a polymer of random structure but a regular polymer composed of identical repeating units, linked to each other and to carbohydrates by dialkyl ether bonds, Figure 1.

In addition to the sulfonation, in which, in acid solution, all eight guaiacylglycerol units in each repeating unit are sulfonated, glycolignin is subject to sulfitolysis. This reaction, which gives rise to low molecular mass glycolignin sulfonic acids, attacks, probably due to steric hindrance, only the terminal repeating units after they have been hydrolytically split off from the fragments of the glycolignin polymer.

Sulfitolysis breaks the bonds between the C_{10} units of the core and the sulfonated guaiacylglycerol units. The C_{10} units of the core are sulfonated and the sulfonated guaiacylglycerol units are liberated. The consecutive liberation of guaiacylglycerol units thus yields a homologous series of eight low molecular mass glycolignin sulfonic acids, the smallest of which consists of the core in which the C_α atoms of the 8 C_{10} units are sulfonated (*2*).

The formation of high and low molecular mass glycolignin sulfonic acids on heating spruce wood in an acid bisulfite solution is illustrated by the series of chromatograms in Figure 2.

The three first chromatograms show that up to a cooking time of 4 hours, the amounts of both high molecular mass glycolignin sulfonates A and low molecular mass glycolignin sulfonates B increased. By this time, most of the glycolignin had dissolved and the amount of high molecular mass glycolignin sulfonates started to decrease owing to depolymerization, which now dominated over dissolution. However, after a cooking time of about 11 hours, the proportion of low molecular mass glycolignin sulfonates, as well as sulfitolysis product C, started to decrease due to recombination with the high molecular mass glycolignin sulfonates. The figure substantiates the view that the high molecular mass glycolignin sulfonates A and the low molecular mass glycolignin sulfonates B are two different groups of glycolignin sulfonates.

Many attempts have been made in the past to isolate lignin in an unaltered state for studies of its structure. However, as this requires splitting of the lignin-carbohydrate bonds, it is not even theoretically possible to extract unaltered glycolignin from wood. Lignin preparations are thus derivatives.

We have avoided the problem of the inaccessibility of unaltered lignin by subjecting a hypothetical glycolignin formula and the rate constants of the reactions that transform glycolignin into high molecular mass glycolignin sulfonates to computer iteration. In this procedure the formula and the reaction rate constants were changed stepwise until the best agreement was obtained between calculated values and the compositions of the corresponding experimentally obtained derivatives.

The correctness of a formula that seeks to describe the constitution of glycolignin inside the bark of the tree cannot be proved. Nevertheless the practical value of such a formula can be determined by the degree of agreement between calculated and experimentally determined data. We shall now, with a few examples, describe the reactions of the repeating unit because the conclusions may be of particular interest.

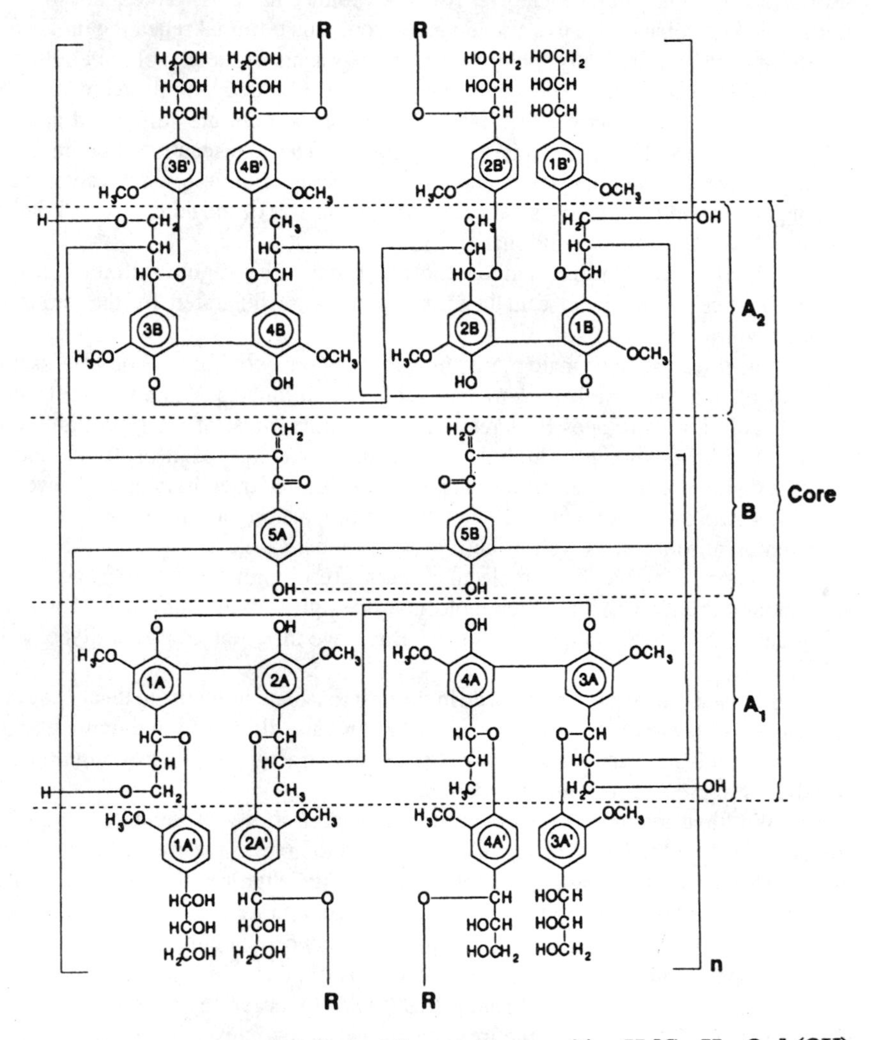

Figure 1. Spruce wood glycolignin. Elemental composition $H_2[C_{178}H_{194}O_{62}]_n(OH)_2$ after hydrolysis of its bonds with carbohydrates, R.

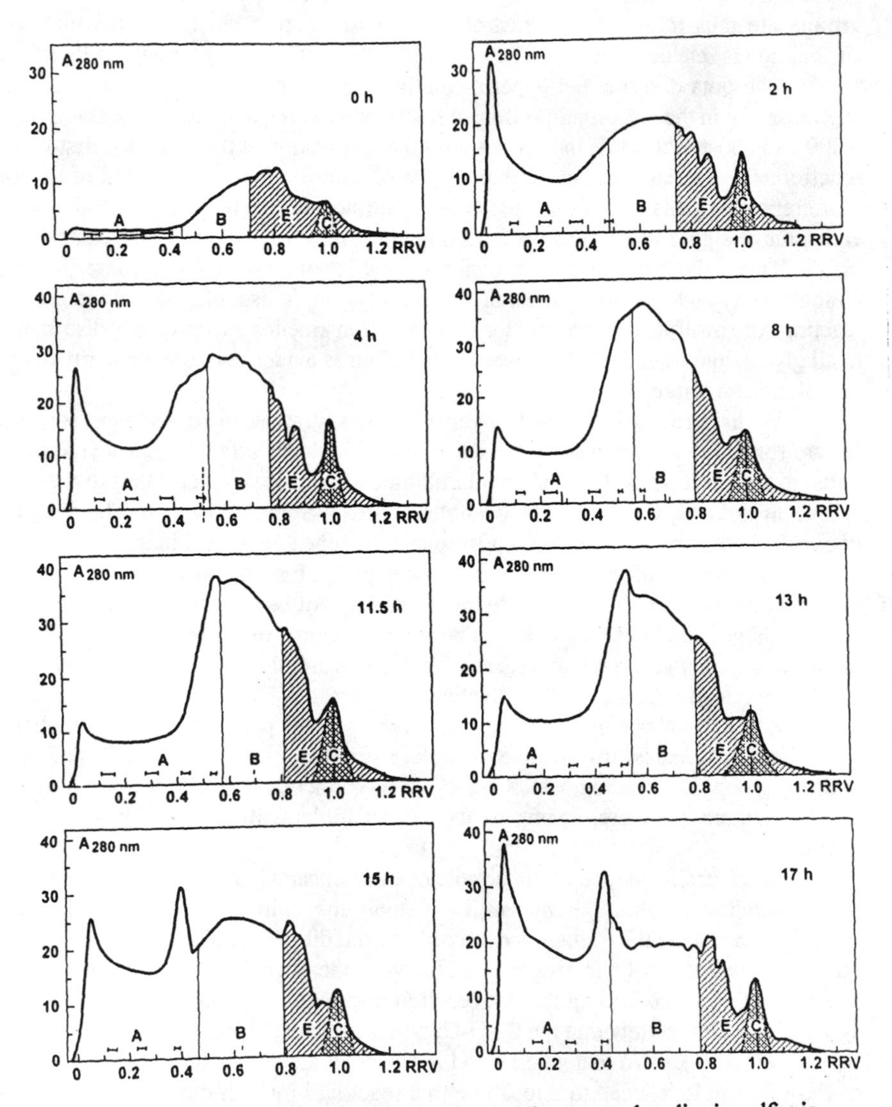

Figure 2. Formation of (A) high, (B) low molecular mass glycolignin sulfonic acids, (C) sulfitolysis product and (E) hemilignin sulfonic acids on heating Norway spruce wood in an acid bisulfite solution.
Column: Sephadex G-50. Eluent: water.

When sulfonation is performed with a nearly neutral bisulfite solution, only the four free benzyl alcohol groups react, whereas in acid solution the four carbohydrate bonds are hydrolyzed in a relatively fast reaction. An ultimate number of eight sulfo groups are thus formed in the hydrolysis and sulfonation reactions, in which the sulfonation is rate determining. The kinetics of the sulfonation is shown in Figure 3.

The dots describe the experimentally obtained number of sulfo groups per repeating unit in the glycolignin sulfonate fractions, covering the molecular mass range 6 800 - 45 000 g/mol, while the curve shows the sulfonation as the predicted first-order reaction. The coincidence shows that all glycolignin fragments composed of two or more repeating units are sulfonated at the same rate and that they approach the same end value of eight sulfo groups per repeating unit irrespective of their molecular mass.

Thus, the repeating unit concept enables us not only to describe the stoichiometry and kinetics of the sulfonation reaction, it also tells us that the groups reacting with bisulfite according to the same reaction kinetics are uniformly distributed in all glycolignin fragments. We regard this finding as evidence of the regularity of the glycolignin structure.

We have found that phenolic hydroxy groups titratable in non-aqueous solution (*3*) are formed in a rearrangement reaction which includes all eight guaiacylglycerol units. In this reaction the C_α-O-C_4 bond linking each guaiacylglycerol unit to the core is broken and a C_α-C_5 bond and a phenolic hydroxy group are formed. The reaction does not change the elemental composition of the repeating unit, Figure 4.

However, of the eight phenolic hydroxy groups formed, only four are titratable in non-aqueous solution. We refer to the finding by Aulin-Erdtman (*4*), according to which a large substituent, especially a non-phenolic one, in the ortho position to the phenolic hydroxy group reduces its acidity. Here again, the experimentally obtained results coincided with the predicted values.

After the rearrangement reaction every second phenylpropane unit in the glycolignin sulfonates, irrespective of molecular mass, carries a titratable phenolic hydroxy group. This finding, which we regard as further evidence of the regularity of the glycolignin structure, seems to us incompatible with the established lignin chemistry.

The four C_β-O-C_4 bonds in the core of each repeating unit are hydrolyzed under acidic conditions, although more slowly than the splitting of C_α-O-C_4 bonds. Hydrolysis results in the formation of an ortho-ortho dihydroxy biphenyl structure but, due to the formation of a hydrogen bond between the phenolic hydroxy groups, only one is titratable. The finding that the reaction does not cause depolymerization is in agreement with the view that the C_β-O-C_4 bonds are present in the core.

It is well known that when wood is heated in an acid bisulfite solution, some of the bisulfite is reduced to thiosulfate in a reaction in which monosaccharides are oxidized to aldonic acids. In our experiments we found that the thiosulfate reacted with glycolignin under formation of two sulfide bonds in each repeating unit, Figure 5.

Demethylation, which attacks an ultimate number of four methoxy groups, is a later reaction than the formation of sulfide bonds, Figure 6.

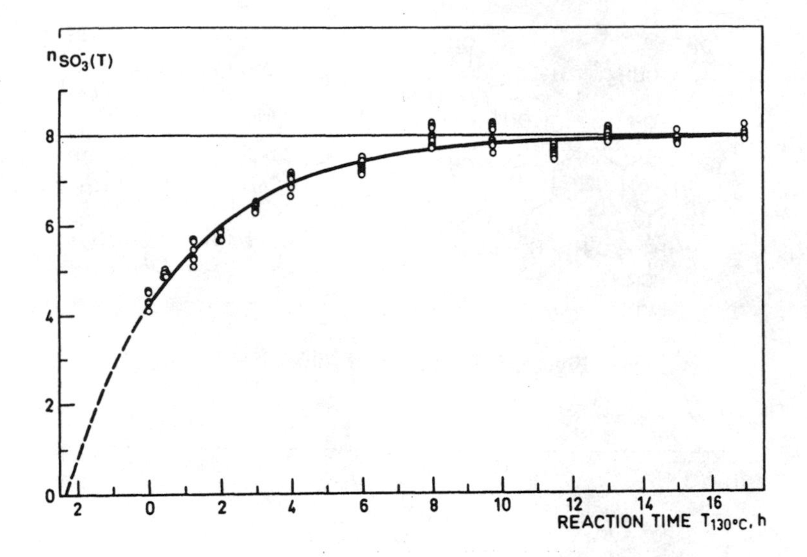

Figure 3. Sulfonation of glycolignin on heating spruce wood in an acid bisulfite solution.

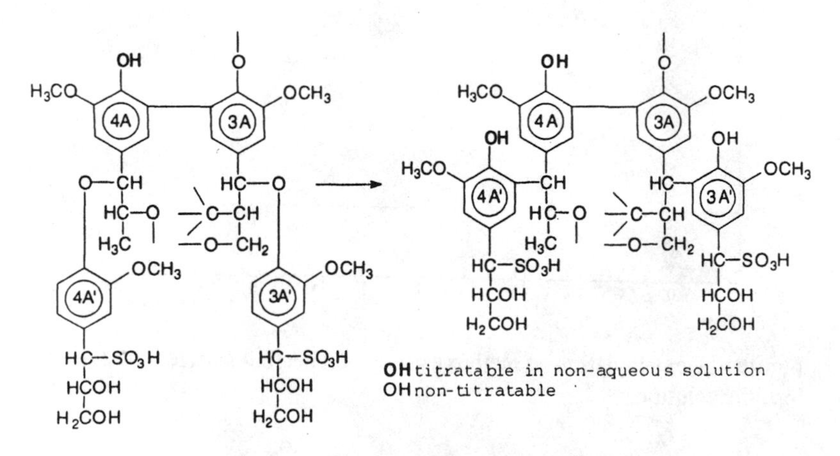

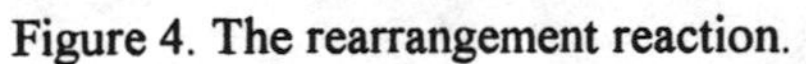

Figure 4. The rearrangement reaction.

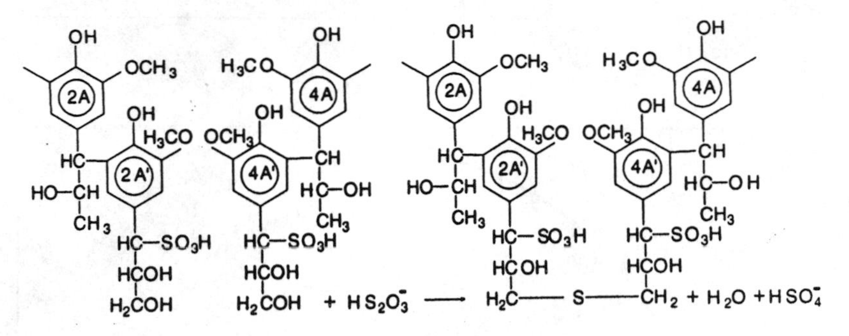

Figure. 5. Formation of sulfide bonds.

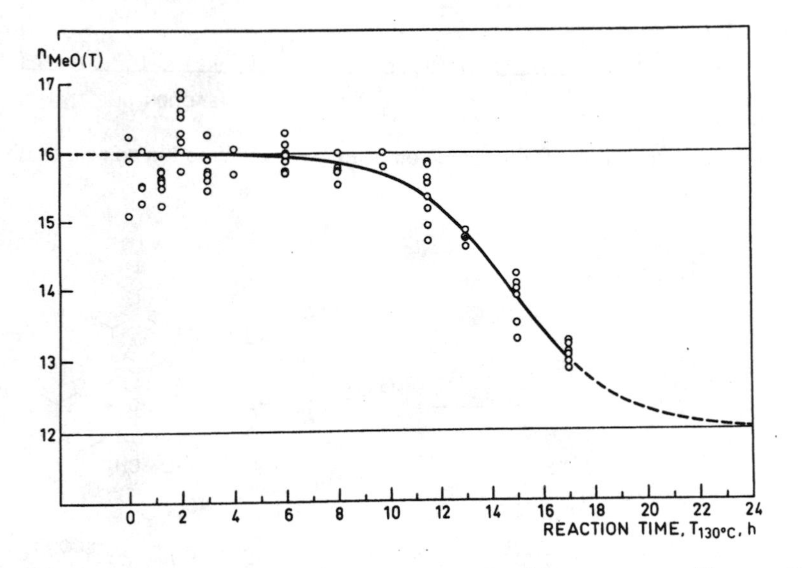

Figure 6. Demethylation of methoxy groups on heating spruce wood in an acid bisulfite solution.

In addition to demethylation, which results in the formation of four catechol groups in units 2A', 4A', 2B' and 4B', four β-keto groups are formed, Figure 7.

The formation of the non-conjugated carbonyl group at the C_β atom is revealed by the formation of an absorption band at about 1705 cm^{-1} (*5*). As carbonyl band formation and demethylation were found to proceed at the same rate, it was concluded that there is a correlation between demethylation and the formation of β-keto groups.

Applied to the formula in Figure 1, and taking into consideration the influence of the end groups and residual bisulfite in the preparations, these reactions quantitatively describe the compositions of the high molecular mass glycolignin sulfonic acids.

Solvolytic Delignification with a Buffered Dioxane-water Mixture

To test the possible universal applicability of the formula in Figure 1, the formula was applied to the description of the glycolignin derivatives obtained from a different set of reactions that takes place when extracted spruce wood meal is heated at 150°C in a mixture of 40 parts per volume of dioxane and 60 parts of a buffer solution of pH 6.8 (*6,7,8*). Based on their solubilities, the dissolved lignins can be separated into three fractions.

It is expedient to assign the lignins that are soluble in n-hexane and those that are soluble in water and in a water-methanol mixture to the group of hemilignins.

While the n-hexane-soluble hemilignin fraction contained no carbohydrates, the finding that the water-soluble lignin fractions, after removal of free carbohydrates, yielded galactose and arabinose on hydrolysis indicates that some of the hemilignins are linked to the arabinose and galactose side chains of arabinoglucuronoxylan and galactoglucomannan.

Unlike both hemilignin fractions, the amount of water-insoluble glycolignin, which did not contain carbohydrates, increased continuously and amounted after 48 hours of heating to 30 percent of the glycolignin in the spruce wood.

The molecular mass distributions of the glycolignin fractions revealed that the glycolignin fragment composed of a single repeating unit has the greatest solubility in the dioxane-buffer mixture. Furthermore, as the molecular mass distribution remained virtually constant, at least up to a heating time of 48 hours, it is obvious that the repeating unit is not degraded when wood is subjected to the reaction conditions employed.

The elemental compositions and the methoxy group contents determined by analysis of the eight glycolignin fractions obtained on heating wood meal for between 0.5 and 48 hours were found to agree with the corresponding calculated values obtained by subjecting the glycolignin, Figure 1 (n = 1), to the splitting off of an ultimate number of 16 molecules of water, Δn_{H2O}:

$$\Delta n_{H2O} = 16(1 - e^{-0.0244T})$$

and to a splitting off of 4 methoxy groups, Δn_{MeO}:

$$\Delta n_{MeO} = 4(1 - e^{-0.0309T})$$

where T is the heating time in hours at 150°C.

Also on heating wood in the dioxane-buffer mixture, glycolignin is transformed by a reaction that leads to the formation of keto groups, as characterized by the formation of an absorption band near 1700 cm^{-1}. It was found that the formation of keto groups is related to the demethylation of the four methoxy groups and to the splitting off of the 16 molecules of water.

The NMR spectrum of each acetylated glycolignin fraction was divided into ranges A - G, of which range D was assigned to the hydrogen atoms of the methoxy groups, Figure 8.

The number of hydrogen atoms in methoxy groups, expressed as a percentage of the total number of hydrogen atoms, equals area D as a percentage of the total area, as shown by the upper expression in Figure 9.

However, the percentage of hydrogen atoms present in the methoxy groups is also given by the lower expression, where $3(16 - \Delta n_{MeO})$ equals the number of hydrogen atoms present in methoxy groups. 198 is the original number of hydrogen atoms according to Figure 1, (n = 1). $2\Delta n_{MeO}$ is the number of hydrogen atoms lost through demethylation of the ultimate number of four methoxy groups and $2\Delta n_{H2O}$ the number of hydrogen atoms lost through the splitting off of the ultimate number of 16 molecules of water.

The agreement between the circles, which represent data based on the NMR spectra, and the curve, which is based on the formula and the determined contents of carbon, hydrogen, oxygen and methoxy groups, reveals that the number of hydrogen atoms n_H corresponding to an area Q of each NMR spectrum can be calculated by means of the following equation:

$$n_H = 3(16 - \Delta n_{MeO})Q/D$$

The NMR spectra can thus be employed for a detailed quantitative description of the reactions. We have found that glycolignin is subject to four reactions when spruce wood meal is heated in the dioxane-buffer solution, Figure 10.

Reaction 1, which does not result in any change in elemental composition or content of functional groups, is revealed by the finding that eight hydrogen atoms cease producing a signal between 1.70 and 0.70 ppm and start producing an increasing signal at the same rate between 2.80 and 2.45 ppm. The reaction was found to correspond to the formation of three ultraviolet absorption bands with maxima at 254, 270 and 288 nm.

It is well known that biphenyl derivatives exhibit conjugation bands in this wavelength range. The absorptivities of these bands depend on the angle between the two aromatic rings. When the rings are co-planar the absorptivities are at a maximum. Twisting about the bond between the rings reduces the conjugation and thus the

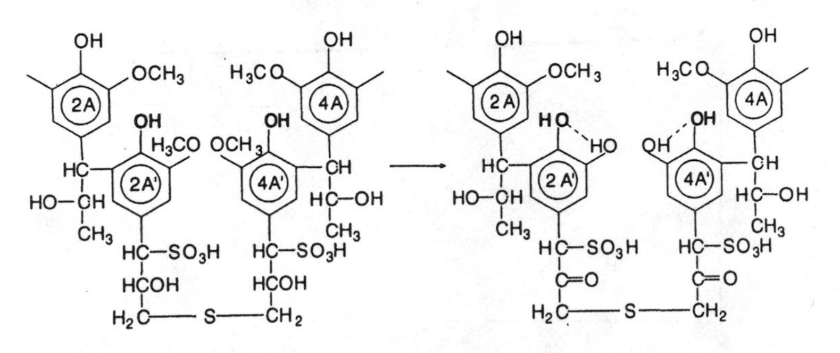

Figure 7. Formation of catechol groups and β-keto groups.

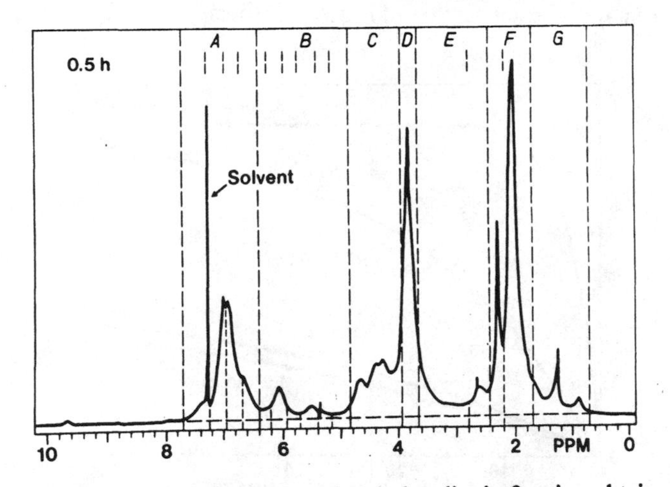

Figure 8. NMR spectrum of the acetylated glycolignin fraction obtained after heating spruce wood in the dioxane-buffer mixture for 0.5 h at 150°C.

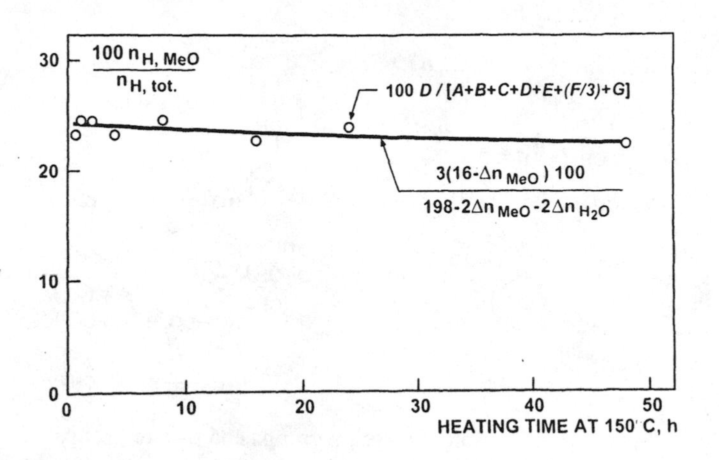

Figure 9. Number of hydrogen atoms in methoxy groups expressed as a percentage of total number of hydrogen atoms.

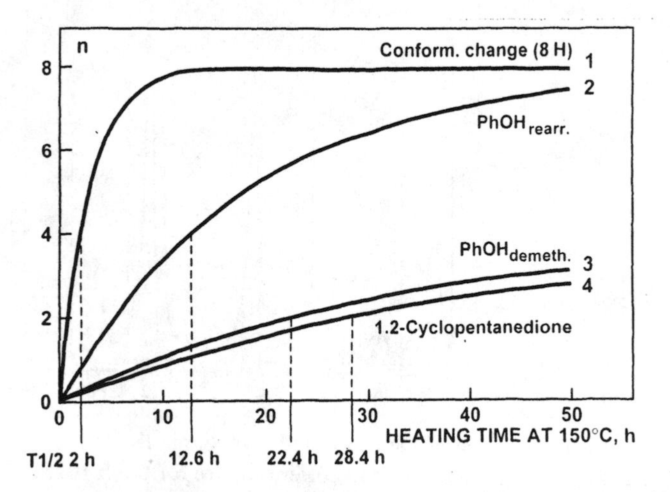

Figure 10. The four reactions of glycolignin on heating spruce wood in the dioxane-buffer mixture.

absorptivities. It seems possible that Reaction 1 reflects a conformation change caused by the cleavage of repeating unit-repeating unit bonds and/or repeating unit-carbohydrate bonds. As Reaction 1 influences the absorptivity at 280 nm it may also have a practical significance.

Reaction 2 is revealed by the loss of eight aromatic hydrogen atoms and the formation of eight phenolic hydroxy groups at the same rate. As the reaction is not accompanied by any change in elemental composition, we believe it is analogous with the rearrangement reaction that takes place during acid bisulfite pulping, Figure 4.

In Reaction 3, which is analogous with the demethylation reaction taking place on extended heating of spruce wood in acid bisulfite solution, four methoxy groups in units 2A', 4A', 2B' and 4B' are split off. In Reaction 4, the ultimate number of 16 molecules of water are split off under formation of four 1,2-cyclopentanedione structures, Figure 11. The finding that Reaction 4, in which all aliphatic hydroxy groups in the eight guaiacylglycerol units disappear and which transforms glycolignin into a virtually hydrophobic polymer, emphasizes the importance of the C_α atoms of the eight guaiacylglycerol units. The rapid formation of the blocking sulfo groups in acid bisulfite pulping prevents not only the formation of 1,2-cyclopentanedione structures but also the recombination of glycolignin with the fiber carbohydrates.

It has been found by many authors that monomeric, dimeric and oligomeric reaction products are formed when wood or milled wood lignin is subjected to ethanolysis, acidolysis, mild hydrolysis or to heating in dioxane-water. Although most of these products were obtained in yields of only a few tenths of a percent of the total lignin, their detection has exerted a great influence on the concept of the nature of lignin. However, none of these compounds were formed from glycolignin in our experiments. We thus find it difficult to avoid the question, 'Are these monomeric, dimeric and oligomeric reaction products derivatives of hemilignins rather than of the glycolignin polymer?' The answer to this question is in our opinion of paramount importance for the understanding of the nature of lignin.

The Structure of Glycolignin

Although we have shown that the constitutional formula in Figure 1 can be used as a basis for quantitative and kinetic descriptions of two different sets of glycolignin reactions, it does not show whether the formula is sterically possible.

Due to the asymmetry of the C_α and C_β atoms in the 16 C_{10} units and to the atropisomerism of the five biphenyl units, the formula corresponds to a theoretical number of 2^{37} stereoisomers. In the following, our aim is to show the number of structures that are sterically possible.

These structures must meet a number of requirements. They must contain acceptable bond lengths, bond angles and torsion angles. They must also meet the requirements for van der Waals correctness, that is the structures must not imply interaction of atoms not bonded to each other. The importance of this requirement is revealed by the fact that the ten phenylpropane units forming the core of the repeating unit are very tightly linked to each other.

The structural description must meet the steric requirements set by the reactions of glycolignin. For example, the benzyl alcohol groups which react with bisulfite must be sterically available to the reaction, and they must also be able to provide space for the bulky sulfo group.

As shown by Goring and his co-workers (*9*), the lignin molecule should be about 2 nm thick. In solution the shape of lignin derivatives should be between an Einstein sphere and a non-freedraining coil.

The shape of the glycolignin polymer must allow the formation of covalent bonds with arabinoglucuronoxylan and/or galactoglucomannan.

Last but not least, the structural formula must reflect the fact that glycolignin preparations as well as all glycolignin degradation products are optically inactive.

To examine the different ways of linking the 10 phenylpropane units of the core together, it is appropriate to subdivide the core, Figure 1, into the two constitutionally identical groups A_1 and A_2 (units 1A, 2A, 3A, 4A and 1B, 2B, 3B, 4B, respectively) and group B comprising the biphenyl unit 5A-5B.

Experiments with space filling models showed that groups A_1 and A_2 can be described by only one structural formula, and its enantiomer, which can clearly be linked to the biphenyl unit 5A-5B, Figure 12.

Further experiments with space filling models revealed that there is only one core structure, and its enantiomer, that meets the requirements set:

Repeating unit *RM*			Repeating unit *SP*		
A_1	B	A_2	A_1	B	A_2
RM	*M*	*RM*	*SP*	*P*	*SP*

In Figure 13 the *SP* repeating unit in the center is linked, both to the left and to the right, by alkyl ether bonds to *RM* repeating units. The figure thus describes the glycolignin polymer as composed of alternating *RM* and *SP* repeating units. As there are symmetry planes through the ether bonds between the repeating units, glycolignin and its degradation products are optically inactive.

Bonds between Glycolignin and Hemicelluloses and between Hemicelluloses and Hemilignins

We have pointed out that the structural description of glycolignin must include its ability to form bonds with carbohydrates. Further modeling experiments revealed that it is possible to form bonds (without strain) between arabinoglucuronoxylan and glycolignin, Figure 14.

In this glycolignin segment, which is composed of four repeating units, each of the four xylan chains is linked by an ether bond to a guaiacylglycerol unit of the repeating units. Eight xylose units of each xylan chain correspond to four repeating

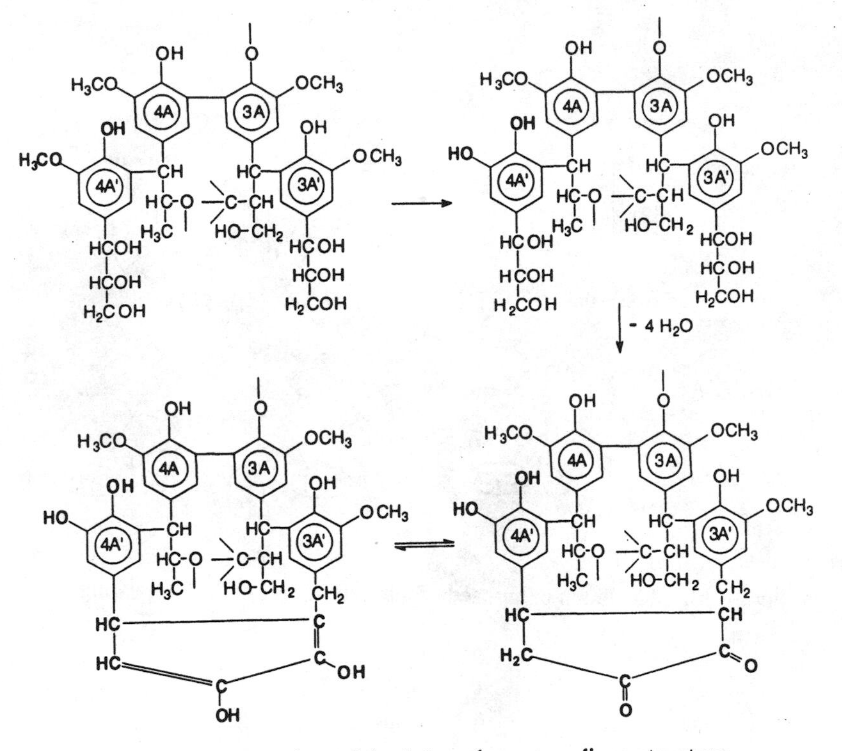

Figure 11. Formation of the 1,2-cyclopentanedione structure.

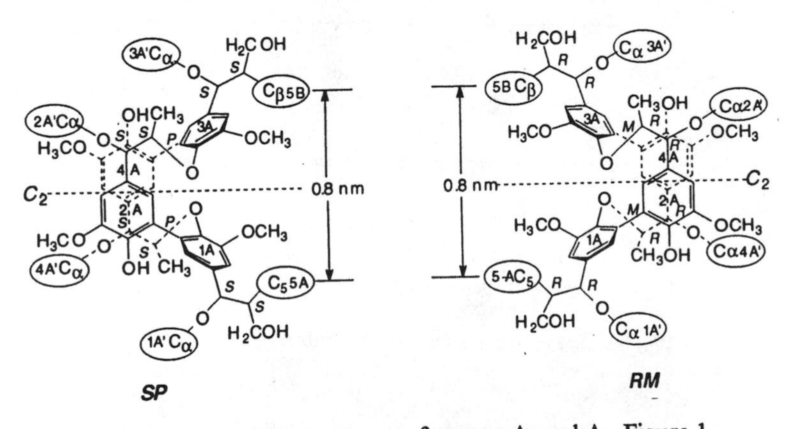

Figure 12. Possible structures of groups A_1 and A_2, Figure 1.

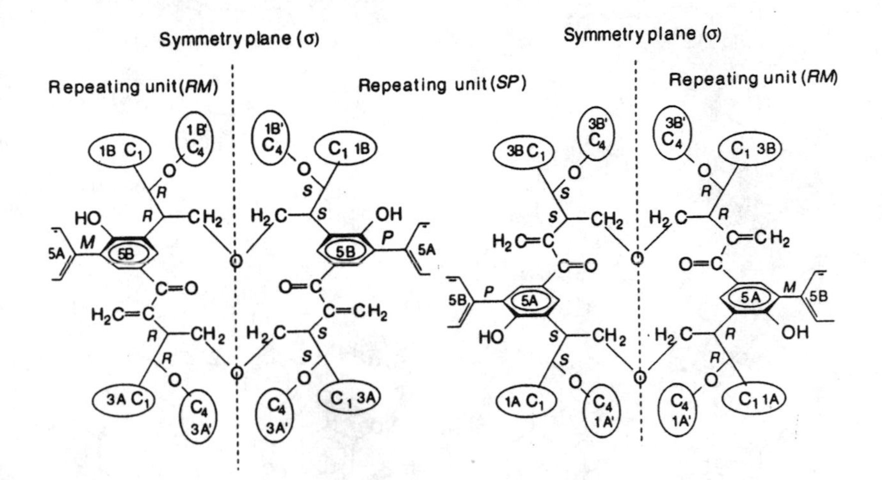

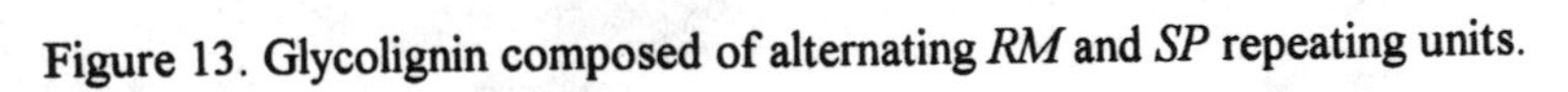
Figure 13. Glycolignin composed of alternating *RM* and *SP* repeating units.

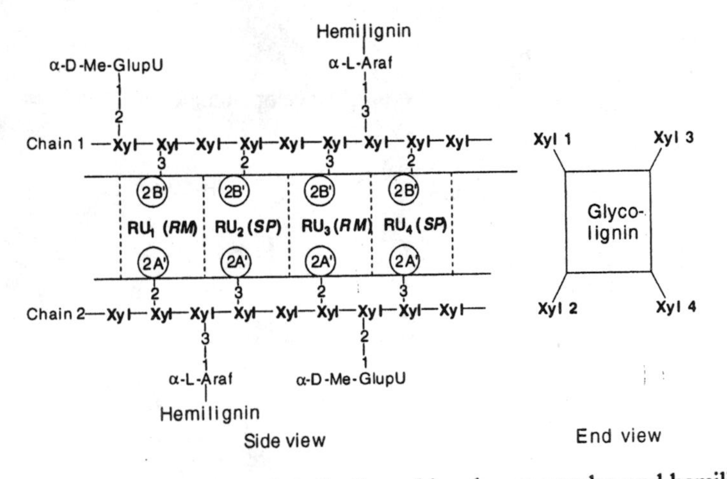

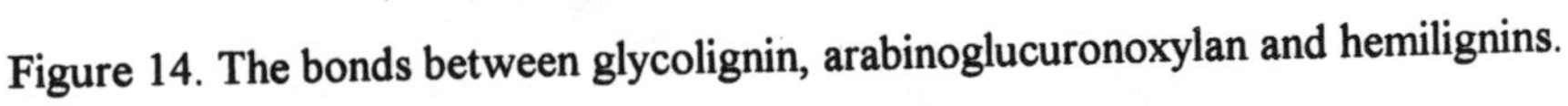
Figure 14. The bonds between glycolignin, arabinoglucuronoxylan and hemilignins.

units. The configurations of the C_α and C_β atoms of the guaiacylglycerol units have little effect on the formation of these bonds. The lack of optical activity, however, requires that the eight guaiacylglycerol units in repeating units *RM* and *SP* are enantiomers.

The figure raises the question as to whether glycolignin and arabinoglucuronoxylan are described as two compounds quantitatively linked to each other. If, as reported, each ten xylose units on average carry two 4-O-methyl-α-D-glucuronic acid groups and 1.3 α-L-arabinose groups, the molecular mass of xylan containing 4 times 8 xylose units is 5 994, while the molecular mass of the glycolignin tetramer is 13 186. The content of 23 percent glycolignin in spruce wood would thus correspond to a content of 10.5 percent of arabinoglucuronoxylan linked to glycolignin. As the content in Norway spruce wood, according to Rydholm (*10*), is 10 percent, it seems possible that the arabinoglucuronoxylan is quantitatively linked to glycolignin.

We have shown that most of the hemilignins are probably monomers and dimers and that some of the hemilignins are linked to the arabinose side chains of arabinoglucuronoxylan and to the galactose side chains of galactoglucomannan.

According to Rydholm (*10*), spruce wood contains 1.9 percent galactan and 0.5 percent araban. If linked to all galactan and araban, a hemilignin dimer with a molecular mass of 344 corresponds to 5.3 percent of the wood, or 18 percent of total lignin. As the hemilignins amount to 15 - 20 percent of total lignin, it is possible that a high proportion of the hemilignins are linked to the arabinose side chains of arabinoglucuronoxylan and to the galactose side chains of galactoglucomannan, which, together with arabinoglucuronoxylan, form nearly 100 percent of the spruce wood hemicelluloses.

Based on the results obtained with the mechanical CPK models, more accurate results were obtained with the aid of the "Chem-X" force field system, developed by Chemical Design Ltd. Oxford, England (*11*). The work, which was performed using a Sigmex 6268 M workstation connected to a VAX 8800 central computer at the Institute of Chemistry, University of Helsinki, made it possible to adjust and to optimize bond distances and bond and torsion angles, to remove non-bonded interactions and thus to confirm that the suggested structure, 2.4 and 3.3 nm in diameter, is sterically possible.

Literature Cited

1. Jensen, W.; Fremer, K-E.; Forss, K. *Tappi* **1962**, *45*, 122 - 127.
2. Forss, K.; Fremer, K-E. *Paperi ja Puu - Papper och Trä* **1965**, *47*, 443 - 454.
3. Forss, K.; Fremer, K-E. *Paperi ja Puu - Papper och Trä* **1978**, *60*, 121 - 127.
4. Aulin-Erdtman, G. *Svensk Papperstidn.* **1953**, *56*, 91 - 101.
5. Hergert, H. In *Lignins, Occurrence, Formation, Structure and Reactions;* Sarkanen, K.V.; Ludwig, C.H., Eds.; Wiley - Interscience: New York, **1971**, pp 267 - 297.

6. Forss,K.; Kokkonen, R.; Sågfors, P-E. In *Lignin, Properties and Materials*; Glasser, W.G.; Sarkanen, S., Eds.; ACS Symposium Series 397; American Chemical Society: Washington DC, **1989**, pp 29 - 41.
7. Forss, K.; Fremer, K-E.; Kokkonen, R.; Sågfors, P-E. *The Reactions of Spruce Wood Lignins on Heating Wood in a Neutral Dioxane-Water Mixture*, Part I, *Dissolution, Composition and Reactions of the Lignins Dissolved,* PSC Communications 42, The Finnish Pulp and Paper Research Institute, **1992**, pp 1 - 58.
8. Forss, K.; Fremer, K-E.; Sågfors, P-E. *The Reactions of Spruce Wood Lignins on* Heating Wood in a Neutral Dioxane-Water Mixture, Part II, *Color Formation and Influence of the Reactions on the Absorption of UV and Visible Light,* PSC Communications 64, The Finnish Pulp and Paper Research Institute, **1994**, pp 1 - 82.
9. Goring, D.A.I. In *Lignin, Properties and Materials,* Glasser, W.G.; Sarkanen, S.; Eds.; ACS Symposium Series 397: American Chemical Society, Washington DC, **1989**, pp 2 - 10.
10. Rydholm, S.A. *Pulping Processes,* Interscience: London, **1965**, p. 96.
11. Davies, E.K.; Murrall, N.W. *Computers Chem.* **1989**, *13*, 149 - 156.

BIOCHEMISTRY

Chapter 3

Monolignol Compositional Determinants in Loblolly Pine: Aromatic Amino Acid Metabolism and Associated Rate-Limiting Steps

Hendrik van Rensburg[1], Aldwin M. Anterola[1], Lanfang H. Levine[2], Laurence B. Davin[1], and Norman G. Lewis[1]

[1]Institute of Biological Chemistry, Washington State University, Pullman, WA 99164–6340
[2] Dynamic Corporation, Mail Code DYN–3, Kennedy Space Center, FL 32899

Phenylpropanoid and phenylpropanoid-acetate pathways are major repositories of organic carbon in vascular plants, affording the related lignins, lignans, suberins, flavonoids, coumarins, and numerous other phenolics. Together these constitute about 30-40% of all organic plant matter. Their formation was central to the evolutionary adaptation of plants to land, this being achieved mainly by elaboration of complex biochemical pathways from phenylalanine and, to a lesser extent, tyrosine. A component of our research program is involved in determining how monolignol formation, leading to lignins and lignans, is controlled in terms of both the ratios and relative amounts of monolignols formed. Thus, loblolly pine (*Pinus taeda* L.) cell suspension cultures were grown under conditions which induced formation of its monolignol precursors, *p*-coumaryl and coniferyl alcohols. Increased phenylalanine availability to the cells resulted in a steady increase in amounts of *p*-coumaryl alcohol relative to that of coniferyl alcohol, until essentially equimolar levels were reached. Monolignol precursor administration experiments and metabolite analysis (pool sizes and end-product accumulation) revealed that there were three operationally slower steps during monolignol formation: these included the overall nitrogen recycling process from phenylalanine, and the two hydroxylation steps affording *p*-coumarate and caffeate, respectively. No evidence was found that cinnamyl alcohol dehydrogenase functions in a rate-limiting capacity, in contrast to previous assertions by others.

The successful transition by plants to a terrestrial environment, from their aquatic algal precursors some 400-450 million years ago (*1*), was accompanied by formation of a large array of plant phenolic compounds. These substances serve a variety of essential physiological functions which include: a means for structural support and water/nutrient conduction; the formation of protective layers that minimize the effects of changes in temperature, humidity, UV-B irradiation and encroachment by pathogens; and generation of hydrophobic layers that prevent uncontrolled water diffusion and sustain a high water potential for active, highly regulated, metabolism. Indeed, all vascular plants, and hence all other dependent life forms, could not exist without these biopolymers, and other lower molecular weight compounds. Among the most abundant of these phenolic derivatives are the distinct metabolic classes of suberins (the aromatic domain thereof), flavonoids, coumarins, furanocoumarins, stilbenes, miscellaneous phenols, condensed tannins and, of special interest for this chapter, lignins and lignans.

Lignans are a large, structurally diverse, class of vascular plant metabolites having a wide range of physiological functions and pharmacologically important properties (*2*). Because of their pronounced biological (antimicrobial, antifungal, and antiviral) (*3-7*), antioxidant (*8,9*), and anti-feedant (*10,11*)) properties, a major role of lignans in vascular plants is to help fight against opportunistic biological pathogens and predators. Lignans were originally classified as phenylpropanoid dimers linked through 8–8′ bonds (*e.g.*, pinoresinol, see Scheme 1 inset) (*12*). However, this definition did not take into account other coupling modes (*e.g.*, 8–*O*–4′ and 8–5′, see Scheme 1 inset). We have since expanded this definition to encompass all of the various interunit linkages, and it is now also well recognized that lignans can exist in higher molecular weight oligomeric forms. The chemistry and biochemistry of lignans have been reviewed elsewhere (*2*).

Lignins constitute 20-30% by weight of all vascular plants, making them the second most abundant organic substances on earth next to cellulose. They are phenylpropanoid heteropolymers interlinked *via* different coupling modes including by 8–*O*–4′ (predominant), 8–5′, 5–5′ and 8–8′ bonds. In contrast to lignans, the lignins mainly have structural roles in plants and function as binding and encrusting materials within certain cell wall types. In so doing, they provide rigidity to plant cell walls and thus help counter the forces of compression acting on the overall plant structure. Indeed, vascular plants could not grow to give the fantastic variety of forms without the "reinforcing" process of lignification. Moreover, their hydrophobic properties give vascular plants the capacity to develop xylem tissues capable of conducting water throughout plants without significant loss due to absorption and evaporation. Lignification is also surmised to be a common response of plants to infection or injury and the wound-healing process of injured plants is often claimed to involve lignification (*13*).

Both lignins and lignans are derived mainly from hydroxycinnamyl alcohols (monolignols). The main purpose of this chapter is to assess our current understanding of the monolignol biosynthetic pathway, in terms of the enzymology involved from phenylalanine to monolignols. This includes a brief description of the enzymatic conversions and the present, but incomplete, state of knowledge regarding the subcellular locations of enzymes associated with the pathway. Additionally, our view of how monolignol amount and composition in lignin is determined is discussed in light of recent results from our laboratory.

Monolignol Biosynthesis

Scheme 1 represents the overall biochemical pathway for monolignol formation which initially involves the conversion of shikimate (**1**) to L-phenylalanine (**6**)/L-tyrosine (**5**). This is then followed by deamination, aromatic ring hydroxylations, CoA ligations and *O*-methylation(s) with consecutive reduction of the cinnamoyl CoA esters (**13, 15, 17**) to afford the monolignols, *p*-coumaryl (**21**), coniferyl (**22**) and sinapyl (**23**) alcohols *via* the intermediate aldehydes (**18-20**) (summarized in Refs. *14,15,16*). Altogether this pathway accounts for about 30-40% of the organic carbon in all plant matter, whose end products represent some of the metabolically most expensive in vascular plants (*17*).

Two central issues often arise when considering plant cells undergoing active monolignol biosynthesis. The first is the metabolic fate of the ammonium ion released, and the second involves identifying the nature of the operationally slower steps which serve as control points during active phenylpropanoid (monolignol) biosynthesis. In addressing these matters, this chapter is broken down into three topics. These include: (i) identification of the nitrogen recycling mechanism able to

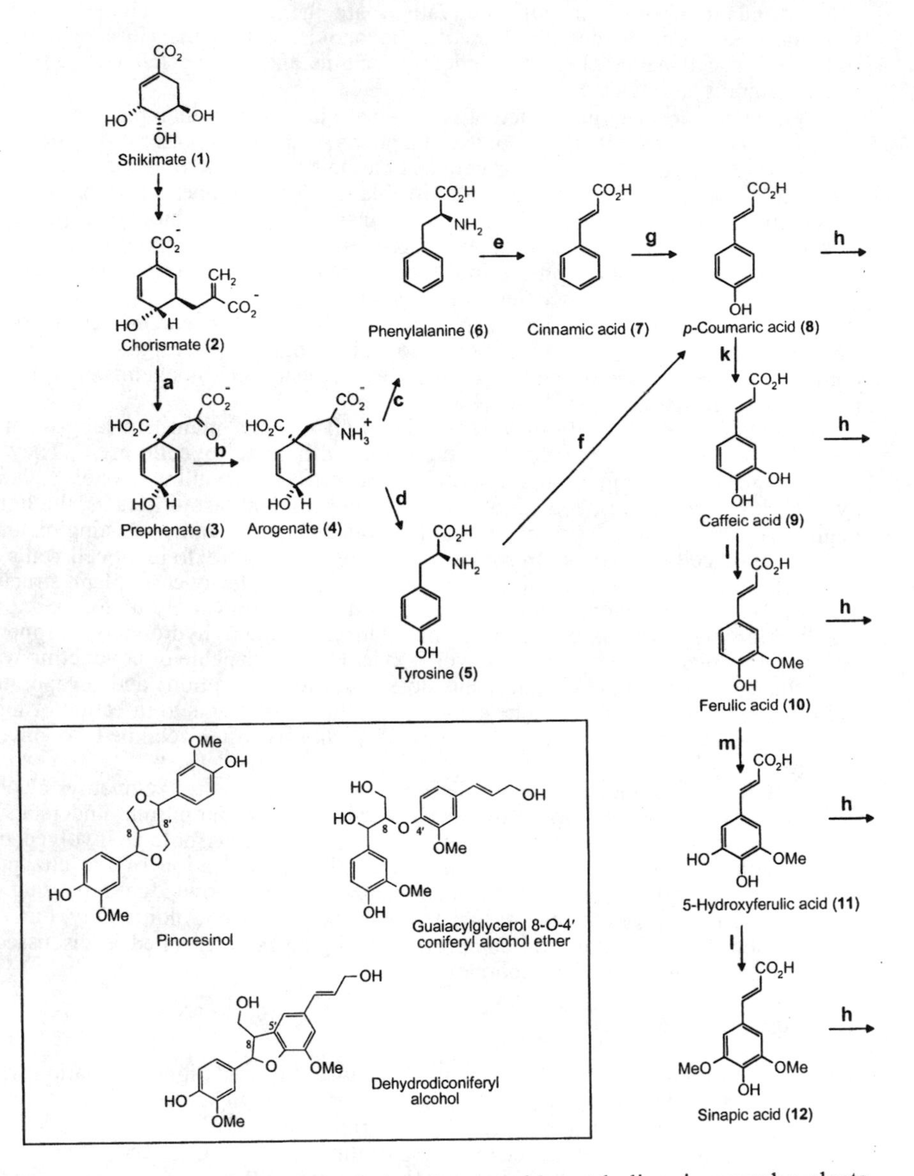

Scheme 1. Simplified denotation of phenylpropanoid metabolism in vascular plants. **(a)** Chorismate mutase, **(b)** prephenate aminotransferase, **(c)** arogenate dehydratase **(d)** arogenate dehydrogenase, **(e)** phenylalanine ammonia-lyase, **(f)** tyrosine ammonia-lyase, **(g)** 4-cinnamate hydroxylase, **(h)** hydroxycinnamoyl CoA ligases, **(i)** cinnamoyl CoA reductase, **(j)** cinnamyl alcohol dehydrogenase, **(k)** *p*-coumarate

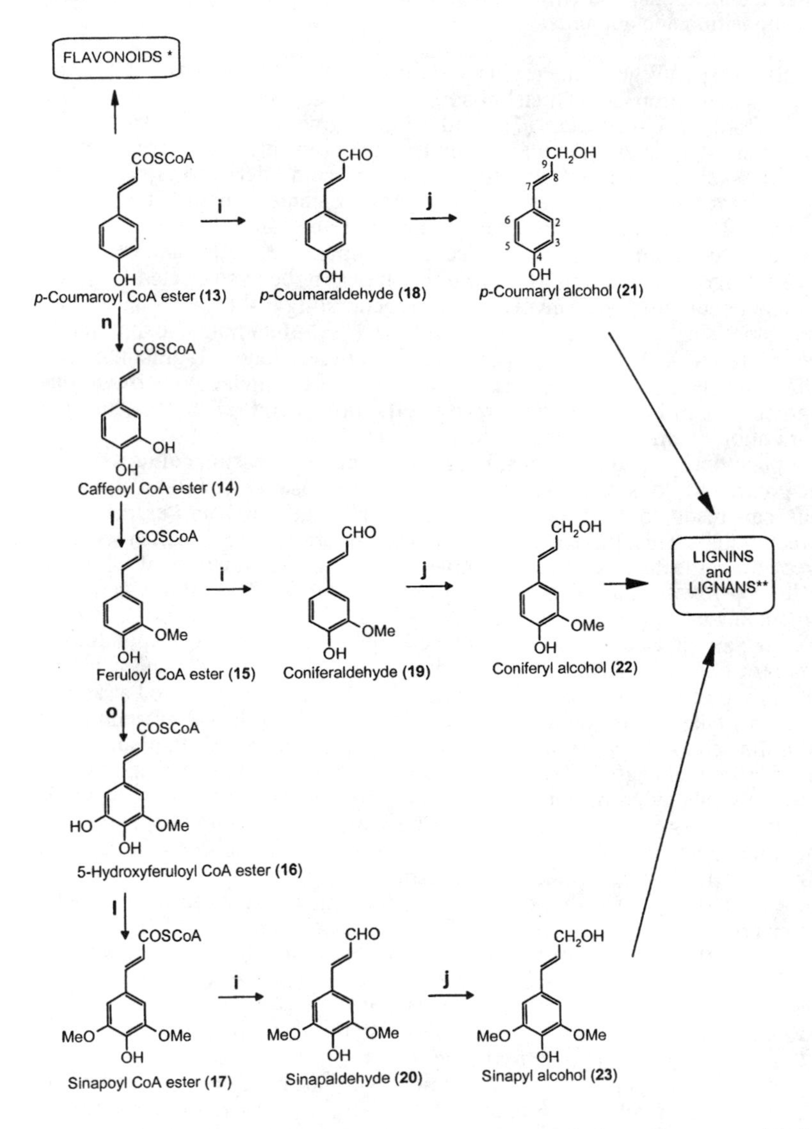

3-hydroxylase, (**l**) hydroxycinnamoyl *O*-methyltransferases, (**m**) ferulic 5-hydroxylase, (**n**) *p*-coumaroyl CoA 3-hydroxylase, (**o**) feruloyl CoA 5-hydroxylse. *: plus acetate pathway for both flavonoids and suberins, **: main source of lignin and lignan skeleta. Inset: Pinoresinol (8–8′, guaiacylglycerol coniferyl alcohol ether (8–*O*–4′) and dehydrodiconiferyl alcohol (8–5′).

sustain phenylpropanoid metabolism; (ii) a brief synopsis of what is currently known thus far about the essential characteristics of each enzymatic step, as well as the tissue and subcellular localizations, and (iii) identification of the limiting steps modulating monolignol composition and amounts.

Nitrogen Availability. When either phenylalanine (Phe) (**6**) or tyrosine (**5**) are conscripted for phenylpropanoid metabolism, the amino group is immediately removed to afford *E*-cinnamate (**7**) and *E*-*p*-coumarate (**8**), respectively. Concomitantly, an equimolar amount of ammonium ion is released for every phenylpropanoid skeleton formed. The cinnamic acid derivatives are then subsequently converted into substances such as lignins, lignans, suberins, flavonoids, and related metabolites. Therefore, it was considered that an effective means of recycling the liberated ammonium ion needed to be operative. Recognition that a mechanism for nitrogen recycling must be in effect has long been suspected (*17*) but, the actual biochemical processes involved for regenerating L-Phe (**6**), have only recently been established *via* the glutamine synthase (GS)/glutamine 2-oxoglutarate aminotransferase (GOGAT) system to generate the amino donor, L-glutamic acid (Glu) (**25**). This was achieved through the investigation of the metabolism of various ^{15}N-labeled precursors in lignifying *Pinus taeda* cell cultures and related systems and involves assimilation of ammonium ion (Scheme 2) (*18-20*).

Active phenylpropanoid metabolism in *P. taeda* cell suspension cultures can be induced when exposed to solutions containing 8% sucrose (*19,21,22*). In such instances, this can result in formation of lignified cell walls and an "extracellular lignin-like precipitate". This, therefore, provided the opportunity to establish whether a nitrogen recycling mechanism was operative following deamination of Phe (**6**). Thus, L-[^{15}N]Phe (**6**) was administered to lignifying *P. taeda* cell cultures and then allowed to metabolize over a four day period. At 24 h, ^{15}N NMR spectroscopic examination of the amino acid extract from the cells revealed signals corresponding to the amide group of glutamine (Gln) (**27**), the α amino groups of Gln (**27**) and Glu (**25**) and the amino group of Phe (**6**) (*19*). Importantly, $^{15}NH_4^+$ was not observed at any of the sampling points taken in the study. Interestingly, the L-[^{15}N]Phe (**6**) present in the soluble pool in the cells was ~90 atom % ^{15}N-enriched at all intervals sampled, but its amount [μg/g fresh weight (gfw)] decreased ~95% as a result of the utilization of its carbon skeleton for phenylpropanoid metabolism. By contrast, the Gln/Glu pools were enriched by only ~43-53 atom % ^{15}N, with both nitrogens of Gln (**27**) labeled in relatively equal amounts; however, the relative amounts (μg/gfw) of Gln/Glu dropped by almost 50% over the duration of the 96 h experiment.

To prove unambiguously that the GS/GOGAT pathway was assimilating the ammonium ion released during active phenylpropanoid metabolism, L-[^{15}N]Phe (**6**) was administered in the presence of the known PAL inhibitor, 0.1 mM L-aminooxy phenylpropionic acid (AOPP) (*23*). As expected, the dominant resonance observed was now that of unmetabolized L-[^{15}N]Phe (**6**), *i.e.*, PAL was substantially inhibited (*19*). Subsequently, experiments were carried out using the glutamine synthetase (GS) inhibitor, 5.0 mM L-methionine-*S*-sulfoximine (MSO). This resulted in only L-[^{15}N]Phe (**6**) and $^{15}NH_4^+$ being observed, indicating that inhibition of glutamine synthase prevented an effective assimilation of $^{15}NH_4^+$, released during Phe (**6**) deamination, into either Gln (**27**), Glu (**25**), or any other amino acid. In addition, treating lignifying *P. taeda* cells with L-[^{15}N]Phe (**6**), in the presence of 5.0 mM azaserine (AZA), an inhibitor of GOGAT (*24*), showed predominantly ^{15}N NMR resonances corresponding to L-[^{15}N]Phe (**6**) and ^{15}N-Gln (**27**). This revealed that further cycling of the nitrogen from L-[^{15}N]Phe (**6**) into the α amino group of Glu (**25**) was greatly reduced. Therefore, the use of specific inhibitors indicate that the primary metabolic fate of the released nitrogen of L-[^{15}N]Phe (**6**) is *via* its assimilation into Gln (**27**) and Glu (**25**) by the action of GS and GOGAT.

The next step was to establish that the ammonium ion, liberated during deamination, was recycled back to Phe (**6**) during active phenylpropanoid metabolism. Consequently, metabolism of $^{15}NH_4Cl$ over a 96 h time period revealed ^{15}N NMR

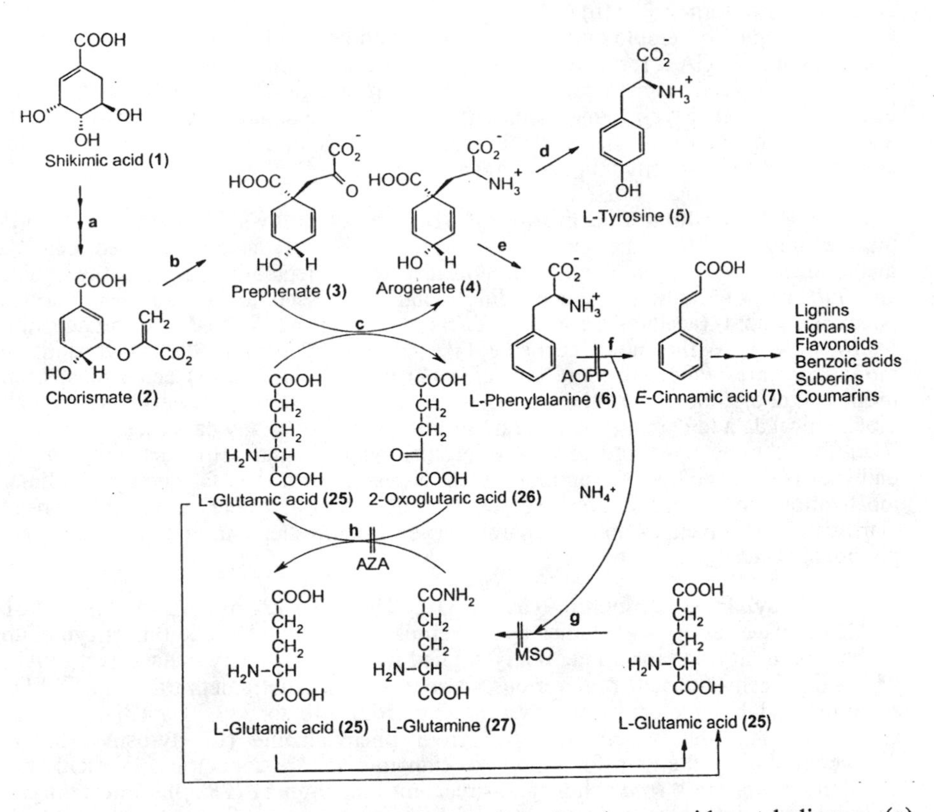

Scheme 2. Proposed nitrogen recycling during phenylpropanoid metabolism. **(a)** Shikimate-chorismate pathway enzymes, **(b)** chorismate mutase, **(c)** prephenate: glutamate aminotransferase, **(d)** arogenate dehydrogenase, **(e)** arogenate dehydratase, **(f)** phenylalanine ammonia lyase (PAL), **(g)** glutamine synthetase **(h)** glutamine 2-oxo-glutarate aminotransferase (GOGAT).

resonances attributable to various amino acids, *i.e.*, Gln, Glu, Ala, Pro, Lys, Orn, γ-aminobutyric acid and NH_4^+. On the other hand, the $^{15}NH_4^+$ ion released from L-[^{15}N]Phe (**6**) during phenylpropanoid metabolism was only incorporated into Gln, Glu, Ala and Ser, *i.e.*, it was apparently not made available for general amino acid/protein synthesis. Additionally, incubation of loblolly pine cell cultures with $^{15}NH_4Cl$, L-[^{15}N]Gln (**27**) or L-[^{15}N]Glu (**25**) in the presence of 0.1 mM of L-AOPP (*23*), gave only a resonance corresponding to L-[^{15}N]Phe (**6**). Taken together, these results prove that the ammonium ion released is transferred to L-Glu (**25**), which then functions as the amino donor to form L-Phe (**6**) (*19*).

In summary, employing ^{15}N-substrates with and without specific inhibitors of PAL, GS and GOGAT, resulted in the discovery of the mechanism for the recycling of nitrogen during active phenylpropanoid metabolism in loblolly pine cell cultures. This efficient nitrogen recycling mechanism (Scheme 2) presumably explains why plants and fungi do not experience any obvious symptoms of nitrogen limitation under conditions of active phenylpropanoid metabolism.

Tissue and Subcellular Localization of Monolignol Pathway Enzymes. Beyond the deamination step, the cinnamic (**7**)/*p*-coumaric (**8**) acid(s) formed can be metabolized *in planta* into the monolignols and hence metabolites, such as the lignins and lignans. [Additionally, depending upon the tissue and/or species, various phenylpropanoid (acetate) derived-products can also be formed *via* branchpoint pathways, *e.g.*, to flavonoids (Scheme 1)]. Surprisingly, all of the conversions in monolignol biosynthesis beyond cinnamic (**7**) and *p*-coumaric (**8**) acids have been proposed, at one time or another, as potential regulatory steps. However, no explicit biochemical data to demonstrate the rate-limiting capacity of any particular step(s) has been forthcoming. Accordingly, a brief historical summary of each step in the pathway is described below, outlining what is also known thus far about subcellular localization, and hence enzyme pathway organization. Where necessary, a brief clarification is given as to why particular enzymatic steps in the pathway have previously been described as "key".

Phenylalanine ammonia-lyase (PAL). The discovery of PAL (TAL) in 1961 (*25,26*) initiated a series of studies by numerous groups which made this enzyme, up to the present day, the most intensively studied in plant secondary metabolism. PAL, which has been obtained from various sources, exists as tetramers of 55 to 85 kDa subunits, and has molecular weights ranging from 240 to 330 kDa (*27*). It was established that the pro-3*S* hydrogen from phenylalanine (**6**) [tyrosine (**5**)] is abstracted during the transformation to cinnamic (**7**) [*p*-coumaric (**8**)] acid and ammonium ion, *via* a *trans* elimination reaction (equation 1) (*28,29*). Interestingly, neither a metal ion nor a co-factor are required for this conversion.

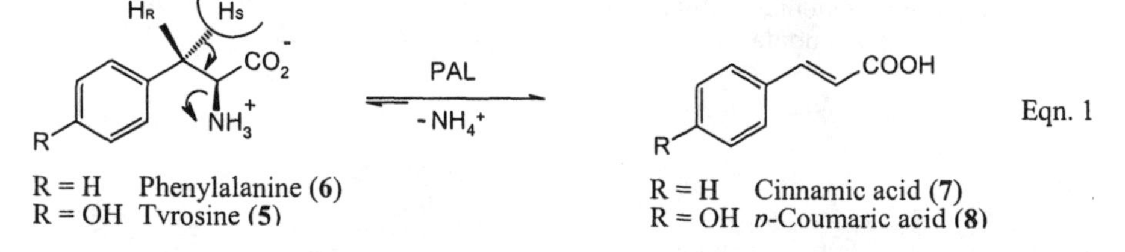

R = H Phenylalanine (**6**)
R = OH Tyrosine (**5**)

R = H Cinnamic acid (**7**)
R = OH *p*-Coumaric acid (**8**)

The considerable attention received by PAL is justified by its strategic position in the metabolic network functioning as a bridge between primary (glycolysis, pentose phosphate, and shikimate pathways) and secondary (phenylpropanoid/acetate pathway) metabolism. Its importance is further magnified by the fact that the phenylpropanoid pathway provides the precursors for the second most abundant biopolymer on earth after cellulose, *i.e.*, lignin. PAL is therefore often considered as an enzyme that catalyzes a "key" step in channeling carbon flux into this colossal metabolic sink.

The evidence provided by various researchers in support of a regulatory role for PAL in lignification is largely based on correlation of induced PAL activity with the production of phenylpropanoid derived metabolites, including lignin. In buckwheat (*Fagopyrum esculentum*) (*30*), asparagus (*Asparagus officionalis*) (*31*), bamboo (*Physollostachys pubescens* and *P. reticulata*) (*32*), *Acer pseudoplatanus* (*33*), and *Coleus blumei* (*34*), PAL activity is high in areas of active lignification. PAL activity is also induced in cases of pathogen attack which involves formation of "lignin-like" material as well as other phenylpropanoid-derived products involved in plant defense. However, no regulatory role for PAL has unequivocally been established.

Many efforts have been made to elucidate the tissue and subcellular distribution of PAL, principally by subcellular fractionation. The results suggested that PAL was not only in the cytoplasm, but was also in the plastids, mitochondria, and microbodies (*27*). Moreover, it has been proposed that in flavonoid metabolism enzyme complexes exist containing PAL, cinnamate 4-hydroxylase, and flavonoid glucosyltransferase which are associated with the endoplasmic reticulum (*35*).

More recently, specific and sensitive immunofluorescence and immunogold labeling techniques, have been utilized to localize PAL at both the tissue and subcellular levels, *e.g.*, in defining that wound-induced PAL in potato cells is predominantly cytoplasmic (*36*). As an additional example, PAL polyclonal antibodies were supplied to us by Whetten and Sederoff (*37*), following immunization of rabbits with a 74-kDa loblolly pine PAL subunit purified by sodium dodecyl sulfate - polyacrylamide gel electrophoresis (SDS-PAGE). These antibodies were then used to study the subcellular localization of PAL in our loblolly pine cell suspension cultures. Initially, the specificity of this PAL antibody was determined by Western blotting experiments (Figure 1). Lane 1 shows the results obtained with a crude cell free extract of the *P. taeda* cells. In addition to the expected 74 kDa subunit, other polypeptides were observed which had been previously reported to be PAL degradation and aggregation products (*38*). Moreover, with the crude protein extract from loblolly pine stem tissue the 74 kDa was again only the major band (Lane 2).

Figure 2 shows the presumed subcellular localization of PAL in the loblolly pine cells. Most of the gold-label was present in the cytosol (Figure 2A), with *much* smaller amounts apparently loosely associated with the plasma membrane (Figure 2B), the significance of which, if any, is not known. Both plastids and mitochondria were apparently free of any immunogold labeling. That this labeling was specific to PAL was established by a number of control experiments. For example, no gold labeling was found when the sections were incubated either with pre-immune serum (Figure 3A) or with protein A-gold complex alone (Figure 3B). These observations are comparable to that reported by Smith *et al* regarding the subcellular localization of PAL in French bean (*38*).

Cinnamate-4-hydroxylase (C4H). C4H is a cytochrome P-450 hydroxylase, which catalyzes the O_2/NADPH-dependent regiospecific hydroxylation shown below, thereby introducing a hydroxy group *para* to the propenoic acid side-chain of cinnamate (**7**) (equation 2) (*39,40*).

C4H is often described as a key enzyme in monolignol biosynthesis. This is mainly because the induction of both its enzymatic activity and mRNA accumulation apparently correlate well with the onset of lignification and the production of various phenylpropanoid metabolites (*41,42*). For instance, in the mesophyll cells of *Zinnia elegans* which are induced to differentiate into lignified tracheary elements, both an increase in C4H activity and the accumulation of C4H mRNAs have been observed

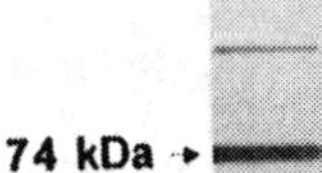

Figure 1. Western immunoblot of PAL showing the specificity of the antiserum raised against the 74kDa PAL from loblolly pine. Lane 1 : purified PAL from loblolly pine. Note the specific band ~74 kDa and other faint bands interpreted as degradation/ aggregation products (*38*). Lane 2 : crude protein extract from loblolly pine mainly visualizing the specific ~74kDa PAL subunit.

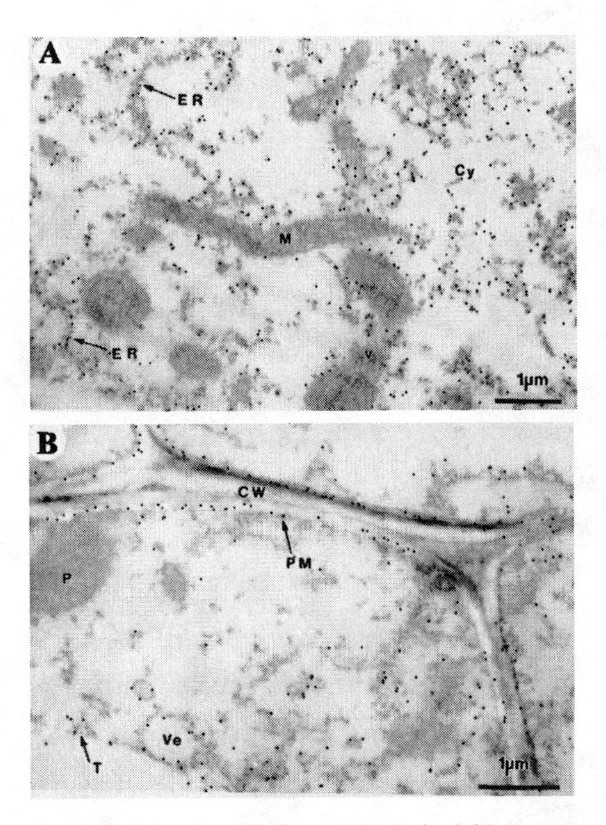

Figure 2. Presumed sub-cellular localization of PAL in loblolly pine suspension culture (2,4-D medium) using immunogold labeling with the PAL antiserum. Note the specific labeling in the cytosol (**A**) and to a much lesser extent along the plasma membrane (**B**) Legends: CW - cell wall; PM - plasma membrane; P - plastid; M - mitochondria; T - tonoplast; Ve - vesicle; Cy - Cytosol; ER - endoplasmic reticulum. (Reproduced with permission from Ref. *16*).

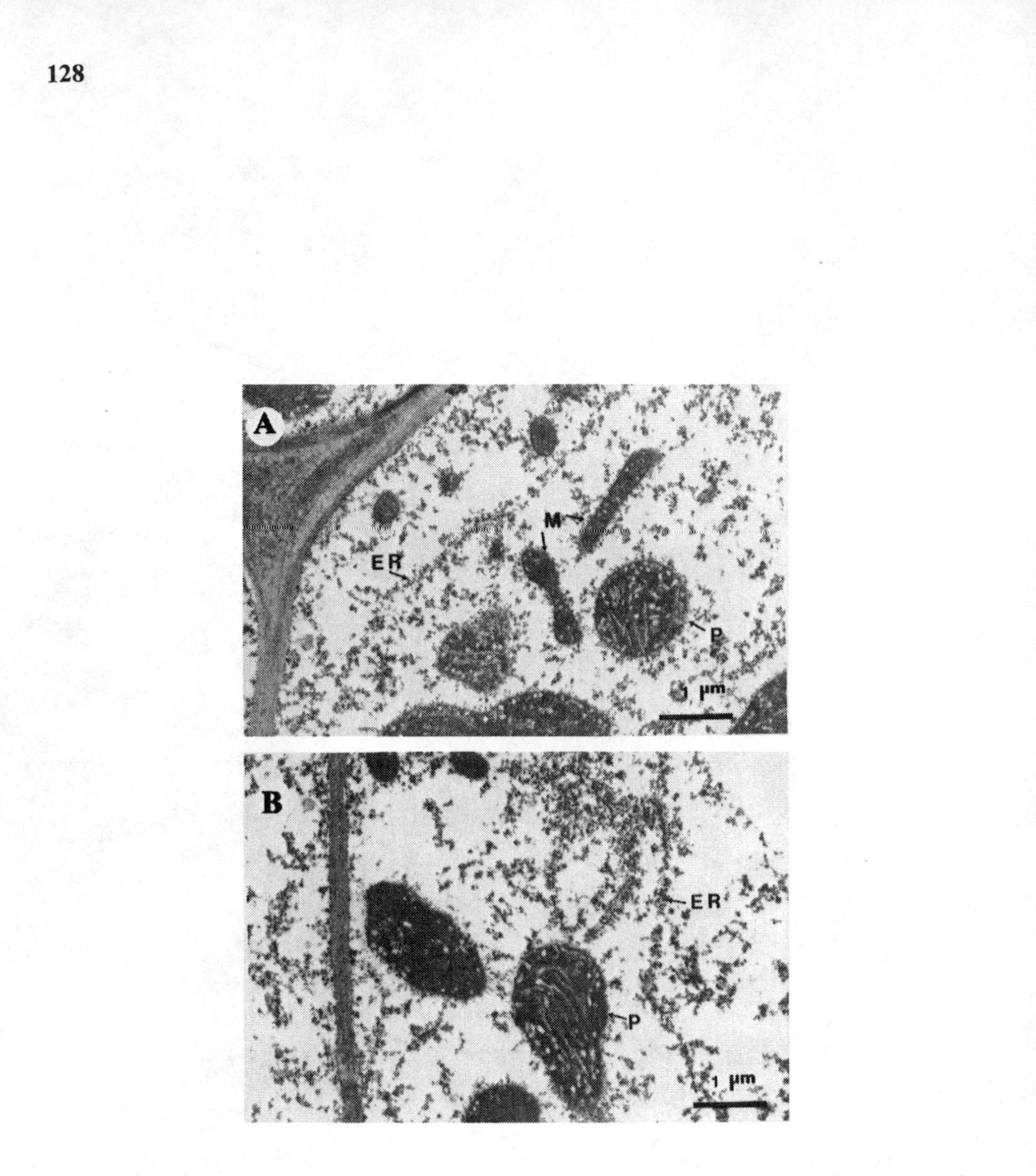

Figure 3. Immunocytochemical controls: (**A**) section incubated with pre-immune serum; (**B**) section incubated with protein A gold alone.

(*41*). Additionally, upon fungal elicitation of alfalfa (*Medicago sativa*) cell cultures, C4H transcripts and enzymatic activity were also found to increase prior to the accumulation of the isoflavonoid, medicarpin, this being a product of the phenylpropanoid/acetate pathway (*42*).

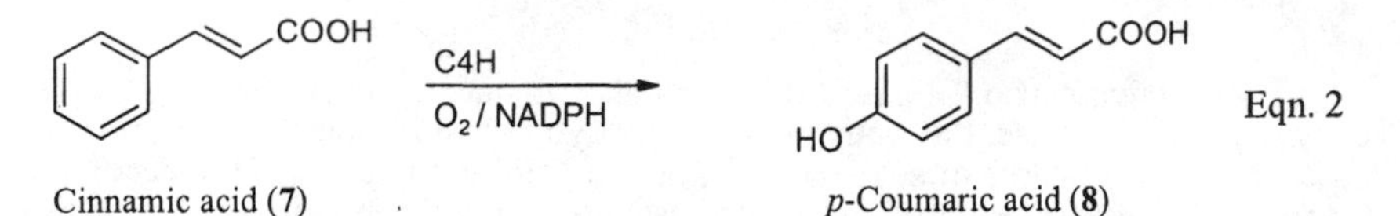

Cinnamic acid (**7**) *p*-Coumaric acid (**8**)

The subcellular localization and tissue-specificity of C4H has received considerable attention, and has been investigated using French bean (*Phaseolus vulgaris*) hypocotyls, *Zinnia elegans* and *Arabidopsis thaliana*, respectively. However, these studies did not specifically correlate C4H localization with precise formation of any specific metabolic class in any definitive manner. In any event, using 8 day old French bean hypocotyls (*38*), immunogold labeling with silver enhancement of C4H polyclonal antibodies was examined. This indicated that C4H was membrane-bound, being localized in both the endoplasmic reticulum and throughout the Golgi stacks. Additionally, tissue-specific immunolocalization of C4H in French bean hypocotyls revealed that C4H is present in young xylem, maturing/mature xylem and in cells adjacent to the metaxylem. C4H was also detected in the epidermal cells following wounding, although no role in lignin deposition was proposed for these cells. On the other hand, when *Zinnia elegans* (*41*) was probed with an antisense C4H mRNA (Northern blots), its expression was observed in stems, roots, flower buds and, to a lesser extent, in leaves. Tissue print hybridization with 4 week old *Zinnia* plantlets also revealed that the first internode contained C4H mRNA in differentiating xylem cells, whereas, in the second internode, it was detected in differentiating xylem tissue and developing phloem fibers. In the third internode, however, it was only present in differentiating xylem of vascular bundles.

Lastly, a GUS reporter gene strategy was utilized to define the tissue-specific expression of C4H in *Arabidopsis thaliana* (*43*). Cotyledons displayed GUS staining throughout their epidermal and mesophyll cells, as well as in ray parenchyma and vascular tissue. GUS staining was also observed in developing primary leaves and was not restricted to the vascular system, whereas the root tissue was intensely stained from the hypocotyl/root interface to just below the root tip. Staining was also noted in both the xylem and sclerified parenchyma of stems. In the flower tissues, it was present in the sepal vasculature below the stigmatic surface and in mature siliques. Again, however, no correlation with any particular metabolic process was established in any definitive way.

Hydroxycinnamoyl CoA ligases. Hydroxycinnamoyl CoA ligases catalyze the stepwise conversion of hydroxycinnamic acids (**8-12**), in the presence of ATP/CoASH, into the corresponding CoA esters (**13-17**) *via* the intermediate AMP derivatives. ATP apparently first binds to the enzyme followed by the substrate, with pyrophosphate (PPi) being released while the adenosyl 5′-acylphosphate still remains bound. Subsequent binding of CoASH is followed by release of CoA ester and AMP (equation 3) (see Ref. *16*).

4-Coumarate:CoA ligase is present as various isoenzymes in different plants, *e.g.*, soybean (*Glycine max*) (*44*), pea (*Pisum sativum*) (*45*), petunia (*Petunia hybrida*) (*46*), parsley (*Petroselinum crispum*) (*47*) and poplar (*Populus* x *euramericana*) (*48*). Of these, several display different substrate specificities suggesting distinct metabolic roles.

$$\text{Ar}\!-\!\text{CH}\!=\!\text{CH}\!-\!\text{COOH} + \text{ATP} + \text{CoASH} \xrightarrow{\text{CoA ligases}} \text{Ar}\!-\!\text{CH}\!=\!\text{CH}\!-\!\text{COSCoA} + \text{AMP} + \text{PPi} \quad \text{Eqn. 3}$$

(Hydroxy)cinnamic acids (**8-12**) CoA esters (**13-17**)

4-Coumarate:CoA ligase (4CL) has also frequently been described as a key enzyme, *e.g.*, because the activation of phenylpropanoid metabolism in fungally challenged cell cultures of soybean (*49*), poplar (*50*) and potato (*51*) is accompanied by induction of 4CL activity and its corresponding gene expression. A comparable correlation was noted in *Zinnia elegans* cells undergoing differentiation into tracheary elements (*52*).

While there have been no reports describing the subcellular localization of 4CL enzymes, the expression of 4CL mRNAs (4CL-1 and 4CL-2) has been quite extensively studied. Thus, expression studies with parsley showed that the 4CL-1 mRNA was very abundant in root tissue, whereas the corresponding 4CL-2 predominated in flowering stem regions (*51*). Similarly, soybean mRNA transcripts encoding different 4CL's, also displayed distinct expression patterns (*53*). On the other hand, 4CL promoter expression studies using transgenic tobacco and parsley (*54,55*) led to the conclusion that there was little difference between the 4CL-1 and 4CL-2 promoter regions, since both were activated by UV and fungal elicitation. GUS-fusion expression of the 4CL-1 promoter was observed predominantly in xylem tissue containing differentiating tracheids and in the primary xylem of axillary buds and developing leaf veins (*56*).

In contrast to distinct 4CL isoforms in parsley (*47*) and aspen (*57*), *Arabidopsis* 4CL was reported as being only a single gene product (*58*). Its mRNA expression in this species was observed 2 days after germination, being localized mainly in the roots with very low levels in the cotyledons. The 4CL promoter activity was induced within 3 days in the developing xylem of the cotyledons, stems and roots, although it was not restricted to vascular tissue. As the plants reached maturity, however, 4CL mRNA expression was noted in bolting stems, maturing leaves and floral buds, whereas promoter expression was highest in the xylem of bolting stems, anthers, flower buds and in developing siliques. As before, no precise correlation of mRNA expression, in terms of formation of specific metabolites, has yet been made; moreover, it will also be instructive to ascertain if 4CL is, or is not, a single gene product in this species.

Cinnamoyl CoA reductase (CCR). CCR catalyzes the four electron, NADPH-dependent reduction of *p*-hydroxycinnamoyl CoA derivatives (**13**), (**15**) and (**17**) to form the corresponding *p*-hydroxycinnamaldehydes (**18**), (**19**) and (**20**) (equation 4) (*59-61*).

Occasionally, this step is described as an entry point to the "lignin branch" of the phenylpropanoid pathway, and accordingly, CCR has been proposed as a key enzyme for lignin biosynthesis (*62,63*). However, these enzymes are not, in fact, lignin-specific since they are also involved in formation of biochemically distinct metabolites, such as the lignans (*2*). Nevertheless, elicitor-induced lignification of cell suspension cultures of *Picea abies* is accompanied by increased levels of CCR activity (*64*), and stimulation of CCR activity has also been observed in the lignification of neoformed calli of bean hypocotyls infected with *Agrobacterium rhizogenes* (*65*).

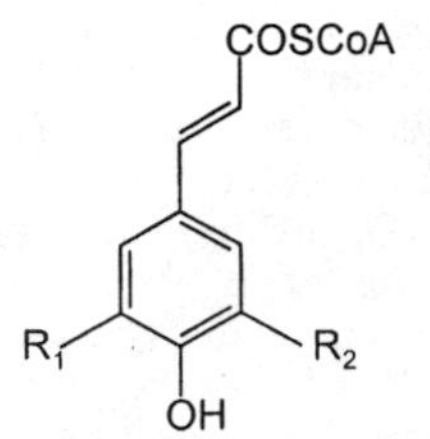

CCR
NADPH

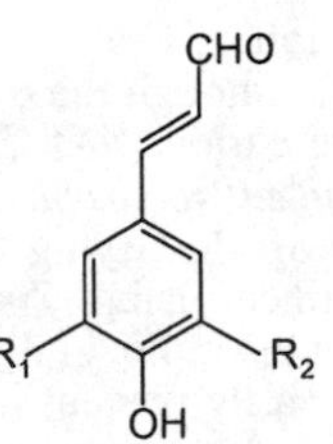

Eqn. 4

$R_1 = R_2 = H$, *p*-Coumaroyl CoA ester (**13**)
$R_1 = H, R_2 = OMe$, Feruloyl CoA ester (**15**)
$R_1 = R_2 = OMe$, Sinapoyl CoA ester (**17**)

$R_1 = R_2 = H$, *p*-Coumaryl aldehyde (**18**)
$R_1 = H, R_2 = OMe$, Coniferyl aldehyde (**19**)
$R_1 = R_2 = OMe$, Sinapyl aldehyde (**20**)

There has been no report thus far describing the determination of the subcellular localization of CCR in any cell type. However, preliminary Northern blot analyses with five year old eucalyptus (*Eucalyptus gunnii*) showed that CCR mRNA transcripts were equally abundant in young stems, roots and leaves (*63*). In 3 month old poplar, signals were only observed in differentiating xylem but not secondary phloem, flower or young periderm tissues. These studies should now be extended to include different stages of plant development and to identify the precise metabolic significance of the particular enzymatic activities in question, *i.e.*, whether they are involved in either lignin or lignan biosynthesis (or both).

Cinnamyl alcohol dehydrogenases (CAD). CAD converts *p*-hydroxycinnamaldehydes (**18**), (**19**) and (**20**) into *p*-coumaryl (**21**), coniferyl (**22**) and sinapyl (**23**) alcohols, respectively, thereby catalyzing the final step in monolignol biosynthesis (equation 5). This two electron hydride transfer reaction is also NADPH-dependent (*66*). As before, CAD is often viewed as being lignin specific, even though its product monolignols are partitioned into various metabolic classes, such as the lignans, dihydrocinnamyl alcohols, lignins and so forth (*2,16*). For example, various plant species, such as western red cedar (*Thuja plicata*), can accumulate coniferyl alcohol derived lignans, *e.g.*, plicatic acid, up to ~20% of its dry weight (*67*).

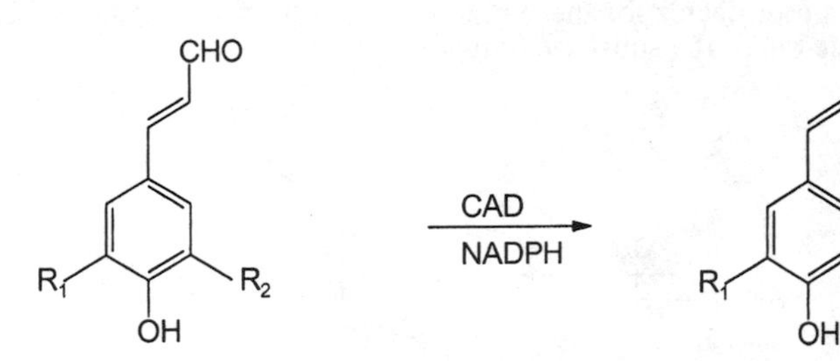

Eqn. 5

$R_1 = R_2 = H$, *p*-Coumaryl aldehyde (**18**)
$R_1 = H, R_2 = OMe$, Coniferyl aldehyde (**19**)
$R_1 = R_2 = OMe$, Sinapyl aldehyde (**20**)

$R_1 = R_2 = H$, *p*-Coumaryl alcohol (**21**)
$R_1 = H, R_2 = OMe$, Coniferyl alcohol (**22**)
$R_1 = R_2 = OMe$, Sinapyl alcohol (**23**)

CAD activity has been demonstrated to be induced in fungal-elicited cell suspension cultures of *Pinus banksiana* (*68*), *Picea abies* (*64*), *Medicago sativa* (*69*), and poplar (*Populus trichocarpa* x *Populus deltoides*) (*70*). On the other hand, transgenic poplar (*71*) and tobacco (*72*) with extensively down regulated CAD levels (~70 and 95%, respectively) did not result in any significant change in lignin content, strongly indicating that this enzyme is not a major control point of metabolic flux to lignin. In contrast, changes in lignin content and composition resulting from the presumed reduction of CAD in a putative mutant loblolly pine (*P. taeda*) have been reported (*73,74*), although the overall measured lignin content was hardly affected. It is therefore quite curious that CAD has been described as a key enzyme in *P. taeda*, particularly in *the entire absence* of any definitive data to establish a regulatory role.

In terms of identifying the subcellular localization of CAD, different results have been obtained using *Zinnia* and poplar (*Populus tremuloides x P. alba*), respectively. For example, in differentiating *Zinnia* tracheids it was concluded that the enzyme was primarily present in secondary wall thickenings and to a lesser extent in the cytosol, particularly in vesicles (*75*). On the other hand, polyclonal antibodies raised against eucalyptus CAD, when applied to four month old poplar plants, indicated the presence of CAD in the cambium, developing xylem cells, ray cells and phloem fibers (*76*). Additionally, CAD promoter expression in tobacco, as visualized by GUS-staining, was higher in roots than in stems and leaves (*77*). In stem sections, however, GUS activity was highest in the regions encompassing the second to the fourth internodes but decreased with maturation of the lignifying cells.

Lastly, when polyclonal antibodies for eucalyptus CAD were used to study 3 month old transformed poplar (*Populus tremuloides x P. alba*) plants, different results were obtained, depending upon whether poplar tissues were first fixed and then stained with β-glucuronidase, or if the sequence of fixation was reversed (*78,79*). Using the first protocol, staining was observed only in ray (parenchyma) cells located between xylem vessels. With the reversed sequence, staining was extremely prominent throughout the vascular cambium/differentiating xylem zone and primary xylem/pith zone as well as in the periderm layer. These findings may be provisionally interpreted as evidence that cells undergoing lignification are themselves fully competent to produce lignin, and that ray parenchyma cells may also be involved in the formation of other metabolites, *e.g.*, lignans.

***O*-Methyltransferases (OMT).** The regiospecific methylation of the hydroxyl group at the *meta*- position can either occur at the level of the hydroxycinnamic acids (**9**) and (**11**) or hydroxycinnamoyl CoA esters (**14**) and (**16**) (see Ref. *16*). These reactions are catalyzed by hydroxycinnamate *O*-methyltransferases (COMT) and hydroxycinnamoyl CoA *O*-methyltransferases (CCOMT), respectively. *S*-Adenosyl methionine (SAM) serves as cofactor for these reactions, transferring its methyl from the sulfur to the phenoxide ion of the substrate (equation 6).

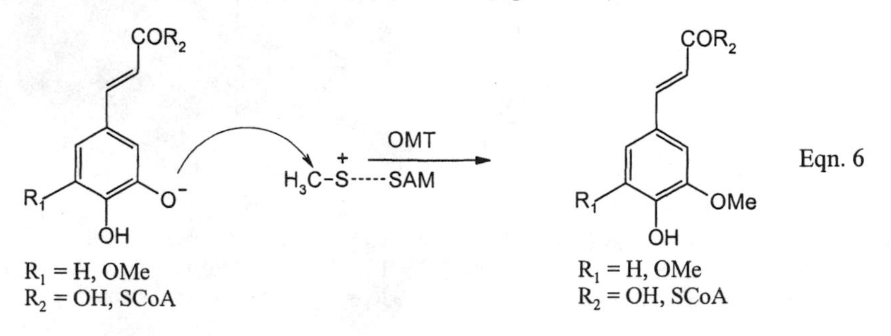

COMT has also been defined as a "key enzyme", for example, because accumulation of its mRNA in *Zinnia elegans* apparently correlates well with the development of lignified tracheids (*80,81*), and both its enzymatic activity and mRNA transcript levels are induced in fungal-elicited alfalfa (*Medicago sativa*) cell suspension cultures (*82*). Moreover, the differences in the specificity of COMT's and 4CL's in gymnosperms and angiosperms have been used to attempt to explain differences in the lignin types of these two groups of land plants (but see later). Additionally, the discovery of the alternative methylation pathway, *i.e.*, *via* the CoA esters (*83*), added more complexity to the role of particular methylation processes in lignin biosynthesis. Indeed, the relative contribution of each methylation pathway is still a subject of much speculation.

Very little is known about the subcellular localization of the hydroxycinnamoyl CoA OMTs. mRNA localization studies in parsley using Northern blot analysis of roots and leaves (*84*) suggested that there was essentially the same level of expression in both tissues. On the other hand, tissue prints of 4-6 week old *Z. elegans* (*81*) sections established that its hydroxycinnamoyl CoA OMT mRNA was mainly located in differentiating xylem and not in developing phloem fibers, whereas, in older internodes, there was extensive transcription in both cell types. With 3-4 year old poplar (*P. tremuloides*), however, mRNA expression was essentially only observed in the stems (*85*), *i.e.*, at the gross tissue level.

Hydroxycinnamic acid OMT localization has only been partially examined with variable results. For example, with aspen (*Populus tremuloides*), Northern blot analysis revealed that hydroxycinnamic acid OMT was mainly present in developing xylem, but was not detectable in leaves (*86*), whereas in poplar (*Populus trichocarpa x P. deltoides*), hydroxycinnamic acid OMT was present in both xylem and leaves (*87*). Using young *Zinnia* internodes, tissue-print hybridization analyses of first and second internodes, as well as immunogold labeling of the stems, showed that its hydroxycinnamic acid OMT was mainly abundant in developing phloem fibers (*81*). Lastly, in maize (*Zea mays*), Northern blot analysis revealed its mRNA to be present in the root elongation zone and in the mesocotyl of differentiated roots mainly (*88*).

Taken together, these data indicate distinct localization patterns for the different hydroxycinnamic acid OMTs and hydroxycinnamoyl CoA OMTs, a matter which now needs to be correlated with the precise metabolic end-products being formed.

Hydroxycinnamate hydroxylases. In the phenylpropanoid pathway, the branch point (see Ref. *16*) can be at the level of either *p*-coumaric acid (**8**) (catalyzed by *p*-coumarate-3-hydroxylase) or *p*-coumaroyl CoA ester (**13**) (catalyzed by *p*-coumaroyl CoA-3-hydroxylase) (*89*). That is, how caffeoyl moieties are actually formed remains an open question at the time of writing (equation 7). A specific cytochrome P-450 membrane associated enzyme in formation of caffeoyl derivatives is probably involved as described in ref. (*16*). However, until the responsible enzyme is purified to homogeneity or its gene cloned and sequenced, the role of the 3-hydroxylation step in the overall regulation of the phenylpropanoid pathway will remain unresolved.

Nevertheless, the hydroxylation step is at a branch point leading to *p*-coumaryl alcohol (**21**) and coniferyl alcohol (**22**), respectively. Accordingly, it is reasonable to hypothesize that it is "key" to the control of lignin monomer composition. However, the sequence of either the protein or gene responsible for this branch point step is not yet known.

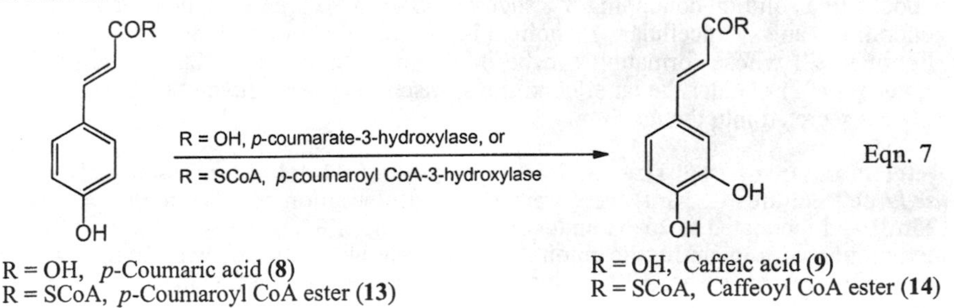

In summary, the preceding empirical account on enzyme localization in the monolignol forming pathway has revealed four noteworthy features. First, caution must be exercised in describing any of the enzymatic steps as being lignin specific since related metabolic processes, for example lignan formation, may also be active even within the so-called lignifying tissue. Second, as secondary xylem maturation and lignin deposition proceed, the cells which form tracheids, vessels and fibers are able to produce the required monolignols, and hence their lignin biopolymers. Third, cinnamyl alcohol dehydrogenase is expressed in many different tissue and cell-types; however, these observations need to be clarified in terms of precise metabolic events involved. The fourth is the striking differences between the localization of the distinct isoenzymes, *e.g.*, hydroxycinnamoyl CoA OMT's and the hydroxycinnamic acid OMT's and also 4CL-1 and 4CL-2, which presumably imply unique metabolic functions for each isoenzyme. Routine techniques are now in place to help establish how these various metabolic processes are differentially initiated, regulated and controlled, and now the major challenge is to examine a single plant species, at all stages of development, in order to ascertain how the phenylpropanoid pathway is regulated.

A much more detailed discussion of tissue and subcellular localization of monolignol pathway enzymes is described elsewhere (*16*).

Metabolic Pool Sizes and Rate-limiting Steps in Monolignol Formation in Loblolly Pine (*Pinus taeda*).

The idea that control of flux through the phenylpropanoid pathway solely resides in one enzyme alone (*i.e.*, a "rate limiting" enzyme) has long been uncertain. As early as 1973, it was suggested that control in the phenylpropanoid pathway resides in groups of enzymes, with PAL exerting primary control and the enzymes catalyzing the subsequent steps (*e.g.*, C4H and 4CL) exerting secondary control (*90*). Yet, in spite of such considerations, virtually every step in the monolignol forming pathway has been referred to as regulatory!

The foregoing remarks thus clearly emphasized the need to analyze the pool levels of different metabolites in the phenylpropanoid pathway under presumed substrate saturating conditions, in order to identify the relative regulatory contributions of each step in the pathway. In this regard, the inducible *P. taeda* monolignol-forming system provided the very *first* opportunity for such a study. The protocol utilized was based, firstly, on the measurement of metabolic pool sizes of various intermediates during active phenylpropanoid metabolism and, secondly, on the determination of the effects of administering several metabolic precursors to monolignol forming cells, *i.e.*, to determine the levels of accumulation within the cells and their effect on subsequent downstream metabolic processes.

In this regard, the *P. taeda* cell suspension cultures serve as an excellent model. They can be induced to undergo active phenylpropanoid metabolism when exposed to a solution containing 8% sucrose. Over a 96 h period, the cells respond by generating an extracellular monolignol-derived dehydropolymerisate, (so-called "lignin like") whose formation can be inhibited by addition of 20 mM KI, a H_2O_2 scavenger (*22*). Under the latter conditions, essentially only the monolignols (**21**) and (**22**) are secreted into the medium.

Determination of Pool sizes of Phenylpropanoid Metabolites. One week old *P. taeda* cell cultures (2.5 ml) were transferred to a solution of 8% sucrose 20 mM KI (25ml) and incubated at 25 °C under continuous light. C_{18} reversed phase HPLC was then employed to quantify the amounts of each metabolite in the cell bathing medium and cellular extracts after 12 h of incubation. Representative metabolic pool sizes of intracellular phenylpropanoid compounds are shown in Table I, and Figures 4A and 4B (lower traces) show the accumulation of *p*-coumaryl (**21**) and coniferyl (**22**)

Table I. Representative intracellular metabolic pool sizes of phenylpropanoid metabolites in monolignol-forming *Pinus taeda* cells (see Ref. *91*).

Metabolite	*Amount in 150 mg dry cells (nmol)*
Cinnamic Acid (**7**)	100.00
p-Coumaric Acid (**8**)	30.00
Caffeic Acid (**9**)	25.00
Ferulic Acid(**10**)	7.00
p-Coumaroyl CoA (**13**)	< 2.50*
Caffeoyl CoA (**14**)	< 2.50*
Feruloyl CoA (**15**)	< 2.50*
p-Coumaryl Aldehyde (**18**)	< 0.25*
Coniferyl Aldehyde (**19**)	< 0.25*
p-Coumaryl Alcohol (**21**)	30.00
Coniferyl Alcohol (**22**)	30.00

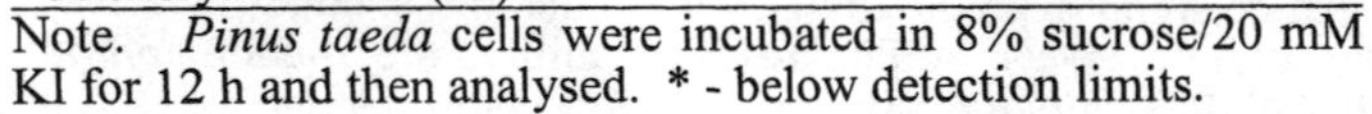

Note. *Pinus taeda* cells were incubated in 8% sucrose/20 mM KI for 12 h and then analysed. * - below detection limits.

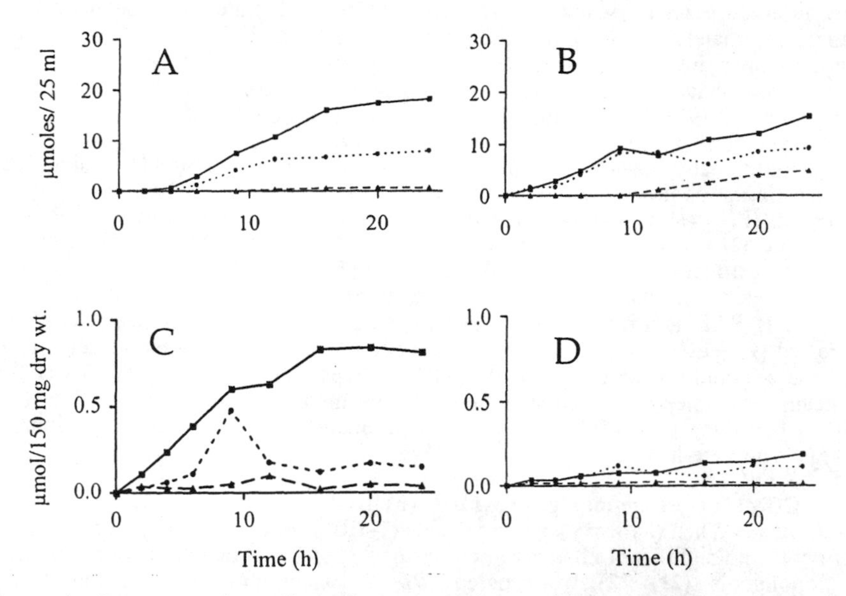

Figure 4. Representative effect of exogenously provided Phe (**6**) on: extracellular accumulation of (**A**) *p*-coumaryl alcohol (**21**) and (**B**) coniferyl alcohol (**22**), and intracellular levels of (**C**) cinnamic acid (**7**) and (**D**) *p*-coumaric acid (**8**) in *P. taeda* cells. Legend: ▲: 0 µmol ; •: 125 µmol ; ■: 1000 µmol Phe (see Ref. *91*).

alcohols secreted into the cell bathing solution. Within the cells, only the acids (**7-10**) and the monolignols (**21**) and (**22**) were detectable, but not the corresponding CoA esters (**13-15**) or aldehydes (**18, 19**) (*91*).

Identification of Rate-limiting Steps in the Pathway. To identify which steps in the pathway might be rate-limiting, various metabolic precursors, namely Phe (**6**), phenylpropanoic acids (*i.e.*, cinnamic (**7**), *p*-coumaric (**8**), caffeic (**9**) and ferulic (**10**) acids), feruloyl CoA ester (**15**), *p*-coumaryl (**18**) and coniferyl (**19**) aldehydes were added at different concentrations to *P. taeda* cells (2.5 ml) exposed to a 8% sucrose solution 20 mM KI (25 ml). For each treatment, cellular pool sizes of all phenylpropanoid metabolic intermediates and the amount of various substances accumulating in the medium (during the 24 h incubation period) were determined by HPLC and LC/MS analyses.

Effect of exogenously supplied phenylalanine to *P. taeda* metabolism. As can be seen in Figures 4A and 4B, increasing phenylalanine (**6**) availability (0 to 1000 μmol) resulted in an approximately 26 and 3 fold increase in extracellular *p*-coumaryl (**21**) and coniferyl (**22**) alcohols released, respectively, within 24 h (*91*). Addition of higher concentrations of Phe (**6**) had no effect on further increasing monolignol levels, indicating that maximum production levels under these conditions were achieved. Significantly, at low levels of Phe (**6**) availability, the *p*-coumaryl alcohol (**21**)/ coniferyl alcohol (**22**) ratio approximated 1 to 8 (Figures 4A and 4B, lower traces). In contrast, at higher levels of exogenously provided Phe (**6**) (up to 125 μmol) differential increases in *p*-coumaryl (**21**) and coniferyl (**22**) alcohol contents resulted in an approximately 1 to 1 monolignol ratio (Figures 4A and 4B, middle traces). This is a particular interesting observation because lignins in compression wood of gymnosperms have a higher *p*-coumaryl alcohol (**21**) content relative to that in "normal" wood (*92*). Perhaps the change in monolignol composition in compression wood results from an increase in Phe (**6**) availability.

Additionally, the increased availability of Phe (**6**) resulted in significant increases in the pool sizes of both cinnamic (**7**) (~16-fold) and *p*-coumaric (**8**) (~10-fold) acids (Figures 4C and 4D). However, the intracellular pool sizes for caffeic (**9**) and ferulic (**10**) acids and *p*-coumaryl (**21**) and coniferyl (**22**) alcohols were not significantly affected, nor did the CoA esters (**13-15**) or aldehydes (**18**) and (**19**) build up to detectable levels. These results are, therefore, contrary to what would be expected if PAL is truly the only rate-limiting enzyme, as has been frequently suggested (*91,93,94*), *i.e.*, if this was the case, cinnamic and *p*-coumaric acids could not have accumulated unless the hydroxylation steps were operationally slower than the deamination step. Furthermore, the very low metabolic pool sizes of Phe (**6**) in actively lignifying cells (*19*) suggests that deamination of Phe (**6**) to form cinnamic acid (**7**) is not rate-limiting.

Effect of exogenously provided (hydroxy)cinnamic acids to *P. taeda* metabolism. When (hydroxy)cinnamic acids (**7-10**) were exogenously provided to *P. taeda* cells, no significant changes in either the intracellular or extracellular amounts of monolignols (**21, 22**) were noted (*91*). Examination of the intracellular constituents, on the other hand, revealed that each exogenously supplied acid accumulated substantially inside the cells, while the levels of other downstream metabolites were not significantly affected. For example, when 40 μmol (hydroxy)cinnamic acids (**7-10**) were individually exogenously administered, the intracellular levels of cinnamic (**7**), *p*-coumaric (**8**) and ferulic (**10**) acids increased ~150, ~155 and ~450 fold, respectively, while caffeic acid (**9**) increased less dramatically (~20 fold) (Figure 5). As can be seen, the (hydroxy)cinnamic acids underwent further metabolism dropping to near basal levels within 12 h, this occurring primarily *via* glucose ester formation. It would, therefore, appear that these acids,

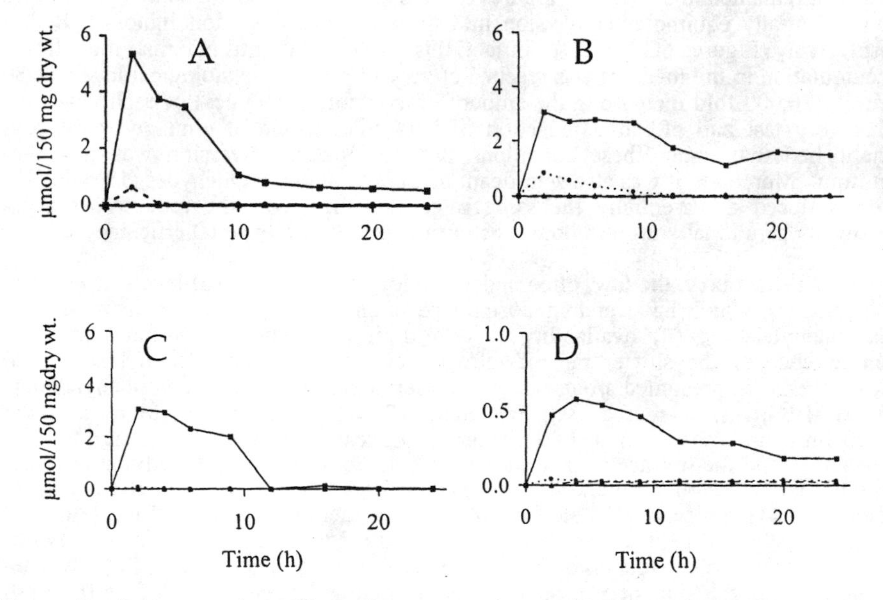

Figure 5. Representative effect of exogenously provided (**A**) cinnamic (**7**), (**B**) *p*-coumaric (**8**), (**C**) ferulic (**10**) and (**D**) caffeic (**9**) acids on their intracellular levels in *P. taeda* cell cultures. Legend: ▲: 0 µmol ; ●: 5 µmol ; ■: 40 µmol (hydroxy)cinnamic acids (see Ref. *91*).

although taken up by the induced *P. taeda* cells, were either not transported to the requisite subcellular locations for further processing and/or were derivatized by way of either a detoxification or a storage mechanism.

Effect of exogenously supplied cinnamyl aldehydes to *P. taeda* metabolites. At no point in the preceding experiments, namely, when the *P. taeda* cells were exogenously provided with phenylalanine (**6**) (up to 1000 μmol) and (hydroxy)cinnamic acids (**7-10**) (up to 40 μmol), did the *p*-coumaryl (**18**) and coniferyl (**19**) aldehydes reach detection levels (0.25 nmol) (*91*). This strongly argues against any regulatory role for this step in monolignol biosynthesis. Indeed, even when *p*-coumaryl (**18**) and coniferyl (**19**) aldehydes were individually administered, at 10 and 30 μmol (*i.e.*, 40,000 and 120,000 fold above the detection limit, respectively) they still remained undetectable (Figures 6A and 6B). This was due to their very rapid and essentially equimolar conversion into the corresponding monolignols (**21, 22**), respectively (Figures 6C and 6D). In fact, this was evident until maximal monolignol accumulation in the medium was reached at very high, non-physiological levels of 80 μmol (320,000 fold increase in the amount of available aldehydes) of each aldehyde. This very fast and effective reduction of very high levels of cinnamyl aldehydes established that under these conditions this enzymatic conversion was not rate-limiting. Moreover, it was also significant that both cinnamyl aldehydes (**18**) and (**19**) were reduced at essentially the same rate, indicating that the reductive process involved operationally utilized both substrates with essentially equal efficiency.

In summary, the low, often undetectable, phenylalanine (**6**) levels in *P. taeda* cell cultures, which have undergone induced phenylpropanoid metabolism, suggest that phenylalanine (**6**) availability is rate limiting. This is assumed to be a consequence of the slower rate of nitrogen cycling and hence phenylalanine (**6**) biosynthesis *via* presumed arogenate (**4**) regeneration. However, when phenylalanine (**6**) availability is increased, both cinnamate (**7**) and *p*-coumarate (**8**) intracellular levels build up, although none of the other downstream metabolites were significantly affected. As the intracellular acid levels rise, however, increased extracellular monolignol secretion occurs. Significantly, at high Phe (**6**) concentrations, the *p*-coumaryl (**21**)/coniferyl (**22**) alcohol ratio in the medium changed from 1 to 8 to approximately 1 to 1. Thus, as cinnamate (**7**) and *p*-coumarate (**8**) levels are elevated intracellularly, the conversion of cinnamate (**7**) into *p*-coumarate (**8**) is initially rate-limiting, until the rate of *p*-coumarate (**8**) formation exceeds that of caffeate (**9**) biosynthesis. Under these conditions, more *p*-coumarate (**8**) is then reduced to the corresponding monolignol (**21**), thus altering the ratio of *p*-coumaryl (**21**) to coniferyl alcohol (**22**) (Scheme 3). Administration of (hydroxy)cinnamic acids (**7-10**) had no significant effect on monolignol (**21,22**) biosynthesis. On the other hand, even very high levels of exogenously supplied cinnamyl aldehydes (**18**) and (**19**) are rapidly and completely converted into the corresponding monolignols (**21**) and (**22**).

Taken together, these data suggest that the monolignol-forming pathway in *P. taeda* cell cultures is not controlled by a single "rate-limiting" step, but rather exhibits *multi-site* regulation. However, it can be concluded that: Firstly, phenylalanine (**6**) availability plays a key role in the phenylpropanoid pathway and is probably regulated in the preceding nitrogen recycling pathway and not during deamination to form cinnamic acid (**7**). Secondly, precisely how cinnamate (**7**) is apportioned into the operationally slower steps affording *p*-coumaric (**8**) and caffeic (**9**) acids, represents an important determinant of how monolignol composition is controlled. Thirdly, reductive transformations of the cinnamyl aldehydes (**18**) and (**19**) to the corresponding alcohols (**21**) and (**22**) are not rate-limiting under these conditions. Indeed, these data explain why down-regulation of CAD has had no significant effect on overall lignin amounts in transgenic plants (*73,74*).

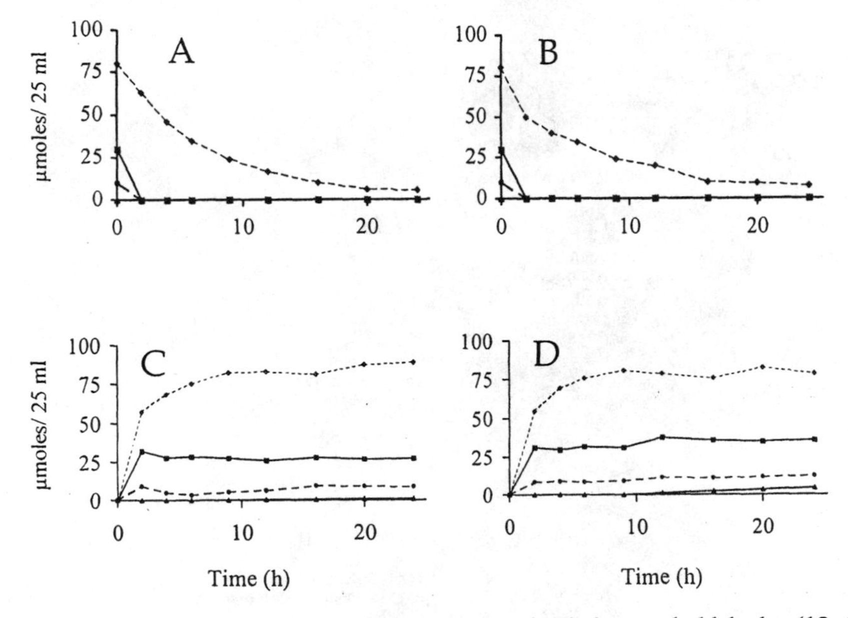

Figure 6. Representative effect of exogenously provided cinnamyl aldehydes (**18, 19**) on extracellular levels of monolignols (**21, 22**) in *P. taeda* cell cultures. Time course of depletion of (**A**) *p*-coumaryl aldehyde (**18**) and (**B**) coniferyl aldehyde (**19**) in the extracellular medium. Time course of formation of (**C**) *p*-coumaryl alcohol (**21**) and (**D**) coniferyl alcohol (**22**) in the extracellular medium. Legend: ▲: 0 µmol; •: 10 µmol; ■: 30 µmol and ♦: 80 µmol cinnamyl aldehydes (see Ref. *91*).

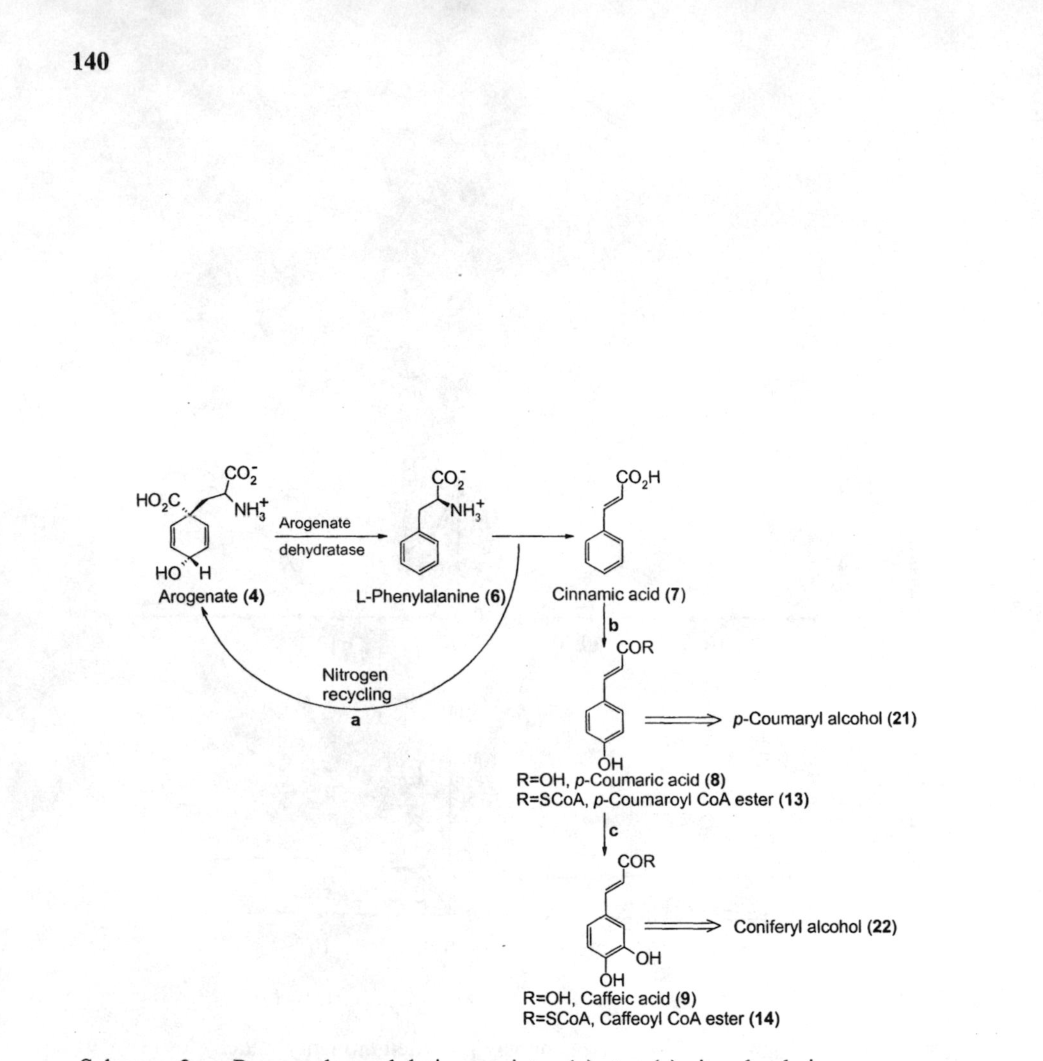

Scheme 3. Proposed modulation points (**a**) to (**c**) involved in gymnosperm monolignol (**21**) and (**22**) biosynthesis in *Pinus taeda*. Modulation points include: (**a**) regeneration of phenylalanine (**6**) *via* nitrogen recycling, (**b**) cinnamate 4-hydroxylase and (**c**) *p*-coumarate 3-hydroxylase.

Acknowledgments

The authors thank the National Aeronautics and Space Administration (NAG100164) and the Lewis B. and Dorothy Cullman and G. Thomas Hargrove Center for Land Plant Adaptation Studies for generous support of this study.

Literature Cited

1. Lewis, N. G., Davin, L. B. In *Isopentenoids and Other Natural Products: Evolution and Function*; Nes, W. D., Ed. ACS Symposium Series: Washington, DC, 1994, Vol. 562, pp 202-246.
2. Lewis, N. G., Davin, L. B. In *Comprehensive Natural Products Chemistry*; Barton, D. H. R., Sir, Nakanishi, K., Meth-Cohn, O., Eds.; Elsevier: London, 1999, Vol. 1, pp 639-712.
3. Hattori, M., Hada, S., Watahiki, A., Ihara, H., Shu, Y.-Z., Kakiuchi, N., Mizuno, T., Namba, T. *Chem. Pharm. Bull.* **1986**, *34*, 3885-3893.
4. Valsaraj, R., Pushpangadan, P., Smitt, U. W., Adsersen, A., Christensen, S. B., Sittie, A., Nyman, U., Nielsen, C., Olsen, C. E. *J. Nat. Prod.* **1997**, *60*, 739-742.
5. Atta-ur-Rahman, Ashraf, M., Choudhary, M. I., Habib-ur-Rehman, Kazmi, M. H. *Phytochemistry* **1995**, *40*, 427-431.
6. Schröder, H. C., Merz, H., Steffen, R., Müller, W. E. G., Sarin, P. S., Trumm, S., Schulz, J., Eich, E. *Z. Naturforsch.* **1990**, *45c*, 1215-1221.
7. Gnabre, J. N., Ito, Y., Ma, Y., Huang, R. C. *J. Chromatogr. A* **1996**, *719*, 353-364.
8. Osawa, T., Nagata, M., Namiki, M., Fukuda, Y. *Agric. Biol. Chem.* **1985**, *49*, 3351-3352.
9. Oliveto, E. P. *Chem. Ind.* **1972** 677-679.
10. Lajide, L., Escoubas, P., Mizutani, J. *Phytochemistry* **1995**, *40*, 1105-1112.
11. Matsui, K., Munakata, K. *Tetrahedron Lett.* **1975**, *24*, 1905-1908.
12. Haworth, R. D. *Annu. Rept. Prog. Chem.* **1937**, *33*, 266-279.
13. Grisebach, H. In *The Biochemistry of Plants*; Stumpf, P. K., Conn, E. E., Eds.; Academic Press: New York, NY, 1981, Vol. 7, pp 457.
14. Lewis, N. G., Yamamoto, E. *Annu. Rev. Plant Physiol. Plant Mol. Biol.* **1990**, *41*, 455-496.
15. Davin, L. B., Lewis, N. G. In *Recent Advances in Phytochemistry*; Stafford, H. A., Ibrahim, R. K., Eds.; Plenum Press: New York, NY. (and references therein), 1992, Vol. 26, pp 325-375.
16. Lewis, N. G., Sarkanen, S., Davin, L. B. In *Comprehensive Natural Products Chemistry*; Barton, D. H. R., Sir, Nakanishi, K., Meth-Cohn, O., Eds.; Elsevier: London, 1999, Vol. 3, pp 618-745.
17. Lewis, N. G., Yamamoto, E. In *Chemistry and Significance of Condensed Tannins*; Hemingway, R. W., Karchesy, J. J., Eds.; Plenum Press: 1989, pp 23-47.
18. Razal, R. A., Ellis, S., Singh, S., Lewis, N. G., Towers, G. H. N. *Phytochemistry* **1996**, *41*, 31-35.
19. van Heerden, P., Towers, G. H. N., Lewis, N. G. *J. Biol. Chem.* **1996**, *271*, 12350-12355.
20. Singh, S., Lewis, N. G., Towers, G. H. N. *J. Plant Physiol.* **1998**, *153*, 316-323.
21. Eberhardt, T. L., Bernards, M., He, L., Davin, L. B., Wooten, J. B., Lewis, N. G. *J. Biol. Chem.* **1993**, *268*, 21088-21096.
22. Nose, M., Bernards, M. A., Furlan, M., Zajicek, J., Eberhardt, T. L., Lewis, N. G. *Phytochemistry* **1995**, *39*, 71-79.
23. Amrhein, N., Gödeke, K.-H. *Plant Sci. Lett.* **1977**, *8*, 313-317.

24. Ronzio, R. A., Rowe, W. B., Meister, A. *Biochemistry* **1969**, *8*, 1066.
25. Koukol, J., Conn, E. E. *J. Biol. Chem.* **1961**, *236*, 2692-2698.
26. Neish, A. C. *Phytochemistry* **1961**, *1*, 1-24.
27. Hanson, K. R., Havir, E. A. In *The Biochemistry of Plants: Secondary Plant Products*; Conn, E. E., Ed. Academic Press: New York, NY, 1981, Vol. 7, pp 577-625.
28. Hanson, K. R., Wightman, R. H., Staunton, J., Battersby, A. R. *J. Chem. Soc., Chem. Commun.* **1971** 185.
29. Ife, R., Haslam, E. *J. Chem. Soc. (C)* **1971** 2818.
30. Yoshida, S., Shimokoriyama, M. *Bot. Mag. Tokyo* **1965**, *78*, 14.
31. Goldstein, L. D., Jennings, P. H., Marsh, H. V. *Plant Cell Physiol.* **1971**, *12*, 657.
32. Higuchi, T. *Agric. Biol. Chem.* **1966**, *30*, 667.
33. Rubery, P. H., Northcote, D. H. *Nature* **1968**, *219*, 1230.
34. Rubery, P. H., Fosket, D. H. *Planta* **1969**, *87*, 54.
35. Hrazdina, G., Wagner, G. J. *Arch. Biochem. Biophys.* **1985**, *237*, 88.
36. Shaw, N. M., Bolwell, G. P., Smith, C. *Biochem. J.* **1990**, *267*, 163.
37. Whetten, R. W., Sederoff, R. R. *Plant Physiol.* **1992**, *98*, 380-386.
38. Smith, C. G., Rodgers, M. W., Zimmerlin, A., Ferdinando, D., Bolwell, G. P. *Planta* **1994**, *192*, 155-164.
39. Russell, D. W., Conn, E. E. *Arch. Biochem. Biophys.* **1967**, *122*, 256-258.
40. Russell, D. W. *J. Biol. Chem.* **1971**, *246*, 3870-3878.
41. Ye, Z.-H. *Plant Sci.* **1996**, *121*, 133-141.
42. Fahrendorf, T., Dixon, R. A. *Arch. Biochem. Biophys.* **1993**, *305*, 509-515.
43. Bell-Lelong, D. A., Cusumano, J. C., Meyer, K., Chapple, C. *Plant Physiol.* **1997**, *113*, 729-739.
44. Knobloch, K. H., Hahlbrock, K. *Eur. J. Biochem.* **1975**, *52*, 311-320.
45. Wallis, P. J., Rhodes, M. J. C. *Phytochemistry* **1977**, *16*, 1891-1894.
46. Ranjeva, R., Boudet, A. M., Faggion, R. *Biochimie* **1976**, *58*, 1255-1262.
47. Lozoya, E., Hoffmann, H., Douglas, C., Schulz, W., Scheel, D., Hahlbrock, K. *Eur. J. Biochem.* **1988**, *176*, 661-667.
48. Grand, C., Boudet, A., Boudet, A. M. *Planta* **1983**, *158*, 225-229.
49. Ebel, J., Grisebach, H. *Trends in Biochem. Sci.* **1988**, *13*, 23.
50. Douglas, C. J., Ellard, M., Hauffe, K. D., Molitor, E., De Sa, M. M., Reinold, S., Subramaniam, R., Williams, F. In *Recent Advances in Phytochemistry*; Stafford, H. A., Ibrahim, R. K., Eds.; Plenum Press: New York, NY, 1992, Vol. 26, pp 63-89.
51. Becker-André, M., Schulze-Lefert, P., Hahlbrock, K. *J. Biol. Chem.* **1991**, *266*, 8551-8559.
52. Church, D. L., Galston, A. W. *Plant Physiol.* **1988**, *88*, 679-684.
53. Uhlmann, A., Ebel, J. *Plant Physiol.* **1993**, *102*, 1147-1156.
54. Hauffe, K. D., Paszkowski, U., Schulze-Lefert, P., Hahlbrock, K., Dangl, J. L., Douglas, C. J. *Plant Cell* **1991**, *3*, 435-443.
55. Douglas, C. J., Hauffe, K. D., Ites-Morales, M.-E., Ellard, M., Paszkowski, U., Hahlbrock, K., Dangl, J. L. *EMBO J.* **1991**, *10*, 1767-1775.
56. Bevan, M., Shufflebottom, D., Edwards, K., Jefferson, R., Schuch, W. *EMBO J.* **1989**, *8*, 1899-1906.
57. Hu, W.-J., Kawaoka, A., Tsai, C.-J., Lung, J., Osakabe, K., Ebinuma, H., Chiang, V. L. *Proc. Natl. Acad. Sci., USA* **1998**, *95*, 5407-5412.
58. Lee, D., Ellard, M., Wanner, L. A., Davis, K. R., Douglas, C. J. *Plant Mol. Biol.* **1995**, *28*, 871-884.
59. Mansell, R. L., Stöckigt, J., Zenk, M. H. *Z. Pflanzenphysiol.* **1972**, *68*, 286-288.
60. Gross, G. G., Kreiten, W. *FEBS Lett.* **1975**, *54*, 259-262.
61. Sarni, F., Grand, C., Boudet, A. M. *Eur. J. Biochem.* **1984**, *139*, 259-265.

62. Goffner, D., Campbell, M. M., Campargue, C., Clastre, M., Borderies, G., Boudet, A., Boudet, A. M. *Plant Physiol.* **1994**, *106*, 625-632.
63. Lacombe, E., Hawkins, S., Van Doorsselaere, J., Piquemal, J., Goffner, D., Poeydomenge, O., Boudet, A.-M., Grima-Pettenati, J. *Plant J.* **1997**, *11*, 429-441.
64. Messner, B., Boll, M. *Plant Cell, Tissue Organ Cult.* **1993**, *34*, 261.
65. Grima-Pettenati, J., Chriqui, D., Sarni-Manchado, P., Prinsen, E. *Plant Sci.* **1989**, *61*, 179-188.
66. Gross, G. G., Stöckigt, J., Mansell, R. L., Zenk, M. H. *FEBS Lett.* **1973**, *31*, 283-286.
67. MacLean, H., Gardner, J. A. F. *For. Prod. J.* **1956**, *6*, 510-516.
68. Campbell, M. M., Ellis, B. E. *Planta* **1992**, *186*, 409-417.
69. Dalkin, K., Edwards, R., Edington, B., Dixon, R. A. *Plant Physiol.* **1990**, *92*, 440-446.
70. De Sa, M. M., Subramaniam, R., Williams, F. E., Douglas, C. J. *Plant Physiol.* **1992**, *98*, 728-737.
71. Baucher, M., Chabbert, B., vanDoorsselaere, J., Pilate, G., Cornu, D., Petit-Conil, M., Monties, B., van Montagu, M., Inze, D., Jouanin, L., Boerjan, W. In *Somatic Cell Genetics and Molecular Genetics of Trees*; Ahuja, M. R., Ed. Kluwer Academic Publishers: Netherlands, 1996, pp 153-158.
72. Halpin, C., Knight, M. E., Foxon, G. A., Campbell, M. M., Boudet, A. M., Boon, J. J., Chabbert, B., Tollier, M.-T., Schuch, W. *Plant J.* **1994**, *6*, 339-350.
73. Ralph, J., MacKay, J. J., Hatfield, R. D., O'Malley, D. M., Whetten, R. W., Sederoff, R. R. *Science* **1997**, *277*, 235-239.
74. MacKay, J. J., O'Malley, D. M., Presnell, T., Booker, F. L., Campbell, M. M., Whetten, R. W., Sederoff, R. R. *Proc. Natl. Acad. Sci.* **1997**, *94*, 8255-8260.
75. Nakashima, J., Awano, T., Takabe, K., Fujita, M., Saiki, H. *Plant Cell Physiol.* **1997**, *38*, 113-123.
76. Samaj, J., Hawkins, S., Lauvergeat, V., Grima-Pettenati, J., Boudet, A. *Planta* **1998**, *204*, 437-443.
77. Walter, M. H., Schaaf, J., Hess, D. *Acta Hort.* **1994**, *381*, 162-168.
78. Feuillet, C., Lauvergeat, V., Deswarte, C., Pilate, G., Boudet, A., Grima-Pettenati, J. *Plant Mol. Biol.* **1995**, *27*, 651-667.
79. Hawkins, S., Samaj, J., Lauvergeat, V., Boudet, A., Grima-Pettenati, J. *Plant Physiol.* **1997**, *113*, 321-325.
80. Ye, Z.-H., Kneusel, R. E., Matern, U., Varner, J. E. *Plant Cell* **1994**, *6*, 1427-1439.
81. Ye, Z.-H., Varner, J. E. *Plant Physiol.* **1995**, *108*, 459-467.
82. Gowri, G., Bugos, R. C., Campbell, W. H., Maxwell, C. A., Dixon, R. A. *Plant Physiol.* **1991**, *97*, 7-14.
83. Pakusch, A.-E., Kneusel, R., Matern, U. *Arch. Biochem. Biophys.* **1989**, *271*, 488-494.
84. Schmitt, D., Pakusch, A.-E., Matern, U. *J. Biol. Chem.* **1991**, *266*, 17416-17423.
85. Meng, H., Campbell, W. H. *Plant. Mol. Biol.* **1998**, *38*, 513-520.
86. Bugos, R. C., Chiang, V. L. C., Campbell, W. H. *Plant Mol. Biol.* **1991**, *17*, 1203-1215.
87. van Doorsselaere, J., Dumas, B., Baucher, M., Fritig, B., Legrand, M., van Montagu, M., Inze, D. *Gene* **1993**, *133*, 213-217.
88. Collazo, P., Montoliu, L., Puigdomenech, P., Rigau, J. *Plant Mol. Biol.* **1992**, *20*, 857-867.
89. Wang, Z.-X., Li, S.-M., Löscher, R., Heide, L. *Arch. Biochem. Biophys.* **1997**, *347*, 249-255.
90. Camm, E. L., Towers, G. H. N. *Phytochemistry* **1973**, *12*, 961-973.

91. Anterola, A. M., van Rensburg, H., van Heerden, P., Davin, L. B., Lewis, N. G. *Biochem. Biophys. Res. Commun.* **1999**, *261*, 652-657.
92. Timell, T. E. *Compression Wood in Gymnosperms*; Springer-Verlag: Berlin, 1986, pp.1338.
93. Jones, H. *Phytochemistry* **1984**, *23*, 1349-1359.
94. Bate, N. J., Orr, J., Ni, W., Meromi, A., Nadler-Hassar, T., Doerner, P. W., Dixon, R. A., Lamb, C. J., Elkind, Y. *Proc. Natl. Acad. Sci. USA* **1994**, *91*, 7608-7612.

Chapter 4

Genetic Engineering of Poplar Lignins: Impact of Lignin Alteration on Kraft Pulping Performances

C. Lapierre[1], B. Pollet[1], M. Petit-Conil[2], G. Pilate[3], C. Leplé[3], W. Boerjan[4], and L. Jouanin[5]

[1]Laboratoire de Chimie Biologique INRA, Institut National Agronomique, INA-PG, F78850, Thiverval-Grignon, France
[2]Centre Technique du Papier CTP, BP 251 F38044, Grenoble cedex 9, France
[3]Station d'Amélioration des Arbres Forestiers, INRA, F45160, Ardon, France
[4]Laboratorium voor Genetika, Vlaams Interuniversitair Instituut voor Biotechnologie VIB, Universiteit Gent, 9000 Gent, Belgium
[5]Laboratoire de Biologie Cellulaire, INRA, F78026 Versailles cedex, France

A poplar clone (INRA 717-1B4) was genetically engineered to down-regulate caffeic acid *O*-methyltransferase (COMT) or/and cinnamyl alcohol dehydrogenase (CAD), which are enzymes of the lignin biosynthetic pathway. One- and two-year old transgenic poplars with severely down-regulated COMT or CAD activity were evaluated for lignin structure and for kraft pulping characteristics. While the subtle lignin alterations caused by CAD depressed activity had a beneficial impact on pulping efficiency, the structural changes induced by COMT down-regulation detrimentally affected the kraft delignification of the poplar trees. These phenomena could be rationally accounted for by the observed alteration of lignin structure. These results point the way to change lignin structure by genetic engineering in order to produce more easily processed trees for the pulp industries.

Currently, there are worldwide efforts to down-regulate the activity of enzymes involved in the lignin monomer biosynthetic pathway as a means to produce plants with altered lignin content and/or structure and improved characteristics when used for pulp and paper production or animal feed (*1*). In this line, we have produced transgenic poplars (*Populus tremula* x *Populus alba*) with depressed caffeic acid/5-hydroxyferulic acid *O*-methyltransferase (COMT) or cinnamyl alcohol dehydrogenase (CAD) activity (*2,3*). Preliminary studies run on three-month old poplars revealed that a severe down-regulation of COMT or CAD was obtained for a few transgenic lines. These lines did not show any change in lignin content relative to the control. No substantial alteration of lignin structure was evidenced in *CAD* transformed poplars which nevertheless displayed a red coloration of the xylem (*3*). Strikingly enough however, when subjected to mild alkaline hydrolysis, the CAD down-regulated poplars released higher levels of benzaldehydes, especially syringaldehyde, while a few percent of the wood lignin were solubilized, relative to control lines.

More importantly, alkaline kraft pulping experiments run on three-month old plants revealed that *CAD* transformed poplar trees were more efficiently delignified than control poplar trees (*3*). In contrast to CAD down-regulated poplars and similar to maize mutants with depressed COMT activity (*4*), COMT down-regulated poplars displayed a drastically altered lignin structure with a drop in the proportion of syringyl (S) units and the incorporation of 5-hydroxyguaiacyl (5-OH G) units in substantial amounts (*2*).

In the present study, in-depth evaluation of lignin structure and pulping characteristics of two-year old poplars with depressed CAD or COMT activity are examined with three objectives: first, to control the stability of transgene expression with age; second, to more reliably evaluate pulping characteristics of CAD or COMT down-regulated poplars, as three-month old trees do not adequately represent older trees with regards to kraft cook; third, to closely evaluate lignin alteration, if any, in *CAD* transformed poplars and to adduce the molecular basis of their specific reactivity towards alkaline treatment.

Phenotype and Lignin Content of Transgenic Poplar Trees with Depressed COMT and CAD Activity

From previous experiments, we selected for this study four transformants with down-regulated CAD or COMT activity : two lines of transgenic trees (ASOMT2B and ASOMT10B) that expressed a part of the *COMT* gene in antisense orientation (*2*); two lines of CAD down-regulated poplar trees obtained either with the full length antisense *CAD* construct (ASCAD21) or with the sense construct (SCAD1) (*3*). In the xylem of these transgenic lines and relative to the control line (clone Institut National de la Recherche Agronomique 717-1B4), we monitored COMT and CAD activity during the growing season in plants ranging from four-month to eighteen-month old (data not shown). CAD activity level was reduced by approximately 70% for the ASCAD21 and SCAD1 lines and relative to control or ASOMT lines. In ASOMT2B and ASOMT10B transformants, COMT activity towards caffeic acid was still more severely reduced by ca. 95% (5% residual activity), relative to control or CAD lines.

In spite of CAD or COMT down-regulation, no phenotypic differences in growth were observed between control and transgenic poplars with antisense *COMT* or *CAD* constructs, grown in the greenhouse or in the field. Similar to three-month old poplars (*3*), two-year old poplars with depressed CAD activity showed a red coloration of the outer part of the xylem, particularly bright in ASCAD21 line. After wood drying, this red coloration turned to a brownish coloration which resisted solvent extraction aimed at removal of extractives.

Although exhibiting a low CAD and COMT activity, the Klason lignin level of transgenic poplars was retained at a similar level as in the control (Table I). Statistical tests run on this series of two-year old poplars and on a larger series of younger poplars (data not shown) showed that the Klason lignin content of ASOMT poplars did not significantly differ from that of control poplar trees. The same tests suggested a tendency for ASCAD21 and SCAD1 poplars to display a slight reduction in Klason lignin. When the oldest branches of two-year old SCAD1 poplars were examined, the same trend could be observed compared with the control. Obviously, more data is needed to confirm the slightly reduced Klason lignin content of ASCAD21 and SCAD1 lines relative to the control.

Table I. Klason Lignin Content (% by weight) of 2-year old Transgenic Poplar Trees with Reduced COMT or CAD Activity, Relative to the Control.

Line	ASCAD21	SCAD1	Control	ASOMT2B	ASOMT10B
Greenhouse	18.7 ± 0.2	17.7 ± 0.3	19.4 ± 0.2	21.1 ± 0.3	20.3 ± 0.2
Field	18.3 ± 0.1	18.5 ± 0.2	19.1 ± 0.1	19.6 ± 0.3	18.5 ± 0.2

For each line, measurements were performed on main stems collected from 5 plants, debarked, ground and extractive-free. These plants were grown in the greenhouse or in the field. Data are averages ± standard error from 3 to 6 independent replicates.

In contrast to the bm3 maize mutant in which the mutation reduced the lignin content (*4*) and similar to COMT down-regulated tobacco (*5*), a depressed COMT activity did not affect the lignin content of the poplar trees. Recently, an alternative methylation pathway has been evidenced in the phenylpropanoid biosynthesis. In this alternative pathway Caffeoyl-CoA-*O*-methyltransferase (CCoAOMT) converts caffeoyl CoA into feruloyl CoA (*6*). The close association of *CCoAOMT* expression with lignification has been demonstrated in the lignifying tissues of several dicot plants (*7*). Therefore, one may assume that CCoAOMT activity compensates for the severe decrease in COMT activity and allows the maintenance of a high lignin level. Moreover, several lines of evidence indicate that, in addition to COMT and CCoAOMT activities, other *O*-methyltransferase activities may occur in the lignin biosynthetic pathway with various substrate specificity (*8-9*). In *CAD* transformed trees and in spite of a decreased CAD activity, the lignin level appeared, at best, slightly reduced or unchanged. A similar trend has been reported for *CAD* null pine mutant (*10*) or CAD down-regulated tobacco (*11*).

From these data taken together, we have ascertained that down-regulation of COMT and CAD activities in a woody plant species with industrial uses does not affect the growth of the tree or the lignin content of the xylem to any substantial extent. Therefore, change in the pulping characteristics of the plants, if any, would essentially rely on the alteration of lignin structure and/or distribution in subcellular or cellular regions. Impact on lignin structure will be discussed in detail, below.

COMT Down-Regulation Induces Severe Structural Alteration in the Lignin of Transgenic Poplar Trees

Lignin structure was evaluated by thioacidolysis, an analytical and routine degradation that is capable of providing a wealth of structural information about native lignins (*12*). Similar to other analytical or industrial lignin degradation processes, the key reaction of thioacidolysis is the cleavage of labile β-O-4 ether bonds which represent major linkages in native lignins (*13*). Guaiacyl (G) and syringyl (S) units only involved in β-O-4 linkages specifically yield thioethylated G and S monomers with an 80% reaction yield, as shown by model experiments (*14*). On this basis, the molar percentage of lignin units only involved in β-O-4 linkages can be readily estimated. When thioacidolysis is performed from CH_2N_2 exhaustively methylated lignins, additional monomers are recovered with methylated phenolic groups. Their relative proportion indicates the frequency of free and methylatable phenolic groups within β-O-4 linked G or S units (*15*). This is an important structural parameter that may affect the efficiency of lignin solubilization and/or fragmentation during kraft cooking, as many studies have shown that free phenolic groups in lignin units markedly facilitate the cleavage of sidechain ether bonds (*16*). A more recent development of thioacidolysis has made possible the determination of lignin-derived dimers that have retained resistant carbon-carbon and diarylether bonds present in the polymer and referred to as lignin condensed bonds. Therefore, the GC analysis of these diagnostic dimers provides information on the nature and relative frequencies of condensed interunit bonds in the lignin (*12*).

On the basis of thioacidolysis yield in G and S monomers, the percentage of lignin units only involved in β-O-4 bonds was monitored as a function of poplar age and line. Within the same line, the lowest content in β-O-4 structures was observed for the more juvenile samples (Figures 1 and 2). This is an indication that lignins deposited at the early stage of wall lignification are enriched in carbon-carbon bonds, in agreement with our current knowledge on the spatio-temporal evolution of lignin pattern (*17*). In the older poplar trees from the control line, more than 60% of the G and S lignin units were found only involved in labile ether structures that readily provided G and S monomers on thioacidolysis. In contrast, the lignins in *COMT* transformed poplars showed a substantial reduction of such labile ether structures relative to the control (reduction by 25 to 35%, Figure 1). In other words, a pronounced impact of COMT depressed activity was found to be an enrichment in carbon-carbon interunit bonds in the lignin polymer.

Consistent with this observation, the distribution of lignin building units in COMT down-regulated samples appeared to be severely affected by the genetic transformation. In agreement with the results reported for bm3 maize mutant with depressed COMT activity (*4*), for analogous three-month old poplars (*2*) and for COMT down-regulated tobacco plants (*5*), the relative importance of S building units in β-O-4 structures was dramatically reduced (Figure 3). Concomitantly to that decrease in S units, 5-OH G units were evidenced in noticeable relative amounts, versus trace amounts in the control. Therefore, the drop in S units and the marked incorporation of 5-OH G units appeared to be a common lignin pattern adopted by different species with depressed COMT activity. It is noteworthy that the capability to provide evidence for 5-OH G units is an attribute of thioacidolysis which preserves labile *o*-diphenolic rings in lignin-derived monomers and does not cause significant demethylation. In contrast, other degradative methods, such as nitrobenzene oxidation, do not preserve 5-OH G units.

The determination of the lignin-derived dimers provided new insight on the specific traits of the lignin in COMT down-regulated poplars. The structure of the main dimers is shown in Figure 4 (compounds ***1-5***). These dimers include representatives from the common condensed interunit linkages of hardwood lignins, namely the 5-5, β-5, β-1, 4-O-5 and β-β linkages (*12*). The frequencies of the

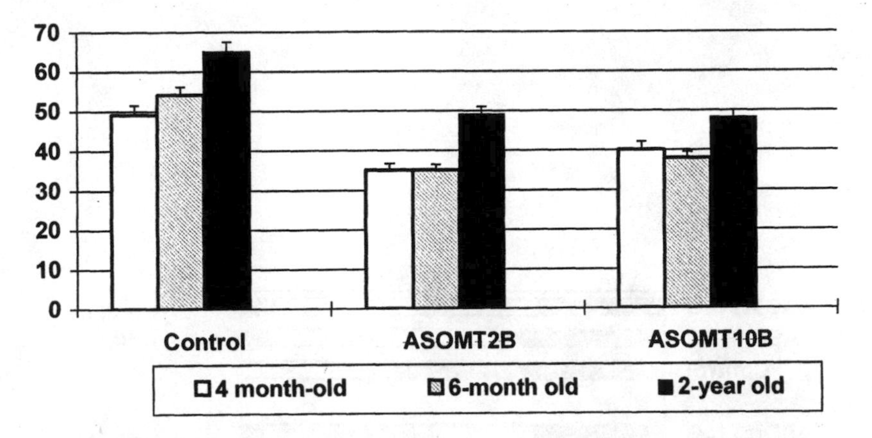

Figure 1. Frequency (% molar) of lignin units only involved in β-O-4 bonds in control and COMT down-regulated poplar trees. These structures are evaluated from the determination of lignin-derived thioacidolysis monomers. The calculation is performed assuming an 80% thioacidolysis yield and a 200 average molecular weight of lignin units. Standard errors between duplicates are in the 3-5% range.

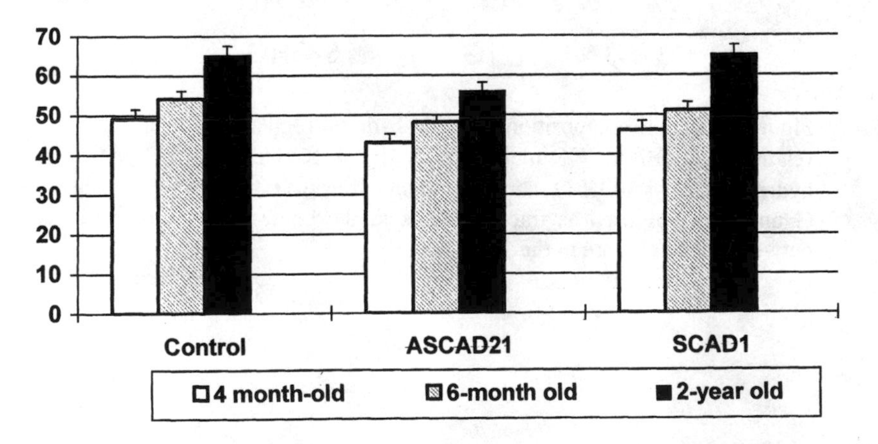

Figure 2. Frequency (% molar) of lignin units only involved in β-O-4 bonds in control and CAD down-regulated poplar trees. For additional comments, see Figure 1.

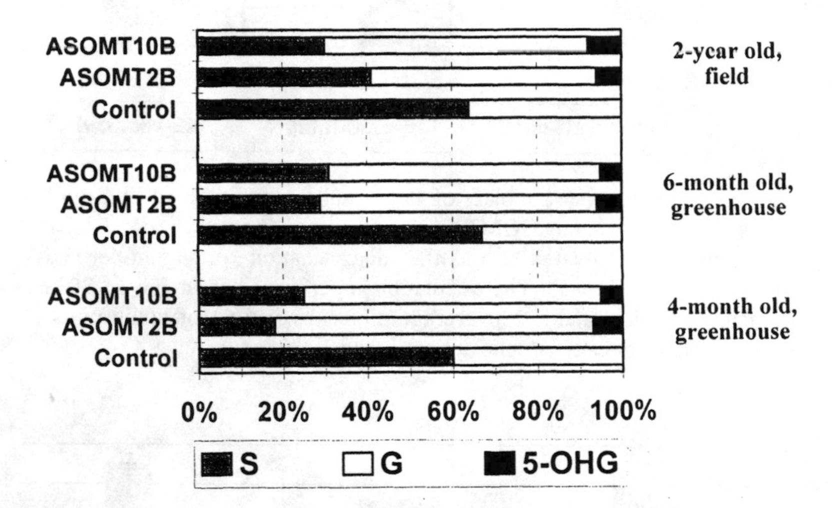

Figure 3. Lignin composition in COMT down-regulated poplar trees : relative proportions (% molar) of Syringyl S, Guaiacyl G, and 5-hydroxyguaiacyl 5-OH G units only involved in β-O-4 bonds. The 5-OH G units are observed as trace amounts in the control. Standard errors between duplicates are in the 2-3% range.

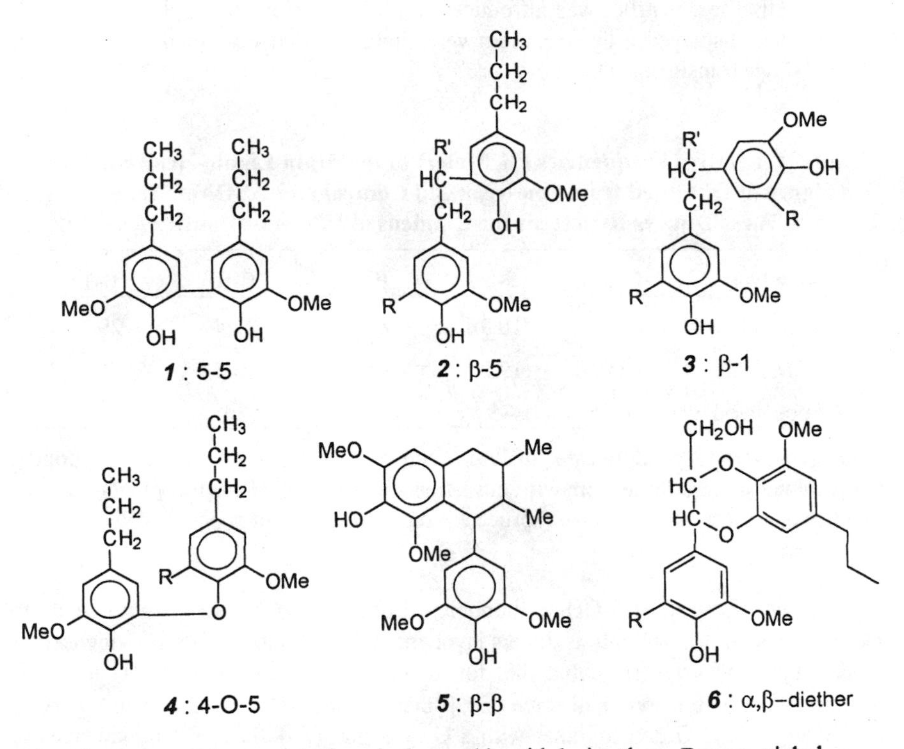

Figure 4. Dimers recovered from thioacidolysis then Raney nickel desulfurization of hardwood lignins (R = H or OMe; R' = H or CH_2OH). The S β-β dimer originates from syringaresinol structures. The α,β-diether dimers ***6*** are specifically found in ASOMT samples.

lignin-derived dimers revealed a dramatic alteration of lignin structure in the *COMT* transformed poplars (Table II). The most striking effect of depressed COMT activity was the twofold increase in 5-5 biphenyl structures. In addition and relative to the control, the proportion of β-5 dimers also increased while that of β-β syringaresinol-derived dimers dropped. Dimers representative of β-1 diarylpropanes and 4-O-5 diphenyl ethers were retained at a similar level. These results are remarkably consistent with what can be expected from the change in S and G relative distribution and further point to the efficiency of the employed technique. Importantly, when *COMT* antisense construct was introduced in *CAD* transformed poplars, the double transformant displayed a lignin pattern very similar to that described herein for the *COMT* single transformants.

Table II. Relative Frequencies (% Molar) of the Main Lignin-derived Dimers (Figure 4) Obtained from One-year Old Control and ASOMT Transgenic Poplars. These Dimers Represent the Condensed Interunit Bonds of the Lignin.

Bond type	4-O-5	5-5	β-5	β-β	β-1
Control	7.5	10.3	28.1	24.4	29.7
ASOMT10B	9.5	19.8	34.2	10.8	25.7
Spruce wood	6	35	33	26	-

Data are average of duplicate analyses with 5-10% standard error. For poplar, measuremenst were done from extractive-free stems collected from 3 plants per line. Data obtained for extractive-free spruce are reported for comparison purpose.

The recovery of 5-OH G monomers from ASOMT plants prompted us to examine whether thioacidolysis dimers involving 5-OH G units could be evidenced as well. Only two dimers could be found in noticeable amounts. Their mass fragmentation pattern was indicative of α,β-diether dimers (compound ***6*** in Figure 4) involving a 5-OH G unit together with a G or S unit (*18*). They were evidenced in minor relative amount from ASOMT poplars (ca. 4% of the main dimers ***1-5*** outlined in Figure 4). Consistent with the proportion of G and S units, the 5-OH G dimer including a G unit was three times more abundant than its analogue including an S unit. We may surmise that 5-OH G units are essentially involved in α,β-diether bonding pattern, in agreement with literature data (*19*). That a small amount of these ether bonds resisted thioacidolysis was not unexpected in view of their survival to other ether-cleaving degradation technique, namely hydrogenolysis at 240°C (*19*). In contrast, GG, GS, and SS dimers diagnostic of traditional β-O-4 bonding patterns were detected as trace amounts or not detected, contrary to recent literature data (*20*).

Even though the lignin content was unchanged in COMT down-regulated poplars, the specific structural traits of the lignin were found dramatically distinct from those of the control lignin. As revealed by thioacidolysis, the dramatic increase in G units and in biphenyl bonds, the reduction in labile β-O-4 linkages, the decrease in syringaresinol structures and in S units were strikingly consistent alterations. These alterations provided poplar lignins with some similarities with softwood samples, as illustrated in Table II. In other words, the expected impact of these structural changes on the kraft pulping characteristics of COMT down-regulated poplars might very likely be detrimental. This effect will be discussed below.

CAD Down-Regulation Induces Subtle but Promising Alteration in the Lignin of Transgenic Poplars

When subjected to thioacidolysis, the lignin of *CAD* transformed poplars yielded similar amounts of G and S monomers as the control. With regards to the β-O-4 content, one may therefore assume that the lignin susceptibility towards industrial processes aimed at ether cleavage was at least preserved. Indeed and similar to control samples, in two-year old *CAD* transformed poplars, approximately 60% of the lignin units were only involved in β-O-4 ether linkages (Figure 2). This amount thereby represented a wealth of potential targets for efficient kraft delignification.

The lignin of *CAD* transformed poplars was comprised of similar proportions of G and S units as that of analogous control samples (Figure 5). This result is consistent with the similarity of thioacidolysis yield as the β-O-4 content has been repeatedly reported to increase with the proportion of S units. The incorporation of cinnamaldehyde units into the lignin of CAD depressed plants has been evidenced from spectrometric characterizations (*10*, *21*), pyrolysis (*11*, *22*), or thioacidolysis (*18*, *23*, *24*). This result has been reported for sorghum mutant (*18*, *21*, *24*), *CAD* null pine mutant (*10*, *21*), and CAD down-regulated tobacco plants (*11*, *23*). With this in view, we evaluated the relative amount of the dithioketal compounds Ar-CHR-CHR-CHR_2 (Ar = G or S ring and R = SEt) which are diagnostic for cinnamaldehyde endgroups. Conclusive support to that claim was given by the results from thioacidolysis of coniferaldehyde or sinapaldehyde which provided these dithioketal derivatives (Lapierre, unpublished data). In addition, G-CHR-CHR-CHR_2 was recovered in elevated amounts from thioacidolysis of synthetic lignins made from coniferaldehyde or of sorghum mutant with depressed CAD activity (*18*). In contrast to the aforementioned examples, *CAD* transformed poplars did not provide higher levels of cinnamaldehyde dithioketals, relative to the control. Interestingly, the recovery yield of S-CHR_2 that represents the dithioketal derivative of syringaldehyde was tenfold higher in the case of ASCAD21 sample relative to the control. This S compound was however recovered in weak relative amounts when compared with the major S thioacidolysis monomers (2% in ASCAD21 sample, versus 0.2% in control). Obviously, we failed to detect significant incorporation of cinnamaldehyde in CAD down-regulated poplars, in contrast to other samples analyzed in our laboratory. These observations suggest that plant species may adopt distinct responses towards CAD depressed activity.

To more closely evaluate lignin alteration in *CAD* transformed poplars, if any, we ran thioacidolysis from exhaustively permethylated samples in order to determine the proportion of free phenolic lignin units involved in β-O-4. The relative frequency of free phenolic groups in lignin is an important trait that may facilitate lignin solubilization and/or fragmentation during kraft cook. In the control lines, ca. 25% of

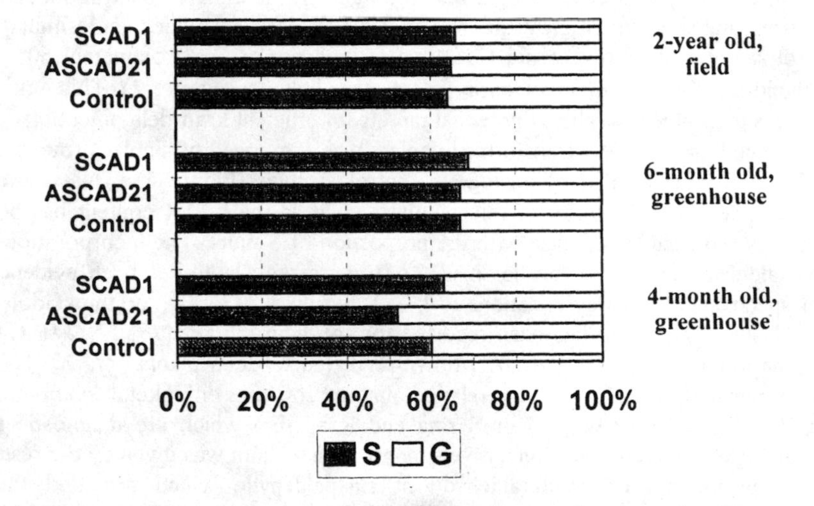

Figure 5. Lignin composition in CAD down-regulated poplar trees : relative proportions (% molar) of Syringyl S and Guaiacyl G units only involved in β-O-4 bonds. Standard errors between duplicates are in the 2-3% range.

G units involved in β-O-4 were terminal units with free phenolic groups (Table III). In contrast, only 3% of the S analogous structures were found to be free phenolic, in agreement with previous studies.

Table III. Frequency (% Molar) of Free Phenolic Groups in β-O-4 linked G and S Lignin Units from Two-year Old Poplars.

Line	ASCAD21	SCAD1	Control	ASOMT10B
G	33.5 ± 1.5	29.8 ± 1.3	26.4 ± 1.3	21.5 ± 1.1
S	4.5 ± 0.2	3.6 ± 0.2	3.0 ± 0.1	2.6 ± 0.2

Data are averages ± standard errors of duplicate thioacidolyses performed on permethylated samples (*15*). For each line , measurements were done from debarked, ground and extractive-free stems collected from 5 plants.

Strikingly, in *CAD* transformed poplars, the levels of free phenolic units were found substantially higher than in the control, not only for the G units but also for the S units. Similar results were obtained for six-month old poplars (data not shown). As the β-O-4 proportions and S levels are the same in CAD transgenic and control lines, the data of Table III clearly pinpoint that CAD down-regulation induces a global enrichment of the lignin in free phenolic groups whereas a reversed tendency was observed for the ASOMT10B line.

The determination of the lignin-derived dimers revealed a higher level of β-1 diarylpropane bonds in *CAD* transformed samples, relative to the control (Table IV). Whatever the biochemical basis of this trait, that the β-1 bonds were observed to increase in *CAD* transformed lignins is consistent with the relative increase in free phenolic groups (Table III). Indeed, diarylpropane structures have been shown to be mainly distributed in lignin as endgroups with free phenolic functions (*25*).

Table IV. Relative Frequencies (% Molar) of the Main Lignin-derived Dimers (Figure 4) Obtained from Two-year Old Control and CAD Transgenic Poplars.

Bond type	4-O-5	5-5	β-5	β-β	β-1
Control	8.2	9.8	26.5	26.0	29.5
ASCAD21	9.6	10.4	23.7	21.8	34.5
SCAD1	8.0	9.6	23.3	23.0	36.1

Data are average of duplicate analyses with 5-10% standard error. Measurements were done from extractive-free stems collected from 5 plants per line.

Similar to *COMT* transformation, CAD down-regulation did not cause any dramatic change in lignin content of transgenic poplars. In contrast to *COMT* transformed lignin which displayed drastic structural change, few differences could be evidenced between the lignin of CAD down-regulated and control poplars. The proportion of β-O-4 linkages or of S units remained at a similarly high level. While we failed to find any increased cinnamaldehyde incorporation in CAD lignins, we observed more syringaldehyde units, albeit in weak total amounts. The most significant change was an enrichment in free phenolic units and in diarylpropane structures. Whatever the biochemical mechanism which may account for that change, more free phenolic units should benefit kraft pulping efficiency, as discussed below.

Impact of COMT and CAD Down-Regulation on Pulping Characteristics of Poplar Trees in Relation with Lignin Structural Alteration

Down-regulation of COMT activity was studied in order to produce plants more easily delignified during kraft cook on the following rationale. Similar to maize mutant (*4*), one could speculate that depressed COMT activity might decrease the poplar lignin content, thereby facilitating kraft delignification. In addition, substitution of some S units methoxylated at C-5 for 5-OH G units hydroxylated at C-5 might increase the level of free phenolic groups and therefore promote kraft delignification (*26*). However, the actual impact of COMT down-regulation was different. There was no change in lignin content. Poplar 5-OH G units were etherified at both phenolic groups. In addition, COMT depressed activity induced a decrease in lignin labile ether bonds and an increase in biphenyl resistant linkages. Not unexpectedly, *COMT* transformed poplars proved substantially less amenable to kraft delignification, as revealed by higher kappa numbers (Table V). Lower brightness levels were reached after an Elemental Chlorine Free (ECF) sequence. The paper strength was the same as for the control sample, as shown by cellulose polymerization degree (DP) and mechanical measurements.

Preliminary pulping assays run on three-month old poplars suggested some improved performance of ASCAD poplars relative to the control (*3*). However, as pulping characteristics of juvenile trees differ from those of mature trees, a more comprehensive evaluation of pulping characteristics on older poplars was done. This evaluation confirmed the promising behavior of *CAD* transformed poplars, especially ASCAD21 line. In two independent experiments run for greenhouse- and field-grown poplars (Table VI), ASCAD21 poplar trees were delignified to a higher extent, relative to the control, as revealed by lower kappa number. This small reduction of pulp kappa number can be of high value because of the large scale of the pulping industries. Other kraft experiments showed that lower alkali charge (15% active alkali) still allowed satisfying delignification of the ASCAD21 line (16.1 kappa number versus 20.5 for the control, G. Toval, personal communication), which suggests that some energy and chemicals could be spared with these transgenic poplars. This easier delignification was not associated with pronounced cellulose degradation. After bleaching, the difference in kappa number between ASCAD21 and control lines was maintained. The final pulp brightness after ECF sequence and the paper strength were found at the same level in control and *CAD* transformed lines.

Table V. Properties of Kraft Pulp Obtained from *COMT* Transformed Poplars

Age	One-year old greenhouse			Two-year old field		
Laboratory cook, 20% act. Alkali, 25% sulfidity, 170°C, liquor/wood ratio = 4						
Line	Control	ASOMT2B	ASOMT10 B	Control	ASOMT2B	ASOMT10 B
Kappa number	27.4	36.8	37.7	17.7	27.4	31.5
Cellulose DP	1880	1940	-	1440	1730	1710
Screened yield	-	-	-	51.3	49.6	-
ECF bleaching sequence O D E D, 10% pulp consistency						
Brightness % ISO	88.1	80.1	83.2	87.9	86.5	84.5
Cellulose DP	1560	1620	1580	1520	1540	1520
Fiber length mm	-	-	-	0.93	0.91	0.90

Experiments were performed on the main stems of several plants (3 to 5 per line), gathered, debarked and cut into chips. Kraft cook was done on ca. 150 g of sample.

Table VI. Properties of Kraft Pulp Obtained from *CAD* Transformed Poplars

Age	Two-year old greenhouse			Two-year old field		
Laboratory cook, 20% act. alkali, 25% sulfidity, 170°C, liquor/wood ratio = 4						
Line	Control	ASCAD21	SCAD1	Control	ASCAD21	SCAD1
Kappa number	18.7	15.6	14.5	17.6	16.3	18.0
Cellulose DP	1900	1910	1830	1440	1670	1670
Screened yield	45.9	47.4	48.8	51.3	50.9	50.9
ECF bleaching sequence O D E D, 10% pulp consistency						
Brightness % ISO	90.6	91.0	91.0	87.9	89.6	86.7
Cellulose DP	1900	1910	1830	1520	1640	1490
Fiber length mm	-	-	-	0.93	0.92	0.93

Experiments were performed as described in Table V.

To further confirm the effect of CAD down-regulation on the efficiency of kraft cook, one-year old poplars grown in the greenhouse and carrying either one (*CAD* single transformants) or two constructs (double transformants with antisense

COMT and *CAD* constructs: ASOMT x ASCAD21; double transformants with sense *COMT* and antisense *CAD* constructs: SOMT x ASCAD21) were collected. CAD activity was found to be similarly reduced in the single and in the double transformants. COMT activity was reduced in ASOMT x ASCAD21 transformants and unchanged in SOMT x ASCAD21 poplars. In agreement with the results of Tables V and VI, lower kappa numbers relative to the control confirmed that the antisense *CAD* construct either in ASCAD21 or in SOMT x ASCAD21 poplars made wood more amenable to kraft delignification (Table VII). A higher kappa number for ASOMT x ASCAD21 pulp relative to the ASCAD21 pulp confirmed that COMT down-regulation decreased the efficiency of the kraft cook. These effects were not related to the lignin content (Table VII), but to the lignin structure. While ASOMT x ASCAD21 transformants displayed the same lignin profile as ASOMT trees, SOMT x ASCAD21 poplars showed the same lignin alteration as ASCAD21 samples.

Table VII. Lignin Content (% Klason Lignin % KL) and Pulping Characteristics of One-Year Old Poplar Single and Double Transformants.

Laboratory cook, 20% act. alkali, 25% sulfidity, 170°C, liquor/wood ratio = 4

Line	Control	ASCAD21	ASOMT X ASCAD21	SOMT x ASCAD21
% KL	17.2	15.6	14.3	15.2
Kappa number	39	13.5	35	17.5
Cellulose DP	2100	1900	1960	2040

For each line, experiments were done on the main stems of several plants (4 to 7 per line), gathered, debarked and ground. Kraft cook was performed on 100 g of sample. The Klason lignin content was determined on the extractive-free ground wood meal.

The structural evaluation of the lignin allowed us to rationally account for the improved pulping characteristics of the ASCAD21 line. The substantial increase in free phenolic lignin units is very likely responsible for the higher susceptibility of ASCAD21 line towards kraft cooking. A support of that hypothesis was afforded from another line, ASCAD52. Compared with ASCAD21, this line displayed a higher CAD residual activity, a lower level of free phenolic groups, and lower pulping performances. These results suggest that a threshold level in free phenolic groups must be reached in order to obtain improved pulping performances.

Conclusions

In the present study, CAD or COMT down-regulated poplars have been obtained which display normal growth and similar lignin content as the control trees. The close evaluation of lignin structure revealed original information and opened new avenues for the rational design of lignin structure through biotechnology techniques.

We have shown that a severely depressed COMT activity in two-year old poplars induces dramatic alterations in the lignins which gives them some similarities with softwood lignins. A higher level of condensed bonds, especially of resistant biphenyl linkages, accounts for an increased resistance towards kraft pulping. A higher lignin condensation degree might be nevertheless beneficial for other wood uses than kraft pulp production.

We have demonstrated that a depressed CAD activity in two-year old poplars induces subtle alterations in the lignin which seem to have a beneficial impact on the pulping characteristics of the tree. While the levels of labile ethers and syringyl units are retained, a substantial increase in lignin units with free phenolic groups and in diarylpropane linkages is observed. This intriguing phenomenon remains to be biochemically addressed. One may speculate that some changes occurred in the conditions prevailing during the polymerization of monolignols. Whatever its biochemical basis, an increase in free phenolic groups makes the lignin more amenable to kraft degradation, thereby facilitating the industrial process. In-depth understanding of the phenomena which control lignin structure and functionality would nevertheless be a challenging task aimed at producing genetically engineered lignin for improved plant uses.

Acknowledgements

This work was done in the framework of European Union Research Programme FAIR-CT95-0424. We sincerely thank F. Legée for the Klason lignin analyses.

References

1. Boudet, A. M. *Trends in Plant Science* **1998,** 3, 67-71.
2. Van Doorsselaere, J.; Baucher, M.; Chognot, E.; Chabbert, B.; Tollier, M.T.; Petit-Conil, M.; Leplé, J.C.; Pilate, G.; Cornu, D.; Monties, B.; Van Montagu, M.; Inze, D.; Boerjan, W.; Jouanin, L. *Plant J.* **1995,** 855-864.
3. Baucher, B.; Chabbert, B.; Pilate, G.; Van Doorsselaere, J.; Tollier, M.T.; Cornu, D.; Monties, B.; Van Montagu, M.; Inze, D.; Jouanin, L; Boerjan, W. *Plant Physiol.* **1996,** 112, 1479-1490.
4. Lapierre, C.; Tollier, M.T.; Monties, B. *C. R. Acad. Sci. Paris, Série III* **1988,** 307, 723-728.
5. Atanassova, R.; Favet, N.; Martz, F.; Chabbert, B.; Tollier, M.T.; Monties, B.; Fritig, B.; Legrand, M. *Plant J.* **1995,** 8, 465-477.
6. Ye, Z.H.; Varner, J.E. *Plant Physiol.* **1994,** 103, 805-816.
7. Ye, Z.H. *Plant Physiol.* **1997,** 115, 1341-1350.
8. Li., L.; Popko, J.L.; Zhang, X-H.; Osakabe, K.; Tsai, C.J.; Joshi, C.P., Chiang, V.L. *Proc. Nat. Acad. Sci. USA* **1997,** 94, 5461-5466.
9. Matsui, N., Fukushima, K.; Yasuda, S.; Terashima, N. *Holzforschung* **1994,** 48, 375-380.
10. MacKay, J.J.; O'Malley, D.M.; Presnell, T.; Booker, F.L.; Campbell, M.M.; Whetten, R.W.; Sederoff, R.R.; *Proc. Nat. Acad. Sci. USA* **1997,** 94, 8255-8260.
11. Halpin, C.; Knight, M.E.; Foxon, G.A.; Campbell, M.M.; Boudet, A.M.; Boon, J.J.; Chabbert, B.; Tollier, M.T.; Schuch, W. *Plant J.* **1994,** 6, 339-350.
12. Lapierre, C.; Pollet, B.; Rolando, C. *Res. Chem. Interm.* **1995,** 21, 397-412.
13. Adler, W. *Wood Sci. Technol.* **1977,** 11, 169-218.
14. Lapierre, C.; Monties, B.; Rolando, C. *Proceedings of the 4th Internat. Symposium Wood and Pulping Chemistry, Paris* **1987,** 2, 431-435.
15. Lapierre, C.; Rolando, C. *Holzforschung* **1988,** 42, 1-4.

16. Gierer, J. *Holzforchung* **1982,** 36, 43-51.
17. Terashima, N.; Fukushima, K.; He, L.F.; Takabe, K. In *Forage Cell Wall Structure and Digestibility;* Jung, H.G.; Buxton, D.R.; Hatfield, R.D.; Ralph, J., Ed.; ASA-CSSA-SSSA, Madison, 1993, pp 247-269.
18. Jacquet, G. *Structure et Réactivité des Lignines de Graminées et des Acides Phénoliques Associés : Développement des Méthodologies d'Investigation* 1997, PhD Thesis Université Aix-Marseille III, pp 1-181.
19. Hwang, B.H.; Sakakibara, A. *Holzforschung* **1981,** 35, 297-300.
20. Ralph, J.; Grabber, J.H. *Holzforschung* **1996,** 50, 425-428.
21. Ralph, J.; MacKay, J.J.; Hatfield, R.; O'Malley, D.M.; Whetten, R.. Sederoff, R.R. *Science* **1997,** 277, 235-239.
22. Pillonel, C.; Mulder, M.M.; Boon, J.J.; Forster, B.; Binder, A. *Planta* **1991,** 185, 538-544.
23. Higuchi, T.; Ito, T.; Umezawa, T.; Hibino, T.; Shibata, D.; *J. Biotechnol.* **1994,** 37, 151-158.
24. Chabbert, B.; Tollier, M.T.; Monties, B. *Proc. 38th Lignin Symp., Kagawa, Japan* **1993,** pp 33-36
25. Gellerstedt, G.; Zhang, L. *Nordic Pulp Paper Res. J.* **1991,** 6, 136-139.
26. Boudet, A.M.; Grima-Pettenati, J. *Molecular Breeding* **1996,** 2, 25-39.

Chapter 5

Lignin–Polysaccharide Interactions in Woody Plants

Richard F. Helm

Department of Wood Science and Forest Products, Fralin Biotechnology Center, Virginia Polytechnic Institute and State University, Blacksburg, VA 24061–0346

Hemicellulose and lignin are theorized to be linked to one another through covalent bonds, and these bonds are thought to limit rapid chemical and biochemical breakdown of woody biomass. After years of study, the mechanism by which lignin and hemicellulose interact with one another is still not clear. This information is crucial to our understanding of wood biosynthesis and the organization of the cell wall three-dimensional matrix. An overview of what is known about these interactions is presented, and bonding patterns are suggested to support the chemical and spectroscopic data reported in the literature.

Many of the products derived from forest biomass are prepared by chemical wood processing. Yields of desired materials from many of these chemical processes are much less than the amount of material theoretically available. This is due, to a great extent, to the way in which the woody cell wall is biosynthesized. Cellulose, hemicellulose and lignin interact with one another through non-covalent and covalent bonds. These interactions interfere with efficient processing, and thus, chemical procedures are needed which decompose some of the desired biopolymers. Improving the yield and efficiency of chemical/biochemical processes would help us to more effectively utilize our wood resources, and also decrease the amount of toxic effluents present in waste streams.

How are we to accomplish this? First we need to understand the regiochemical organization of the wood cell wall matrix. However, in order to understand the cell wall matrix, we need to determine how the individual polymeric components come together to establish the three-dimensional biopolymer composite. When we fully understand the way in which the woody cell wall is constructed on the chemical level, we can begin ascertaining rational methods for improved chemical processing and/or genetic manipulation.

The Formation of Lignin-Polysaccharide Bonds in Woody Plant Cell Walls.

The theory that is used, almost without exception, to describe the formation of lignin-carbohydrate (LC) bonds in lignified tissues, centers on the quinone methide formed during lignin biosynthesis. The free radical coupling of a monolignol and a lignin oligomer as shown in Figure 1, affords an intermediate quinone methide. It has been suggested that these reactive intermediates undergo nucleophilic attack to form either α-hydroxy compounds, α-alkyl/aryl ethers, α-esters or α-glycosides (*1,2*).

This theory is under revision. In a series of investigations on the structure of forage cell walls (*3,4*), it has been shown that *p*-coumaric acid is attached to corn lignin (and other grass lignins) exclusively at the γ-position (Figure 2A). Thus the

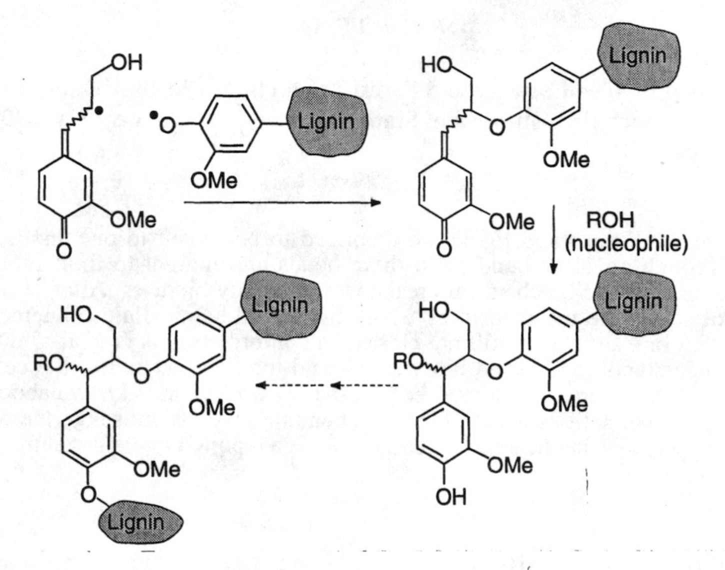

Figure 1. The coupling of two free radicals generates an intermediate quinone methide. Nucleophilic attack followed by additional free radical condensations leads to incorporation of the nucleophile into the lignin macromolecule.

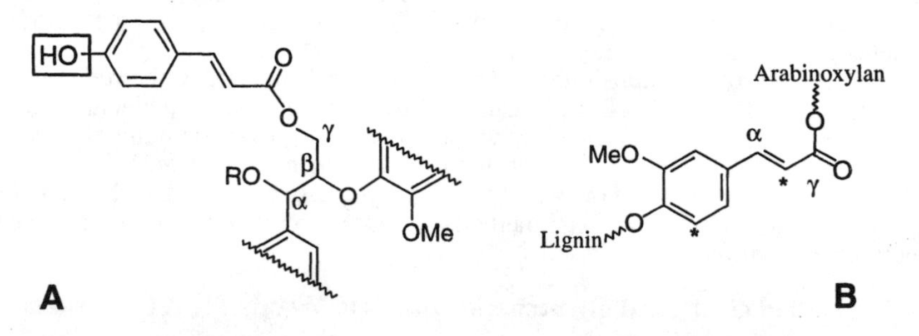

Figure 2. Two modes of hydroxycinnamic acid incorporation into the cell wall of forages. A, *p*-coumaric acid attached at the γ-position of lignin, without cross-coupling. B, ferulic acid is capable of cross-linking between arabinoxylans and lignin. The asterisks indicate other potential sites of lignin attachment.

ester bond is located at a "non-quinone methide" lignin hydroxyl—a process that requires a discrete biological process. The coumarate moiety possesses a free phenolic hydroxyl, which does not participate in the lignification process. This is in contrast to the other predominant cinnamic acid present in forage cell walls, ferulic acid. Ferulates can undergo free radical condensation with lignin and be fully incorporated into the lignin structure (*5*). As it has been amply shown that ferulic acid is esterified to arabinoxylans in forages through the 5-position of α-L-arabinofuranosyl moieties (*6,7*), herein lies an important mechanism for lignin-hemicellulose covalent interactions in forages. Ferulic acid is esterified to arabinoxylans through the carboxyl group, and the free phenolic hydroxyl allows for free radical formation leading to incorporation into the lignin matrix (Figure 2B).

How does this relate to woody cell wall structure? The studies with forages clearly indicate that biochemical processes are involved in the placement of *p*-coumaric acid on the lignin polymer, and the attachment of ferulic acid to arabinoxylans. Ferulate is subsequently incorporated into the lignin polymer whereas *p*-coumarate is not. It would seem logical to assume then that if forages undergo biochemically-controlled processes for the formation of lignin-polysaccharide interactions, a controlled process would also occur in wood. Therefore, the quinone-methide theory may not adequately describe LC formation in wood.

The chemical reactivity of the lignin quinone methide has been the subject of several investigations (*8-12*). One of the overriding points of all of these studies is that the lignin quinone methide is somewhat unreactive towards nucleophiles such as carbohydrate hydroxyls. The reactivity of the quinone methide towards carboxylic acids (uronic acids) is much higher. This reactivity is in contrast to the reported types of major LC linkages in wood. In the case of softwood and hardwood glucuronoxylans, studies indicate that the predominant linkage is an alkali-stable ether linkage, not an ester linkage through the uronosyl moiety (*2,13,14*). Could a biological process be controlling this bond formation?

NMR spectroscopy has been extremely useful in ascertaining the predominant linkages in lignin isolates. Brunow and and others (*15,16*) have listed the major and minor interunit linkages, and these are shown in Figure 3. As can be seen from a perusal of the structures, α-alkyl ethers are not shown—efforts to locate such structures have largely been unsuccessful. Indeed, of all the structures indicated the only one that could possibly involve a carbohydrate is Structure **G**, where an alkyl group is attached to the γ-position of a β-O-4 unit. This concept will be revisited later in the section entitled "Points to Ponder."

Do Different Wood Species and Wood Types Have Different Linkages? It is emphasized here that the choice of substrate can have profound influence on the presence and type of LC bonds found. Using temperate zone hardwoods as an example, their principal hemicellulose is an *O*-acetyl-(4-*O*-methylglucurono)xylan, comprising 15-30% of the weight of these woods (*17*). The (1→4) β-D-xylopyranosyl backbone carries occasional substitutions at the two position by 4-*O*-methyl-α-D-glucopyranosiduronic acid, as well as randomly distributed acetyl groups. Approximately 7 acetate groups are present for every 10 xylose units, and there is approximately 1 uronosyl moiety for every 7 xylose units. It is interesting to note that this structure and composite makeup is quite conserved for all temperate zone hardwood species. This implies xylan plays an integral role in hardwood cell wall structure.

Obst has investigated the LC bonds present in aspen; about 20% were labile to base, implying an ester linkage (*14*). This is in accordance with the observation that an almost quantitative yield of glucuronoxylan can be obtained from treating aspen wood meal with 24% KOH (*18*). The same base extraction can be done with paper birch (*Betula papyrifera*), providing a glucuronoxylan in 80% yield (*17*). These isolated materials have minimal lignin content (<2%) and only differ from the native material by the lack of acetate groups. Peeling is minor in the preparation, as the presence of an α-(1→2) uronosyl moiety has the net effect of stopping the reaction.

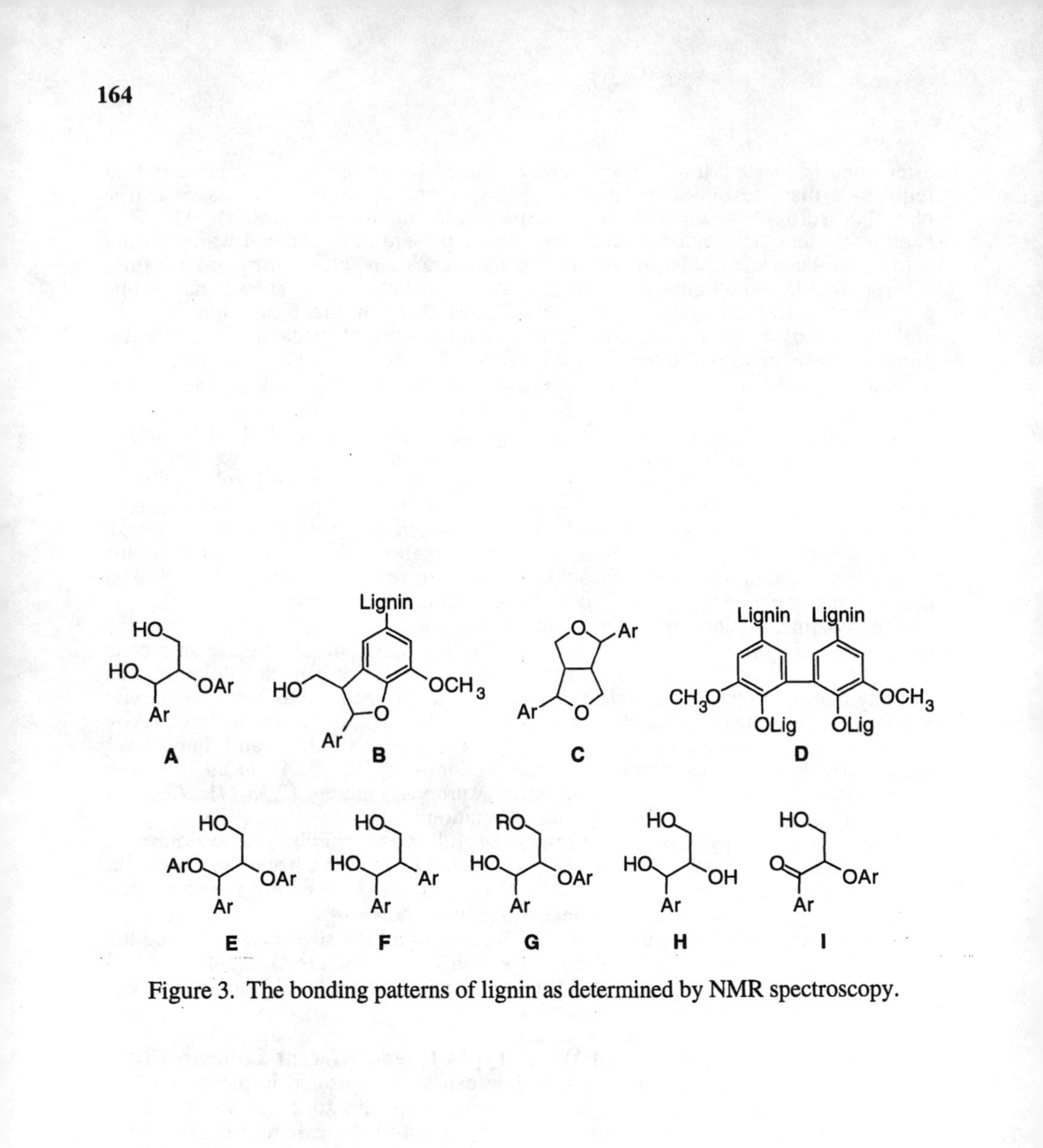

Figure 3. The bonding patterns of lignin as determined by NMR spectroscopy.

The high hemicellulose yield seems to imply that there are relatively few alkali-stable bonds between lignin and hemicellulose in these species.

In contrast to poplar and birch, American elm (*Ulmus americana*) has a 19% glucuronoxylan content. Treatment of this milled wood with 24% KOH affords a 37% yield of the total glucuronoxylan (*17*). Most hardwood species are intermediate in this extractability; poplar is the most extractable and elm the least. Are non-ester linkages more prevalent in species such as elm and lacking in the aspens and birches? If this is the case, in searching for LC bonds, one should keep in mind that different wood species may have different levels and types of LC bonds.

Another point to bear in mind is the differences between normal and reaction woods. Tension wood is found in hardwoods, typically on the top side of branches. Compression wood is found in softwoods on the bottom side of branches. Tension wood is under tension and compression wood is under compression. As wood creep would be undesirable in these locations, one could expect that the concentration of LC bonds would be higher in reaction woods than in normal woods. However, only a few studies have investigated this possibility. The concentration of LC bonds is higher in compression wood than normal wood and the bonds appear to be associated with galactans (*14*). Interestingly, tension wood is low in lignin and high in cellulose and galactan relative to normal wood, and compression wood is high in lignin and galactan but low in cellulose. Clearly, in searching for LC bonds in native woods, one must consider the heterogeneity of woody plants when deciding on a substrate.

Methods of Isolation and Analysis.

The choice of substrate to be investigated is not the only issue to be considered when investigating LC bonds. The methods employed in isolating and characterizing the LC fractions are also critically important in determining the nature and scope of these bonds.

Isolation of Lignin-Carbohydrate Complexes (LCCs). The key to any work in this area is the isolation of LCCs representative of those in native tissue. Several methods have been used, although all rely initially on the preparation of solvent-extracted wood meal and subsequent milling. The use of milled wood is obviously somewhat of a compromise as one wishes to increase the surface area of the sample for subsequent treatments, and this leads to some mechanical degradation. Once the milled wood has been obtained, strategies for isolation of the LCCs differ in that one either uses direct solvent extraction with dioxane:water (*19*) or a treatment with cellulase (*20,21*). The most recent methods described for the isolation of native LCCs utilize direct hot water extraction of a dioxane:water treated residue (*19,22*). Yields from beech and pine have been reported to be 3.6 and 9.3%, respectively, of the dioxane:water treated material (*22,23*). In the study of Japanese beech (*Fagus crenata*), Imamura *et al.* showed that this water-soluble LCC (named LCC-WE) had a 80% carbohydrate content, a 10% uronic acid content, and a 7.6% lignin content (*23*). Assuming the uronic acid is 4-*O*-methyl-D-glucuronic acid, this uronic acid was the third most abundant carbohydrate in the extract behind xylose (48%) and glucose (12%). The published NMR spectrum, however, shows negligible uronosyl signals, including the 4-*O*-methyl group (61 ppm).

Treatment of milled wood with commercial cellulase preparations prior to dioxane:water extraction improves the overall recovery of milled wood enzyme lignin. In the classic study of Chang *et al.* (*24*), it was shown that cellulase-treated milled wood provided a much higher yield of dioxane-water soluble lignin, termed milled wood enzyme lignin (MWEL). Carbohydrate contents in this fraction were in the 4.0-9.0% range for both sweetgum and spruce. More than half of the lignin could be dissolved in dioxane:water. This is still considered the most effective way of isolating native lignin.

Part of the problem with using commercial cellulases for pretreatment is that the preparations contain contaminating enzymes. Glycosidases could potentially cleave glycosidic linkages between lignin and carbohydrate, and esterases may cleave any uronosyl-lignin bonds that may be present. However, cellulase preparations which are relatively free of contaminating activity are now commercially available. Use of this enzyme complex (*25*) would help to insure that no loss of these potential LC linkages would occur, while also improving subsequent purification steps.

An enriched sample of LCCs can then be obtained by treating solvent-extracted milled wood with a cellulase preparation and subsequently with dioxane:water. With a majority of the lignin and cellulose removed what remains is hemicellulose, unsolubilized lignin, LCCs and potentially some undegraded cellulose. To further enrich the LCCs fractions that are high in carbohydrate content, the dioxane-water insoluble residue can be extracted with a polar solvent such as water, DMF or DMSO. The LCCs are solubilized along with a portion of the hemicellulose. The use of hot water is advocated by Azuma and others (*19*) although a higher yield of LCCs (with higher molecular weights) can be extracted with DMSO (*2,19*). If one desires to obtain as much material as possible which is indicative of native LC structures, without resorting to conditions which may afford cleavage of some LC bonds, the use of DMSO is warranted. In addition, we have found that several model LCCs are susceptible to rearrangement and/or degradation in aqueous solvents (*26*).

The material solubilized by DMSO is the fraction that contains the majority of the LCCs. The major components of this mixture are non-lignin bonded hemicelluloses and the LCCs. Further enrichment of the LCCs can be accomplished either by chromatography into acidic, neutral and lignin-rich fractions by ion-exchange chromatography (*19, 27-30*), or by enzymatic degradation of the hemicellulosic component with purified enzymes (*31,32*). While the chromatographic route is useful for the isolation of three different fractions, it would appear that the more effective method of concentrating the LCCs would be by enzymatic depolymerization of the non-bonded hemicellulose regions first. In the case of hardwood hemicelluloses, enzymatic breakdown of the *O*-acetyl-(4-*O*-methyl-glucurono)xylans is required.

In order to break this hemicellulose down by selective enzymatic means, one needs a series of purified enzymes. An acetyl xylan esterase cleaves the acetate groups to expose the hemicellulose (*33*). This glucuronoxylan is then susceptible to depolymerization by endo-xylanases (*34*). The endo-xylanases are not completely capable of cleaving the polymer down to its constitutive monomers; an α-glucuronidase is needed to cleave off the uronosyl groups, and a β-xylosidase is needed to break the xylo-oligomers down the monomer D-xylose. Thus at least four enzymes are needed by any organism wishing to break down this hemicellulose (*35*). For the purposes of enzymatic breakdown of cellulase treated wood, a purified enzyme system is required that has acetyl xylan esterase, endo-xylanase, and α-glucuronosidase activities. β-Xylosidase activity is undesirable as this would cleave any potential glycosidic bonds that may be present. Unfortunately, obtaining preparative amounts of α-uronidase is difficult, due to its low abundance and fragile nature (*36-38*).

Recent studies have attempted to investigate the LCCs present in aspen steam-exploded wood by the use of selective enzymatic breakdown (*31,32,39*). The aqueous fraction from the thermomechanically-treated aspen was purified by size-exclusion chromatography and treated with acetyl xylan esterase, an endo-xylanase and a β-mannanase (*32*). It was noted that hydrolysis was not complete and monosaccharides were not produced. It was then theorized that the LCCs went under conformational changes which restricted complete enzymatic breakdown. The lack of monosaccharides is not surprising as a xylosidase was not present. The conformational changes limiting degradation is an important finding, but this could be due to the nature of the substrate, steam-exploded wood. The mechanism of lignin degradation during steam explosion has been suggested to be free radical-based, with depolymerization and condensation reactions occurring (*40,41*). Loss of LC bonds has also been documented (*42*). A more highly condensed LCC structure would be less amenable to solubilization in water after removal of a portion of the hemicelluloses. Regardless of whether this occurs or not, one must consider that as carbohydrates are removed, the solubility of the remaining lignin-rich fraction in water will decrease.

This discussion of LCC isolation has shown that numerous attempts have been made to analyze LCCs in wood. From these efforts, it can be seen that the most successful approach for hardwood LCC isolation is by the preparation of solvent-extracted ball milled material which is submitted to a cellulase treatment with a purified enzyme mixture. Dissolution of MWEL can be accomplished with dioxane:water, and the carbohydrate-rich LCCs can be obtained from the insoluble fraction by dissolution in a polar solvent such as DMSO. The solubilized LCCs are then further enriched by an enzymatic treatment with acetyl xylan esterases and endo-xylanases. The resulting lignin-carbohydrate containing materials are then purified by chromatography and subjected to analysis.

Analysis of LCCs. While the techniques needed to isolate LCCs are fairly straightforward and reproducible, methods for analysis of LCCs are still in the developing stages. The approaches taken can be separated into two groups, those studies interested in alkali-labile bonds and those interested in alkali-stable bonds. An overview of the results obtained to date are shown in Table 1 and discussed below.

TABLE 1. Overview of LC Linkage Data

Source/Substrate	Method	Carbohydrate Bonds Implicated	Ref.
	ALKALI-LABILE LINKAGES		
Beech/MWL	DDQ/Titration	Uronosyl esters	*23*
Pine/MWL	DDQ/Titration	Uronosyl esters	*2*
Pine/MWEL	Base treatment	Uronosyl esters	*14*
Pine Compression/ MWEL	Base treatment	Uronosyl esters	*14*
Aspen/MWEL	Base treatment	Uronosyl esters	*14*
Spruce/MWEL	Base treatment	Uronosyl esters	*14*
Spruce/MWL	Base treatment	Uronosyl esters	*44*
	ALKALI-STABLE LINKAGES		
Pine/Base Treated MWEL	Methylation Analysis	Primary hydroxyls	*20*
Pine Compression/ Base Treated MWEL	Methylation Analysis	Primary hydroxyls	*20*
Spruce/MWL + glycanases	Methylation Analysis	Primary hydroxyls	*45*
Beech/MWL + cellulase	DDQ	Xylose	*19*
Pine/MWL + cellulase	DDQ	Primary hydroxyls	*19*
Pine/MWL + cellulase	DDQ	Primary hydroxyls/Xylose	*2*
Tropical legume/MWL	DDQ	Xylose	*43*
	COMBINED STUDIES LINKAGES		
Spruce/MWL + glycanases	Acid/Base Treatments	Primary hydroxyls/ Uronosyl esters	*46*
Aspen/MWL + xylanase	Acid/Base Treatments	Primary hydroxyls/ Uronosyl esters	*47*

Alkali-labile bonds are the easiest bonds to study since they presumably uronsyl-linked esters that are selectively released by saponification. Obst characterized the alkali labile linkages of several woods and estimated that about 10-20% of the LC bonds in wood were alkali labile (*14*).

Watanabe's group has been instrumental in the development of DDQ (2,3-dichloro-5,6-dicyano-1,4-benzoquinone) procedures for the detection and quantitation of LCC ester and ether bonds attached at either the α- or conjugated γ-positions of lignin (*2,13,43*). Lignin-carbohydrate bonds at other locations *would not be quantitated.* A recent modification of the procedure has been recently published to quantitate the frequency of α- and/or conjugated γ-esters in *Fagus crenata* (*23*). In this modification, the isolate was ethylated with diazoethane to protect the free phenolic hydroxyl groups on the lignin portion of the LCC. The material was then acetylated and submitted to the DDQ oxidation in refluxing CH_2Cl_2/2M aq. TFA. The released α- and conjugated γ-uronate moieties were then quantitated by comparing the neutralizing capacity of the DDQ-treated material relative to the untreated material. We have found in our studies on lignin-uronate esters that migration of the uronosyl moiety between the α- and γ-positions occurs under neutral aqueous conditions (*26*). This process is accelerated by the presence of acid, and favors the γ-position (Figure 4). If migration occurs as a natural process in wood, and DDQ does not oxidize unconjugated γ-esters, then the DDQ method could seriously underestimate the importance of this linkage in native tissue.

The analysis of alkali-stable linkages is usually performed either with DDQ or by methylation analysis (a protocol typically used for carbohydrate analysis). Neither of these methods are optimal, and they both suffer from one drawback or another. However, it is interesting to note that the methods typically indicate that the primary hydroxyls of either pentofuranosyl residues or hexopyranosyl residues are involved in these linkages. It is only the DDQ method that indicates xylopyranosyl units linked either at the 2- or 3-hydroxyls.

NMR evidence for lignin carbohydrate bonds is still lacking. One can clearly observe signals for both polymers in NMR spectra, but long range correlations between resonances assigned to lignin and carbohydrate have yet to be made. There are probably several reasons for this, most importantly relating to the stereochemistry of lignin and polysaccharides. Since lignin is formed in a chiral environment, one can envision some regiochemical organization, but with respect to chirality, only a slight enantiomeric excess. Thus the attachment of lignin (basically racemic) to polysaccharide (chiral) will generate diastereoisomerism and, in the case of NMR, signal "smearing." This makes confident assignments difficult, especially when appropriate models are not available. It is hoped that in the near future the application of more sophisticated NMR techniques (digital oversampling, pulse-field gradients, etc.) will ameliorate this problem (*3*).

Points to Ponder.

Efforts to fully describe the chemical nature of LC bonds will require the application of new techniques as well as the use of unconventional strategies. In an effort to provide some new thinking on the subject, a theory is offered to mechanistically explain γ-alkyl ethers. This is followed by a cursory mention of some of the new techniques that may provide an insight into this old question.

As the structural bonding patterns determined by NMR indicate only one possible site of attachment of carbohydrates to lignin (γ-alkyl ether), one must consider how such a linkage could be formed. The hemicelluloses of woody plants contain relatively large amounts of 4-*O*-methyl D-glucopyranosyl uronic acid moieties glycosidically linked to the xylan backbone. While uronic acids are relatively stable to acid, in alkaline environments they can rearrange to hexenuronic acids. Interestingly, this process is analogous to the mechanism of glycosidic bond cleavage for carbohydrates linked to the 4-position of uronic acids (Figure 5). If such a hexenuronic acid were to form, it would be susceptible to nucleophilic attack at the 4-position. Could the purpose of the 4-*O*-methyl group on uronic acids in xylans be to provide a leaving group for eventual connection to lignin? Although this is speculation, it is clearly a mechanistic possibility for the formation of LC bonds during kraft cooking.

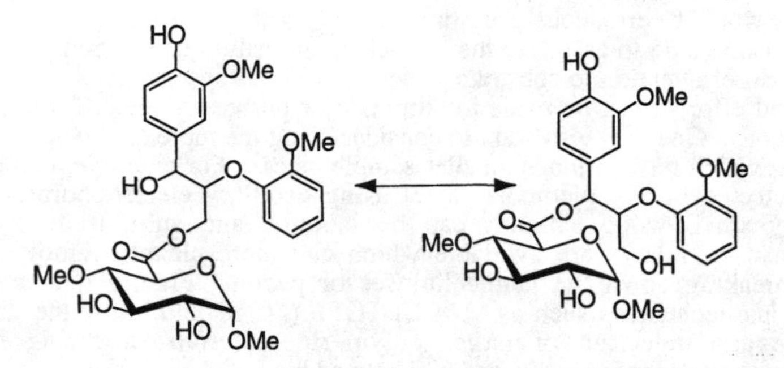

Figure 4. Uronosyl migration between the α- and γ-positions of a lignin model dimer.

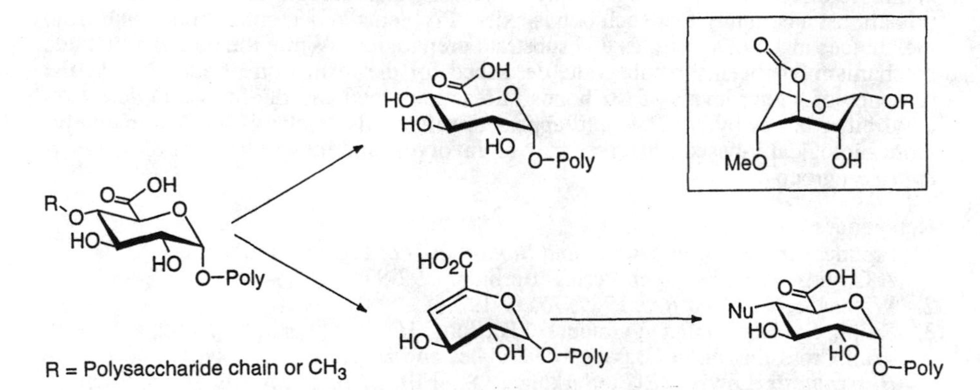

Figure 5. Pathways for modification of 4-*O*-substituted glucuronic acid. Removal of the 4-*O*-substitution can lead to a hexenuronic acid which can undergo nucleophilic substitution to provide a 4-linked structure.

Consideration must also be given to the chemistry of uronic acid groups. In an effort to generate a series of model compounds that depict potential LC bonds, several lignin-uronic acid esters were made (*26*). During their isolation and characterization, it was observed that a migration between the α- and γ-positions could occur in the presence of water, with the γ-position being thermodynamically favored. If such is the case, even if an α-ester was formed via a quinone methide, it may eventually migrate to the γ-position. Thus any method which is only looking at α-linkages may not detect it. Another novel reaction of uronic acids is their ability to form internal 6,3-lactones. Such methods which rely on titration may over-estimate LC bonds, as the lactone would be erroneously attributed to LC bonds.

What can we do to ascertain the regiochemical nature of LC bonds? First, one must pay closer attention to substrate selection. Woods species need to be carefully chosen, and efforts must be made to eliminate or purposely select for reaction and normal woods. One may also want to consider using the more sophisticated analysis techniques which rely on much smaller sample sizes. For example, carbohydrates can be detected on the picomolar level using capillary electrophoresis (*48,49*). Extremely small wood samples can be isolated and submitted to analysis. Recombinant cellulases are available which can more cleanly remove cellulose without breaking down the hemicelluloses or pectin. Finally the newer mass spectroscopic techniques such as MALDI-TOFF (*50*) should allow the desorption, ionization and detection of larger polymeric materials, allowing for mass spectroscopic evidence for lignin-carbohydrate adducts.

Summary

While no direct evidence has been presented for lignin-carbohydrate bonds in wood, it is almost a certainty that such bonds exist. Problems in detecting them come from the choices made in substrates and substrate preparation. While the quinone methide mechanism has been the sole route described for the formation of such bonds, the presence of higher levels of LC bonds in reaction wood and the failure to detect α-alkyl ether bonds by NMR would appear to refute this mechanism. Alternatively, more biologically-based processes may be involved, and may operate through the γ-hydroxyl group.

References

1. Freudenberg, K., *Constitution and Biosynthesis of Lignin*, Freudenberg K.; Neish, A.C., Eds. 1968, Springer-Verlag: Berlin. pp. 92-97.
2. Watanabe, T., *Wood Res.* **1989**, *76*, 59-123.
3. Ralph, J.; Hatfield, R.D.; Grabber, J.H.; Jung, H.G.; Quideau, S.; Helm, R.F., Cell Wall Cross-linking in Grasses by Ferulates and Diferulates, in *Lignin and Lignan Biosynthesis* Lewis, N.G.; Sarkanen, S., Eds. (American Chemical Society, Washington, DC, 1998) pp. 209-236.
4. Ralph, J.; Hatfield, R.D.; Quideau, S.; Helm, R.F.; Grabber, J.H.; Jung, H.-J. G., *J. Am. Chem. Soc.* **1994**, *116*, 9448.
5. Ralph, J.; Quideau, S.; Grabber, J.H.; Hatfield, R.D., *J. Chem. Soc., Perkin Trans. 1* **1994**, 3485.
6. Himmelsbach, D.; Hartley, R.D.; Borneman, W.S.; Poppe, L.; van Halbeek, H., *Magn. Reson. Chem.* **1994**, *32*,158.
7. Ishii, T., *Carbohydr. Res.* **1991**, *219*,15.
8. Ralph, J.; Young, R.A., *J. Wood Chem. Technol.* **1983**, *3*, 161.
9. Brunow, G.; Sipilä, J.; Mäkelä, T., *Holzforschung* **1989**, *43*, 55.
10. Sipilä, J.; Brunow, G., *Holzforschung* **1991**, *45*, 275.
11. Sipilä, J.; Brunow, G., *Holzforschung* **1991**, *45(Suppl.)*, 3.
12. Sipilä, J.; Brunow, G., *Holzforschung* **1991,** *45(Suppl.)*, 9.
13. Watanabe, T., Ohnishi, J.; Yamasaki, Y.; Kaizu, S.; Koshijima, T., *Agric. Biol. Chem.* **1989**, *53*, 2233.
14. Obst, J.R., TAPPI **1982**, *65(4)*, 109.

15. Kilpeläinen, I.; Ämmälahti, E.; Brunow, G.; Robert,D., *Tetrahedron Letters* **1994**, *35*, 9267.
16. Kilpeläinen, I.; Sipilä, J.; Brunow, G.; Lundquist, K.; Ede, R.M., *J. Agric. Food Chem.* **1994**, *42*, 2790.
17. Timell, T.E., *Adv. Carbohydr. Chem.* **1964** *19*, 247.
18. Jones, J.K.N., Purves, C.B. ; Timell, T.E., *Canad. J. Chem.*, **1961**, *39*, 1059.
19. Azuma, J., *Modern Methods of Plant Analysis*, Linskens, H.F.; Jackson, J.F., Eds. Vol. 10, Springer-Verlag: Berlin, 1989; pp. 100-129.
20. Minor, J., *J. Wood Chem. Technol.* **1982**, 2, 1.
21. Minor, J., *J. Wood Chem. Technol.* **1991**, *11*,159.
22. Watanabe, T.; Azuma, J.; Koshijima, T., *Mokuzai Gakkaishi* **1987**, *33*, 798.
23. Imamura, T.; Watanabe, T.; Kuwahara, M.; Koshijima, T., *Phytochemistry* **1994**, *37*, 1165.
24. Chang, H.-M.; Cowling, E.B.; Brown, W.; Adler, E.; Miksche, G., *Holzforschung* **1975**, *29*, 153.
25. Henrissat, B., *Cellulose* **1994**, *1*, 169.
26. Li, K.; Helm, R.F., *J. Agric. Food Chem.* **1995**, *43*, 2098.
27. Azuma, J.; Takahashi, N.; Koshijima. T., *Carbohydr. Res.* **1981**, *93*, 91.
28. Watanabe, T., Azuma, J.; Koshijima, T., *Mokuzai Gakkaishi* **1985**, *31*, 52.
29. Azuma, J.; Takahashi, N.; Isaka, M.; Koshijima, T., *Mokuzai Gakkaishi* **1985**, *31*, 587.
30. Takahashi, N.; Koshijima, T., *Wood Sci. Technol.* **1988**, *22*, 177.
31. Johnson, K.G.; Overend, R.P., *Holzforschung* **1992**, *46*, 31.
32. Ross, N.W.; Johnson, K.G.; Braun, C.; MacKenzie, C.R.; Schneider, H., *Enzyme Microb. Technol.* **1992**, *14*, 90.
33. Biely, P.; Puls, J.; Schneider, H., *FEBS Letters*, **1985**, *186*, 80.
34. Elegir, G.; Szakacs, G.; Jeffries, T.W., *Appl. Environ. Microb. 1994*, *60*, 2609.
35. Jeffries, T.W., *Biodegradation of lignin and hemicelluloses*, C. Ratledge, Ed. Kluwer Academic, Amsterdam, 1994, pp. 233-277.
36. Ishihara, M.; Shimizu, K., *Mokuzai Gakkaishi* **1988**, *34*, 58.
37. Puls, J.; Granzow, C., *Enzyme Microb. Technol.* **1987**, *9*, 83.
38. Fontana, J.D.; Gebara, M.; Blumel, M.; Schneider, H.; MacKenzie, C.R.; Johnson, K.G. *Methods in Enzymology* **1988**, *160*, 560.
39. Johnson, K.G.; Overend, R.P., *Holzforschung* **1991**, *45*, 469.
40. Tanahashi, M.; Karina, M.; Tamabuchi, K.; Higuchi, T., *Mokuzai Gakkaishi*, **1989**, *35*, 135.
41. Overend, R.P.; Chornet, E., *Phil. Trans. R. Soc. London, A* **1987**, *321*, 523.
42. Ebringerova, A.; Kosíkova, B.; Kacuráková, M. *Drevársky Vyskum* **1994**, *3*, 23.
43. Watanabe, T.; Karina, M.; Sudiyani, Y.; Koshijima, T.; Kuwahara, M., *Wood Res.* **1993**, *79*, 13.
44. Lundquist, K.; Simonson, R.; Tingsvik, K., *Svensk Papperstidn.* **1983**, *86*, R44-47.
45. Iversen, T., *Wood Sci. Technol.* **1985**, *19*, 243-251.
46. Eriksson, O.; Lindgren, B.O., *Svensk Papperstidn.* **1977**, *79*, 59-63.
47. Joseleau, J.-P.; Gancet, C., *Svensk Papperstidn.*, **1981**, *84*, R123-127.
48. Rydlund, A.; Dahlman, O., *J. Chromatogr. A* **1996**, *738*, 129.
49. Chen, F.-T. A.; Evangelista, R.A., *Anal. Biochem.* **1995**, *230*, 273-280.
50. Cancilla, M.T.; Penn, S.G.; Lebrilla, C.B., *Anal. Chem.* **1998**, *70*, 663-672.

Chapter 6

Structural Heterogeneity of Hardwood Lignin: Characteristics of End-Wise Lignin Fraction

Keko Hori and Gyosuke Meshitsuka

Laboratory of Wood Chemistry, Department of Biomaterial Sciences, Graduate School of Agricultural and Life Sciences, The University of Tokyo, 1-1-1 Yayoi, Bunkyo-ku, Tokyo 113–8657, Japan

The authors have investigated the structural heterogeneity of lignin, in particular, heterogeneous distribution of various types of linkages and aromatic structures. A β-0-4 rich endwise lignin fraction was found in earlier studies in the low molecular weight part of birch periodate lignin and a phenylcoumaran rich fraction in TMSil-depolymerized periodate lignin. In this study, a periodate lignin from diazomethane-methylated wood meal was partially depolymerized by stepwise TMSil treatment and the low molecular weight fractions obtained were characterized in comparison with those from milled wood lignin (MWL). It was found that the low molecular weight fractions, particularly acetone soluble fractions from periodate lignin, were rich in β-0-4 linkages and syringyl units. This was a confirmation of the endwise type lignin fraction in a polymer lignin.

Lignin structure contains various types of linkages and aromatic nuclei, and their heterogeneous distributions have been reported for different wood components, such as vessels, wood fibers, ray cells, and so on. As it has been demonstrated in DHP preparations, the bonding pattern of lignin is expected to be highly dependent on lignification conditions. Endwise type and bulk type lignins are understood to be formed by slow and rapid supply of monomers respectively to the lignification site (1). The highly condensed nature of compound middle lamella lignin should be attributed to the latter condition.

The heterogeneous distribution of different aromatic structures and linkage types was also examined by the isolation of cell wall fractions, such as middle lamella and secondary wall (2-4). Another interesting point is the structural heterogeneity in a lignin molecule. In our earlier study (5), a β-0-4 and syringyl rich fraction was found in a low molecular weight fraction of birch periodate lignin. This fraction had a molecular weight comparable to birch milled wood lignin, and more than 85% of the aromatic structures were estimated to be syringyl units. Although very small amount of β-β and β-1 type linkages were confirmed, β-5 (phenylcoumaran) type linkage were not found by 2D-NMR spectroscopy. This fraction was predominantly composed of β-0-4 linkages and was a typical endwise lignin fraction. Although obtained from the low molecular weight lignin fraction, this was the first experimental evidence of the endwise lignin fraction.

In order to discuss the structural heterogeneity in the high molecular weight fraction of lignin, a periodate lignin from birch wood was partially depolymerized by trimethylsilyl iodide (TMSiI), and the low molecular weight lignin fraction formed by the treatment was examined (6). It was interesting to note that a phenylcoumaran rich fraction was obtained at a very low yield by TMSiI treatment. Contribution of phenylcoumaran type structures in this fraction was estimated to be at least 45%. Since TMSiI was known to cleave α- and β-ether bonds under very mild condition (7-9), the above fraction must be liberated from the polymer lignin by ether bond cleavage.

In this study, in order to further investigate the structural heterogeneity of polymer lignin, the high molecular weight fraction of a periodate lignin prepared from diazomethane methylated birch wood meal was degraded gradually by repeated TMSiI treatments. The low molecular weight fractions thus formed were characterized to find the contributions of endwise type and bulk type lignin fractions in a polymer lignin. Periodate lignin first proposed by Purves (10) is an isolated lignin prepared almost quantitatively from wood meal without artificial condensation but with some oxidation at free phenolic units. Therefore, if the oxidative structural changes of lignin can be avoided by some pre-treatments, it will be a desirable lignin preparation. For this reason, a periodate lignin prepared from diazomethane methylated wood meal was used as a lignin sample in this study.

The same stepwise depolymerization procedure was applied also to a birch MWL to see its structural heterogeneity and thereby discuss its origin in the cell wall. MWL has long been used for lignin structural studies as a standard lignin preparation with negligible structural modification. It should be considered, however, that the yield of MWL is generally not high enough to be representative of all lignin in wood. In our earlier study, total yield of MWLs from birch wood meal by stepwise milling for 216 hr, was less than 30% of the lignin in the original wood meal (11). Although the yield may be increased under different milling conditions, it is not realistic to prepare MWL quantitatively. Another drawback of MWL is that its structural heterogeneity depends on yield. It has been previously reported (11) that the chemical structure of MWL extracted at the initial milling stage was quite different from the fractions extracted later. The former fraction was quite rich in guaiacyl units and was highly condensed. When subjected to nitrobenzene oxidation, the initial fraction has a syringaldehyde to vanillin ratio of 1.0 as opposed to a ratio of 2.59 for the total lignin. Therefore, it is very important to keep in mind that MWL originates from a special part of lignin in the wood cell wall. Based on separate studies on lignin from the differentiating xylem (12-17), the authors concluded that MWL from hardwood, at least the initial fraction, originated from the compound middle lamella.

Experimental

Preparation and Treatment of Periodate Lignin. The scheme for the preparation of periodate lignin fractions is shown in Figure 1. Pre-extracted birch wood meal was methylated with diazomethane for 1 week at room temperature until the yellow color of the diazomethane-diethyl ether solution became stable. This methylated wood meal was then used for the preparation of periodate lignin according to the previous study (5). The periodate treatment was repeated to improve the lignin purity, which was found to be 83.4% by the Klason lignin determination method.

The periodate lignin was pre-extracted with dioxane/water (9/1). The extracted solution was evaporated to dryness and acetylated with acetic anhydride/pyridine at room temperature for 1 day. The reaction mixture was poured into cold water and extracted with dichloromethane/acetone (2/1). The organic layer was washed with hydrochloric acid and water, dried over Na_2SO_4, and evaporated to dryness. Then the lignin fraction was completely extracted with methanol and then acetone to give PIL-M1 and PIL-A1.

The fraction of the periodate lignin which was unextracted with aqueous dioxane was acetylated with acetic anhydride/pyridine under the previous conditions. The reaction mixture was poured into cold water and acetylated lignin was collected as precipitate by centrifugation. The acetylated lignin was treated with TMSiI in dry chloroform at 0°C for 3hr. In this case, the amount of TMSiI required was about one twentieth of the full treatment of MWL in the earlier study (8). After this treatment, the reaction was stopped by adding pyridine and water according to the previous study (9). The reaction mixture was filtered and the precipitates washed with a small amount of chloroform. The filtrate and the washings were combined and washed with hydrochloric acid and water to remove HI and pyridine from the reaction mixture. This filtrate and the precipitates were combined, evaporated to dryness, and extracted to obtain the newly formed low molecular weight fractions of PIL-M2 and PIL-A2, respectively. These TMSiI treatment procedures were repeated three more times to obtain PIL-M3 to PIL-M5 and PIL-A3 to PIL-A5, respectively, as shown in Figure 1.

Preparation and Treatment of Milled Wood Lignin. The scheme for the preparation of MWL fractions is shown in Figure 2. MWL was extracted with 90% aqueous dioxane from birch wood meal ball-milled for 72hr and then purified under ordinary conditions. MWL was acetylated by the same procedure as for periodate lignin. Acetylated MWL was extracted with methanol to obtain ML-M1. The

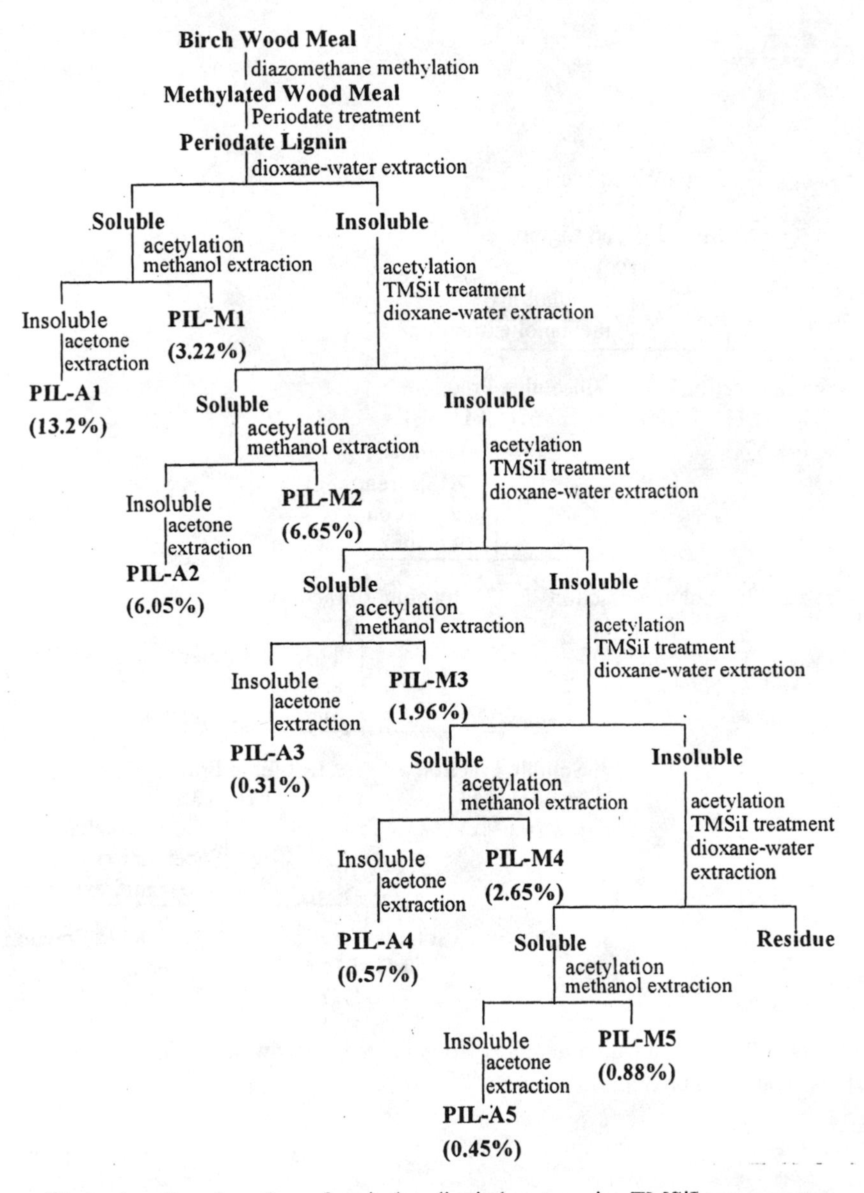

Figure 1. Fractionation of periodate lignin by stepwise TMSiI treatment.
* Yield: baed on original periodate lignin.

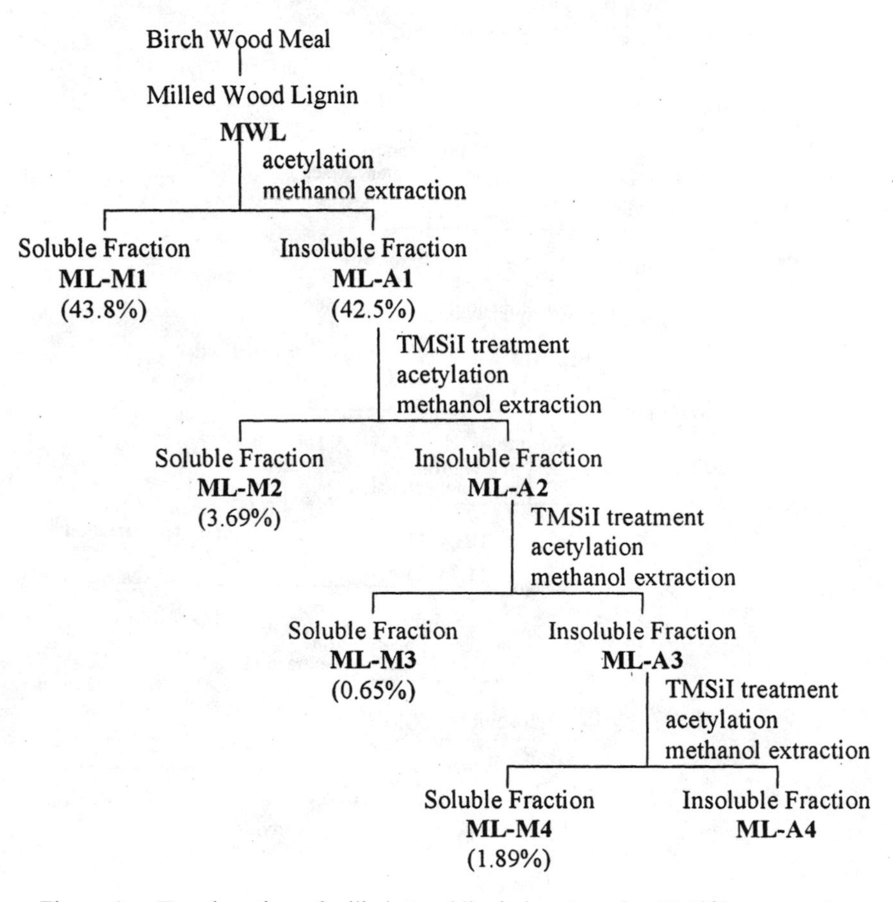

Figure 2. Fractionation of milled wood lignin by stepwise TMSiI treatment.
* Yield: baed on original MWL.

insoluble fraction (ML-A1) was treated with the same amount of TMSiI as for periodate lignin. The newly formed low molecular weight fraction was extracted with methanol after acetylation. These procedures of TMSiI treatment were repeated two more times as shown in Figure 2.

Chemical Analyses of Various Lignin Fractions. Acetylated periodate lignin fractions and acetylated MWL fractions after TMSiI treatment were analyzed by ^{1}H-NMR, GPC and alkaline nitrobenzene oxidation.

^{1}H-NMR spectra of 5mg acetylated lignin samples dissolved in 0.5ml $CDCl_3$ were recorded on a Bruker AC300 spectrometer and their chemical shifts were referenced to $CHCl_3$. The data matrix was 32k and 512 or 1024 scans were accumulated.

Molecular weight distribution of lignin samples were measured by HPLC system of Showa Denko Co. Ltd. and GPC columns of Shodex KF-801, 802, 803 for periodate lignin fractions or Shodex KF-801, 802 for MWL fractions. The eluent was THF and its flow rate was kept at 1ml/min.

Alkaline nitrobenzene oxidation of lignin samples was conducted in a 10ml volume stainless steel autoclave under the following conditions: sample weight 5-10mg, 2M NaOH 4ml, nitrobenzene 0.25ml, 170°C, 2hr. After the reaction, ethyl vanillin was added to the reaction mixture as the internal standard and the mixture was extracted with dichloromethane mainly to remove nitrobenzene and its reaction products. The reaction mixture was then acidified with hydrochloric acid and extracted with dichloromethane followed by diethyl ether. These solutions were combined and evaporated to dryness. Phenolic reaction products were determined by GLC as trimethylsilyl ethers. The conditions for GLC were as follows: columns Neutrabond-1 (0.25mm x 30m), columns temperature 150-280°C, 5°C/min, carrier gas He, injection temperature 280°C, detector FID. The results of alkaline nitrobenzene oxidation are shown as the average of two or three determinations.

Results and Discussion

Structural Characteristics of Periodate lignin Fractions. Periodate lignin used for the previous studies (5,6) was prepared from birch wood meal without protecting the free phenolic hydroxyl groups, so some structural changes must have occurred during its preparation. This might have caused the very low yield of periodate lignin itself and the lack of an endwise type lignin fraction in the low molecular weight fractions obtained after the partial cleavage of ether bonds by TMSiI treatment. Therefore, in this study, birch wood meal pre-methylated with diazomethane was used for the preparation of periodate lignin.

The periodate lignin fraction used for TMSiI treatment was the high molecular weight fraction, since the low molecular weight fraction originally present in the periodate lignin was removed by aqueous dioxane extraction.

Since the high molecular weight fraction of acetylated periodate lignin was sparingly soluble in the chloroform used as the solvent in the TMSiI treatment, dissolution was accomplished under heterogeneous conditions. As shown in Figure 1, yields of the low molecular weight fractions were 7.38 % as PIL-A series and 12.1 % as PIL-M series, for a total of 19.5 % based on original periodate lignin. The highest yields in these fractions were observed for A2 (5.5 %) and M2 (5.1 %), then gradually decreased to A5 (0.45 %) and M5 (0.88 %). Therefore, it was assumed that an increased yield of the low molecular weight fraction would not be obtained with additional TMSiI treatments. It appears that most of the lignin fraction which was the source of the low molecular weight fraction in the TMSiI treatment, was already liberated from the polymer lignin by this stage.

Structural characterization of each low molecular weight lignin fraction obtained by the stepwise TMSiI treatment was based on ^{1}H-NMR spectra and alkaline nitrobenzene oxidation. Syringyl unit to guaiacyl unit ratio (S/G ratio) was calculated based on the number of the aromatic protons in the corresponding, regions in the ^{1}H-NMR spectra. Those were 6.2ppm to 6.9ppm for syringyl protons and 6.6ppm to 7.1ppm for guaiacyl protons, and calculated by image analysis. Figure 3 shows the ^{1}H-NMR spectrum of PIL-A2. It is obvious that this lignin fraction is quite rich in syringyl unit and β-0-4 type structure (Hα: 5.9-6.1 ppm). It is difficult to confirm phenylcoumaran type structure (Hα: 5 18-5.74 ppm) in this spectrum. Although there must be some condensed guaiacyl units, their contribution in this syringyl and β-0-4 rich lignin fraction should be quite low. Therefore, for S/G ratio calculation, two protons in syringyl units and three protons in guaiacyl unit were assumed. S/G ratios are shown in Table I and Figure 4, together with S/V ratio obtained by nitrobenzene oxidation. It is obvious that these two ratios are generally high for the PIL-A series compared with the PIL-M series. PIL-A2 had remarkably high S/G and S/V ratios (7.45 and 5.94, respectively). These high values for PIL-A2 may indicate that about 90% of the structural units in this fraction are syringyl types. Fractions in the PIL-M series had S/G and S/V ratios similar to those of original PIL and residue (finally insoluble fraction). Therefore, the syringyl-rich parts in the original periodate lignin were preferentially liberated as low molecular weight fractions of the PIL-A series by the stepwise TMSiI treatment.

As far as other linkage types in these low molecular weight fractions, contribution of pinoresinol (β-β) type linkages (Hα: 4.7 ppm, Hβ: 3.1 ppm) was almost negligible. Although contribution of diaryl ether and biphenyl type linkages were not evaluated, the PL-A series is quite likely to be the endwise type lignin fraction. This was a confirmation of the endwise type lignin fractions in a polymer lignin.

Molecular weight distributions of these fractions evaluated by Gel Permeation Chromatography (GPC) are shown in Figures 5 and 6. The PIL-A series showed two peaks or shoulders at Mw around 18000 and 8000, judging, from polystyrene standards. Although this was not true for all of the fractions, PIL-A2 had a sharp peak at the latter position together with a shoulder at the former. PIL-A4 showed a

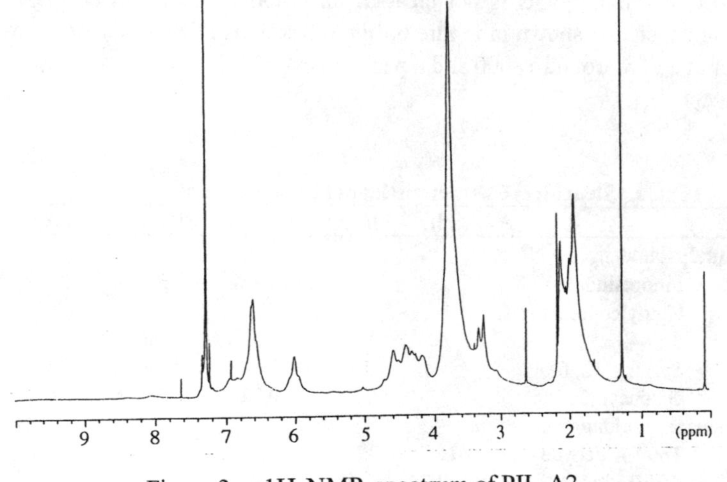

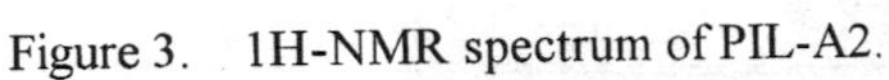

Figure 3. 1H-NMR spectrum of PIL-A2.

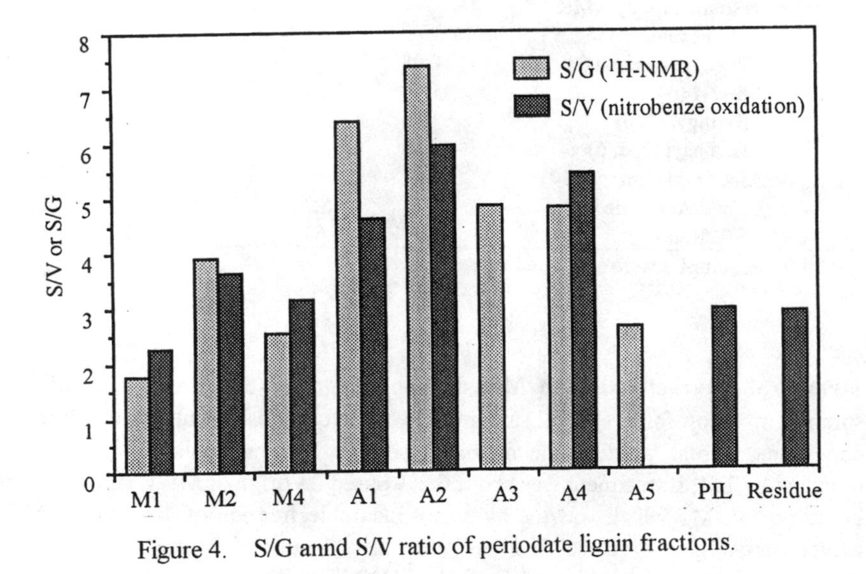

Figure 4. S/G annd S/V ratio of periodate lignin fractions.

very sharp peak at the former position and only a slight shoulder was observed at the latter position. Since the PIL-A series has the structural characteristics of endwise type lignin as mentioned above, the molecular weight shown here indicated the possible size of the endwise lignin fraction in polymer lignin. An endwise lignin fraction of Mw 18000 indicates that about one hundred units are mostly linked by β-O-4 bonds as shown in Figure 7. The reason these two peaks or shoulders appear in these GPC chromatograms is not clear at this moment and will be discussed in a following paper. As shown in Figure 6, lignin fractions in PIL-M series show a weak shoulder at a Mw around 18000 and a peak at Mw 5500.

Table I. Structural Characteristics of Periodate Lignin Fractions

	PIL	PIL-A1	PIL-A2	PIL-A3	PIL-A4	PIL-A5
Structural Abundance by NMR						
Pinoresinol (β)		+	+	±	±	±
Phenylcoumaran (α)		±	+	+	+	+
β-O-4 (α)		1.00	1.00	1.00	1.00	1.00
Syringyl (2, 6)		1.86	1.64	1.48	1.73	1.36
Guaiacyl (2, 5, 6)		0.29	0.22	0.33	0.36	0.52
Nitrobenzene Oxidation						
Total Aldehyde (%)	15.7	11.9				
S/V Ratio	2.92	4.62	5.94		5.43	
		PIL-M1	PIL-M2		PIL-M4	Residue
Structural Abundance by NMR						
Pinoresinol (β)		+	+		+	
Phenylcoumaran (α)		0.05	+		0.03	
β-O-4 (α)		1.00	1.00		1	
Syringyl (2, 6)		1.46	1.46		1.32	
Guaiacyl (2, 5, 6)		0.82	0.37		0.52	
Nitrobenzene Oxidation						
Total Aldehyde (%)		8.1	15.2			14.6
S/V Ratio		2.27	3.63		3.14	2.88

*All lignin samples were acetylated before analyses.

Structural Characteristics of Milled Wood Lignin. Since acetylated MWL is soluble in chloroform, TMSiI treatment was accomplished under homogeneous conditions. Total yield of the methanol soluble low molecular weight fractions obtained by TMSiI treatment was about 6.2% based on original MWL and about 15% based on ML-M1 which was the methanol insoluble fraction of the original MWL. Molecular weights of these fractions were always around Mw 3000 at the peak position except for a very low value of ML-M1 (Figure 8).

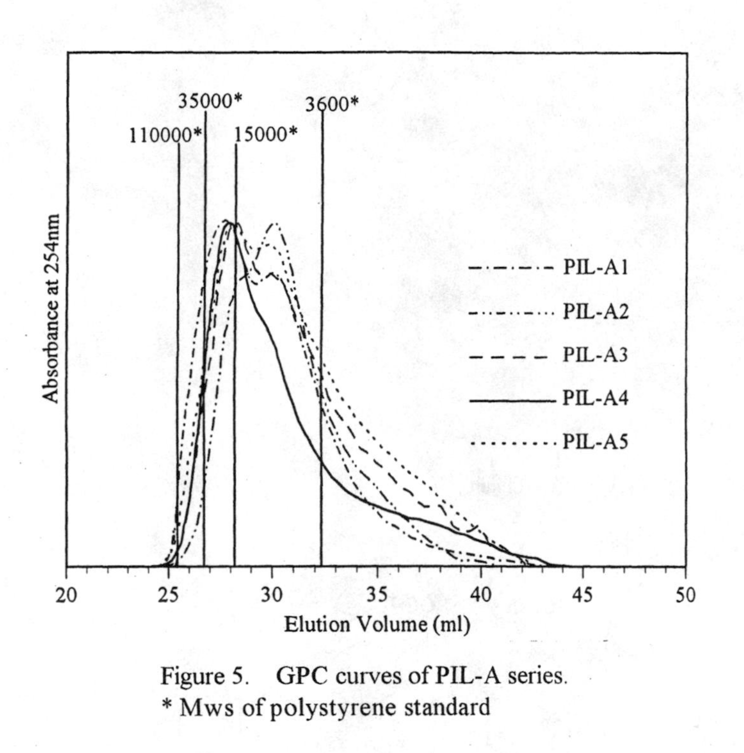

Figure 5. GPC curves of PIL-A series.
* Mws of polystyrene standard

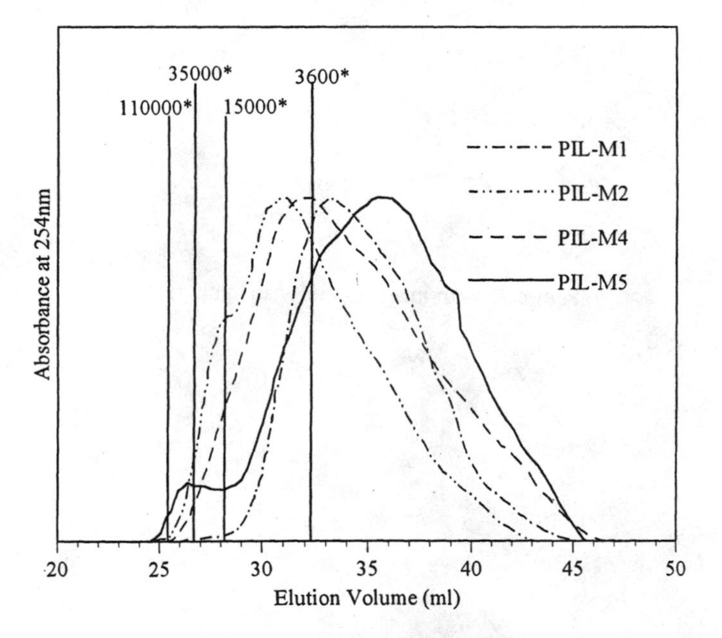

Figure 6. GPC curves of PIL-M series.

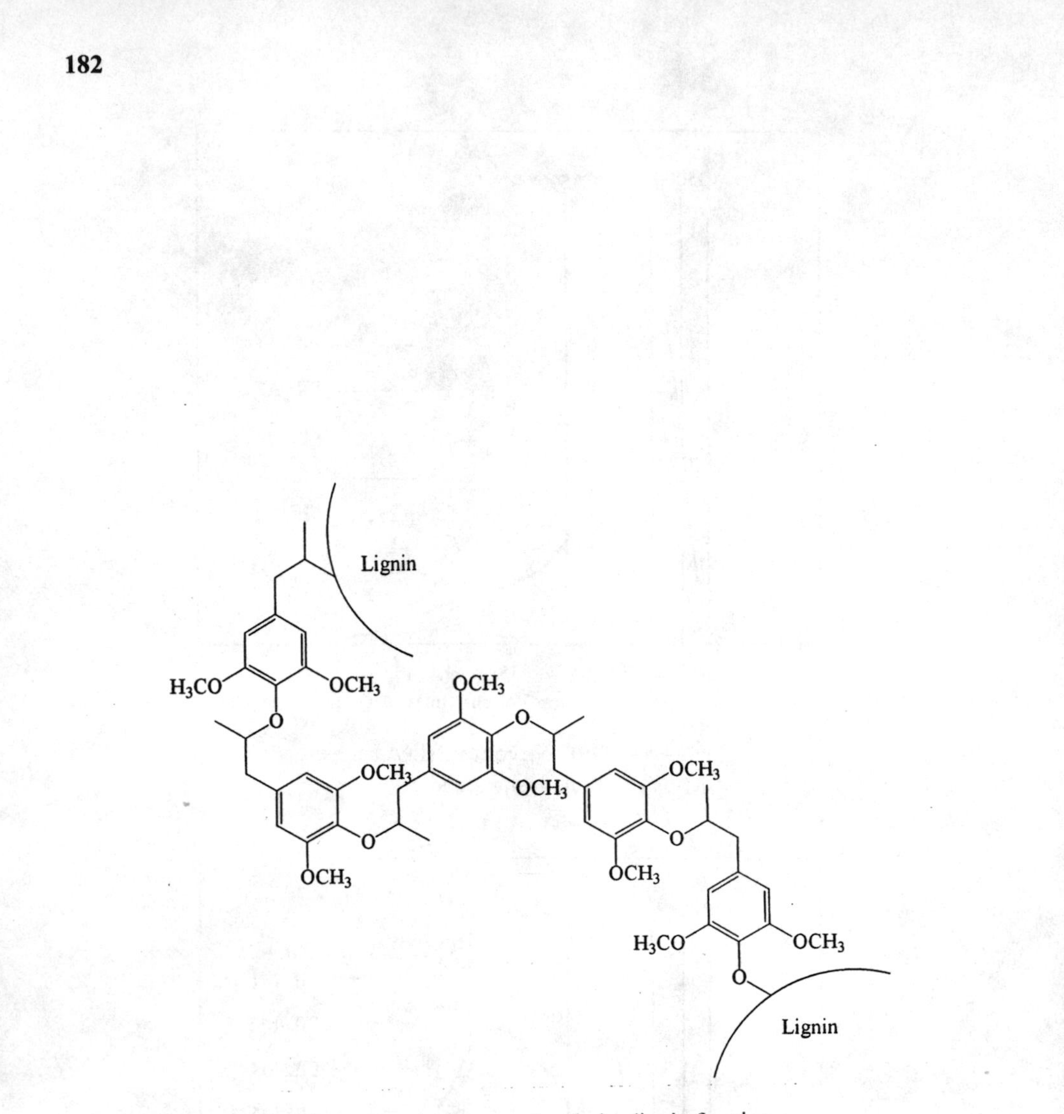

Figure 7. An image of endwise lignin fraction.

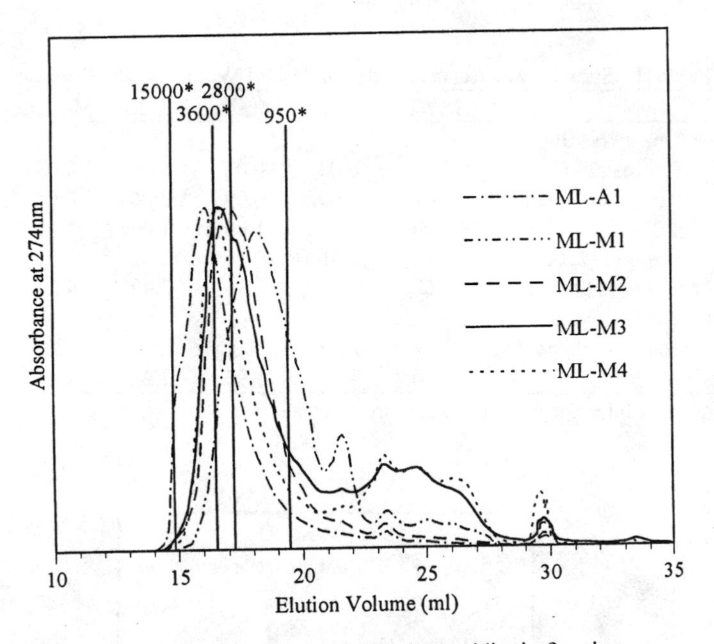

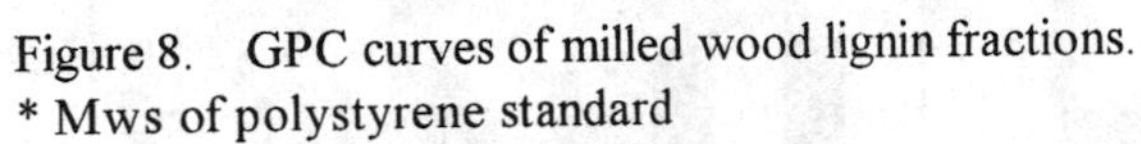
Figure 8. GPC curves of milled wood lignin fractions.
* Mws of polystyrene standard

Structural characteristics of methanol soluble fractions (ML-M2 to ML-M4) formed by partial ether bond cleavage of MWL using TMSiI are shown in Table II and Figure 9. It is important to note that any one of these fractions does not show obviously different structural characteristics from the original MWL itself, although there are some differences between samples. Signals attributed to the β-proton of the pinoresinol structure and the α-proton of phenylcoumaran structures were always found in ^{1}H-NMR spectra of these fractions. In MWL, it is difficult to find an endwise type lignin fraction like the case of PIL-A series. The S/V ratio by nitrobenzene oxidation did not show clear differences between the original MWL and MWL fractions, although total aldehyde yields were scattered for unknown reasons.

Table II. Structural Characteristics of Milled Wood Lignin Fractions

	MWL	ML-AI	ML-MI	ML-M2	ML-M3	ML-M4
Relative Area by NMR						
Pinoresinol (β)	0.01	0.03	0.02	0.04	0.03	0.03
Phenylcoumaran (α)	0.09	0.07	0.09	0.06	0.03	0.14
β-0-4 (α)	1.00	1.00	1.00	1.00	1.00	1.00
Syringyl (2, 6)	2.17	2.46	2.25	2.85	2.56	2.22
Guaiacyl (2, 5, 6)	2.67	3.37	4.06	2.42	4.24	2.76
Nitrobenzene Oxidation						
Total Aldehyde (%)	25.4	20.4	20.0	26.3	23.1	18.0
S/V Ratio	2.10	2.00	1.90	2.22	1.86	2.02

*All lignin samples were acetylated before analyses.

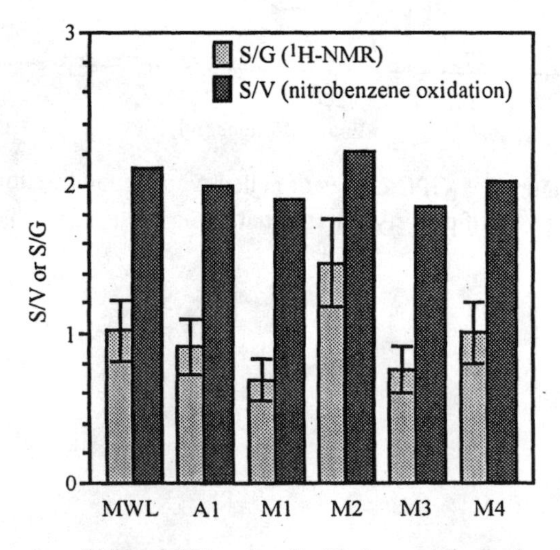

Figure 9. S/G and S/V ratio of milled wood lignin fractions.

In other words, it may be reasonable to say that MWL has a relatively homogeneous structure and does not contain a typical endwise type fraction. This may support our earlier conclusion that hardwood MWL basically originates from the compound middle lamella in wood.

Conclusions

1. Endwise lignin fractions were confirmed in the polymer fraction of birch wood periodate lignin by stepwise cleavage of α- and β- ether bonds by TMSiI treatments.
2. Possible size of an endwise lignin fraction was estimated to be one hundred lignin structural units. β-0-4 bonds in this fraction link most of the units.
3. Birch milled wood lignin has a homogeneous structure and does not contain an endwise lignin fraction.

Reference

1. Sarkanen, K. V. In *Lignins:* Sarkanen, K. V., Ludwig, C. H.; Eds.; Wiley-Interscience, New York, **1971**, pp 95
2. Iwamida, T.; Sumi, Y.; Nakano, J. *J. Japan Tappi* **1975**, *29*, 324.
3. Cho, N.-S.; Lee, J.-Y.; Meshitsuka, G.; Nakano, J. *Mokuzai Gakkaishi* **1980**, *26*, 527.
4. Hardell, H.-L.; Leary, G. J.; Stoll, M.; Westermark, U. *Svensk Papperstidn.* **1980**, 83, 71.
5. Fukagawa, N.; Meshitsuka, G.; Ishizu, A. *J. Wood Chem. Technol.* **1992**, *12*, 91.
6. Nakahata, A.; Kim, Y.-S.; Meshitsuka, G. *Proceedings of 8th ISWPC, Helsinki* **1995**,*vol.1*, 123.
7. Meshitsuka, G.; Kondo, T.; Nakano, J. *J. Wood Chem. Technol.* **1987**, *7*, 161.
8. Makino, S.; Meshitsuka, G.; Ishizu, A. *Mokuzai Gakkaishi* **1990**, *36*, 460.
9. Fujino, K.; Meshitsuka, G.; Ishizu, A. *Mokuzai Gakkaishi* **1992**, *38*, 956.
10. Wald, W. J.; Ritchie, P. F.; Purves, C. B. *J. Am. Chem. Soc.* **1947**, *69*, 1371.
11. Lee, Z.-Z.; Meshitsuka, G.; Cho, N.-S.; Nakano, J. *Mokuzai Gakkaishi* **1981**, *27*, 671.
12. Meshitsuka, G.; Nakano, J. *J. Wood Chem. Technol.* **1985**, *5*, 391.
13. Eom T.-J.; Meshitsuka, G.; Nakano, J. *Mokuzai Gakkaishi* **1987**, *33*, 576.
14. Eom T.-J.; Meshitsuka, G.; Ishizu, A.; Nakano, J. *Mokuzai Gakkaishi* **1987**, *33*, 716.
15. Eom T.-J.; Meshitsuka, G.; Ishizu. A.; Nakano. J. *Cellulose Chem. Technol.* **1988**, *22*, 211.
16. Eom T.-J.; Meshitsuka, G.; Ishizu, A. *Mokuzai Gakkaishi* **1988**, *35*, 820.
17. Kim, Y.-S.; Meshitsuka, G.; Ishizu, A. *Mokuzai Gakkaishi* **1994**, *40*, 407.

Chapter 7

Molecular Orbital Calculations on the Interaction of Veratryl Alcohol with the Lignin Peroxidase Active Site

Thomas Elder[1] and David C. Young[2]

[1]School of Forestry, Auburn University, Auburn, AL 36849
[2]Alabama Research and Education Network, Nichols Research Corporation, Auburn, AL 36849

Ab initio calculations for the lignin peroxidase active site and veratryl alcohol have been performed at the STO-3G level using the unrestricted Hartree-Fock approximation for open-shell species. These results are interpreted in terms of edgewise interactions between the catalyst and substrate.

During the recent past, considerable interest has been focused on the white rot fungus *Phanerochaete chrysosporium* and the lignin peroxidase enzyme (E.C.1.11.1.) thereof, as a catalyst for lignin oxidation. Such reactions have various potential industrial applications including biobleaching of pulp and the removal of chlorinated organics from waste waters *(1)*.

To date, three x-ray diffraction solutions of the lignin peroxidase structure have been deposited with the Brookhaven Protein Data Bank, with accession numbers pdb1lga *(2)* (Figure 1a), pdb1llp and pdb1qpa. The active sites of all three structures are quite similar, with the only consistent geometric differences occurring in the vinyl pendant groups.

Lignin peroxidase, with an iron protoporphyrin IX prosthetic group coordinated to a histidine (Figure 1b), initially reacts with hydrogen peroxide to produce compound I *(3)*. Due to the isolated nature of the lignin peroxidase active site and the size of the lignin macromolecule, veratryl alcohol, a secondary metabolite of *Phanerochaete chrysosporium* has been proposed as a mediator that undergoes an initial enzymatic oxidation and subsequently oxidizes the lignin polymer *(4)*. Compound I, a cation radical with three unpaired electrons, is proposed to oxidize veratryl alcohol by a single electron transfer mechanism resulting in the veratryl alcohol cation radical and Compound II, which is neutral with two unpaired electrons, as summarized in Figure 2a. Furthermore, the geometry of interaction between the

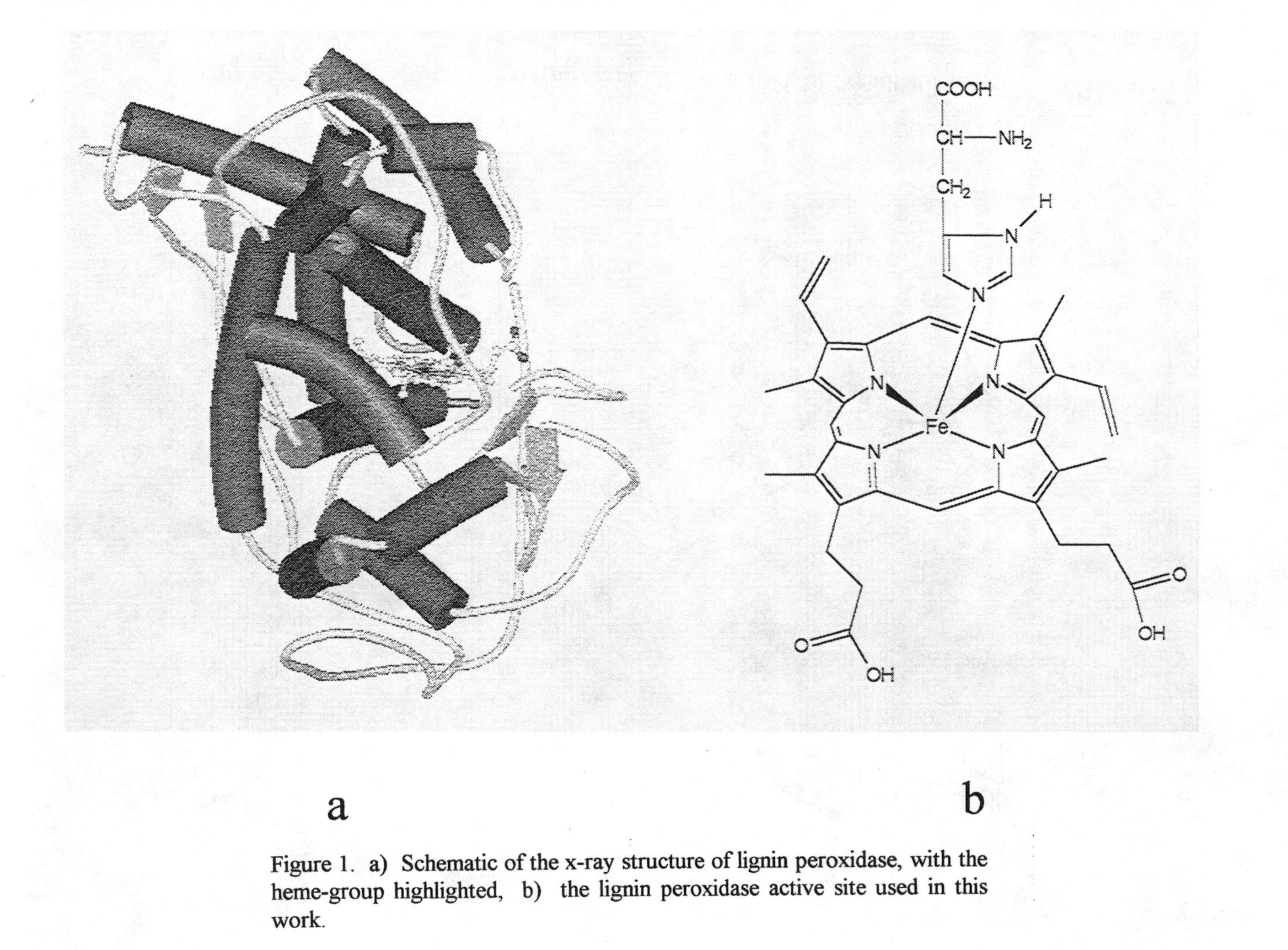

Figure 1. a) Schematic of the x-ray structure of lignin peroxidase, with the heme-group highlighted, b) the lignin peroxidase active site used in this work.

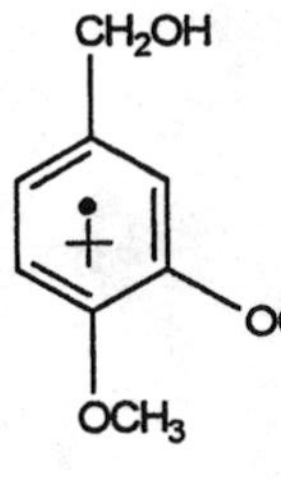

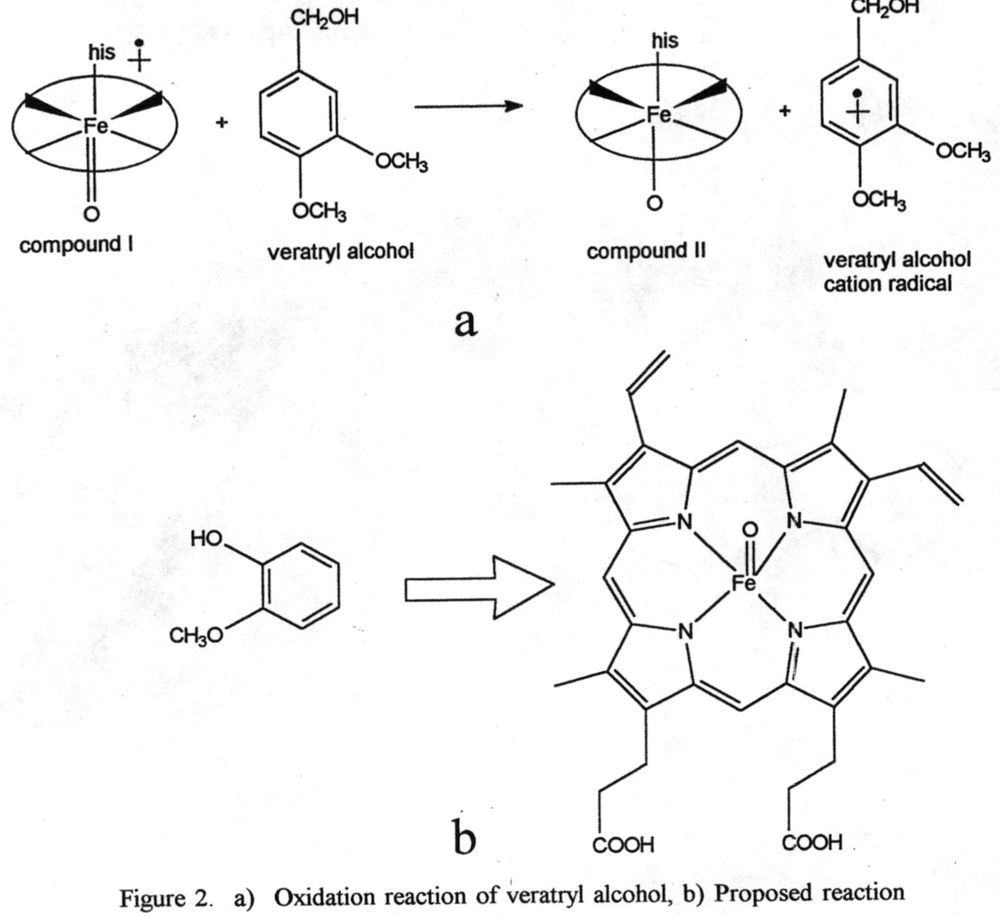

Figure 2. a) Oxidation reaction of veratryl alcohol, b) Proposed reaction geometry from reference *(5)*.

substrate and active site has been discussed in studies of cytochrome C peroxidase and guaiacol *(5)*, wherein a co-planar, edgewise approach has been proposed (Figure 2b).

Due to the general biological significance of the metalloporphyrins and the enzyme systems in which they occur, including lignin peroxidase, these compounds have been extensively studied using the methods of theoretical chemistry. This has been reviewed for the iron porphyrins by Loew *(6)*, with particular attention to the varying oxidation states that may be exhibited, the calculation of electronic absorption spectra, and dynamic aspects of binding to the iron complexes. The heme group has also found application as a model system for the active site of peroxidase enzymes *(7, 8)* with complementary work reported for a number of substrate molecules *(9,10)*. Work on the active site of peroxidase enzymes has involved cytochrome P450 *(11-13)* and, notably, the electronic absorption spectra, spin densities, and proton transfer reactions of cytochrome *c* peroxidase *(14-16)*. The calculations have been largely executed using a version of INDO (intermediate neglect of differential overlap), a semi-empirical molecular orbital method, that has been specifically parameterized for transition metals and the prediction of spectral properties *(17)*.

Recently, results from density functional theory (DFT) calculations on a peroxidase model have been reported by Kuramochi et al. *(18)*. This work characterized the electronic structure of compounds I and II, using a triple-ζ basis set, and calculated Mössbauer spectrum parameters that agreed well with experiment. Related research has applied restricted Hartree-Fock and complete active space SCF calculations to iron porphyrin-pyridine complexes *(19, 20)*.

Using the crystallographic structure for the lignin peroxidase active site, the current work reports on unrestricted Hartree-Fock (UHF) calculations for the open-shell compounds I and II, and the veratryl alcohol cation radical at ab initio levels of approximation. The UHF method constructs orbitals describing the α and β electrons of free radicals separately and is a more general solution to the electron density problem than the restricted calculations *(21)*. The UHF calculations are more rapid to perform than the restricted methods, but may suffer from the problem of spin contamination (i.e. spin states other than the one specified may be mixed into the solution, possibly leading to energies that are too low and unrealistic spin densities *(22)*).

Methods

Given the structural similarities among the three available lignin peroxidase structures, the structure of Poulos et al. *(2)* was selected as basis for the current work. The heme and coordinated histidine (His-176), were extracted for subsequent calculations.

As a first step, an oxygen was added to the iron center of the active site and the whole structure was relaxed using the Universal Force Field *(23)*. This was done to relieve any crystal packing forces that might have been distorting the structure, and to provide a reasonable Fe-O distance. The optimized structure was found to be not dissimilar to the original, maintaining slight puckering and non-planarity within the porphyrin system. Next single-point UHF/STO-3G calculations were performed, for compound I (charge=+1, multiplicity=4) and compound II (charge=0, multiplicity=3). In order to verify the identification of a reasonable electronic ground state, and insure an open-shell structure with minimal spin contamination, the wave functions were varied by exchanging selected molecular orbitals representing the α-electrons. STO-3G calculations were also done for the neutral, closed-shell veratryl alcohol (Figure 1b) and the cation radical, using HF and UHF methods respectively.

For examinations of the interaction between veratryl alcohol and the lignin peroxidase active site, the reaction geometry proposed by DePillis *(5)* was approximated, and distances between the groups were set at about 5 and 10 Å. Charge was maintained at +1 and the multiplicity was kept at 4. Single-point UHF/STO-3G calculations were performed on these complexes.

All molecular orbital calculations were performed using Gaussian 94, Revision E2 *(24)* on a Cray-C90, maintained in Huntsville, Alabama by the Alabama Supercomputer Authority. Unichem, licensed from Oxford Molecular, running on a Silicon-Graphics Indigo2 workstation in the Department of Chemistry at Auburn University, was used as a graphical interface.

Results

Compound I. The properties of the various states of compound I, identified by exchanging the specified molecular orbitals are as indicated in Table 1. While the exchange of the HOMO and LUMO (orbitals 207 and 208) gave the lowest energy, not unexpectedly, an increase in the spin contamination was also found. Furthermore, exchanging 206 with 208 resulted in unrealistic values for both charge and spin density. (The theoretical spin density for this compound is 3.75).

Table I. Properties of Compound I

	E(UHF) (hartrees)	$\langle S^2 \rangle$	Charge Fe	Charge O	Spin density Fe	Spin density O
original state	-3,661.97327	4.76	1.25	-0.50	4.70	-0.85
207 exchanged with 208	-3,662.04833	19.61	1.25	-0.50	4.73	-0.87
207 exchanged with 209	-3,661.91515	5.03	1.19	-0.47	0.91	1.05
206 exchanged with 208	-3,661.95976	17.44	1.27	-0.03	-0.93	1.77

Both the initial calculation and the 207-209 exchange resulted in lower spin contamination, with the latter state having a slightly higher energy. Based on Mulliken charges (Table 1, Figure 3a,b), these two states are quite similar. In both cases, the iron possesses a large partial positive charge, while the coordinated oxygen has a large partial negative charge. These values are in qualitative agreement with the literature. Given these results, however, it must be borne in mind that charge calculations are extremely dependent on the basis set that is used *(25)*. The current values are presented only for internal comparison. Furthermore, it has been documented that the use of more sophisticated basis sets does not lead to converging Mulliken charges *(25)*, such that the use of higher levels of theory offers no readily apparent advantage over the minimal basis set used in the current work.

It can be seen, however, that there are striking differences between these two states with respect to the spin density (Table 1, Figure 4a,b). Restricted open-shell calculations on related model compounds resulted in values of 1.10-1.91 for iron and 0.06-0.89 for the oxygen, with most of the values toward the upper end of this range *(19,20)*. Unrestricted density functional theory calculations gave values of 1.15-1.24 and 0.98-0.99, for the iron and oxygen, respectively. Based on these results, it would appear that the state arising from the exchange of orbitals 207 and 209 is more consistent with the literature, indicating an appreciable level of spin density on the oxygen. Furthermore, as can be seen from Figure 4b, this state also has substantial spin density on the meso-carbons, possibly supporting the edgewise reaction geometry proposed in the literature *(2,5)*. Preliminary restricted open-shell calculations on the lignin peroxidase active site have resulted in spin densities that are in accord with the literature, and will be reported elsewhere.

Concomitant with the range of spin densities reported for iron porphyrin compounds is a controversy over the nature of the ground state of compound I, as manifested by orbital symmetry. The two states, designated a_{1u} and a_{2u}, in the porphyrin literature *(6)*, have been found to be energetically similar (a difference of 0.15eV has been reported from density functional calculations), and sensitive to substituents on the porphyrin ring, and the nature of the axial ligand *(6)*. The orbital plots from the current calculations (Figure 4c,d) are consistent with the previously reported symmetries, such that the original, lower energy, structure is a_{2u}, while the state identified by 207-209 exchange, is a_{1u}.

Compound II. The reduction of compound I, by the abstraction of an electron from veratryl alcohol results in compound II, with an energy of –3661.96781 hartree. An electronic state was identified with a spin contamination value of 2.13, which compares favorably with the theoretical value of 2.0. Perhaps owing to the low $\langle S^2 \rangle$ value, the spin densities for the iron (0.97) and oxygen (1.01) compare favorably with the recent literature (Yamaguchi et al. 1997). Graphically, it can be seen that the spin density, while concentrated on the Fe=O bond, is also present on the porphyrin ring (Figure 5b), albeit to a smaller extent than in Figure 3b. This might be interpreted in terms of the lower number of unpaired electrons in compound II. As also might be expected, the partial positive charges on the meso-carbons are smaller upon reduction (Figure 5a). A perturbation in the HOMO can also be seen (Figure 5c).

Veratryl alcohol. Semi-empirical results have been reported for the electronic structure of both veratryl alcohol and its cation radical *(26)*. In the interest of consistency and in order to allow direct comparisons with the porphyrins, however, STO-3G ab initio calculations have been completed and will be reported here. Results for Mulliken charge and orbital plots are shown in Figure 6a,b and c. As was the case for the proposed reaction channel in compound I, veratryl alcohol generally exhibits small partial negative charges, with the exceptions being positions 3 and 4 with methoxyl substituents. Although the partial positive charges involved are small, invoking coulombic forces to account for the interactions of the reactants might be questionable. It can be seen, however, that veratryl alcohol has a large, well

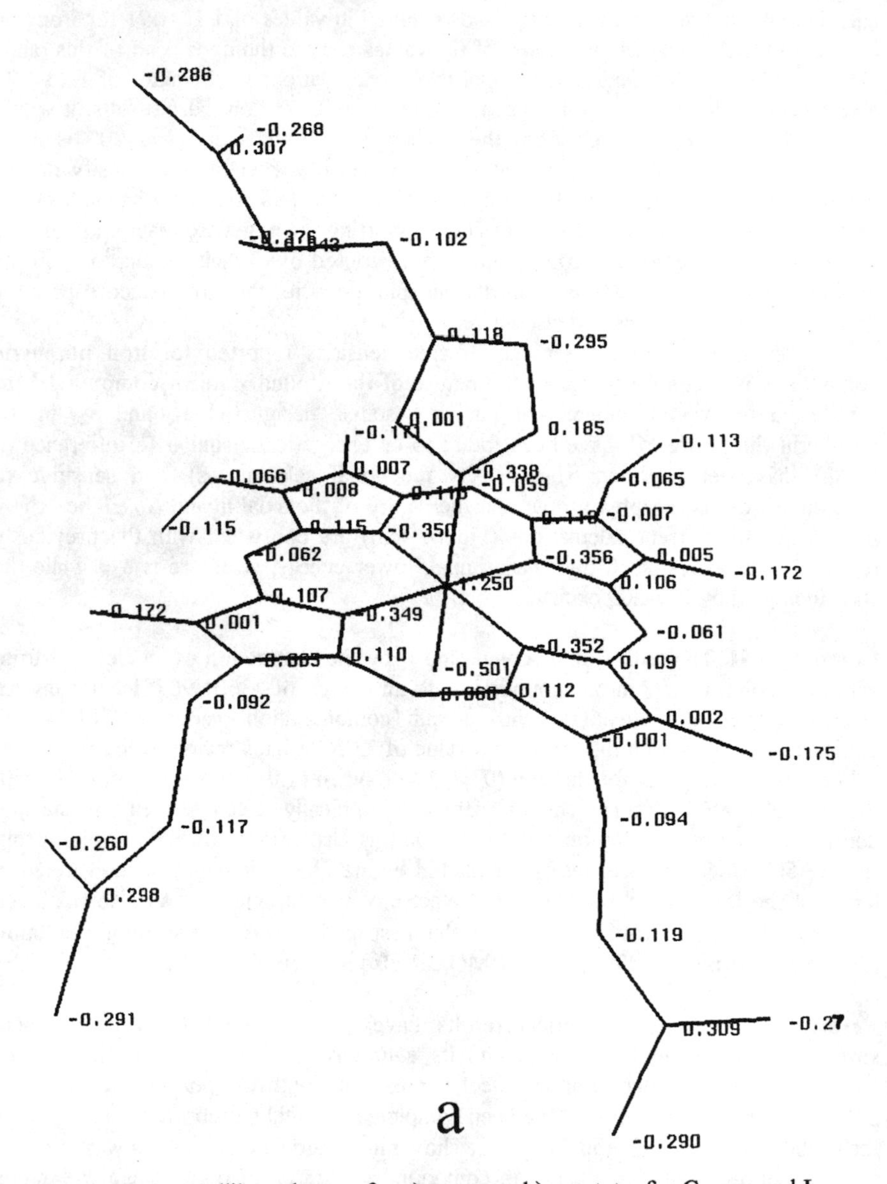

Figure 3. Mulliken charges for a) a_{2u} state, b) a_{1u} state, for Compound I.

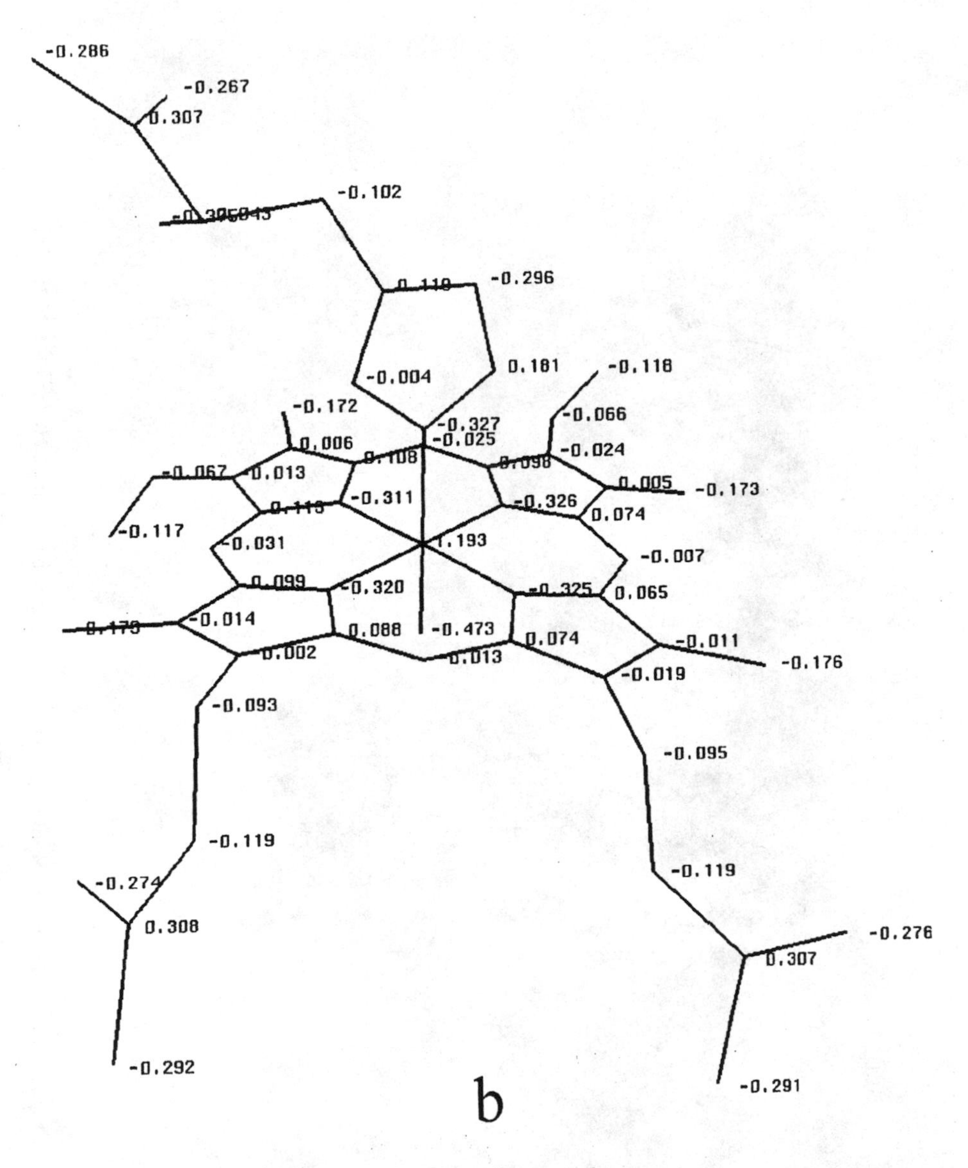

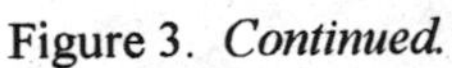

Figure 3. *Continued.*

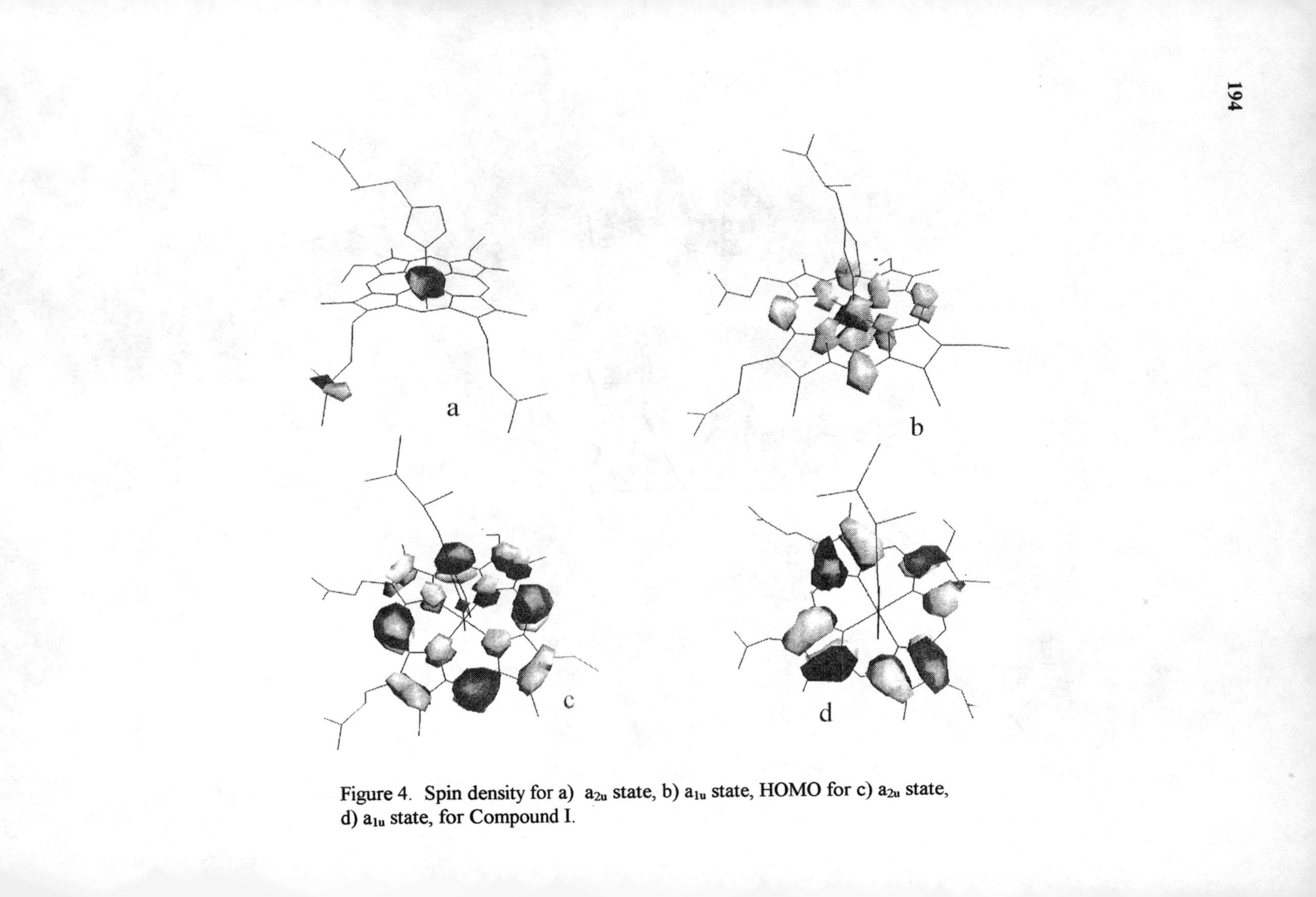

Figure 4. Spin density for a) a_{2u} state, b) a_{1u} state, HOMO for c) a_{2u} state, d) a_{1u} state, for Compound I.

distributed HOMO, that may be capable of overlap with the porphyrin. The ab initio energy was found to be –565.11094 hartree

Veratryl alcohol cation radical. The Mulliken charges found for the veratryl alcohol cation radical are similar to the ground state (Figure 6d), and as has been reported in the previous semi-empirical work, the SOMO (singly occupied molecular orbital) (Figure 6e) of the cation radical is similar to the HOMO of the ground state. The spin density of the cation radical is as might be expected from resonance structures, with the bulk of the unpaired density present on C-1, C-3, C-5 and the C-4 oxygen (Figure 6f), and a UHF energy of –564.98216746 hartree.

Interactions between lignin peroxidase active site and veratryl alcohol. The structures used to study the interaction of veratryl alcohol and the heme active site are as indicated in Figures 7 and 8. Using the a_{2u} state for compound I, the oxidation reaction is found to be exothermic at a level of 0.13424 hartrees (84.234 kcal/mole). In comparing the charge differences as the veratryl alcohol approaches the heme, there appear to be few large changes among the positions that are getting closer to each other. It is interesting to note, however, that the coordinated oxygen has gone from having a large partial negative charge to a slight partial positive charge as the separation decreases (Figure 7a, 8a). There are obvious changes occurring as a function of distance in both the spin density (Figures 7b and 8b) and spatial distribution of the HOMO (Figures 7c and 8c). The increase in unpaired density on the veratryl alcohol at the smaller separation may be indicative of the electron loss that occurs in the oxidation reaction. Also, at the closest approach the HOMO plot shows a distortion towards the veratryl alcohol.

Conclusions

In the current work, the lowest energy state for the lignin peroxidase active site is found to exhibit the a_{2u} symmetry that has been previously identified and reported in the literature for related compounds. Furthermore, when the catalytic site and veratryl alcohol substrate are in close proximity, an increase in spin density can be observed within the veratryl alcohol, indicative of the single electron transfer reaction.

From a methodological standpoint, the results for the lignin peroxidase active site, obtained by the use of the UHF approximation, are the basis for a comparison with other theoretical methods. It is apparent from this work that the transition metals, with partially filled d-orbitals and range of possible oxidation states, are intrinsically difficult to address using theoretical methods. The identification and verification that a specified electronic state has been found for an organometallic complex is a non-trivial undertaking. Furthermore, for the compounds in question in this paper, the literature is replete with results that vary with computational methods. While this is initially disturbing, it should be no more surprising than the well-accepted differences in molecular weights of polymers that are found by different analytical methods. Ongoing work will be concerned with a systematic comparison of theoretical methods and structure-activity relationships for varying catalysts.

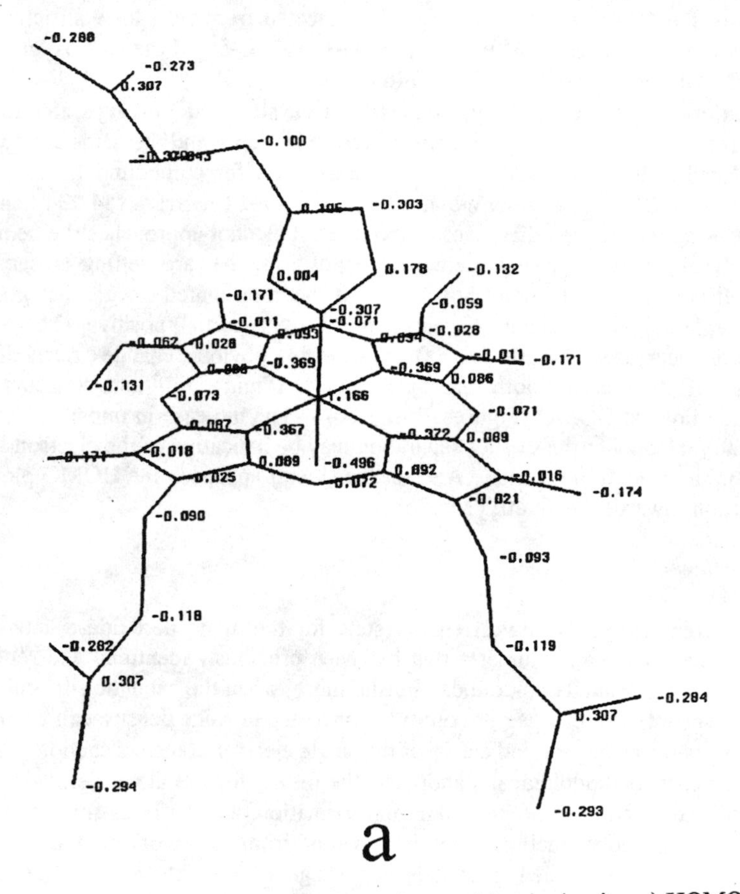

Figure 5. Compound II a) Mulliken charges, b) spin density, c) HOMO.

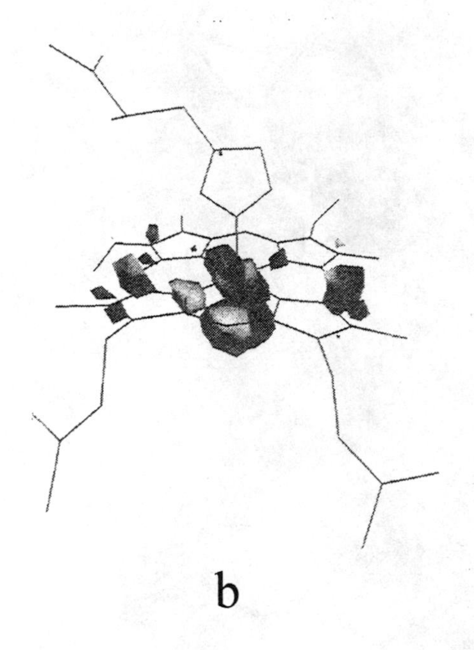

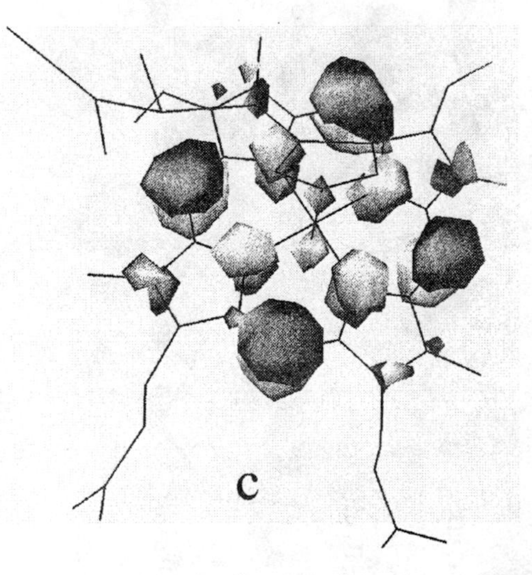

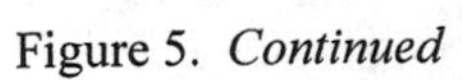

Figure 5. *Continued*

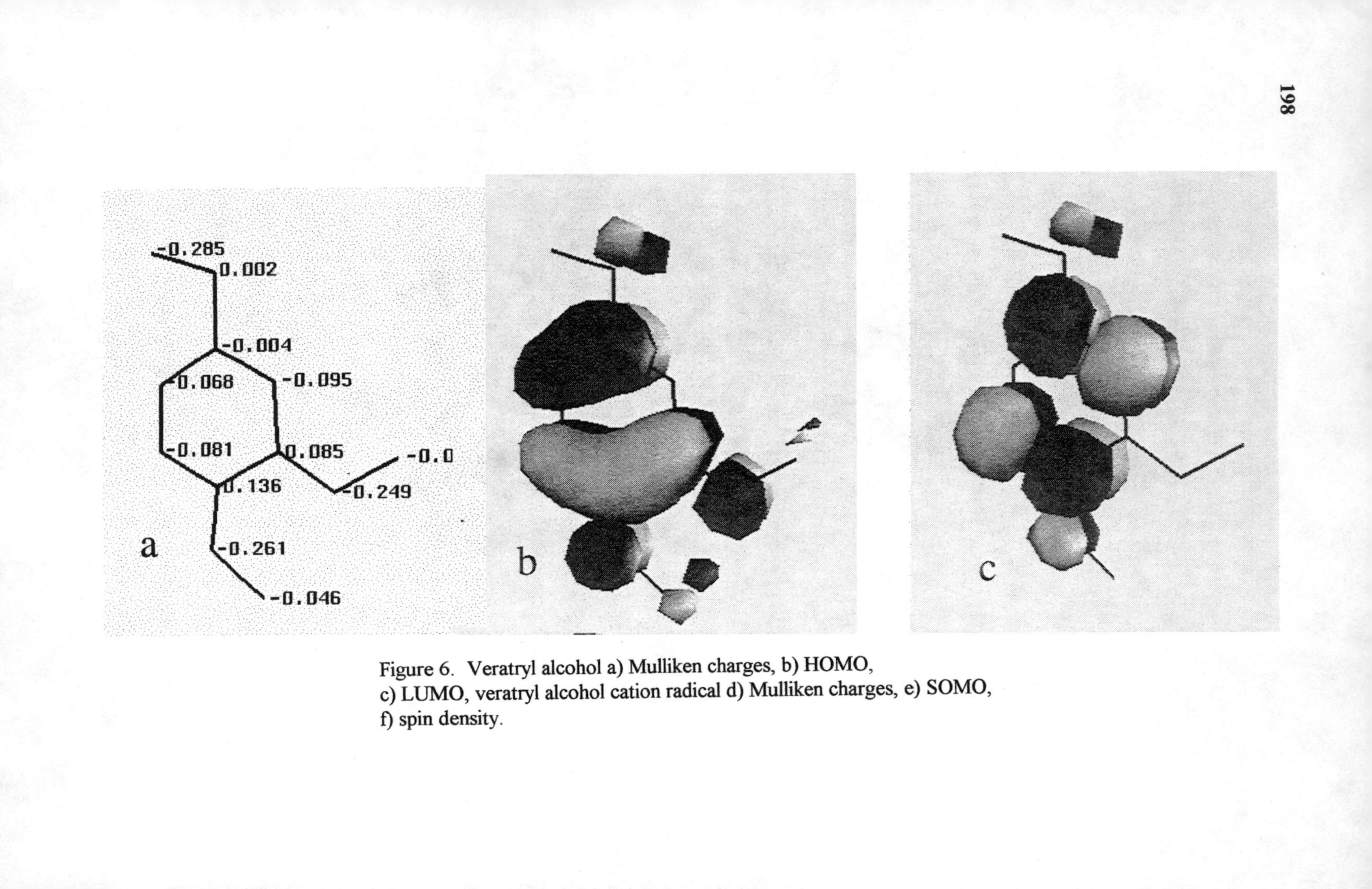

Figure 6. Veratryl alcohol a) Mulliken charges, b) HOMO, c) LUMO, veratryl alcohol cation radical d) Mulliken charges, e) SOMO, f) spin density.

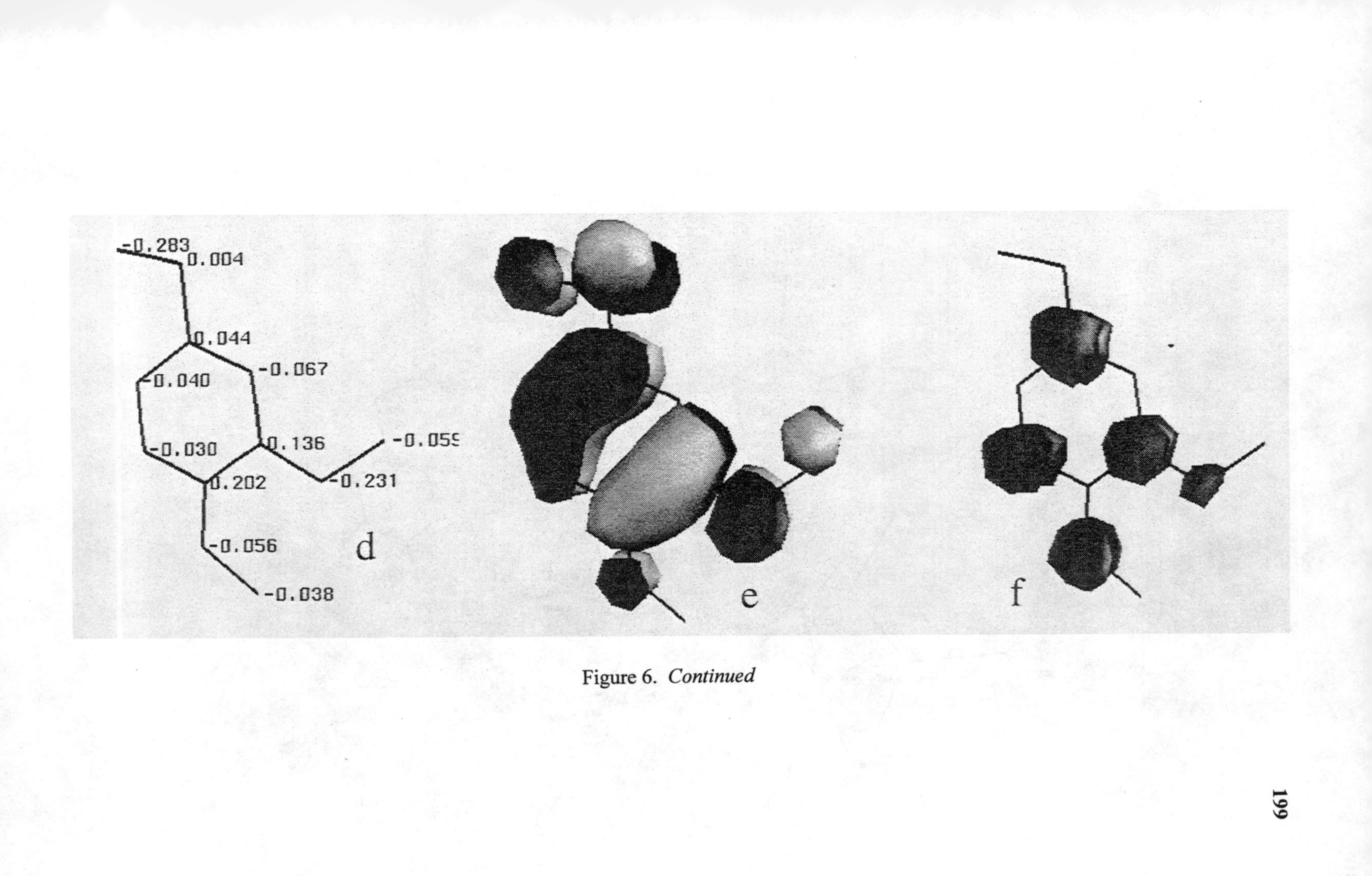

Figure 6. *Continued*

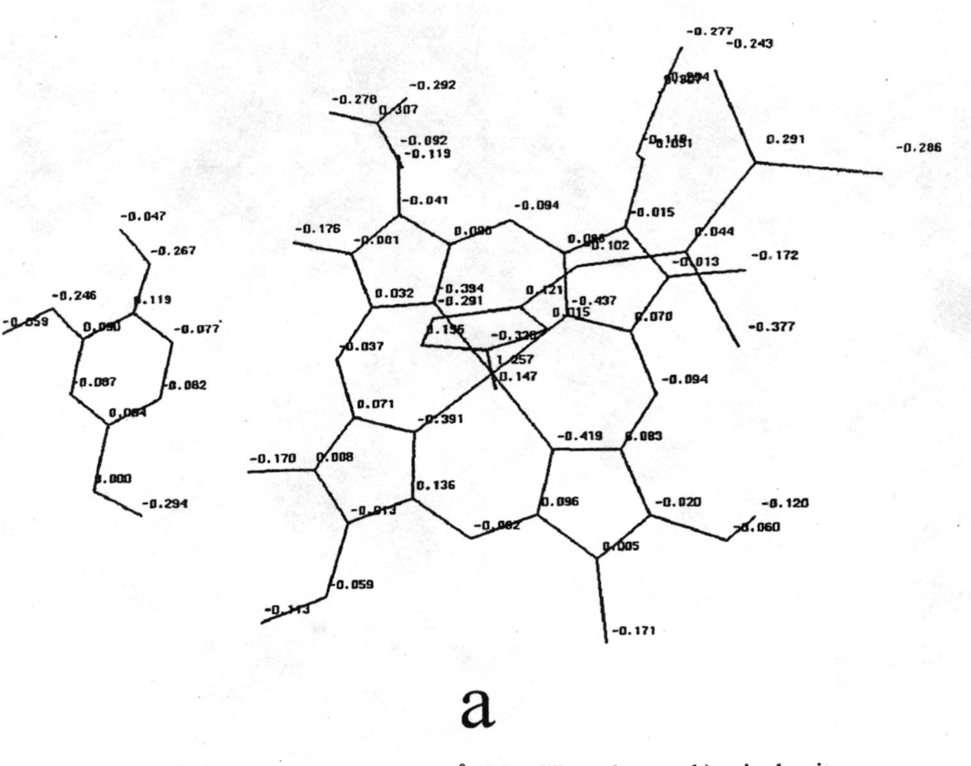

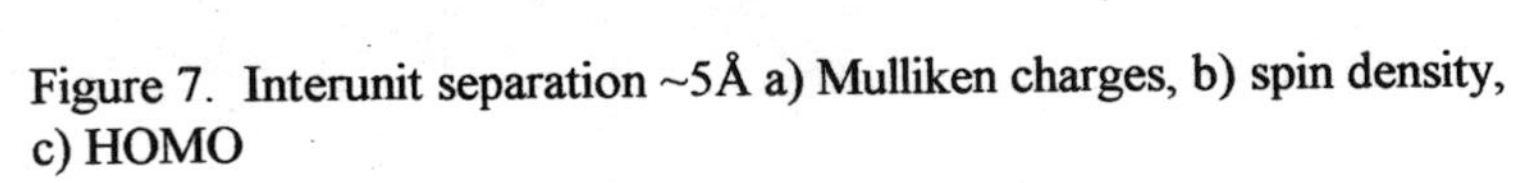

Figure 7. Interunit separation ~5Å a) Mulliken charges, b) spin density, c) HOMO

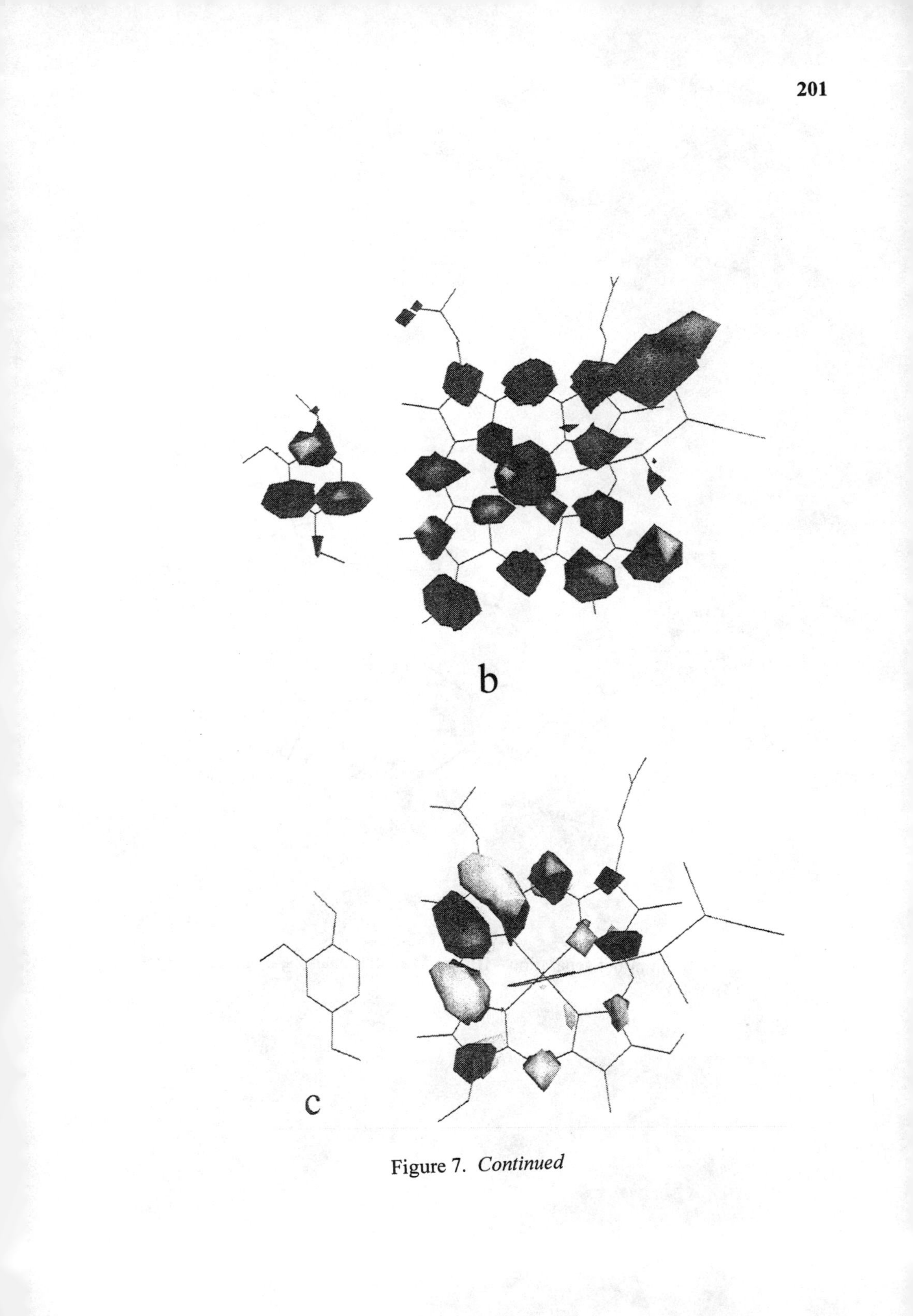

Figure 7. *Continued*

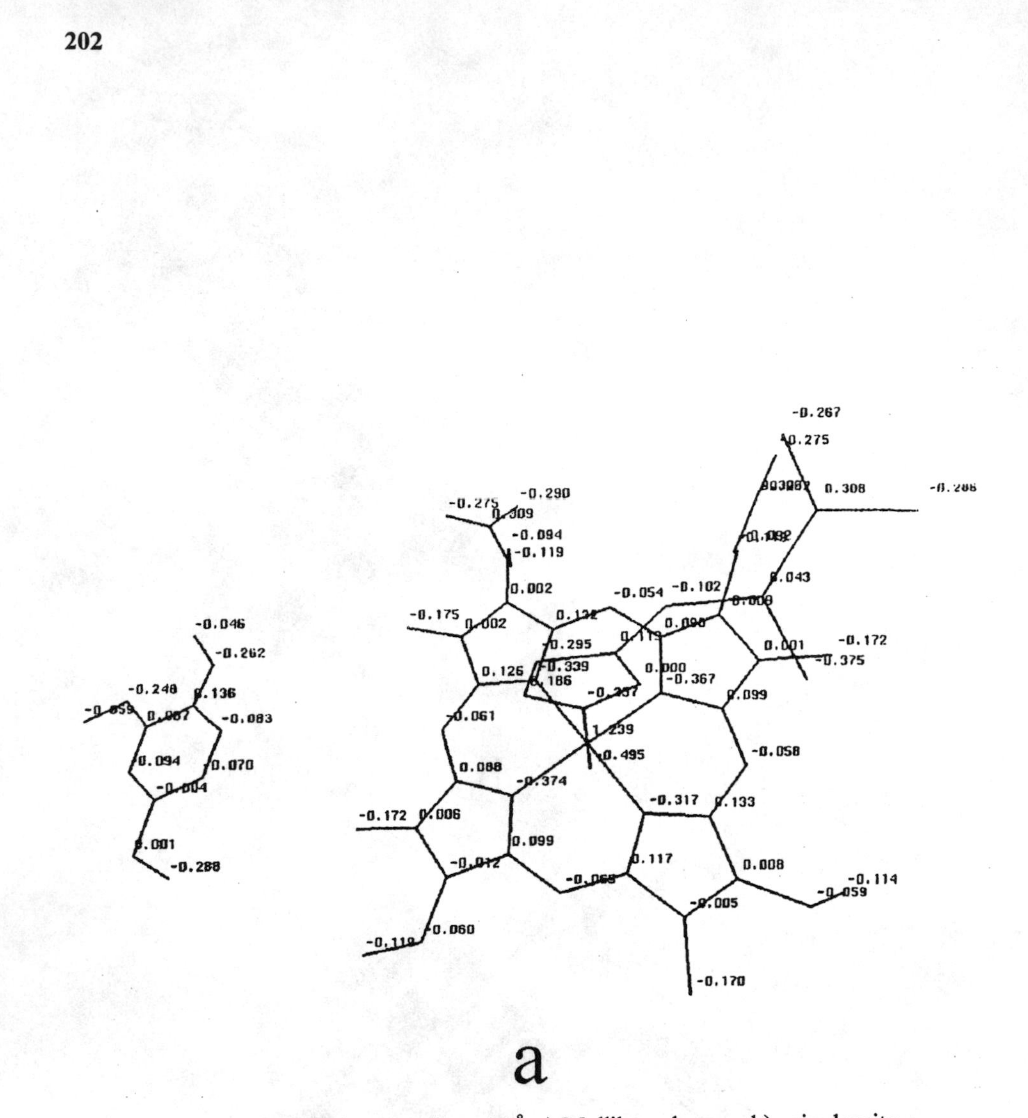

Figure 8. Interunit separation ~10Å a) Mulliken charges, b) spin density, c) HOMO.

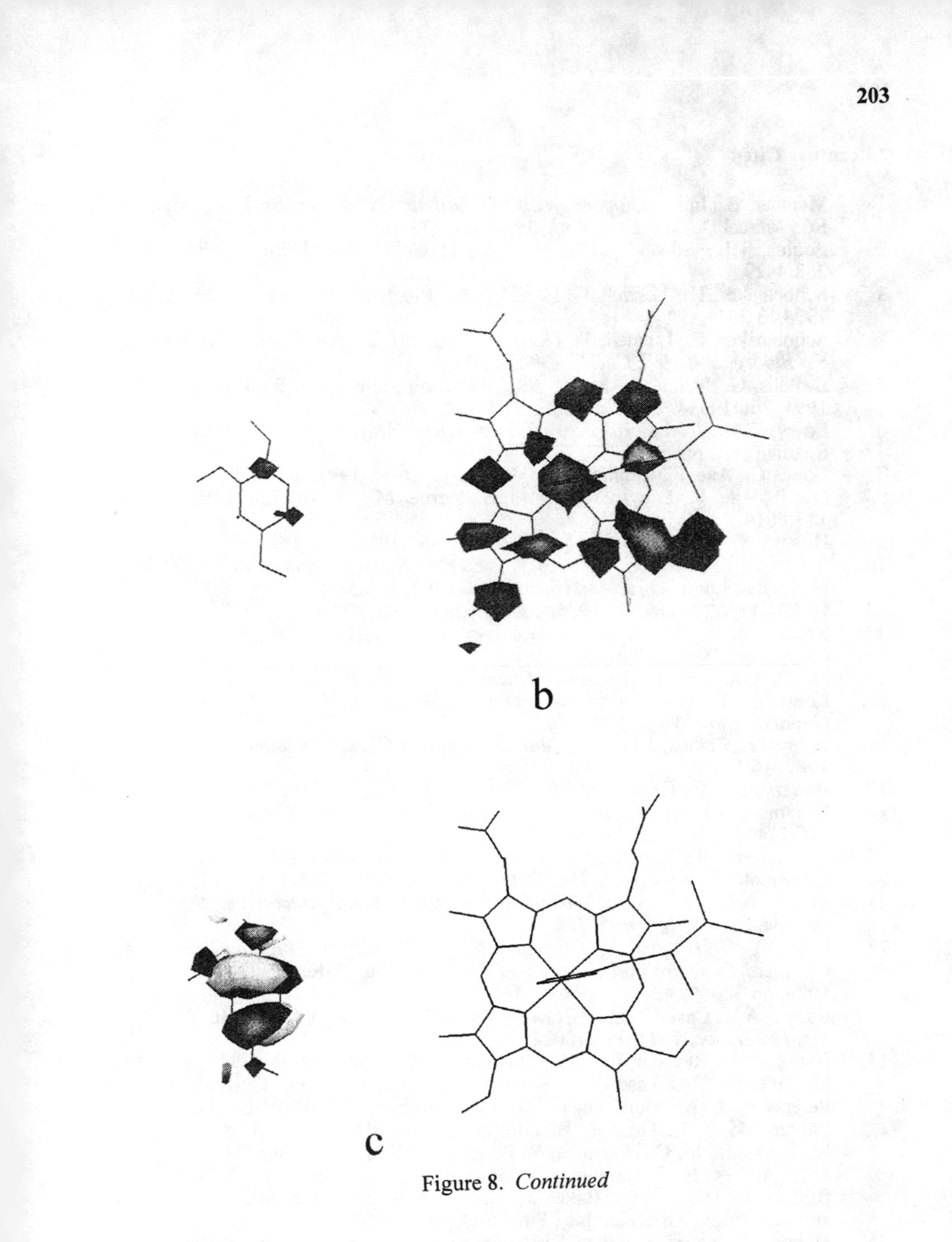

Figure 8. *Continued*

Literature Cited

1. Meunier, B. In *Metalloporphyrins in Catalytic Oxidations*, Sheldon, Roger A, Ed., Marcel Dekker, New York, 1994, pp. 133-155.
2. Poulos, T.L., Edwards, S.L., Wariishi, H.,Gold, M., *J. Biol. Chem.* **1993**, 268:4429.
3. Schoemaker, H., Lundell T., Hatakka, A., Piontek K. *FEMS Microbiol. Rev.* **1994**,13:321.
4. Schoemaker, H., Lundell, T., Floris, R., Glumoff, T., Winterhalter, K., Piontek, K. *Bioorg. and Med. Chem.*, **1994**, 2:509.
5. DePillis, G., Sishta, B., Mauk, A.G., Ortiz de Montellano, P., *J. Biol. Chem.*, **1991**, 266:19334.
6. Loew, G. In *Iron Porphyrins,* Lever A and Gray H. Eds, Addison-Wesley, Reading MA, pp. 3-87.
7. Loew, G., Axe, F., Collins, J., Du, P., *Inorg. Chem.* **1991**, 30:2291.
8. Du, P., Axe, F., Loew, G., Canuto, S., Zerner M., *J. Am. Chem. Soc.* **1991**, 113:8614.
9. Hosoya, T., Fuji, T., Ogawa S., *J. Theor. Biol.*, **1983**, 100:283.
10. Sakurada, J., Sekiguchi, R., Sato, K., and Hosoya, T., *Biochem.* **1990**, 29:2093.
11. Harris, D., Loew, G., *J. Am. Chem. Soc.*, **1993,** 115:5799.
12. Goldblum, A., Loew, G., **1985,** *J. Am. Chem. Soc.* 107:4265.
13. Korzekwa, K., Trager, W., Gouterman, M., Spangler, D., Loew G., *J. Am. Chem. Soc.* **1985,** 107:4273.
14. Du, P., Loew, G., 1991, *J. Phys. Chem.* **1991,** 95:6379.
15. Loew, G., Collins, J., Chantranupong, L., Waleh, A., *Int. J. Quantum Chem. Quantum Biol. Symp.*, **1987,** 14:75
16. Loew, G., Collins, J., Axe, F., *Int. J. Quantum Chem. Quantum Biol. Symp.*, **1989,** 16:199.
17. Anderson, W.P., Edwards, W.D., Zerner, M.C., *Inorg, Chem.* **1986,** 25:2720.
18. Kuramochi, H., Noodleman, L., Case, D.A., *J. Am. Chem. Soc.* **1997,** 119:11442.
19. Yamamotoa, S., Teroka, J., Kashiwagi, H., *J. Chem. Phys.* **1988,** 88:302.
20. Yamamoto, S., Kashiwagi, H., *Chem. Phys. Lett.* **1988,** 145:111.
21. Leach, A.R., *Molecular Modelling-Principles and Applications,* Longman, Harlow, UK, 1996, pp. 78-79.
22. Clark, T., *A Handbook of Computational Chemistry-A Practical Guide to Chemical Structure and Energy Calculations*, Wiley-Interscience, New York, 1985, pp. 98-99.
23. Rappe, A.K. Casewit, C.J., Colwell, K.S., Goddard III, W.A., Skiff, W.M., *J. Am. Chem. Soc.* **1992,** 114:10024.
24. Gaussian 94, Revision E.2, M. J. Frisch, G. W. Trucks, H. B. Schlegel, P. M. W. Gill, B. G. Johnson, M. A. Robb, J. R. Cheeseman, T. Keith, G. A. Petersson, J. A. Montgomery, K. Raghavachari, M. A. Al-Laham, V. G. Zakrzewski, J. V. Ortiz, J. B. Foresman, J. Cioslowski, B. B. Stefanov, A. Nanayakkara, M. Challacombe, Y. Peng, P. Y. Ayala, W. Chen, M. W. Wong, J. L. Andres, E. S. Replogle, R. Gomperts, R. L. Martin, D. J. Fox, J. S. Binkley, D. J. Defrees, J. Baker, J. P. Stewart, M. Head-Gordon, C. Gonzalez, and J. A. Pople, Gaussian, Inc., Pittsburgh PA, 1995.
25. Hehre, Warren J., *Practical Strategies for Electronic Structure Calculations,* Wavefunction, Irvine, California, pp199-204.
26. Elder, Thomas, *Holzforshung* **1997**, 51:47.

Chapter 8

Lignin Influence on Angiosperm Sapwood Susceptibility to White-Rot Fungal Colonization: A Hypothesis

Tor P. Schultz and Darrel D. Nicholas

Forest and Wildlife Research Center and Forest Products Laboratory, Box 9820, Mississippi State University, Mississippi State, MS 39762–9820

It has long been recognized that angiosperm sapwood in nature is usually degraded by white-rot fungi. This susceptibility to white-rot fungi is generally believed to be due to the structure and concentration of angiosperm lignin. However, an explicit explanation as to why lignin structure and/or concentration makes a particular wood vulnerable to white-rot fungal colonization and subsequent degradation has apparently never been given. We propose that phenolic groups in wood, such as those present in the lignin or heartwood extractives, can act as free radical scavengers (antioxidants) to inhibit the various white-rot free radical degradative mechanisms. Consequently, the presence of a relatively high phenolic "density", such as that in gymnosperm sapwood or angiosperm heartwood, may retard white-rot colonization. Conversely, white-rot fungi may rapidly colonize wood with a relatively low free phenolic content, such as angiosperm sapwood. Once a wood is colonized, subsequent antagonistic factors known to exist among fungi may give the advantage to the principal fungus present rather than fungi with fast decay rates (based on rates measured using monocultures in the laboratory). The complex structure of angiosperm wood where different cell types have different amounts and types of lignin – and consequently different phenolic "densities" – influences the susceptibility of angiosperm wood to initial white-rot colonization and, perhaps, also the subsequent decay rate. In addition to the free phenolic "density" and amount and bioactivity of any extractives present other factors undoubtedly also affect the decay resistance of a particular wood.

Introduction

In this paper we propose that sapwood susceptibility to white-rot fungal colonization and subsequent degradation is related to phenolic "density" [with the decay resistance of heartwood related to both this factor and the bioactivity of the extractives]. Consequently, wood with a low phenolic "density", such as angiosperm sapwood, is more vulnerable to white-rot colonization and subsequent degradation than wood which has a high phenolic "density" such as gymnosperm sapwood or angiosperm heartwood. This hypothesis is developed in part from our earlier work on the natural durability of angiosperm heartwood where we suggested that angiosperm heartwood extractives play a dual role in natural durability: extractives have some limited fungicidal activity and are also excellent free radical scavengers and thus inhibit the various fungal free radical degradation pathways (*1*).

This article is a discussion of our tentative hypothesis and consequently does not have experimental results followed by discussion and conclusions. Instead, an effort is made to review prior work in this area and discuss how some of these results support our hypothesis.

Discussion

Large volumes of wood/lumber are treated with various biocides [wood preservatives] to form a product resistant to fungal and insect degradation. The major wood preservatives used today in the US are CCA [chromated copper arsenate], pentachlorophenol and creosote. Recent environmental concerns have resulted in these biocides being scrutinized by various governmental agencies, and use of pentachlorophenol and/or CCA is restricted or banned in many countries. Consequently, many organizations are trying to develop alternative, environmentally-benign wood preservatives.

The heartwood of some trees has considerable resistance against fungi. As part of our effort to develop environmentally-benign wood preservatives, we have studied the role which extractives, particularly stilbenols, play in the natural durability of angiosperm heartwood (*1, 2*). Particularly intriguing to us was recent work (*3*) which reported that white-rot fungi are believed to initially disrupt xylem cell walls using free radicals to increase the pore size (*4*) so that the relatively large extracellular fungal enzymes can then penetrate the cell wall. Backa's et al. report, combined with the common knowledge that extractives are excellent antioxidants and dispersed throughout the cell wall, suggested that extractives may prevent the cell wall from being perturbed or "opened up" by free radicals during the initial attack by white-rot fungi. The above reasoning and related experimental work (*5*) led us to propose (*1*) that extractives protect angiosperm heartwood against white-rot fungal degradation by: A) extractives having some fungal activity; and B) angiosperm phenolic extractives being effective antioxidants and protecting wood from fungal free radical oxidative-type degradations.

It is well known that angiosperm sapwood in nature is highly susceptible to white-rot fungal degradation. While the exact reason for this vulnerability is not known, research has indicated that the different types of lignin in angiosperm and gymnosperm sapwoods, and possibly also the amount of lignin, are significant factors (*6, 7*). However, the exact reason why lignin type influences the susceptibility of a particular wood to white-rot colonization and subsequent degradation has not been proposed. In reviewing these reports we wondered if the presence of phenolic groups, which should be antioxidants, may influence the susceptibility of a particular sapwood to white-rot degradation. This susceptibility would be related to our earlier hypothesis (*1*) that the natural durability of angiosperm heartwood may be influenced by fungicidal extractives which are also antioxidants. Since almost all free phenolic groups present in normal sapwood are due to the lignin component, we decided to see if data in prior studies indicate that a low phenolic "density" in sapwood is related to the ease of degradation by white-rot fungi.

Highley (*7*) found that the rate of degradation of sweetgum (*Liquidambar styraciflua* L.) sapwood by a white-rot fungus [*Coriolus versicolor* (L.) Quél.] was significantly faster than for southern pine (*Pinus* spp.). The lignin in normal gymnosperm xylem is composed almost entirely of guaiacyl (G) units while temperate angiosperm lignin is normally composed of both guaiacyl and syringyl (S) units. This difference influences the lignin structure, including the free phenolic content. Lignin in normal gymnosperm xylem is reported to have 15-30 phenolic groups per 100 lignin units as compared to 10-15 phenolic groups in angiosperm lignin (*8*). In addition, angiosperm xylem typically has a slightly lower lignin content than gymnosperms. These two factors, a lower lignin content and lower phenolic content per unit of lignin present, suggest that phenolic "density" may be lower for angiosperm than gymnosperm sapwood. A rough estimate of the phenolic density can be calculated for sweetgum and southern yellow pine sapwood given a lignin content of 20.1 and 28.0%, respectively (*7*), and averaging the free phenolic content from the above to give an estimated 13 and 23% free phenolic groups per 100 lignin units. Based on this, pine will have about 35 mMoles and sweetgum about 13 mMoles of phenolic groups per 100 grams of sapwood. The significantly lower level -- less than half -- of phenolic groups in sweetgum as compared to pine sapwood supports our tentative hypothesis that the free phenolic "density" may be correlated to resistance to white-rot fungal degradation.

While angiosperm sapwood has a low phenolic "density", the heartwood of many angiosperm species has extractives which are mostly phenolic in nature. These extractives would be expected to increase the free phenolic content of the wood and impart enhanced resistance to white-rot fungi. For example, the highly-durable heartwood of osage orange [*Maclura pomifera* (Raf.) Schneid.] (*2*) was found to have 9.9 wt.% extractives and all the extractives identified had three or more phenolic groups. From the compounds identified and the total amount present we can estimate that osage orange heartwood has about 145 mMoles of phenolic groups per 100 grams of wood **due only to the extractives**, or approximately an order of magnitude greater than that due to the lignin. Thus, the amount of extractives present in a particular wood has been suggested to be a good indicator of durability (*2*). A high level of compounds which have some fungicidal activity in addition to being antioxidants will further protect heartwood against decay fungi.

Our hypothesis is dependent on the proposition that the free phenolic groups of lignin can act as free radical scavengers or antioxidants. Phenolic plant extractives are well known to be excellent free radical scavengers (*9*). The stilbenol resveratrol which is found in red wine and many woods is one classic example; the extensive studies on utilizing quercetin obtained from Douglas-fir bark [*Pseudotsuga menziesii* (Mirb.) Franco] as a commercial antioxidant is another. While the antioxidant properties of lignins have been studied less, several references have reported that lignins are antioxidants (*10, 11, 12*).

When we use the term "density", we do not suggest that dense wood is more durable. Rather, we imply that the number of phenolic groups per unit mass of cell wall substance (or sample mass, since all wood cell wall substance is reported to have the same specific gravity) is related to a wood's resistance to white-rot colonization. It has been suggested that decay susceptibility is inversely related to specific gravity (*13*). However, the differences reported were relatively minor (decay ranged from about 16 to 19.5 wt. %) and may be related to the amount of summerwood, as the authors suggested. [An external reviewer pointed out that the researchers plotted decay susceptibility (% oven dry weight loss) versus specific gravity. He noted that the negative slope may be due to the fact that high density wood has a higher initial weight. Plotting the mass loss per unit volume versus the specific gravity gave a plot with a slight positive slope which suggested to the reviewer that the higher density samples did not have greater durability.] Furthermore, decay rates, not susceptibility, were measured. Many researchers have used decay rates as an indication of decay resistance. We are unsure, however, if small differences in decay rates indicates a difference in susceptibility to fungal attack, especially considering the inherent error in laboratory decay tests. If density is related to white-rot decay resistance, then the sapwood of a very dense angiosperm such as osage orange should be more resistant than aspen sapwood (specific gravity of 0.76 and 0.35, respectively). This is not the case -- normal sapwood of all angiosperms has very little decay resistance.

In further support of our hypothesis, Srebotnik et al. (*14*) found that a white-rot fungus degraded a synthetic lignin much faster when the phenolic groups were methylated as compared to the same lignin which had not been methylated. In addition, the decay rates of three angiosperm sapwoods (*7*) by *C. versicolor* (sweetgum > white oak [*Quercus alba* L.] > ceibo [*Erythrina crista-galli* L.]) appear to be related to their approximate free phenolic "density" based on the lignin content and the atypical lignin structure of ceibo wood [mainly G] as compared to the other two angiosperm woods.

We are still unsure, however, of the relationship between decay rate versus susceptibility to colonization for untreated sapwood samples. Colonization is, of course, necessary before degradation can occur. Thus, it could be possible that white-rot fungi can colonize angiosperm sapwood faster than brown-rot fungi. Once colonization has occurred, antagonism (*15, 16*) between white- and brown-rot fungi may give the upper hand to the fungus which colonized the wood first rather than the fungus with the fastest decay rate (where decay rate is the mass loss per unit time for a fungal culture on a particular wood sample under laboratory conditions with no other fungus present).

Examining only the macro phenolic content of a particular wood is misleading, of course. The complex structure of angiosperms as compared to gymnosperms has also been suggested to influence decay susceptibility to white-rot fungi (*6, 17*). Obst et al. (*6*), in an excellent study, found that the decay susceptibility of different cell types in the sapwood of seven angiosperms generally followed the pattern of fibers > vessels > ray parenchyma. As discussed by Obst et al., angiosperm fiber cell walls are believed to be composed of mainly S lignin, vessel cell walls of G lignin, and ray cells of S/G lignin. The lignin concentration in birch (*Betula* spp.) for the secondary cell wall of these three cell types has been reported to be 19, 27 and 27% lignin for fibers, vessels, and ray cells, respectively (*8*). It would be expected that the phenolic content of lignin would be in the order of G > G/S > S. Thus, the finding of Obst et al. that fibers were much more susceptible to white-rot fungi than vessels would agree with our hypothesis, since vessels should have a higher phenolic content than fibers based on a higher lignin concentration (27 vs. 19%) and the lignin type (G vs. S), respectively. However, the high decay resistance of ray parenchyma cells, which have a high lignin content (27%) but with the lignin probably made up of G/S units, is surprising since we would predict that ray parenchyma would have a decay resistance between that of fibers and vessels. One possibility the authors suggested is that some extractives may have been present in these sapwood samples. Extractives would influence the results since the food reserves held in the sapwood parenchyma cells can often be quickly converted -- within hours for some angiosperms -- into antimicrobial metabolites (phytoalexins) such as phenolic extractives once a tree is cut (*18, 20*). The parenchyma cells will remain viable for an extended period until the log is dried, steamed or fumigated.

The heterogenous distribution of the lignin units in the cell wall (*21*) may also affect colonization by white- and brown-rot fungi. Specifically, white-rot fungi gradually erode cell walls from the lumen inwards [the initial attack occurs at the S_3 layer] while the entire cell wall is uniformly degraded by brown-rot fungi. It has been reported (*21*) that the S_3 and S_2 cell layers in angiosperms contain a high proportion of uncondensed S-type lignin units [i.e. high levels of ether linkages and consequently few phenolic groups] as compared to the lignin in compound middle lamella which contains larger amounts of condensed G and *p*-coumaryl alcohol elements. Thus, the layers of the cell wall which is first degraded by white-rot fungi may be the easiest layers **for white-rot fungi to degrade**, and consequently these fungi may have a relatively easy start colonizing a cell. Conversely, brown-rot fungi degrade the entire cell wall and thus experience both condensed and uncondensed lignin. This heterogenous lignin distribution may result in white-rot fungi being able to colonize angiosperm wood faster than brown-rot fungi.

We have concentrated on white-rot fungi, but brown-rot fungi are also reported to initially degrade wood by generating free radicals (*19, 22*). This might suggest that angiosperm sapwood in nature should be attacked and degraded by both brown- and white-rot fungi. We believe that the susceptibility of angiosperm sapwood in nature to white-rot fungi is simply due to white-rot fungi colonizing some or all of angiosperm sapwood faster than brown-rot fungi. Once a wood is colonized, subsequent antagonistic factors may give the advantage to the fungus present at the high level rather than fungi with fast decay rates (based on decay rates measured using monocultures in the laboratory). We believe this rapid colonization may be due to the low phenolic "density" of angiosperm sapwood. Conversely, white-rot fungi are initially inhibited when colonizing wood with a relatively high phenolic "density". For example, Highley (*7*) reported that the white-rot fungus *C. versicolor* degraded the sapwood of two typical angiosperms (sweetgum and white oak) much faster than three gymnosperms (southern pine, white spruce [*Picea glauca* (Moench.) Voss.], and western hemlock (*Tsuga heterophylla* (Raf.) Sarg.]. In addition, we and others (T. Amburgey, personal communication) have observed that angiosperm sapwood is degraded by white-rot fungi in an above-ground test much faster than similar gymnosperm samples are degraded by brown-rot fungi. Furthermore, when we added the antioxidant BHT to aspen (*Populus* spp.) sapwood, at a level which would approximately double the phenolic "density", followed by inoculation with the white-rot fungus *Trametes* (formerly *Coreolus*) *versicolor* (L.) Quél. in the soil-block test, we observed some protective effect **but only during the early stages of incubation** (Table 1). This relatively short protected period against white-rot fungi, however, may result in a "window of opportunity" for brown-rot fungi. (We measured strength rather than weight loss in this test since strength loss is a more accurate indication of incipient decay). Alternatively, it may be possible that the type and mode of action of free radicals generated by brown- and white-rot fungi during the initial colonization stage are different and, thus, that the mechanisms(s) by which wood is initially perturbed differ. Another possibility might be that syringyl lignin units are simply more susceptible to free radical oxidation/degradation than guaiacyl units (*23*). Consequently, free radical attack would mainly affect angiosperm wood and it's syringyl-guaiacyl lignin rather than normal gymnosperm wood which has guaiacyl lignin. Finally, it is well known that white-rot fungi degrade wood by radical mechanisms. Thus, the degradation rate of white-rot fungi may be more affected by a high phenolic "density" than that of brown-rot fungi.

Interestingly, phenolic syringyl units would be expected to be a better antioxidant than a free phenolic quaiacyl. (The second *o*-methoxyl would confer beneficial steric and electronic properties). Indeed, this is supported by data from a recent study (*12*). However, the higher phenolic quantity in gymnosperm is likely more important than the faster antioxidant rate but lower phenolic level in angiosperms (quantity vs. quality).

Table 1. Average percent compression strength loss for aspen sapwood samples either untreated or treated with 5% BHT, then inoculated with *T. versicolor* and tested in the soil-block test for 2-, 3-, and 4-weeks incubation. The results are the average of five replicates.

	Avg. % Strength Loss		
	Incubation Time, Weeks		
Treatment	2	3	4
Control	70.0	87.3	89.1
5% BHT	34.2	56.7	78.6

We need to emphasize that even if our tentative hypothesis is correct, other factors probably also influence angiosperm sapwood susceptibility to white-rot colonization. For example, it is well known that white-rot fungi prefer wood with a higher moisture content than brown-rot fungi. The higher hydrophobicity of gymnosperm extractives, such as the terpenoid-based extractives found in pine sapwood and heartwood, would make gymnosperm wood relatively water repelling. Other decay susceptibility factors could be differences in the structure, amount and accessability of the hemicelluloses, the fungicidal activity, microdistribution and hydrophobicity of any extractives present, antagonism between different fungi, and other factors -- some as yet unknown. Synergism or antagonism between two or more factors is also possible. Whatever the cause(s), after studying natural durability for some time we believe that multiple factors affect natural durability for both gymnosperms and angiosperms and that most wood extractives have very poor fungicidal activity when compared to commercial biocides. It is also interesting that our radical hypothesis [pun intended] suggests that wood components protect wood by a general, nonspecific mechanism. It is likely that if an extractive were to inhibit a specific part of a critical fungal metabolic pathway, and thus was highly fungicidal, a decay fungus would rapidly (in an evolutionary time frame) mutate into a strain unaffected by the extractive. A nonspecific defense mechanism such as our antioxidant hypothesis, by contrast, may be difficult for fungi to overcome by mutation.

Extractives or other antioxidants may protect the wood cell wall from being disrupted or perturbed by free radicals during an initial attack by white-rot fungi (*3*). If the wood is disrupted in some manner such as by autohydrolysis, a previous attack by white-rot fungi, a limited lignin degradation by some chemical means (*7*) or photodegradation as discussed above, then we would predict that adding nonbiocidal antioxidants to an already-disrupted xylem would have no or minimal influence on later white-rot susceptibility. Thus, we are unable to test our hypothesis by methylating the lignin phenolic groups of a wood block, since methylation would require a pretreatment to make the cell walls permeable. Furthermore, we are unsure of the relationship between colonization, decay susceptibility, and decay rate for untreated sapwood samples. Finally it should be noted that resistance to white-rot fungi does not imply that a wood will also be resistant to brown- or soft-rot fungi.

As mentioned in the beginning, our hypothesis is only tentative. We have presented these thoughts so that these ideas may lead to a better understanding of wood decay and natural durability.

Conclusion

Based on prior articles of angiosperm sapwood susceptibility to white-rot fungal degradation we suggest that:

> The resistance of sapwood to white-rot fungal colonization and subsequent degradation is related to a wood's phenolic "density", with the phenolic groups acting as antioxidants. In normal sapwood essentially all phenolic groups are in the lignin and thus phenolic "density" is related to lignin concentration and structure.

This antioxidant function prevents the fungal-generated free radicals from initially perturbing the wood structure and, consequently, thwarts the relatively large extracellular fungal enzymes from penetrating into and degrading the wood substance. Once the cell wall has been disrupted by some means, either biological, physical or chemical, this antioxidative protective mechanism may no longer help prevent white-rot colonization.

Acknowledgments

Funding was provided by a USDA-competitive grant and the state of Mississippi. Discussions with Dr. K. Kirk, USDA-Forest Products Lab and Dr. T. Nilsson, SLU, Sweden, were much appreciated. Approved for publication as Journal Article No. FP102 of the Forest and Wildlife Research Center, Mississippi State University.

References

1. Schultz, T.P.; Nicholas, D.D.; Minn, J.; McMurtrey, K.D.; Fisher, T.H. IRG Paper WP 98-30172, **1998**, and references therein.
2. Schultz, T.P.; Harms, W.B.; Fisher, T.H.; McMurtrey, K.D.; Minn, J.; Nicholas, D.D. *Holzforschung* **1995**, *49*, 29, and references therein.
3. Backa, S.; Gierer, J.; Reitberger, T; Nilsson, T. *Holzforschung* **1993**, *47*, 181.
4. Flournoy, D.S.; Paul, J.A.; Kirk, T.K.; Highley, T.L. *Holzforschung* **1993**, *47*, 297.
5. Schultz, T.P.; Nicholas, D.D. U.S. Patent 5,730,907. **1998**.
6. Obst, J.; Highley, T.L.; Miller, R.B. *Biodeterioration; Research 4*; Plenum Press: New York, NY; **1994**, Vol. 4 pp. 357-374.
7. Highley, T.L. *Can. J. For. Res.* **1982**, *12*, 435, and references therein.
8. Sjostrom, E. *Wood Chemistry: Fundamentals and Applications*, 2nd Ed., Academic Press: New York, NY. **1993**, pp. 83.
9. Larson, R.A. *Phytochem.* **1988**, *27*, 969, and references therein.
10. Han, D.N.S. In *Wood and Cellulosic Chemistry*; Han, D.N.S. and Shiraishi, N., eds; Marcel Dekker, New York; **1991**, chp. 11.
11. Schmidt, J.A.; Rye, C.S.; Gurnagul, N. *Polymer Degrad. Stability* **1995**, *49*, 291.
12. Barclay, L.R.C.; Xi, F.; Norris, J.Q.; In *J. Wood Chem. Tech.*; **1997**, *17*, 73.
13. Schmidtling, R.C.; Amburgey, T.L. *Holzforschung* **1982**, *36*, 159.
14. Srebotnik, E.; Jensen, K.A., Jr.; Hammel, K.E. *Proc. Natl. Acad. Sci.* **1994**, *91*, 12794.
15. Holmer, L.; Stenlid, J. *FEMS Microbiol. Ecology*, **1993**, *12*, 169.
16. Owens, E.M.; Reddy, C.A.; Grethlein, H.E. *FEMS Microbiol. Ecology*, **1994**, *14*, 19.
17. Faix, O.; Mozuch, M.D.; Kirk, T.K. *Holzforschung* **1985**, *39*, 203.
18. Hillis, W.E. In *Heartwood and Tree Exudates*, Springer-Verlag; New York, NY **1987**.
19. Hirano, T.; Tanaka, H.; Enoki, A.; *Holzforschung*, **1997**, *51*, 389.
20. Chen, Y-r; Schmidt, E.L.; Olsen, K.K. *Wood Fiber Sci.*, **1998**, *30*, 18, and references therein.
21. Lewis, N.G.; Davin, L.B. In: *Lignin and Lignan Biosynthesis*; Lewis, N.G.; Sarkanen, S., Eds; ACS Symp. Series: Washington, DC, **1998**, Vol. 697, pp. 344, also see pp. 188.
22. Backa, S.; Gierer, J.; Reitberger, T.; Nilsson, T. *Holzforschung* **1992**, *46*, 61.
23. Ander, D. *Holzforschung* **1996**, *50*, 413.

Analysis

Chapter 9

Classification of Lignin According to Chemical and Molecular Structure

Wolfgang G. Glasser

Biobased Materials and Recycling Center, Department of Wood Science and Forest Products, Virginia Polytechnic Institute and State University, Blacksburg, VA 24061

The commercial trade in isolated lignins generated in conjunction with the fractionation of biomass into constitutive components, including pulp and paper and saccharification/fermentation processes, requires the adoption of standard analytical techniques by both suppliers and users of such products. Simple, reliable, quantitative, and widely accepted techniques include elemental analysis, functional group analysis, analytical degradations, various spectroscopic techniques, molecular weight analysis, and thermal analysis. These techniques are capable of elucidating the chemical composition of repeat units; the nature of intermonomer bonds; molecular weights and weight distributions; and the potential for inter- and intramolecular interactions. Several aspects of this classification methodology are reviewed.

The huge potential of lignin as an underutilized (or wasted) raw material has been pointed out repeatedly (1-3). Although lignin's potential is not entirely unexploited (lignin sulfonates, for example, constitute a major resource in markets for ionic surfactants, drilling mud additives, dye stuff dispersants, etc., see refs. 4,5) the acceptance of non-sulfonated lignins in markets for structural polymers and materials has been slow at best. Despite the fact that numerous studies have pointed repeatedly to lignin's potential contribution to the physical properties of structural thermosets and thermoplastics, only a very small fraction of the lignin separated from wood (or biomass) finds its way today into engineered structural materials. One of the reasons for this apparent reluctance results from the absence of well-defined, standard analytical techniques that might be adopted by both suppliers and users of lignins in markets for structural polymers. Since most lignins are cogenerated in the process of pulp and paper making, and since this separation is optimized for pulp production and paper properties, lignin is usually viewed as a material of significant non-uniformity and variability in both chemical and molecular structure. It is this perceived variability in combination with the absence of widely accepted and understood, quantitative, fast and simple analytical techniques suitable for general adoption in quality control procedures which are widely practiced in relatively non-specialized laboratories, that present the most significant obstacles to lignin utilization in engineered structural materials. It is the objective of this review to draw attention to analytical methodologies qualified to overcome this obstacle.

Lignins are separated from lignified ("woody") biomass by technologies that involve partial depolymerization and/or partial derivatization, and combinations thereof (6-9). These depolymerization and modification chemistries and their underlying reactions are well understood. However, this understanding has often resulted from work with lignin-like model substances, and it has ignored macromolecular properties and molecular interaction parameters. It is the extent to which each individual depolymerization and/or modification reaction has taken place within a certain native lignin, the structure of which is inherently variable with tree age, tree location within the ecology of a forest, cell type, etc. (10), that presents the lignin analyst with the challenge of quantitation. One can distinguish between four somewhat interrelated tasks for lignin analysis: (a) chemistry of the basic repeat unit(s); (b) chemistry of intermonomer bonds; (c) molecular size/weight considerations; and (d) inter/intramolecular interactions.

The following is to summarize widely accepted and understood, fast and simple, quantitative, and (hopefully) universally available analysis techniques in non-specialized laboratories that are qualified to shed light on these four points. Excellent reviews dealing with several details of the structure analysis of lignin have recently appeared elsewhere (11-13).

The Chemistry of Monomeric Repeat Units

The chemistry of lignin's monomeric repeat units, its "monolignols," normally relies on the determination of elemental composition in conjunction with the determination of methoxy groups. Standard C, H, N, (S?), and OCH_3-analysis allows the formulation of basic C_9-unit structures in accordance with the suggestion of Freudenberg (14) (Figure 1). C_9-structures provide useful information on the overall composition of monomeric repeat units (15,16). They also provide information about the overall composition of lignin in terms of the specific mixture of *p*-OH cinnamyl alcohols, *p*-coumaryl alcohol, coniferyl alcohol, and sinapyl alcohol (Fig. 1).

Additional information on the chemical composition of the average, basic repeat unit can be obtained by optional functional group analysis (17,18). This is particularly relevant in relation to phenolic and total hydroxyl groups, and carbonyl groups (18). Widely accepted methods of phenolic OH-determination involve aminolysis (19), and H-NMR spectroscopy of acetylated lignin derivatives (20-25), and, to a lesser extent and with less reliability, UV-spectroscopy (17,18).

Elemental Analysis Results

C, H, N, S, OCH_3

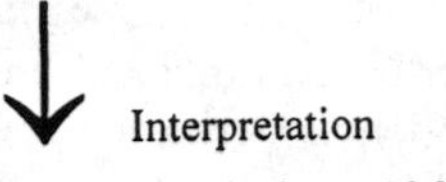

Interpretation

(Freudenberg, 1968)

Mixture of *p*-OH cinnamyl alcohols	$C_9H_{(10-i)}O_2[OCH_3]_i$
DHP from *p*-coumaryl alcohol	$C_9H_{8.07}O_2[H_2O]_{0.40}$
Softwood MWL	$C_9H_{7.92...8.83}O_{2.37...2.88}[OCH_3]_{0.92...0.96}$
Beech MWL	$C_9H_{7.49..8.50}O_{2.53...2.95}[OCH_3]_{1.39...1.46}$

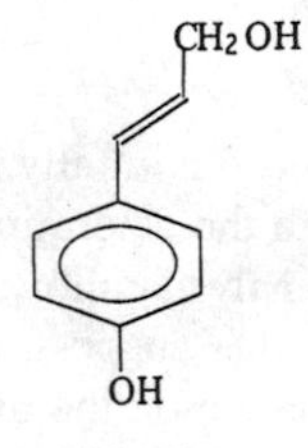

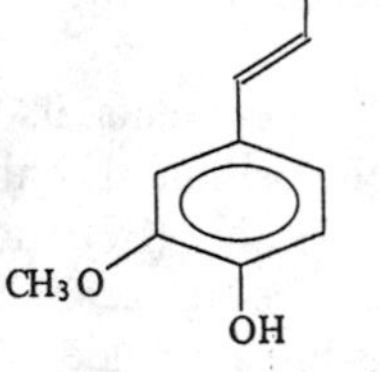

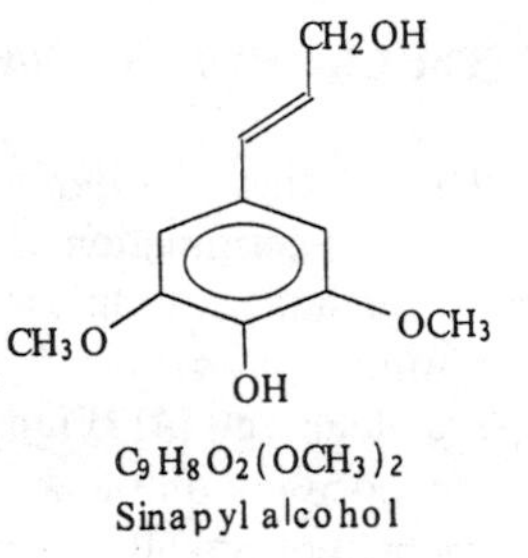

Fig. 1 Schematic conversion of primary elemental and functional group analysis results into lignin repeat unit (phenylpropane, C_9-unit)-structures according to Freudenberg (14). Data from Sakakibara (15).

The Chemistry of Intermonomer Bonds

The most widely accepted methods for defining the intermonomer bonds present in lignin involve various types of analytical degradation procedures (2,13,14,15,16,26, 27,28). Although the final determination regarding the suitability of degradation methods as opposed to spectrophotometric analysis for intermonomer bond identification remains outstanding, it is primarily analysis by three alternative analytical degradation methods that has shed light on the quantitative distribution of lignin's many intermonomer bonds. These methods are discussed in the following.

The degradation of lignin by alkaline solutions of potassium permanganate has been the subject of much research in the laboratories of Karl Freudenberg in the 1950's and 60's (14,29). Permanganate oxidation of lignin generated the first fragments qualified to provide insight into the intermonomer bonding of softwood lignins. The method has later been refined by Miksche et al. (30,31), and by Glasser et al. (27,28,32,33), among others (see ref. 34). The most refined protocol (27,28) involves the sequential treatment of isolated lignin preparations with diethylsulfate (for phenolic OH-protection), alkaline cupric oxide (for depolymerization), dimethylsulfate (for protection of phenolic OH groups formed by cleavage of alkyl-exyl ethers), alkaline permanganate and H_2O_2 (for degradation of aliphatic sidechains to carboxylic acid groups), and eventual methylation with diazomethane (for separability by GC) (Figure 2). A typical gas chromatogram, and the most important monomeric and dimeric fragments, are shown in Figure 3. The protocol typically generates between 20 and 40% by weight of mono- and dimeric degradation products depending on lignin source and nature of pretreatment. An elaborate but widely accepted data interpretation technique, which takes account of stepwise yield reduction in relation to chemical treatment involved, has been devised by Miksche et al. (30,31). Adoption of this interpretive technique provides accountability for the intermonomer bonds of between 60 and 70% of all monomeric units (16).

The definition of (a) a "hydrolysis ratio," which measures the molar ratio of (uncondensed) C_9-units originally present with free phenolic OH groups to the sum total of all monobasic, monomeric benzoic acid units present in the degradation product mixture; and (b) that of a "condensation factor" relating the molar ratio of benzoic acids to dimers or that of the monobasic to dibasic monomers (i.e., ratio of benzoic to phthalic and iso-phthalic acids, BA/PA), has helped provide classification criteria for various lignins (27,28). Lignins with low hydrolysis ratios (< 0.4) are recognized as polymers primarily bonded by alkyl-aryl ether bonds, and those with high hydrolysis ratios (>0.6) are rich in carbon to carbon bonding (Table I). Conversely, high BA/PA ratios typically signify lignins with a high proportion of alkyl aryl ether bonds compared to C-C bonds; and those lignins that have low BA/PA-ratios are mostly C-C-linked and have few (remaining) alkyl aryl ether linkages (Table I). Not surprisingly, lignins with intermonomer bonding closely resembling their native structure reflect a high degree of alkyl-aryl ether bonding; and lignins isolated following significant depolymerization catalyzed by alkali or acid reveal lesser or greater degrees of ether-cleaving depolymerization and

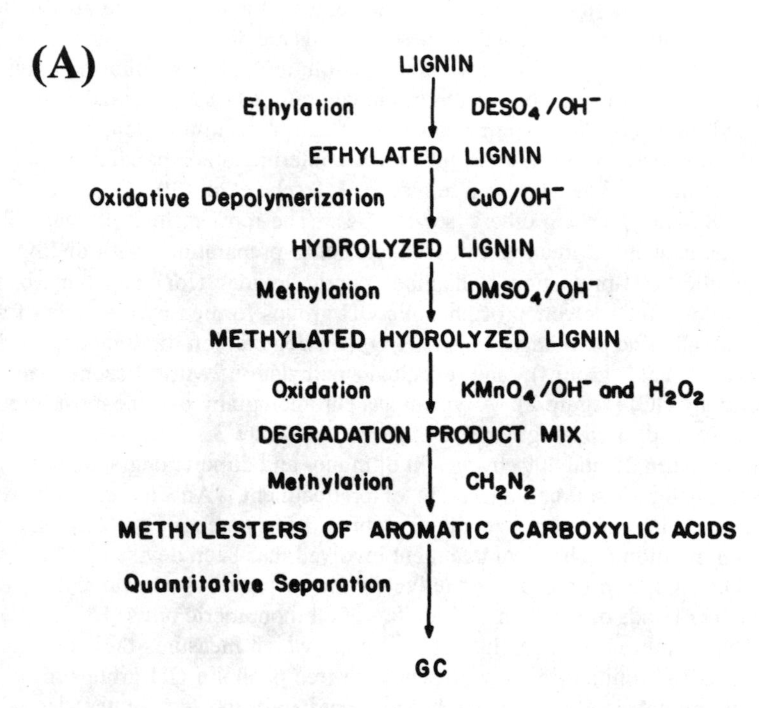

Fig. 2 (A) Reaction pathway for lignin analysis by degradative permanganate oxidation; and (B) Typical gas chromatogram of a mixture of monomeric and dimeric permanganate oxidation products using the degradation protocol illustrated in Fig. 2A. The sample represents a hardwood lignin. (Reproduced with permission from ref. 27. Copyright 1983 by American Chemical Society.)

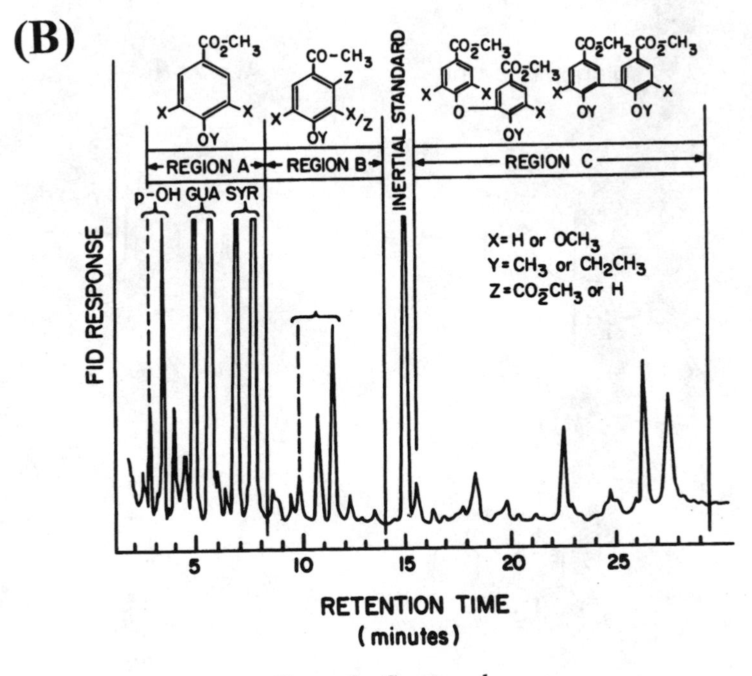

Figure 2. *Continued*

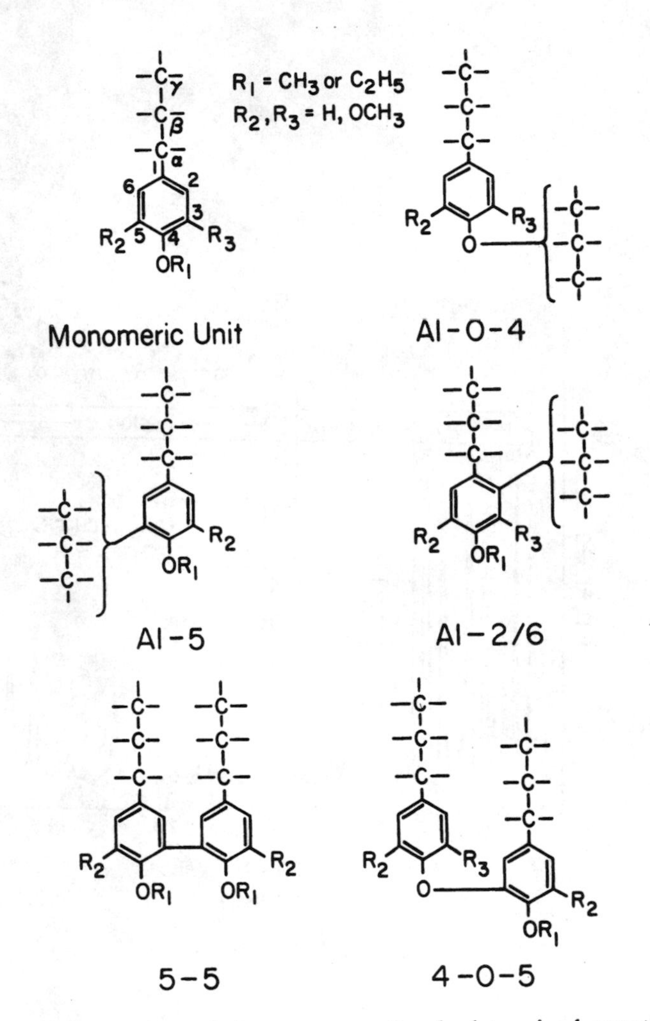

Fig. 3 Types of (softwood) intermonomer bonds determined quantitatively by permanganate oxidation on the basis of the distribution of monomeric and dimeric fragments found in the gas chromatogram of Fig. 2B. (Reproduced with permission from ref. 27. Copyright 1983 by American Chemical Society.)

Table I. Intermonomer Bond Types of Isolated Lignins (Adopted from ref. 16,27,28).

Lignin Types	Hydrolysis Ratio[1)]	BA/PA[2)]
MWL - softwood - hardwood	0.28-0.32 0.22	3.7 – 4.8 13-16
Kraft Lignin (pine)	0.83	2.1
Acid Hydrolysis Lignin (2% H_2SO_4, > 220°C, 10-60s) - Softwood - Hardwood	 0.91 0.79	 0.72 0.39
Steam Explosion Lignin - Straw - Hardwood	 0.81 0.5 – 0.8	 2.0 5.2 – 8.1
Organosolv Lignin - Softwood - Hardwood	 0.87 0.2 – 0.9	 2.1 1.6 – 15

[1)]Molar ratio of uncondensed C_9-units originally present with free phenolic OH groups to the sum total of all monobasic, monomeric-benzoic acid units. This ratio expresses alkyl-aryl ether content (i.e., inverse relationship).
[2)]Molar ratio of benzoic to phthalic and isophthalic acid units. This ratio decreases with increasing degree of condensation.

condensation, respectively (Table I). These prevalent structural features can be expected to have a significant impact on the utilization potential of isolated lignins in structural polymers.

Thioacetolysis involves the sequential treatment of lignin (isolated or in wood) with thioacetic acid followed by treatment with alkali and Raney Nickel (for desulfurization) (35,36). While thioacetolysis of hardwood lignin was found to generate as much as two-third (by weight) monomeric and dimeric degradation products (35), and this was substantially in excess of all previously available degradation methods, this method failed to find the universal acceptance required of a standard protocol. The emergence of an alternative degradation technique, with fewer steps and equal quantifiability, may have been responsible for the low acceptance rate.

Thioacidolysis, first advocated in the early 1980's, provided for the treatment of lignin with ethanethiol/BF_3 (37). The desulfurized (with Raney Nickel) reaction product mixture typically consists of between 1/3 and 2/3 of monomeric and dimeric degradation products per average phenylpropane repeat unit (i.e., >1,000 to <3,000 μmoles/g of lignin) (38-41). High yield and preservation of the propanoid sidechain are principal advantages of this method when it is applied to unmodified (esp. un-depolymerized) lignins. Some additional general advantages and disadvantages of the two most prominent degradation procedures are given in Table II.

Table II. Comparison of Two Major Analytical Degradation Methods.

Degradation Procedure	Advantage	Disadvantage
Permanganate Oxidation:	Many bonds broken (incl. ether and C-C bonds)	Multi-step protocol
	Applicable to all lignins (incl. fully hydrolyzed)	Elaborate data treatment
	High yield following data correction for sidechain degradation (60-70% monomers and dimers)	Low yield of primary degradation products due to yield loss during the many procedural steps (usually 20-40% by wt. of monomers and dimers (27))
Thioacidolysis:	Effective alky-aryl ether cleavage	Smell/odor
	High yield of primary degradation products (30-50% by wt. monomers and dimers)	Not applicable to depolymerized lignins
	Simple protocol and data treatment	Complex mixture of products

It is unlikely, however, that any one of these laborious and skill-requiring degradation methods will find universal acceptance before spectroscopic (primarily ^{13}C-NMR) methods become qualified to quantitatively distinguish between main intermonomer bonds. Until such time, the detailed analysis of intermonomer bonding in lignin will likely remain confined to a few selected lignin chemistry laboratories.

Molecular Size/Weight Considerations

The overwhelming importance of lignin depolymerization as the basis for delignification during paper production has spurred considerable interest in molecular weight determination techniques for many years. Since normal solution viscosimetry remained decidedly uninformative (42), greatest focus has always been on size exclusion (SEC) or gel permeation chromatography (GPC) (43,44). Today's state-of-art involves commercially available GPC-systems operating at high pressure, with multiple columns packed with gels of different pore sizes, in non-aqueous, non-ionic solvents, and with multiple detection devices (45-49). These capital-intensive systems are typically supported by complex software that has found universal acceptance. After many decades of debate over bimodal and non-uniform molecular weight distributions (44,50), current state-of-art systems are indeed capable of producing molecular weight distributions that resemble those obtained with other (esp. man-made) polymers (Figure 4). The molecular weight analysis of lignins is complicated by the branched character of lignin, its poor solubility in non-aqueous solvents, and its fluorescence properties that interfere with light scattering for analysis (51). However, lignin derivatization for the purpose of solubility

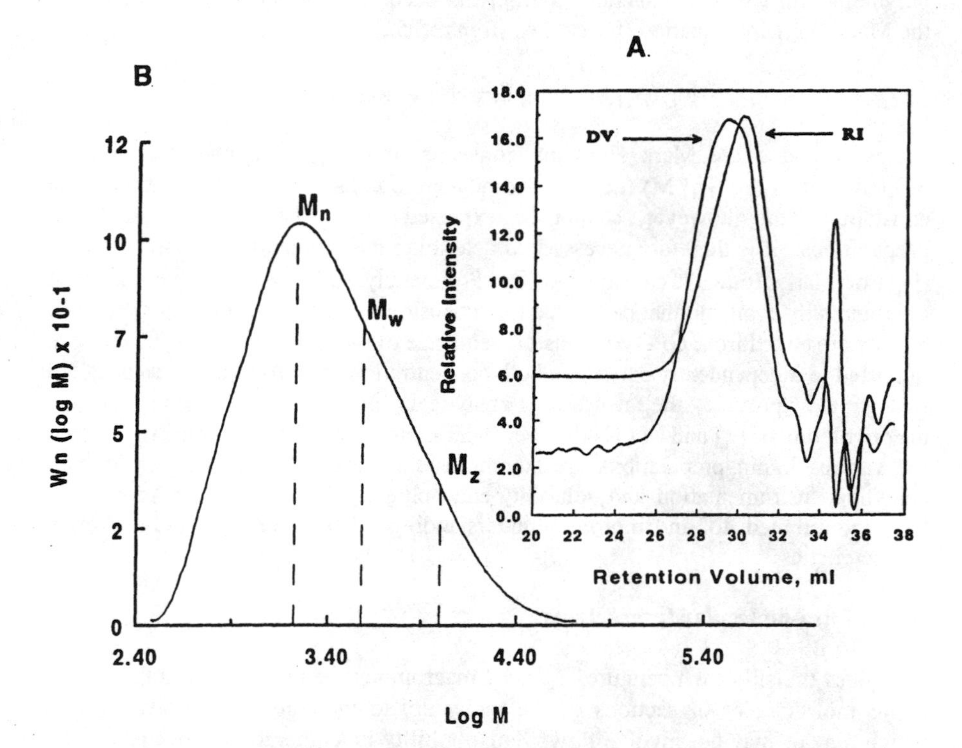

Fig. 4 Typical molecular weight distribution of a (THF-soluble) lignin derivative as revealed by HPLC-GPC using sequential eluent monitoring by refractive index and differential viscosity. The primary analysis data (inset) are converted from a volume-basis to absolute molecular weight (M)-basis using a universal calibration generated by Unical software (Viscotek Inc.). (Reproduced with permission from ref. 49. Copyright 1993 by Marcel Dekker, Inc.)

enhancement in solvents normally employed with GPC analysis (i.e., THF), and the wide acceptance of universal calibration, have resulted in a widely accepted, standardized molecular weight determination methodology (49).

As a consequence of the fact that GPC-analysis is based on separating molecules according to their hydrodynamic volume (defined as the product of intrinsic viscosity, [η], and absolute molecular weight, M), the analysis of branched polymers with variable branching coefficients becomes complicated. According to the Mark Houwink equation (form taken from ref. 52),

$$\log([\eta]\,M) = \log K + (1 + a)\log M \qquad (1)$$

where K and a are Mark Houwink constants, the log M-parameter is directly proportional to log ([η] M) (i.e., the hydrodynamic volume) only if K and a remain constant. This, however, cannot be expected to be the case for all lignin preparations. It is therefore necessary to determine independently intrinsic viscosity ([η]) or Mark Houwink constants (a,K). Fortunately, GPC-detectors have become commercially available that determine the intrinsic viscosities of molecular fractions as they are eluted from GPC columns. A sequence of two detectors in a GPC system, qualified to independently determine both concentration and differential viscosity "in real time," provide the necessary analytical information qualified for the deconvolution of [η] and M (47-49). Pertinent molecular weight parameters reported for various lignin preparations are summarized in Table III. Advances in both analytical instrumentation and solubility-enhancing lignin modification techniques have contributed to an improved understanding of lignin's molecular weight characteristics.

Inter-/Intra-molecular Interactions

The glass transition temperature, T_g, of a macromolecule reflects the ability of an entire molecule, or of sections of the molecule, to undergo translational motion which may or may not involve flow. This mobility is temperature dependent. The cooperative mobility of molecular segments having a length of 40 to 50 C-atoms

Table III. Molecular Weight Parameters of Lignin Acetates (Adopted from refs. 47-49).

Lignin Types	Log K	a	[η] (dLg^{-1})
Kraft Lignin			
- softwood	-2.1...-2.3	0.24...0.29	0.05...0.08
- hardwood	-2.4	0.33	0.06
Organosolv Lignin			
- hardwood	-2.1	0.21...0.24	0.04...0.07
Steam Explosion Lignin			
- hardwood	-2.4	0.35	0.07
- bagasse	-1.8...-2.4	0.17...0.25	0.06...0.08
Hydroxypropyl Lignin			0.03...0.05

results in molecular wriggling and jumping that is responsible for changes in several physical properties (53). The property most widely employed for the determination of the transition between the glassy and rubbery state (i.e., T_g) involves heat capacity, C_p. This preponderance of C_p-based T_g-analysis methods is mostly the result of instrumental preference. Other physical properties may have equal T_g-resolving power.

Glass transition temperatures are influenced by such factors as free volume between polymer chains; the existence and abundance of attractive forces between molecules (and this obviously is related to solubility); the freedom of molecular side groups, branches, and segments to rotate around intermonomer bonds; chain stiffness; and chain length. The most widely accepted method for determining T_g-values of isolated lignins involves differential scanning calorimetry, DSC (54). Whereas DSC is highly accommodating towards liquids, powders, and films, the alternative method involving dynamic mechanical thermal analysis (DMTA) requires test specimens that are capable of transmitting mechanical forces across an unsupported, temperature-controlled space, usually a short distance (55). Dielectric thermal analysis (DETA) has been finding interest by adhesive chemists primarily for the purpose of analyzing cure behavior of thermosets (56). An analysis of T_g-values reported for various types of lignins (Table IV) reveals that lignin has its lowest T_g when left unmodified in wood (57-60). All types of isolation protocols seem to

Table IV. T_g – Values of Lignins

Sample	T_g (°C)	Method	Reference
Lignin in Wood			
- Hardwood	65-85	DMTA	Olsson, Salmen, 60
		DMTA	Kelley et al., 59
- Softwood	90-105	DMTA	Olsson, Salmen, 60
		DMTA	Kelley et al., 59
MWL			
- Softwood	138	TMA[1)]	Sakata, Senju, 65
	150	Powder collapse[2)]	Takamura, 66
	160	DSC	Glasser et al., 27
- Hardwood	110-130	DSC	Glasser et al., 27
Periodate Lignin	193	Powder collapse[3)]	Goring, 67
Kraft Lignin	124[4)]	DSC	Hatakeyama et al., 68
	174	TMA[1)]	Sakata, Senju, 65
	169	DSC	Glasser et al., 27
Organosolv Lignin	91-97	DSC	Glasser et al., 27
Steam Explosion L.	113-139	DSC	Glasser et al., 27

[1)]Flow tester (Shimadzu Seisakusho Co. Ltd.)

[2)]Powder shrinking and discoloration observed using a magnifying glass.

[3)]Powder collapse indicated by plunger movement.

[4)]Measured as onset of ΔC_p-transition.

result in a lesser or greater increase in T_g-value. Although not rigorously explored, this increase must be suspected to be related to factors involving chemical structure, functional group (especially phenolic OH) content, intermonomer bond type, and molecular size. The very broad molecular weight distributions of many isolated lignins, and the preponderance of phenolic OH-functionality that invites strong hydrogen bonding effects, are supposedly primarily responsible for difficulties with the determination of T_gs of lignins. T_g-values are often subject to extreme subtlety (i.e., indistinct inflections of the C_P vs. temperature-curve) and to lack of reproducibility (i.e., disappearance of an inflection, often an enthalpy relaxation (61,62) following the first temperature scan through T_g). These problems are normally overcome by derivatization involving esterification (acetates) or etherification (esp. methylation or hydroxypropylation) (63,64). An example of a T_g-determination of a lignin acetate by DSC is illustrated in Fig. 5.

The application of thermal analysis to the evaluation of lignin's intra and intermolecular interaction potential vis-à-vis thermosetting and thermoplastic polymer systems has received increasing attention in the recent past. After all, the incorporation of lignin into man-made polymer and materials systems is one of the principal objectives of any quantitative lignin structure determination effort.

The analysis of lignin-containing thermosetting resins, phenolics, epoxies, polyurethanes, and acrylates, has been studied in numerous adhesives and adhesion laboratories (69). Recent work on epoxy and polyurethane systems involving lignin has attempted to interpret the process of molecular lignin incorporation into infinite networks by adopting Gillham's time-temperature-transition (TTT)-model diagram (70-75). The TTT-diagram (Figure 6) provides a conceptual framework for expressing the process of network formation on a time-scale (70). By isothermally heating a curable resin, molecular transitions can be detected that are related to gelation (i.e., the formation of an infinite network structure from low molecular weight precursors that represents a state which exists below T_g) followed by a vitrification event (i.e., the formation of a more densely crosslinked network structure whose T_g has reached the isothermal cure temperature of a polymer that has undergone glass formation). The gelation and vitrification phenomena have been displayed in resin-impregnated glass fabrics by DMTA (71) (or by other dynamic mechanical thermal analysis methods, such as vibrational braid analysis (70)). By using either temperature or modulus-dependent models, TTT-diagrams have been derived for lignin-containing epoxy and polyurethane systems (72-74). It was observed that in those systems in which a crosslinkable lignin component is converted from liquid (resin) state to vitrified state at such low degrees of conversion that the rest of the resin (i.e., the lignin-free or low lignin-containing component) still exists in the pregel-state, a cured resin structure emerges that resembles a two-phase, glass-filled material in contrast to a thermoset with infinite molecular architecture. This is illustrated in Figure 7. The contribution an isolated lignin can make towards the performance of a thermosetting resin system obviously is related to its chemical structure and reactivity, its molecular architecture (size and branching), and its solubility and compatibility with other components in the resin system. Premature vitrification of the lignin-rich component, at a time when the lignin-free component is still in a state of low degree of crosslinking, represents a condition that inevitably

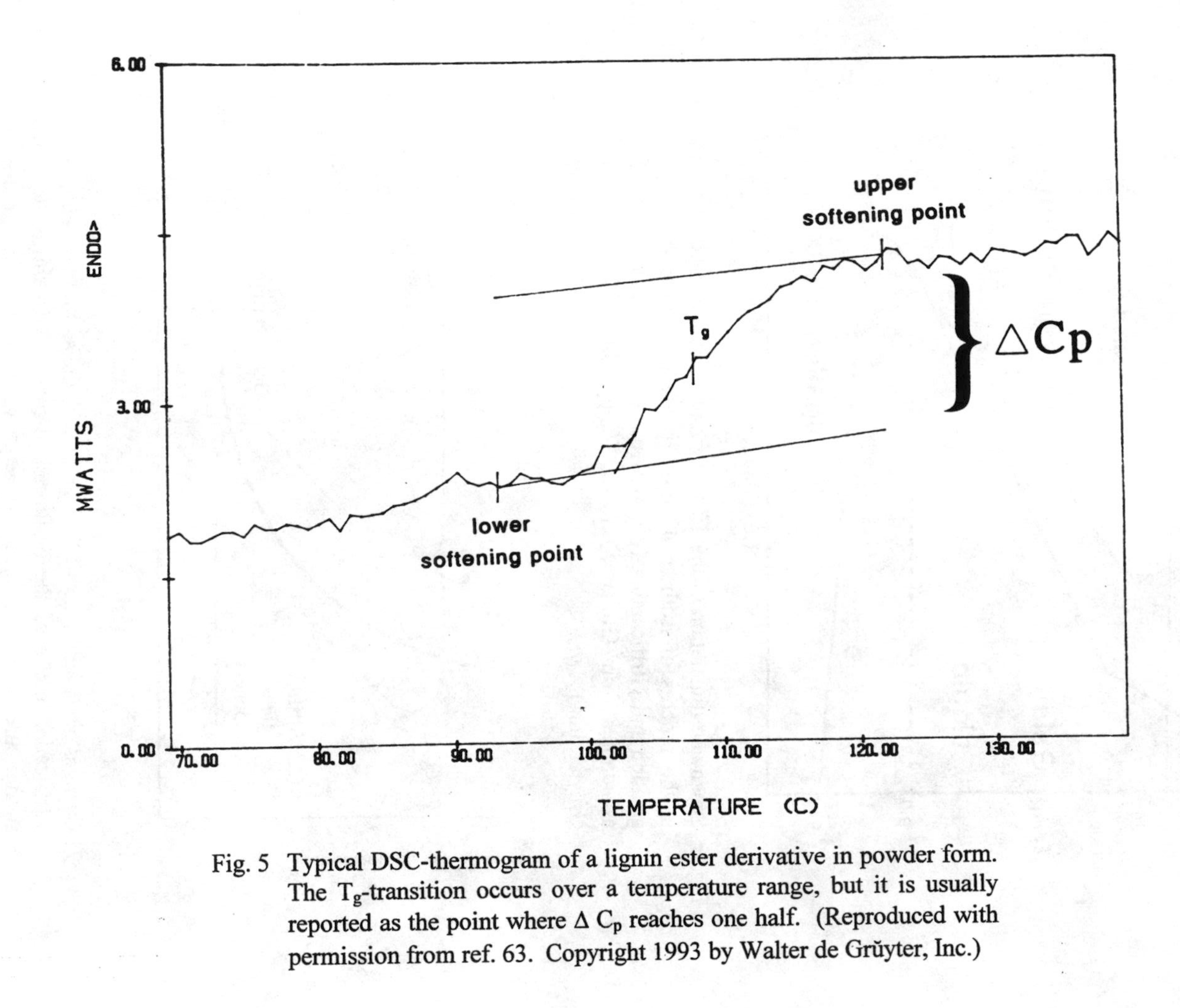

Fig. 5 Typical DSC-thermogram of a lignin ester derivative in powder form. The T_g-transition occurs over a temperature range, but it is usually reported as the point where ΔC_p reaches one half. (Reproduced with permission from ref. 63. Copyright 1993 by Walter de Grüyter, Inc.)

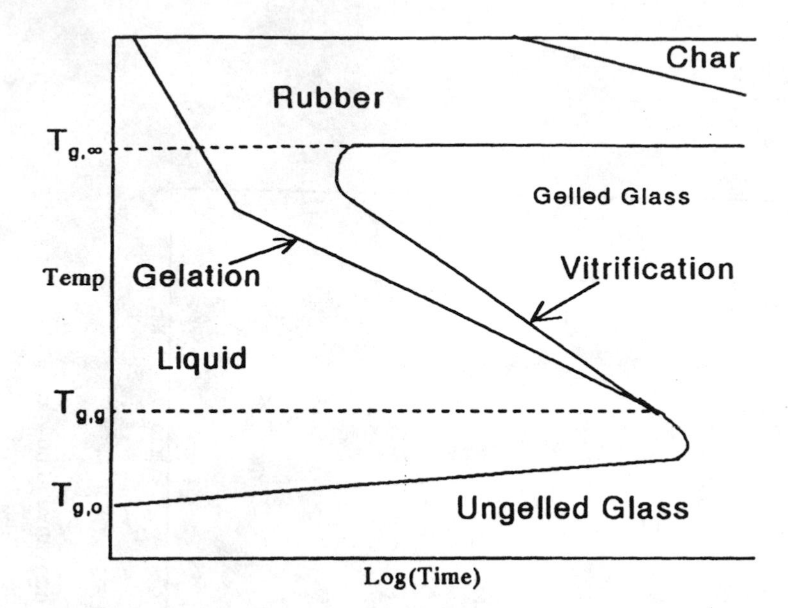

Fig. 6 Schematic illustration of the TTT-diagram revealing important events in the process of isothermal cure: phase separation (not shown), gelation and vitrification. (Reproduced with permission from ref. 74 as adopted from Gillham, ref. 70. Copyright 1997 by Walter de Grŭyter, Inc.)

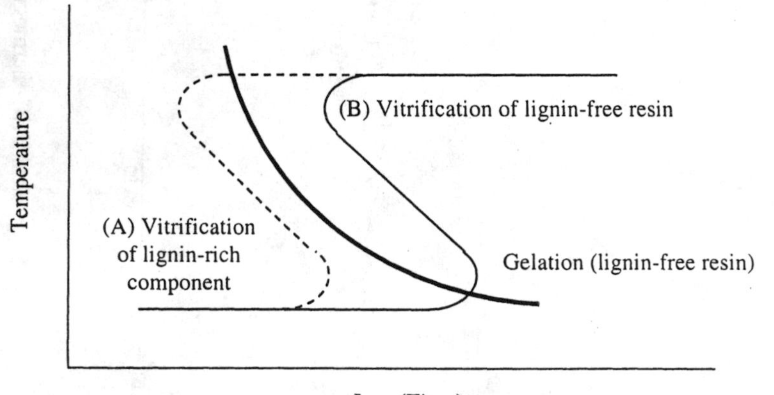

Fig. 7 Changes observed in the TTT-diagram with the addition of lignin: the vitrification curve of the lignin-rich component is shifted towards shorter cure times, often ahead of the gelation curve of the lignin-poor component. This results in glassy inclusions in the uncured resin that behave like glass-like filler particles without contributing to adhesive strength. The extent of shift depends on chemical compatibility

promotes early phase separation of rèsin components and formation of a glass-filled resin structure. If intramolecular attractive forces within a lignin-rich component overwhelm intermolecular attractions vis-à-vis lignin-free resin molecules, a glass-filled composite is produced. Chemical compatibility between the lignin and non-lignin components is therefore of paramount importance to the performance of a lignin-containing thermoset. Lignin chemically modified in such a way as to raise the compatibility with other resin components therefore reveals superior network-forming characteristics. This has been examined for the case of several different types of hydroxyalkyl lignin derivatives in polyurethane resins (76). It was observed that only the most soluble lignins, those being compatible with the other resin-forming components, were capable of sustaining a homogeneous network structure.

The potential contribution isolated lignins and lignin derivatives can make towards the engineering properties of melt-processed thermoplastic materials has been examined in numerous blend studies (77-90). These have involved solution and/or melt-blended mixtures of lignin and lignin derivatives with PVA (78), PVAc (78), LDPE (79), PP (89) EVA (79), PMMA (77), PCL (86), PS (88), PVC (86), cellulose esters (80-83), and even cellulose (90). These studies typically have focused on the ability of a given lignin or lignin derivative to interact with another macromolecule having a linear backbone, on the molecular level. Molecular level thereby requires the dispersion of lignin at a scale below one micron. The relationship between phase dimension and glass transition behavior (by DMTA) is illustrated in Figure 8. If the phase dimensions of the enclosed (i.e., minor) polymeric phase are very large in comparison to the resolving power of the analytical procedure used (i.e., the probe size), distinct phase incompatibility is detected (91). If, conversely, the phase dimensions become much smaller than the probe size of the analytical method a compatible blend results that exhibits a single glass transition located between the T_g of the respective molecular parents. If probe size and phase dimension are about equal, a semi-compatible material structure emerges that exhibits a hybrid thermal behavior with an intermediate T_g (Fig. 8). Thus, most thermoplastic blend systems have been evaluated using DMTA-methodology in relation to T_g-migrations. Although strong intermolecular interactions have been observed with blends of lignin with cellulose (90) and cellulose esters (80-83), most man-made polymers revealed such positive interactions only after lignin modification (85-88). Obviously, the abundance of H-bonds present in lignin leads to an imbalance between intramolecular secondary bonds and the potential for forming intermolecular attractions. Following modification partial, but often substantial, intermolecular attractions have been established for mixtures of lignin with most man-made polymers, especially those with some polar functionality (78,79, 85-88). This is illustrated for blends of lignin derivatives with PVC (Figures 9,10) in which lignin was progressively compatabilized with PVC by chemical modification (87). Thermal analysis results were found to be supported by transmission electron microscopy (Fig. 9) (86). The magnitude of phase separation, and the dimensions of the included phase within a continuous phase, have thereby been recognized to be the consequence of the balance between intra- vs. intermolecular forces. This has laid the foundation for understanding the structure-property relationship between isolated

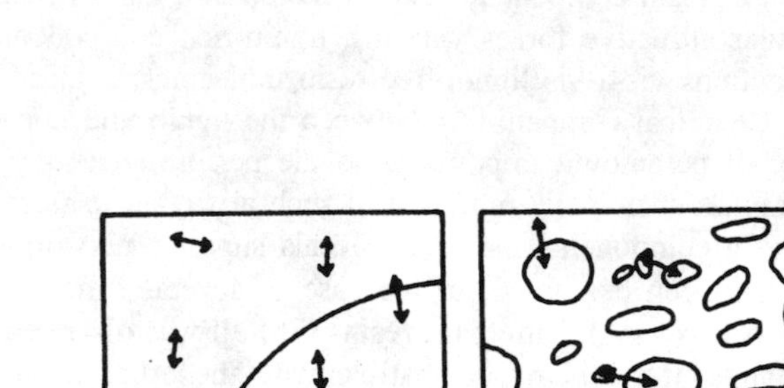
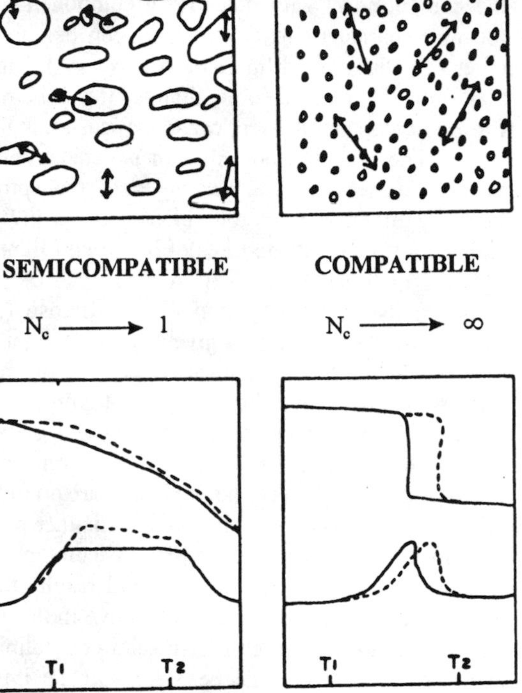

Fig. 8 Illustration of phase-separated vs. compatible blends in relation to dimensional parameters of the analytical method: if the resolution of the phase-probing method is much smaller than the phase dimensions (left), incompatibility is revealed; and, conversely, if the resolution is much larger than the phase dimensions (right), compatibility manifests itself in a single T_g-transition. (Reproduced with permission from ref. 91. Copyright 1976 by John Wiley & Sons, Inc.)

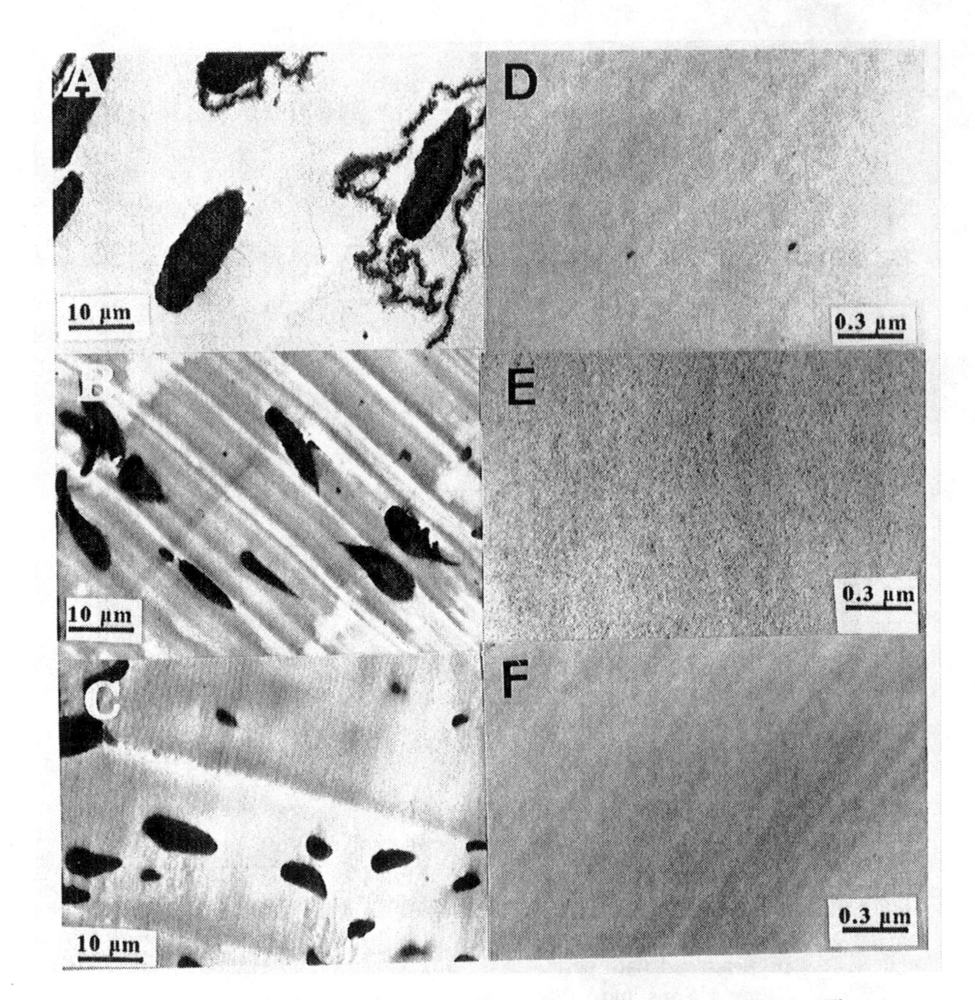

Fig. 9. Blends of lignin (derivative) and PVC as examined by TEM. The solvent-cast film samples consisted of mixtures of PVC with (unmodified) lignin (10%, A; 20%, B; 40%, C) or with lignin-caprolactone copolymer (10%, D; 20%, E; 40% F). Note the difference in magnification between the A to C-series and the D to F-series. (Reproduced with permission from ref. 86. Copyright 1994 by John Wiley & Sons, Inc.)

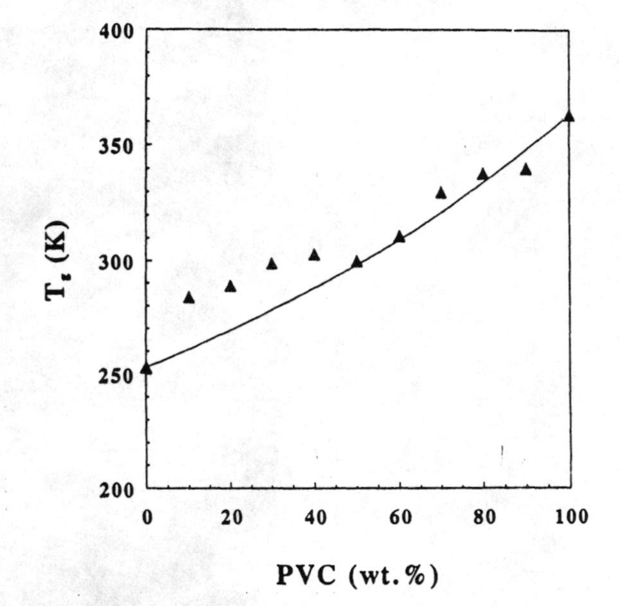

Fig. 10. The glass transition behavior of the solvent-case blends of lignin-caprolactone copolymers with PVC illustrated in Fig. 9, D to F. The recorded single T_g-values are intermediate to those of the blend components, and they obey the Fox-equation (solid line). (Reproduced with permission from ref. 86. Copyright 1994 by John Wiley & Sons, Inc.)

lignin derivatives and synthetic polymers in thermoplastic multi-component material systems.

Conclusions

The quantitative description of the chemical and molecular structures of isolated lignins has advanced substantially in the past decades. Lignin preparations can now be quantitatively described and classified in terms of the chemistry of their monomeric repeat units; the chemistry of the major intermonomer bonds; their molecular weight parameters; and several aspects of their inter- and intramolecular interactions. Isolated lignin needs no longer be considered as the *bête inconnue*. Simplicity and field-wide acceptance as well as universal adaptability and method-availability all are viewed as being essential for the adoption of underutilized isolated lignins in structural material systems.

Literature Cited

1. Glasser, W. G. "Lignin," Chapter 2 in Pulp and Paper—Chemistry and Chemical Technology, 3rd ed., Vol. I, J. P. Casey, ed., John Wiley & Sons, **1980**, pp. 39-111.
2. Glasser, W. G. and S. S. Kelley. "Lignin," Enc. Polym. Sci. Eng., Vol. 8, John Wiley & Sons, Inc., **1987,** pp. 795-852.
3. Chum, H. L., S. K. Parker, D. A. Feinberg, J. D. Wright, P. A. Rice, S. A. Sinclair, and W. G. Glasser. "The Economic Contribution of Lignins to Ethanol Production from Biomass." SERI/Technical Report 231-2488, **1985**, 86 pg.
4. Northey, R. A., Chapter in Emerging Technologies for Materials and Chemicals from Biomass, R. M. Rowell, T. P. Schultz, and R. Narayan, eds., ACS Symp. Ser. **1992**, 476, 164-175.
5. Gargulak, J. D., and S. E. Lebo. ACS Symp. Ser., in press.
6. Gierer, J. Holzforschung, **1982,** 36(2):55-64.
7. Gierer, J. Wood Sci. Technol., **1986**, 20, 1-33.
8. Gierer, J. Holzforschung, **1990**, 44, 387-394.
9. Gellerstedt, G., and E. Lindfors. Holzforschung **1998**, 38, 151-158.
10. Lewis, N. G., and M. G. Paice, eds. Plant Cell Wall Polymers—Biogenesis and Biodegradation. ACS Symp. Ser. **1989**, 399, 676 pg.
11. Hon, D. N.-S., and N. Shiraishi, eds., Wood and Cellulosic Chemistry, Marcel Dekker, Inc., New York, **1991**, 1020 pg.
12. Jung, H. G., D. R. Buxton, R. D. Hatfield, J. Ralph, eds., Forage Cell Wall Structure and Digestibility, American Society of Agronomy, Inc., Madison, WI, **1991**, 794 pg.
13. Lin, S. Y. and C. W. Dence, eds., Methods in Lignin Chemistry, Springer-Verlag, New York, **1992**, 578 pg.
14. Freudenberg, K. and A. C. Neish. Constitution and Biosynthesis of Lignin. Springer Verlag New York, Inc., **1968**, 123 pg.

15. Sakakibara, A., Chemistry of Lignin. Chapter 4 in Wood and Cellulosic Chemistry, D. N.-S. Hon and N. Shiraishi, eds., Marcel Dekker, Inc., **1991**, pp. 113-175.
16. Glasser, W., H. Glasser, and N. Morohoshi. Macromolecules **1981,** 14, 253-262.
17. Lai, Y. Z. Chapter in Methods in Lignin Chemistry, Y. S. Lin and C. V. Dence, eds., Springer Verlag Berlin, **1992**, 423-434.
18. Faix, O., B. Andersons, D. S. Argyropoulos, D. Robert. "Quantitative Determination of Hydroxyl and Carbonyl Groups of Lignins-An Overview." Proc. 8th Interntl. Symp. Wood Pulping Chem., **1995**, Vol. I, pp. 559-566.
19. Mansson, P., B. Samuelsson. Holzforschung **1983**, 37, 143-146.
20. Ludwig, C. H., B. J. Nist, J. L. McCarthy. J. Amer. Chem. Soc. **1964,** 80:1186-1196.
21. Ludwig, C. H., B. J. Nist, J. L. McCarthy. J. Amer. Chem. Soc. **1964**, 80:1196-1202.
22. Lenz, B. L. Tappi **1968**, 51:511-519.
23. Lundquist, K. Acta Chem. Scand. **1979**, B33:27-30.
24. Lundquist, K. Acta Chem. Scand. **1980**, B34:21-26.
25. Glasser, W. G., C. Barnett, T. Rials, V. Saraf. J. Appl. Polym. Sci. **1984**, 29, 1815-1830.
26. Glasser, W. G. and H. R. Glasser. Paperi ja Puu **1981**, 63(2):71-83.
27. Glasser, W. G., C. A. Barnett, P. C. Muller, K. V. Sarkanen. J. Agric. Food Chem. **1983**, 31(5):921-930 (1983).
28. Glasser, W. G., C. A. Barnett, Y. Sano. Appl. Polym. Symp. **1983**, 37:441-460.
29. Freudenberg, K., C.-L. Chen. Chem. Ber. **1967**, 100: 3683-3688.
30. Larsson, S., G. E. Miksche. Acta Chem. Scand. **1967**, 21:1970-1971 (1967).
31. Eriksson, M., S. Larsson, G. E. Miksche. Acta Chem. Scand. **1973**, 27:127 -140.
32. Morohoshi, N., W. G. Glasser. Wood Sci. Technol. **1979**, 13(3):165-178.
33. Morohoshi, N., W. G. Glasser. Wood Sci. Technol. **1979**, 13(4):249-264.
34. Gellerstedt, G. Chapter in Methods in Lignin Chemistry, S. Y. Lin and C. W. Dence, eds., Springer-Verlag, New York, **1992**, p. 322-333.
35. Nimz, H., and K. Das. Chem. Ber. **1971**, 104, 1871-1876.
36. Nimz, H. "Analytical Methods in Wood, Pulping and Bleaching Chemistry," Proc. 8th Interntl. Symposium on Wood and Pulping Chemistry, Gummerus Kirjapaino OY, Jyvaskyla, Finland, **1995**, Vol. I, 1-31.
37. Lapierre, C., and B. Monties. Holzforschung **1986**, 40(2):113-118.
38. Lapierre, C. Holzforschung **1988**, 42(1):1-4.
39. Lapierre, C., B. Pollet, B. Monties. Holzforschung **1991**, 45(1):61-68.
40. Rolando, C., B. Monties, and C. Lapierre. Chapter in Methods in Lignin Chemistry, S. Y. Lin and C. W. Dence, eds., Springer-Verlag, New York, **1992**, pp. 334-349.

41. Lapierre, C. Chapter in Forage Cell Wall Structure and Digestibility, H. G. Jung, D. R. Buxton, R. D. Hatfield, J. Ralph, eds., ASA, Inc., Madison, WI, **1993**, pp. 133-166.
42. Goring, D. A. I., Chapter in Lignins-Occurrence, Formation, Structure and Reactions. K. V. Sarkanen and C. H. Ludwig, eds., Wiley Interscience, New York, **1971**, p. 695-768
43. Gupta, P. R., and J. L. McCarthy. Macromolecules 1, **1968**, 236.
44. Forss, K. G., B. G. Stenlund, and P. Sagfors. Appl. Polym. Symp. **1976,** 28:1185-1194.
45. Gellerstedt, G., Chapter in Methods in Lignin Chemistry, S.Y. Lin and C. W. Dence, eds., Springer-Verlag, Berlin and Heidelberg, **1992**, p. 487-497.
46. Himmel, M. E., K. Tatsumoto, K. K. Oh, K. Grohmann, D. K. Johnson, H. L. Chum. Chapter in Lignin-Properties and Materials, W. G. Glasser and S. Sankanen, eds., ACS Symp. Ser. **1989**, 397, 82-99.
47. Siochi, E. J., M. A. Haney, W. Mahn, and T. C. Ward. Chapter in Lignin-Properties and Materials. W. G. Glasser and S. Sarkanen, eds. ACS Symp. Ser. **1989**, 397, 100-108.
48. Siochi, E. J., T. C. Ward, M. A. Haney, W. Mahu. Macromolecules **1990**, 23, 1420-29.
49. Glasser, W. G., V. Dave, C. E. Frazier. J. Wood Chem. Tech. **1993**, 13(4), 545-559.
50. Bolker, H. I. Natural and Synthetic Polymers, Marcel Dekker Inc., New York, **1974**, pg. 577-621.
51. Pla, F. Chapter in Methods in Lignin Chemistry, S. Y. Lin and C. W. Dence, eds., Springer-Verlag, Berlin and Heidelberg, **1992**, p. 498-508.
52. Young, R. J., and P. A. Lovell. Introduction to Polymers. 2nd edition, Chapman & Hall, London, **1991**, p. 214.
53. Wunderlich, B., Chapter in Thermal Characterization of Polymeric Materials, E. A. Turi, ed., Academic Press, New York, **1981**, p. 92-234.
54. Irvine, G. M. TAPPI J. **1984**, 67(5):118-121.
55. Wendlandt, W. W. and P. K. Gallagher. Chapter in Thermal Characterization of Polymeric Materials, E. A. Turi, ed., Academic Press, New York, **1981**, p. 1-90.
56. Wolcott, M. P. and T. G. Rials. For. Prod. J., **1995**, 45(2), 72-77.
57. Salmen, L. Temperature and Water-Induced Softening Behavior of Wood Fiber -Based Materials. Dissertation, The Royal Institute of Technology, Stockholm, **1982**.
58. Salmen, L., J. Mater. Sci. **1984**, 19(9), 3090.
59. Kelley, S. S., T. G. Rials, W. G. Glasser. J. Materials Sci. **1987**, 22:617-624.
60. Olsson, A.-M. and L. Salmen. Chapter in Viscoelasticity of Biomaterials, W. G. Glasser and H. Hatakeyama, eds., ACS Symp. Ser. **1992**, 489, 133-143.
61. Hatakeyama, T., and F. X. Quinn. Thermal Analysis-Fundamentals and Applications to Polymer Science. J. Wiley & Sons, New York, **1994**, 81-86.
62. Rials, T. G., and W. G. Glasser. J. Wood Chem. Technol. **1984**, 4, 331-345.
63. Glasser, W. G., and R. K. Jain. Holzforschung **1993**, 47(3):225-233.
64. Jain, R. K., and W. G. Glasser. Holzforschung **1993**, 47(4), 325-332.

65. Sakato, I. and R. Senju. J. Appl. Polym. Sci. 19, **1975**, 2799-2810.
66. Takamura, N. J. Japan Wood Res. Soc. **1968**, 14, 75-79.
67. Goring, D. A. I. Pulp and Paper Mag. Can., **1963** (Dec.), T517-527.
68. Hatakeyama, H., K. Iwashita, and G. Meshitsuka. J. Japan Wood Res. Soc. **1975**, 21, 618-623.
69. Lewis, N. G., and T. R. Lantzy. Chapter in Adhesives From Renewable Resources, R. W. Hemingway and A. H. Conner, eds., ACS Symp. Ser. No. 385, Washington, DC, **1989**, pg. 13-26.
70. Gillham, J. K. Chapter in Structural Adhesives – Developments in Resins and Primers. A. J. Kinloch, ed. Elsevier Applied Sci. Publ., London, **1986**, 1-27.
71. Hofmann, K., and W. G. Glasser. Thermochimica Acta **1990**, 166, 169-184.
72. Hofmann, K., and W. G. Glasser. J. Adhesion **1993**, 40, 229-241.
73. Hofmann, K. and W. G. Glasser. Macromol. Chem. Phys. **1994**, 195, 65-80.
74. Toffey, A. and W. G. Glasser. Holzforschung **1997**, 51, 71-78.
75. Kinloch, A. J. Chapter in Developments in Structural Adhesives, A. J. Kinloch, ed., Elsevier Appl. Sci., Oxford, **1986**, pg. 127-162.
76. Rials, T. G. and W. G. Glasser. Holzforschung **1986**, 40, 353-360.
77. Ciemniecki, S. L. and W. G. Glasser. Polymer **1988**, 29, 1021-1029.
78. Ciemniecki, S. L. and W. G. Glasser. Polymer 29, **1988**, 1030-1036.
79. Glasser, W. G., J. S. Knudsen, and C.-S. Chang. J. Wood Chem. Technol. **1988,** 8(2),221-234 (1988).
80. Rials, T. G. and W. G. Glasser. J. Appl. Polym.-Sci. **1989**, 37, 2399-2415.
81. Rials, T. G. and W. G. Glasser. Wood Fiber Sci. **1989**, 21(1):80-90.
82. Demaret, V. and W. G. Glasser. Polymer 30(3), **1989**, 570-575.
83. Rials, T. G. and W. G. Glasser. Polymer **1990**, 31, 1333-1338.
84. Kelley, S. S., T. C. Ward, and W. G. Glasser. J. Appl. Polym. Sci. **1990**, 41, 2813-2828 (1990).
85. De Oliveira, W. and W. G. Glasser. Macromolecules **1994**, 27, 5-11.
86. De Oliveira, W. and W. G. Glasser. J. Appl. Polym. Sci. **1994**, 51, 563-571.
87. De Oliveira, W. and W. G. Glasser, Polymer **1994**, 35(9), 1977-1985.
88. De Oliveira, W. and W. G. Glasser. J. Wood Chem. Technol. **1994,** 14(1), 119-126.
89. Kosikova, B., V. Demianova, and M. Kacurakova. J. Appl. Polym. Sci. **1993**, 47, 1065-1073.
90. Glasser, W. G., T. G. Rials, S. S. Kelley, and V. Dave. ACS Symp. Ser. No. 688, **1998**, 265-282.
91. Kaplan, D. S. J. Appl. Polym. Sci. **1976**, 20, 2615-2629.

Chapter 10

An Overview of Chemical Degradation Methods for Determining Lignin Condensed Units

Y.-Z. Lai, H. Xu[1], and R. Yang

Empire State Paper Research Institute, Faculty of Paper Science and Engineering, State University of New York College of Environmental Science and Forestry, Syracuse, NY 13210

Although a variety of chemical degradation methods can be used to characterize the chemical nature of lignocellulosic substrates, it is still a great challenge to reveal quantitatively all type of lignin units in situ, especially the diphenylmethane (DPM)-type structures. Recent studies on lignin DPM-model dimers indicate that they were rather reactive under both nitrobenzene and permanganate oxidation conditions including the formation of oxidized dimers characteristic of the parent units. The extent to which these dimeric products maybe used to evaluate lignin condensation reactions is discussed

Lignin in general plays a negative role in the chemical utilization of lignocellulosic materials. It must be modified and either partially or totally removed depending on the desired quality of final products. The aromatic groups in lignin exist in either phenolic or etherified form with the phenolic units accounting for less than 15% of wood lignin in situ (*1*). In alkaline pulping, the etherified lignin components are gradually degraded to phenolic structures resulting in dissolution (*2*). The last 5-10% of residual pulp lignin is known to be very resistant to alkaline degradation and generally requires a multiple bleaching sequence for a complete delignification (*2-4*).

[1]Current address: Union Camp Corporation, P.O. Box 3301, Princeton, NJ 08543–3301

Despite extensive studies (*5, 6*), the chemical nature of residual lignin which causes resistance to degradation is still not fully understood notably; the role of lignin condensation reactions in alkaline pulping play in the process. Major challenges encountered in lignin analysis (*7*) are the lack of a specific procedure for a quantitative isolation of lignin in a pure and unaltered form as well as the limitation of analytical methods which can reveal quantitatively all types of lignin units in situ.

Figure 1 indicates the major substitution patterns of guaiacyl nuclei that may occur in softwood or technical lignins. These include the uncondensed **1**, β-5 **2**, 5,5' **3** and 4-0-5 **4**, - linked units. Additionally, the diphenylmethane (DPM)-type units 5-7 maybe formed during acidic or alkaline treatments, and are generally thought of being absent in the wood lignin except small amounts of the α-6 type **6**. Although a variety of chemical degradation methods (*8*) have been used to determine the nature of lignin units in situ, most of the product yields were not quantitative and can only be used to analyze certain types of condensed units. Also, none of the existing methods is capable of determining specifically DPM-type structures.

This paper reviews briefly the principle and limitations of available degradation methods used for revealing the lignin condensed units in situ, as well as our recent attempts to develop specific procedures for determining the DPM-type structures.

Acidolysis, Thioacidolysis and DFRC Methods

Acidolysis is conducted in a 9:1 dioxane-water mixture containing 0.2 M HCl (*9*), whereas thioacidolysis (*10*) is a solvolysis in ethandiol with boron trifluoride etherate. Both procedures are effective in the cleavage of α and β-aryl ether linkages, and have been used to determine the content of uncondensed β-0-4 aryl ether structures in wood or pulp (*11, 12*) lignins. The thioacidolysis method, however, has the advantage of providing simpler products in higher yields (Figure 2). In addition to monomeric products, several dimeric products representing the β-5, 5,5', and β-1 linkages have also been identified by acidolysis or a combined thioacidolysis and desulfurization technique (*13*).

Recently, a "DFRC" method based on Derivatization (with acetyl bromide) Followed by Reductive Cleavage (in an acidic medium) and acetylation has been proposed to determine the uncondensed β-0-4 units (*14*). This procedure, based on the yield of 4-acetoxycinnamyl acetate **12**, is quite similar to the thioacidolysis method. It is evident that all these acidic procedures determine mainly the uncondensed β-0-4 structures, and can only provide an indirect estimation of the condensed units.

Nitrobenzene Oxidation

Figure 3 illustrates the typical alkaline nitrobenzene oxidation products from the guaiacyl units of Norway spruce wood lignin (*15*). Traditionally, the yield of vanillin

Uncondensed

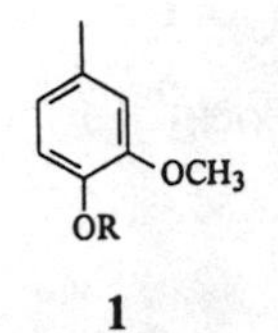

Condensed Units in the Wood Lignin

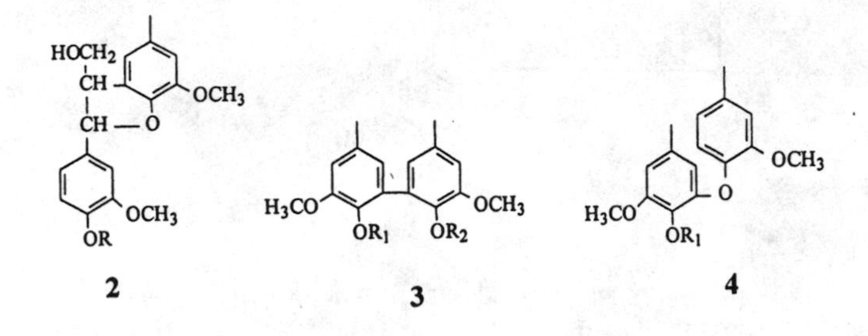

Diphenylmethane Types Condensed Units

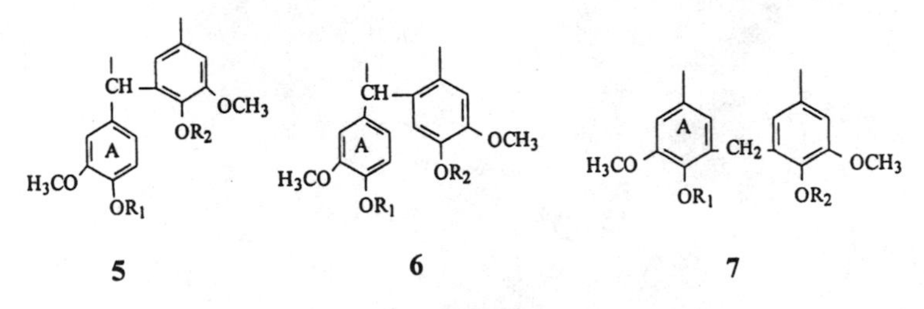

Figure 1. Major types of guaiacyl nuclei.

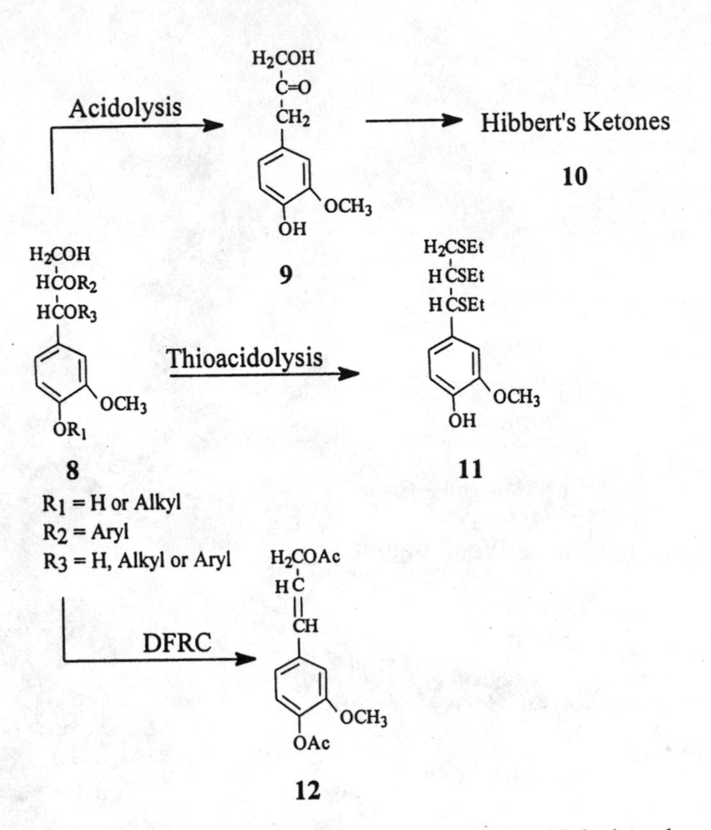

Figure 2. Major products from acidolysis, thioacidolysis,and "DFRC" methods.

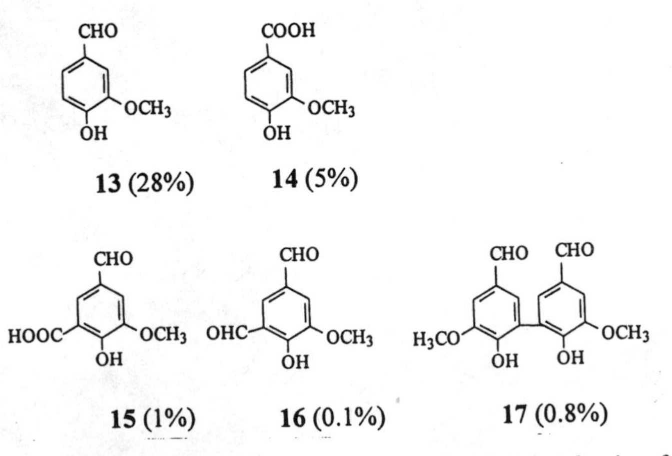

Figure 3. Nitrobenzene oxidation products from guaiacyl units of Norway spruce wood lignin (*15*).

13 and vanillic acid **14** has been used to estimate the content of uncondensed guaiacyl units (*16, 17*). These products, however, may be also produced to a certain extent from the DPM units as determined in lignin model compound experiments (*18, 19*). On the other hand, 5-carboxyvanillin **15** and 5-formylvanillin **16** are derived from the C-5 condensed guaiacyl units, whereas dehydrodivanillin **17** is from the biphenyl units. Since the yield of nitrobenzene oxidation products from uncondensed structures is considerably affected by chemical nature of the side chain units, it is not entirely quantitative. Also, the yield of condensed products **15-17** is relatively small, and therefore this method serves only as a qualitative identification procedure.

Permanganate Oxidation

Permanganate oxidation involves an initial methylation or ethylation of the phenolic hydroxyl groups followed by sequential oxidation with permanganate and then hydrogen peroxide to yield a variety of mono- and di-carboxylic acid derivatives (*20*). Figure 4 illustrates the yield proportion as percentages of total oxidation products identified from Norway spruce wood lignin. The relative amount of veratric acid **18** is frequently used as an indication of the uncondensed guaiacyl units. This acid, however, may also be derived partially from the DPM units as determined through lignin model compound reactions (*21, 22*).

Although the isohemipinic **19** and metahemipinic acid **20** acids are derived from C-5 and C-6 condensed units respectively, the quantitative contribution of lignin sub-structures to these acids is still not fully understood. For example, model compound reactions (*20*) indicate that the source of isohemipinic acid **19** may include, in addition to β-5 related structures **2**, a mono-phenolic biphenyl unit or an incomplete alkylation of phenolic biphenyl units **3** (Figure 1). On the other hand, the dimeric acids **21** and **22** are derived from the 4-0-5 and biphenyl structures, respectively.

The proportion of dicarboxylic acids **19-22** has been used to estimate the extent of lignin condensed units among the phenolic structural units (*23, 24*). However, this procedure reveals little information on the nature of the etherified lignin component, which accounts for the bulk of wood or pulp lignins (*1, 5*). Recent findings (*25*) suggest that the phenolic and etherified components of wood lignins differ significantly in chemical characteristics as reflected in the yield of nitrobenzene oxidation products. The phenolic lignin units appears to be appreciably less condensed than the etherified counterpart.

Nucleus Exchange Reactions

The phenyl nucleus exchange technique (*26*), developed originally by Funaoka and Abe (*27, 28*), is based on the degradation of lignin by boron trifluoride in the presence of excess phenol. They observed that the uncondensed and α-5 diphenylmethane model compounds are much more reactive than other types of lignin model units and can be selectively oxidized to guaiacol **23** and catechol **24** (Figure 5). This contention, however, was recently questioned by Chan et al. (*29*).

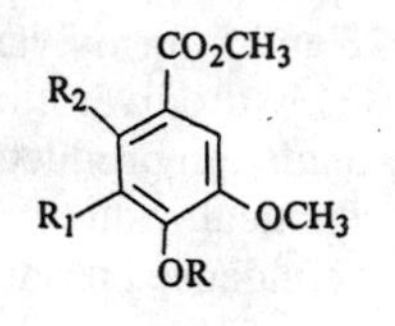

18 $R_1 = R_2 = H$ (67%)
19 $R_1 = CO_2H, R_2 = H$ (10%)
20 $R_1 = H, R_2 = CO_2H$ (8%)

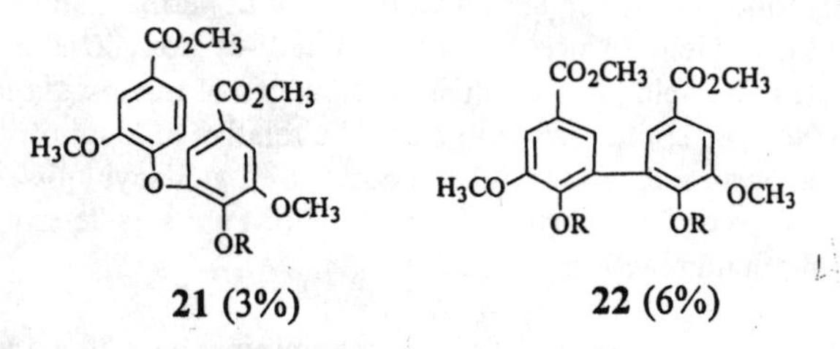

Figure 4. Permanganate oxidation products from guaiacyl units of Norway spruce wood lignin (*20*).

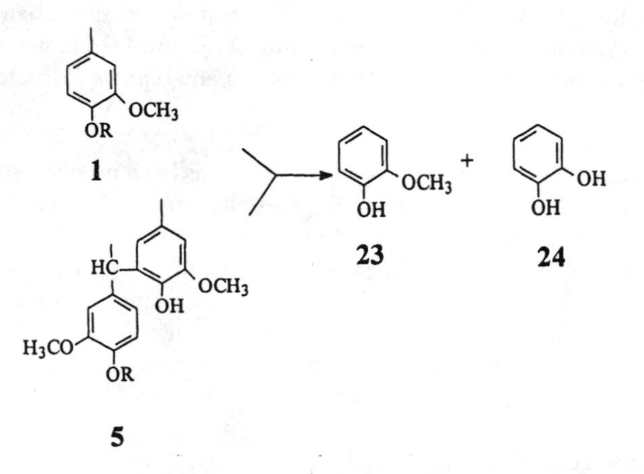

Figure 5. Major products from nucleus exchange reaction of uncondensed and α-5 diphenylemethane-type guaiacyl units (*26*).

There are no apparent explanations for these inconsistent findings. However, it appears that the reaction conditions used by Chan et al. (*29*) were much more drastic than that of Funaoka judging from the absence of guaiacol in the reaction mixture. Funaoka et al. (*26-28*) reported that the yield of guaiacol accounted for more than 30% of the total products. The extent to which the selectivity of nucleus exchange reactions toward different lignin units may vary with reaction conditions merits a careful evaluation.

Determination of Condensed Units

Although a variety of chemical degradation techniques have been used to identify certain types of condensed units, these methods, as noted earlier, are not suitable for a direct and quantitative determination of total condensed structures. Permanganate oxidation is the only method able to measure C-5 and C-6 condensed units of the phenolic type. This method, however, is unable to differentiate between the specific types of C-5 condensed structures, and also does not measure etherified lignin structures. Additionally, no existing procedures are available for identification of DPM-type units.

Attempts to Develop Methods for Detecting DPM units

The reactivity of guaiacyl DPM-model dimers was examined under both nitrobenzene (*19*) and permanganate (*22*) oxidation conditions as part of our studies aimed at understanding the significance of DPM-type condensed units in residual kraft pulp lignin (5, 30). These unsubstituted DPM dimers were shown to be quite reactive under oxidative conditions resulting in the formation of monomeric and dimeric products. The dimeric products as shown in Figures 6 and 7 are characteristic of the corresponding parent DPM units.

Nitrobenzene oxidation. A common reaction of the α-5 **26**, α-6 **27**, and α-1 **28** dimers is an oxidation of the α-methylene groups leading to the formation of diguaiacylketone derivatives **31, 33** and **34**, which were obtained in 5-28% yield (Figure 6). In contrast, the methylene group of 5,5-DPM dimer **29** was fairly resistant to oxidation. This dimer gave two 5,5-DPM aldehydes **35** (5%) and **36** (53%) corresponding to the oxidation of one and two methyl groups, respectively. Similar reactions were also observed for the α-5 and α-6 dimers (**26** and **27**) giving the corresponding diguaiacylmethane monoaldehyde **30** (29%) and **32** (11%), respectively.

Permanganate oxidation. Major products from the permangante oxidation of unsubstituted guaiacyl α-DPM model dimers were the diguaiacylketone compounds **37-41** (36-47%) (Figure 7). Thus, there is a close similarity between permanganate and nitrobenzene oxidation on the reaction of unsubstituted DPM dimers as evident in the formation of the diguaiacylketone products. The feasibility of using these products to detect the DPM units in modified lignins merits a further study.

Detection of DPM Units in Lignin. Theoretically, the formation of diguaiacylketone derivatives from the nitrobenzene and permangante oxidation

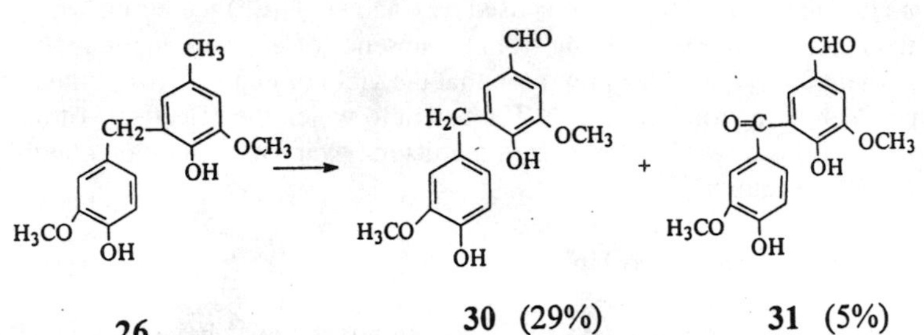

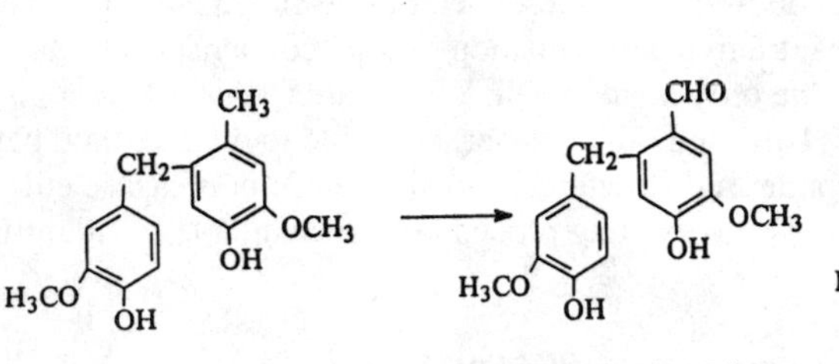

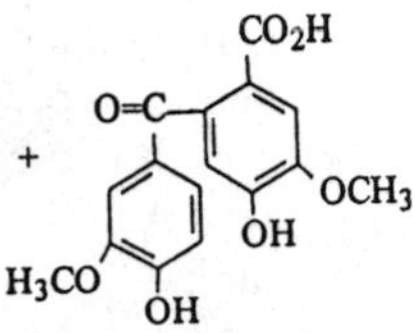

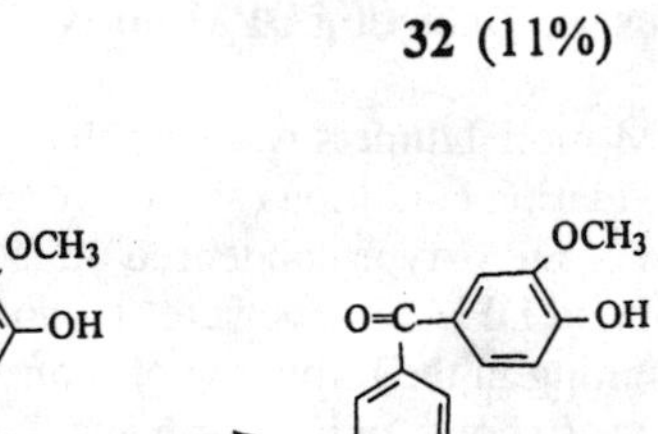

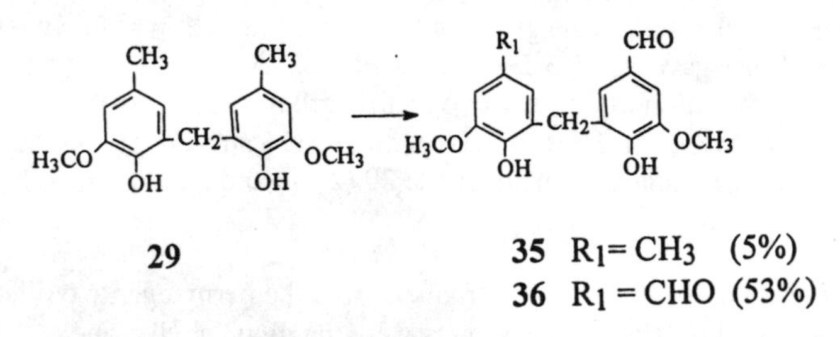

35 $R_1 = CH_3$ (5%)
36 $R_1 = CHO$ (53%)

Figure 6. Nitrobenzene oxidation products characteristic of the parent diphenylmethane units conducted in 2M NaOH at 170°C for 3h (*19*).

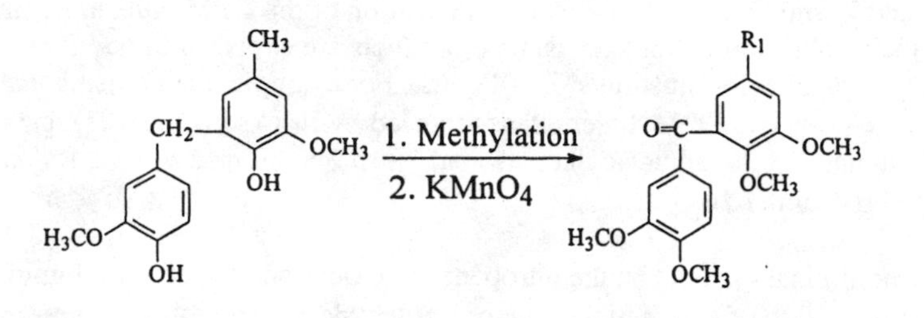

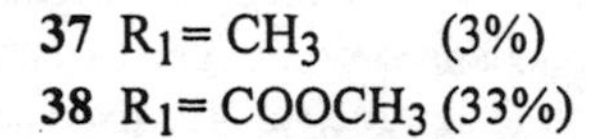

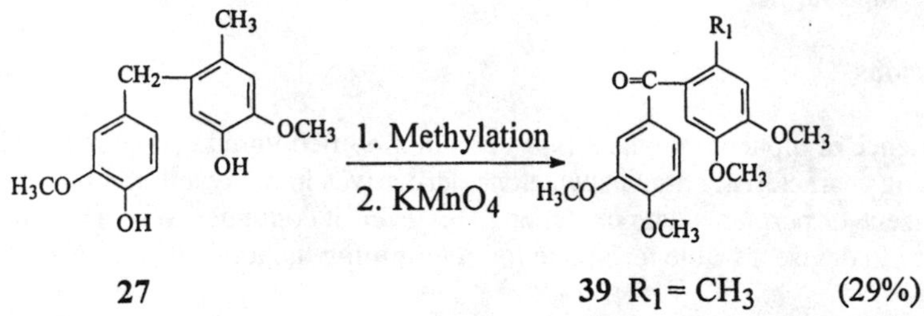

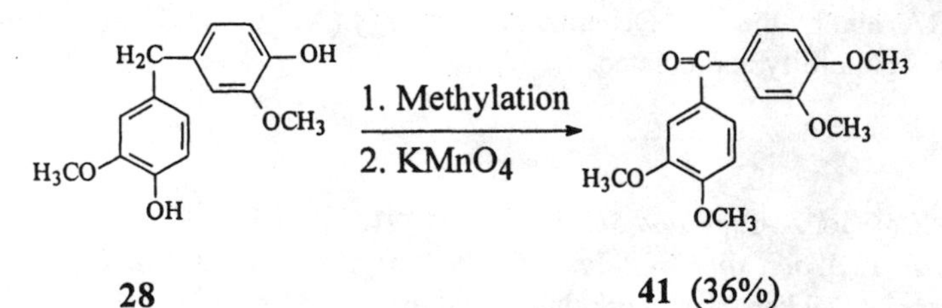

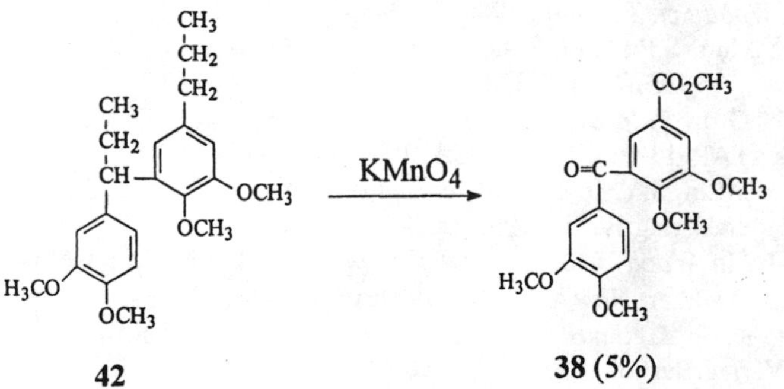

Figure 7. Permanaganate oxidation products characteristic of the parent diphenylmethane units (*22*).

(Figures 6 and 7) should provide a direct identification of the DPM units in lignin. However, the yield of these diguaiacylketone products are expected to be substantially reduced with substituted DPM units. For example, the permanganate oxidation of an ethyl α-5 DPM dimer **42**, as reported by Erickson et al. (*21*), gave only small amounts of the diguaiacylketone acid **38** (5% as compared 33% for an unsubstituted α-5 dimer **26**.

Our preliminary results on the nitrobenzene oxidation of kraft pulp lignin has detected the formation of α-5 diguaiacylketone aldehyde **31** indicating the presence of α-5 condensed units. The extent to which the nitrobenzene and permanganate oxidation techniques maybe used as analytical tool for condensed units are currently being pursued further.

Conclusions

The presence of diphenylmethane-type units in modified lignin can be proved by identifying characteristic diguaiacylketone derivatives in the reaction mixture on nitrobenzene or permanganate oxidation. However, it continues to be a major challenge to devise a viable technique for determining the total condensed units.

Acknowledgments

The financial support of this study by the Empire State Paper Research Associates, Inc., (ESPRA) and by the U.S. Department of Energy (Award No. DE-FC07-96 ID134438) is gratefully appreciated.

References

1. Lai, Y.-Z.; Guo, X.-P., *Wood Sci. Technol.* **1991**, *25*, 467-472.
2. Lai, Y.-Z. In *Wood and Cellulosic Chemistry*; Hon, David N.-S., Shiraishi, N., Eds.; Marcel Dekker: New York, NY, **1990**, pp 455-523.
3. Gierer, J., *Wood Sci. Technol.*, **1985**, *19*, 289-312.
4. Gierer, J., *Wood Sci. Technol.*, **1986**, *20*, 1-33.
5. Lai, Y.-Z.; Mun, S.P.; Luo, S.-G.; Chen, H.-T.; Ghazy, M.; Xu, H.; Jiang, J.E.; *Holzforschung*, **1995**, *49*, 319-322.
6. Gellerstedt, G. In *Pulp Bleaching Principles and Practice*; Dence, C.W.; Reeve, D.W., Eds.; TAPPI Press: Atlanta, GA, 1996, pp 93-111.
7. Dence, C.W.; Lin, S.Y. In *Methods in Lignin Chemistry*; Lin, S.Y.; Dence, C.W.; Eds.; Springer-Verlag:Berlin, 1992, pp 3-19.
8. Chen, C.-L. In *Wood Structure and Composition*, Levin, M.; Goldsteins, I.S., Eds.; Marcel Dekker: New York, NY **1991**, pp 183-261.
9. Lundquist, K. In *Methods in Lignin Chemistry*; Lin, S.Y.; Dence, C.W., Ed.; Springer-Verlag:Berlin, 1992, pp 289-300.
10. Rolando, C.; Monties, B.; Lapierre, C. In *Methods in Lignin Chemistry*; Lin, S.Y.; Dence, C.W.; Eds.; Springer-Verlag: Berlin, 1992, pp 334-349.
11. Gellerstedt, G.; Lindfors, E.L. *Svensk Papperstidn.* **1984**, *87*, R61-R67.
12. Pasco, M.F.; Suckling, D. *Holzforshung* **1994**, *48*, 504-508.
13. Lapierre, C.; Brigitte, P.; Monties, B. *Holzforschung* **1991**, *61*, 61-68.

14. Lu, F.; Ralph, J. In *Proceedings of 9th International Symposium on Wood and Pulping Chemistry*; Montreal, 1997, pp L3-1 - L3-4.
15. Leopold B. *Acta Chem. Scand.* **1952**, *6*, 38-39.
16. Chen, C.-L. In *Methods in Lignin Chemistry*; Lin. S.Y.; Dence, C.W.; Eds.: Springer-Verlag: Berlin, 1992, pp 301-321.
17. Chang, H.-M.; Allan, G.G. In *Lignins*; Sarkanen, K.V.; Ludwig, C.H.; Eds.; Wiley-Interscience: New York, 1991, pp 301-321.
18. Chan, F.D.; Nguyen, K.L.; Wallis, A.F.A. *J. Wood Chem. Technol.* **1995**, *15*, 329-347.
19. Xu, H.; Lai, Y.-Z. *Holzforschung* **1998**, *52*, 51-56.
20. Gellerstedt, G. In *Methods in Lignin Chemistry*; Lin, S.Y.; Dence, C.W., Eds; Springer-Verlag: Berlin, 1992, pp 322-333.
21. Erickson, M.; Larsson, S.; Miksche, G.E. *Acta. Chem. Scand.* **1973**, *27*, 127-140.
22. Meguro, S.; Xu, H.; Lai, Y.-Z. *Holzforschung*, **1998**, *175*, 175-179.
23. Glasser, W.G. *Svensk Papperstidn.* **1981**, *84*, R25-R32.
24. Glasser, W.G.; Barnett, C.A.; Sano, Y. *J. Appl. Polym. Symp.* **1983**, *37*, 441-460.
25. Chen, H.-T.; Funaoka, M.; Lai, Y.-Z. *Wood Sci. Technol.* **1997**, *31*, 433-440.
26. Funaoka, M.; Abe, I.; Chang, V.L. In *Methods in Lignin Chemistry*; Lin, S.Y.; Dence, C.W., Eds.; Springer-Verlag: Berlin, 1992, pp 369-386.
27. Funaoka, M.; Abe, I. *Mokuzai Gakkaishi* **1983**, *29*, 781-788.
28. Funaoka M.; Abe, I. *Wood Sci. Technol.* **1987**, *21*, 261-279.
29. Chan, F.D.; Nguyen, K.L.; Wallis, A.F.A. *J. Wood Chem. Technol.* **1995**, *15*, 473-491.
30. Lai, Y.-Z.; Funaoka, M.; Chen, H.-T. *Holzforschung* **1994**, *48*, 355-359.

Chapter 11

Using Raman Spectroscopy to Identify Chromophores in Lignin–Lignocellulosics

U. P. Agarwal and R. H. Atalla

U.S. Department of Agriculture FS, Forest Products Laboratory, Madison, WI 53705

Raman spectroscopy is being increasingly applied to study lignin and lignin-containing materials. One of the strengths of Raman is that it is highly sensitive to chromophoric structures. Various color-generating species are present in small quantities and are not easily isolated; therefore, the method is ideally suited to the study of chromophores *in situ*. Chromophore contributions can be easily detected because, except for the 1600 cm^{-1} band, which is due to lignin, the region where chromophores most strongly contribute (1500-1750 cm^{-1} region) is devoid of strong contributions from lignocellulosic components. Several chromophore lignin-models including quinones and stilbenes were studied and their Raman contributions were identified. This knowledge was then applied in the studies of photoyellowed thermomechanical pulp (TMP), bleached TMP, and acid-hydrolyzed TMP. In all cases, Raman studies provided important information on presence or absence of various chromophore groups in pulps.

The development of near-IR FT Raman spectroscopy has made it possible to acquire good quality spectra of most lignocellulosic materials [*1*]. Moreover, with the availability of commercial instruments Raman analysis of such materials has become routine [*2*]. In our laboratory, Raman spectroscopy has been used to obtain important information on a number of lignocellulosic materials [*3-6*]. Considering that Raman spectroscopy is complementary to IR, both techniques should be used to obtain full information on the vibrational spectra.

Raman spectroscopy has some capabilities that are not available in IR spectroscopy [*7*]. For instance, the capability to detect a particular group or structure at very low concentration using resonance Raman, conjugation and/or surface enhancement effects. The major advantage is that when one of these effects (which are molecular-structure specific) is present in a sample, Raman spectroscopy becomes a highly sensitive analytical tool. In contrast, in IR the same structures can only be detected at much higher concentrations. Moreover, since some Raman effects can be introduced in a sample, such capabilities are quite versatile. For example, resonance Raman effect depends upon the wavelength of excitation; therefore, it is possible to induce this effect by choosing an appropriate excitation wavelength. We have recently used this approach to study TMPs. Contributions of chromophores in the spectra of TMPs were determined using two different wavelengths of excitation [*8*].

Sensitivity to Chromophores

It has been reported that Raman intensity depends upon several factors [7,9]. In addition to being a function of the Raman instrument and the concentration of a component in a sample, effects like π-conjugation, resonance/pre-resonance Raman, and surface-enhanced Raman are important [*7*]. Depending upon the type of effect and the degree to which a particular effect is present, Raman intensity can be enhanced up to a million times. In the studies of lignocellulosics, intensity enhancement due to π-conjugation and pre-resonance Raman effects has been detected [*10*]. For example, when the Raman spectrum of black spruce was obtained using the 514.5 nm excitation, as much as 86% of the intensity of the 1595 cm^{-1} band of lignin could be assigned to conjugation and pre-resonance Raman effects [*10*].

To gain a better understanding of the dependence of Raman intensity on molecular structure, especially in the case of lignin substructures, a study was undertaken using lignin models. Some of these models can be classified as chromophores (e.g., stilbene and coniferaldehyde). Models that were (to varying degrees) and were not π-conjugated to the aromatic ring were selected (Fig. 1, models 1-19). These model structures are named in Table I.

Spectra of model compounds in CH_2Cl_2 at the same molar concentration were obtained using the 514.5 nm laser excitation. Results of this study are summarized in Fig. 2 where relative Raman intensity of certain model-bands (peak heights ratioed to peak height at 1420 cm^{-1}; the latter band is due to the solvent) is plotted for various models. For the phenyl-group band at 1595 cm^{-1}, the intensities were corrected for the number of phenyl groups per molecule. However, for the band representing a C=C and/or C=O group, except for model 9 intensity correction for models was not needed (other models contain only one C=C and/or C=O group per molecule). In the case of *o*- and *p*- quinones (models 15 and 16), respectively, the intensities of 1559 and 1670 cm^{-1} bands were used. It is worth noting that in quinones, the coupling of C=C and C=O stretch modes some times makes the assignment of the Raman bands difficult [*11*].

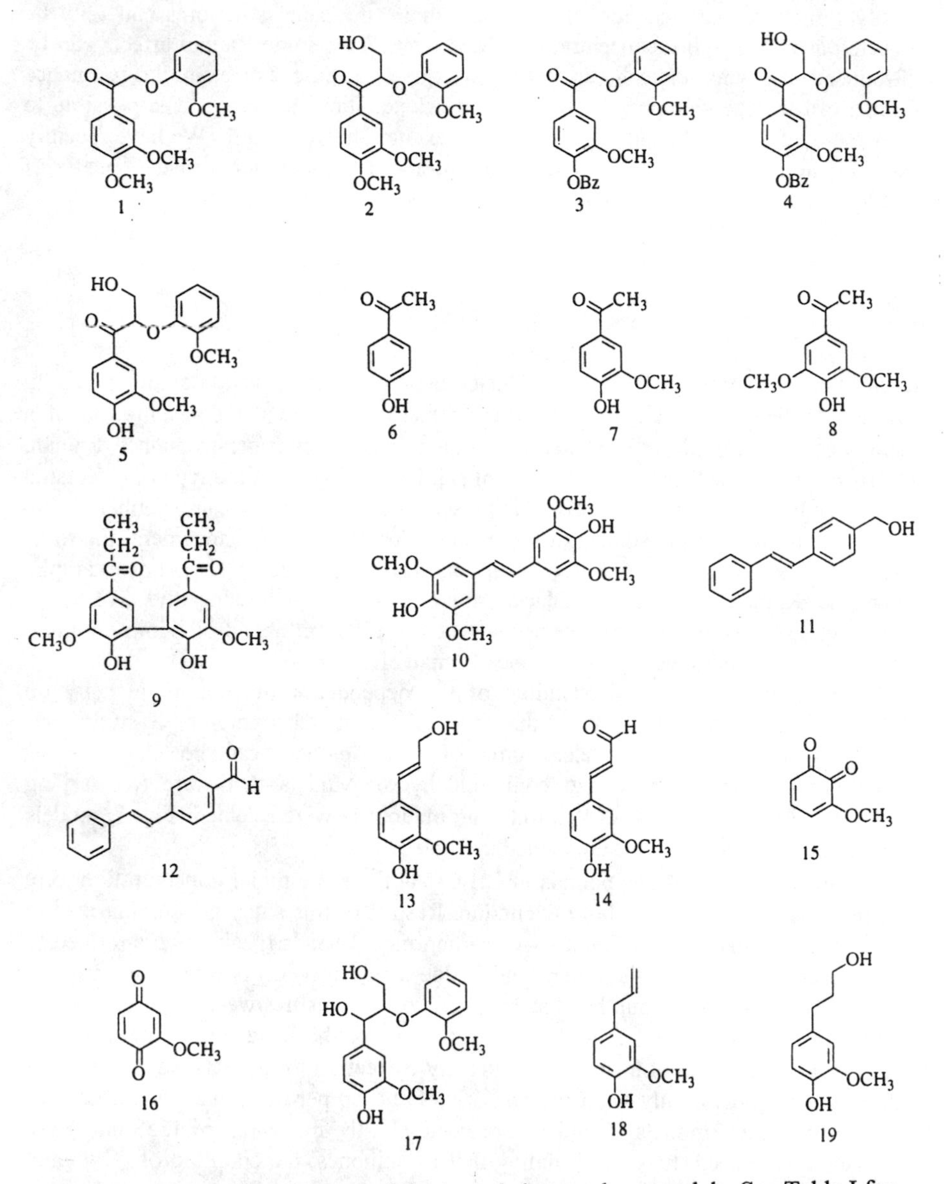

Figure 1. Molecular structures of lignin and chromophore models. See Table I for names of the models.

Table I. Model compounds shown in Fig. 1

Model	Name
1	1-(3,4-dimethoxy phenyl)-2-(2-methoxy phenoxy) ethanone
2	1-(3,4-dimethoxy phenyl)-3-hydroxy-2-(2-methoxy phenoxy) propan-1-one
3	1-(3-methoxy-4-benzoxy phenyl)-2-(2-methoxy phenoxy) ethanone
4	1-(3-methoxy-4-benzoxy phenyl)-3-hydroxy-2-(2-methoxy phenoxy) propan-1-one
5	1-(3-methoxy-4-hydroxy phenyl)-3-hydroxy-2-(2-methoxy phenoxy)-propan-1-one
6	4-hydroxy acetophenone
7	4-hydroxy-3-methoxy acetophenone
8	3,5-dimethoxy-4-hydroxy acetophenone
9	5,5'-bi-(4-hydroxy-3-methoxy propiophenone)
10	syringyl stilbene
11	stilbene methanol
12	stilbene carboxaldehyde
13	coniferyl alcohol
14	coniferaldehyde
15	3-methoxy *o*-quinone
16	3-methoxy *p*-quinone
17	1-(3-methoxy-4-hydroxy phenyl)-2-(2-methoxy phenoxy) propane-1, 3-diol
18	4-allyl-2-methoxy phenol
19	1-(3-methoxy-4-hydroxy phenyl)-propan-3-ol

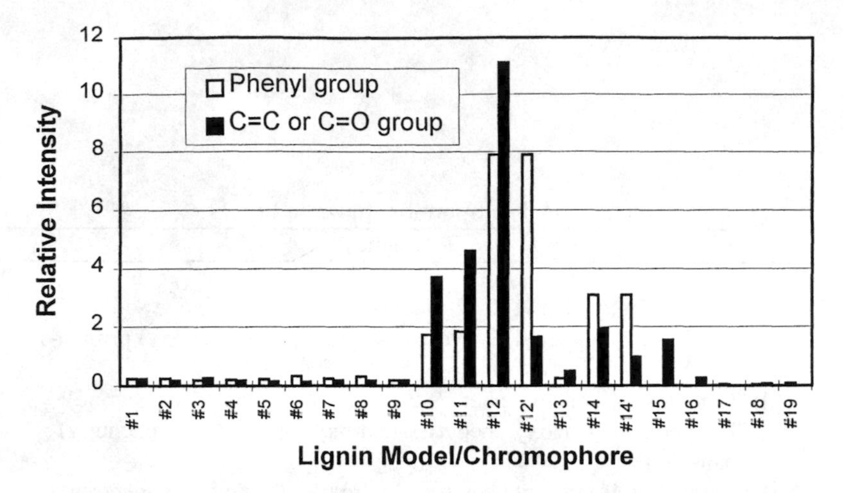

Figure 2. Relative Raman intensities (in 514.5 nm excited spectra) of lignin model compounds (for phenyl, C=O, and C=C groups). Key: see Fig. 1 and Table I. For models 12 and 14, which contain both C=C and C=O groups, #12' and #14' solid bars represent intensities of C=O stretching mode.

From Fig. 2 note that the intensity differences between models were quite dramatic and depended upon the molecular structure. Moreover, for models wherein both the phenyl and C=C or C=O group modes were present the intensity of both modes was enhanced due to the conjugation effect. Considering that the Raman frequency of the phenyl group is similar among various models, the phenyl band is unlikely to be used for characterizing a model. What is needed is a band that is distinct. This requirement is satisfied by the band representing the C=C and/or the C=O bond vibrations (see Table II). The pattern of intensity variation for this vibration (Fig. 2) indicated that different chromophores have different detection sensitivities in Raman. For the models studied, the order of detection, from high to low, was stilbene carboxaldehyde (#12) > stilbene methanol (#11) > syringyl stilbene (#10) > coniferaldehyde (#14) > *o*-quinone (#15) > coniferyl alcohol (#13) > *p*-quinone (#16) > α-keto containing lignin monomers and dimers (#1 - #8) and α-keto biphenyl model (#9) > unconjugated lignin models (#17 - #19).

Raman Spectra

Of the Raman spectra obtained, spectra of *p*-quinones (Fig. 3), methoxy *o*-quinone (Fig. 4), stilbene methanol (Fig. 5), coniferaldehyde (Fig. 6), and methoxy coniferyl alcohol (Fig. 7) are shown in the 850-1850 cm^{-1} region. The displayed wavenumber interval is the region where most prominent features of chromophore models were detected. For the models whose spectra are reported here, the most useful Raman frequencies are listed in Table II.

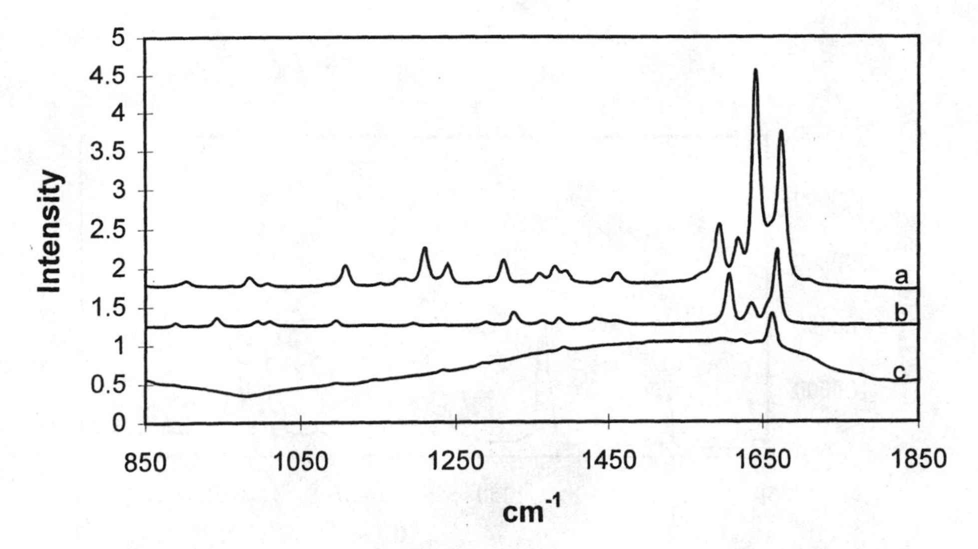

Figure 3. FT Raman spectra of 2-methoxy 1,4-quinone (a), and 2,3-dimethoxy-5-methyl 1,4-quinone (b), and methyl 1,4-quinone (c). The C=O bond vibrations in *p*-quinones have peaks in the region 1665-1680 cm^{-1}.

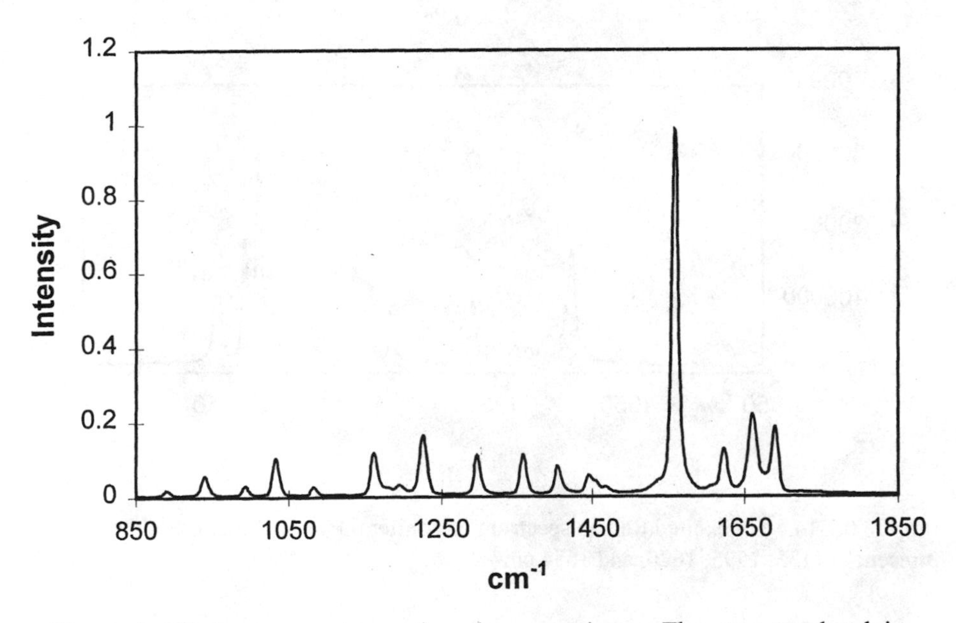

Figure 4. FT Raman spectrum of methoxy *o*-quinone. The strongest band is present at 1559 cm^{-1}.

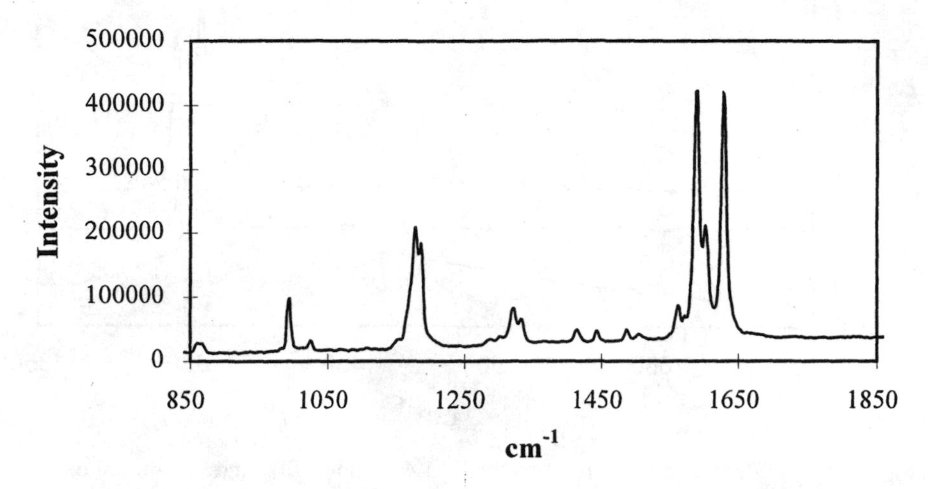

Figure 5. 514.5 nm excited Raman spectrum of stilbene methanol. The C=C band of stilbene is detected at 1635 cm^{-1}.

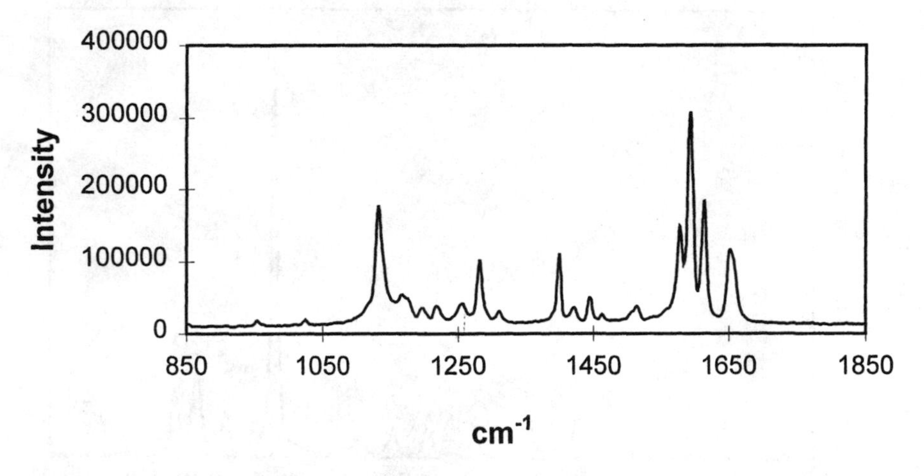

Figure 6. 514.5 nm excited Raman spectrum of coniferaldehyde. Most useful bands are present at 1135, 1595, 1620, and 1654 cm^{-1}.

Table II. Useful Raman frequencies of chromophores[a]

Chromophore	Frequency, cm^{-1}
o-Quinone	1559
Coniferaldehyde	1135, 1620, 1654
Stilbene	1635
Coniferyl alcohol	1654
p-Quinone	1670

[a]Except for the 1135 cm^{-1} band of coniferaldehyde, the bands are due to C=C or C=O groups.

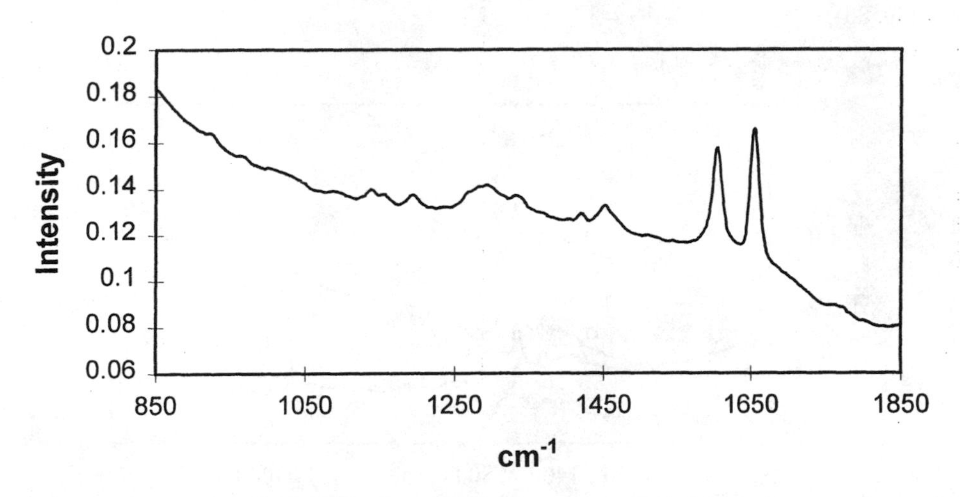

Figure 7. FT Raman spectrum of 4-methoxy coniferyl alcohol. Prominent peaks are detected at 1600 and 1654 cm^{-1}.

From Table II it is clear that, among the models studied, the peak positions of the strongly Raman active modes have vibrational frequencies that are clearly distinguishable. This means that if more than one chromophores was to be present in a sample they all are likely to be detected. Moreover, if on the basis of single-peak-analysis a chromophore group, listed in Table II, was not clearly shown to be present, additional Raman frequencies could be used to facilitate an analysis (Fig. 3-7). This latter approach was used to detect coniferaldehyde structures in mechanical pulps [*12*]; in addition to the feature at 1654 cm^{-1}, coniferaldehyde bands at 1620 and 1135 cm^{-1} were used in the analysis.

Applications

Photoyellowing. Light-induced yellowing of lignin-rich mechanical pulps is an area of research that continues to be challenging. Because of the complexity of the phenomenon, however, only limited progress has been made. Recently, using Raman spectroscopy, important information was obtained [*5,13*]. Of the many results obtained in our laboratory using Raman, one showed that the chromophores responsible for brightness loss are mostly *p*-quinones [*13*]. More specifically, when thermomechanical pulps were photoyellowed, a new Raman feature at 1675 cm^{-1} was detected [*5*]. In addition to the pulps, the 1675 cm^{-1} feature was present in pulp extracts (yellow) (Fig. 8a, 8b, and 8c). Subtraction of the extracted pulp-spectrum from the unextracted pulp-spectrum clearly showed that the substance removed contributed to the 1675 cm^{-1} band (Fig. 8d).

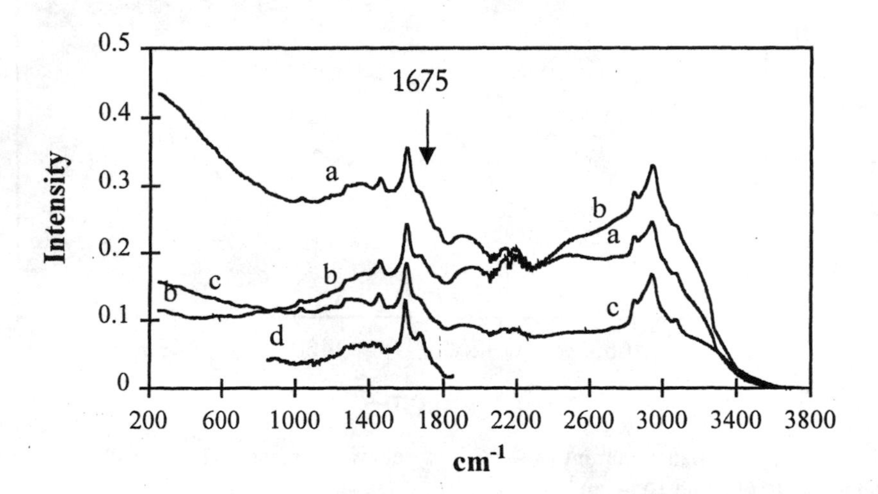

Figure 8. FT Raman spectra of materials extracted from photoyellowed thermomechanical pulps; (a) unbleached, (b) borohydride bleached, (c) diimide reduced, and (d) difference of spectra obtained before and after methanol extraction of a yellowed TMP sample. All spectra show a band at 1675 cm^{-1} which is related to photoyellowing.

Raman spectroscopy played a key role in assigning the 1675 cm^{-1} band to *p*-quinones. When the spectral contribution of yellow chromophores were compared with those of *o*- and *p*- quinones and of Fremy's salt-treated TMP (Fig. 9), it became clear that the contribution at 1675 cm^{-1} was due to *p*-quinone structures [*13*]. This conclusion was also supported by IR analysis of these samples. In Raman spectroscopy, when *o*-quinones and samples containing this functionality are analyzed a band near 1559 cm^{-1} is detected (Fig. 9, spectra d and e). In contrast, for *p*-quinone functionality, a Raman band is expected near 1675 cm^{-1} (Fig. 9, spectra b and c and Table II). Using this information, the band observed at 1675 cm^{-1} in yellowed pulps was assigned to *p*-quinones. The possibility of the 1675 cm^{-1} band being due to a carbonyl bond stretch (other than *p*-quinone) was ruled out by the fact that this band was quite weak in the IR spectra [*5*]. Carbonyl bond stretch-modes are strong in IR and weak in Raman. On the contrary, in the TMP study, the 1675 cm^{-1} band was strong in Raman and weak in IR.

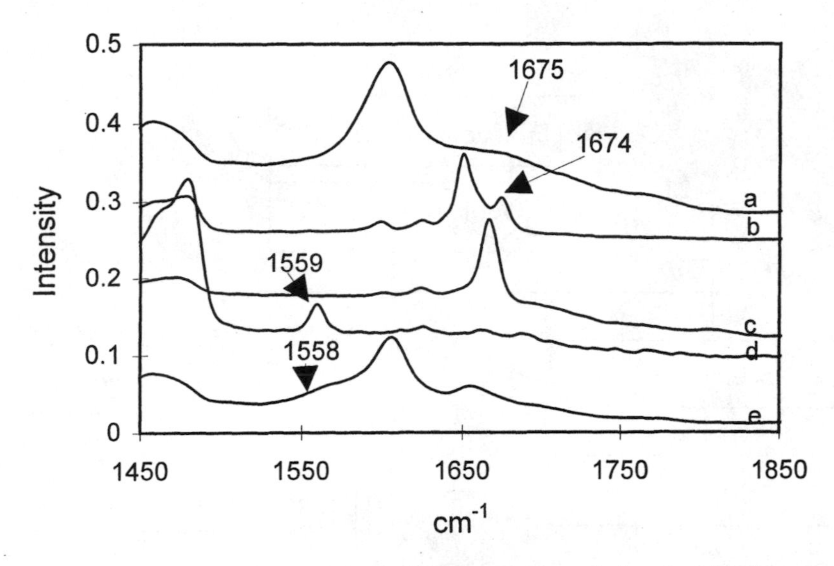

Figure 9. FT Raman spectra of (a) photoyellowed borohydride-bleached thermomechanical pulp, (b) 3-methoxy *p*-quinone, (c) methyl *p*-quinone, (d) 3-methoxy o-quinone, and (e) Fremy's salt-treated, borohydride-bleached TMP. Important band positions are annotated.

Using a conventional Raman system, laser-induced fluorescence-quenching evidence suggested that both *p*-quinone and hydroquinones were present in mechanical pulp [*13*]. This provided further support that hydroquinone/*p*-quinone redox system was involved in photoyellowing.

Bleaching of Thermomechanical Pulp (TMP). To understand which chromophores are affected when mechanical pulps are bleached using reductive or oxidative bleaching agents, a Raman study was undertaken. Spruce TMP was bleached in accordance with the sequence outlined in Fig. 10. After each bleaching step, the Raman spectrum of the bleached pulp was obtained. This spectrum was compared with the one obtained after the previous bleaching step in the sequence (Fig. 10). In addition, the spectrum of each bleached TMP was compared with that of unbleached TMP. Pulp brightness is listed in the flow diagram in Fig. 10.

The most prominent changes in the Raman spectrum occurred when the pulp was bleached for the first time with H_2O_2 (Fig. 11, plot c). The resultant changes showed peaks at 1135, 1595, 1622, and 1663 cm^{-1}. These band positions were similar to those reported for the borohydride bleached TMP [*5*]. This suggested that in both reductive and oxidative bleaching the nature of the changes was similar.

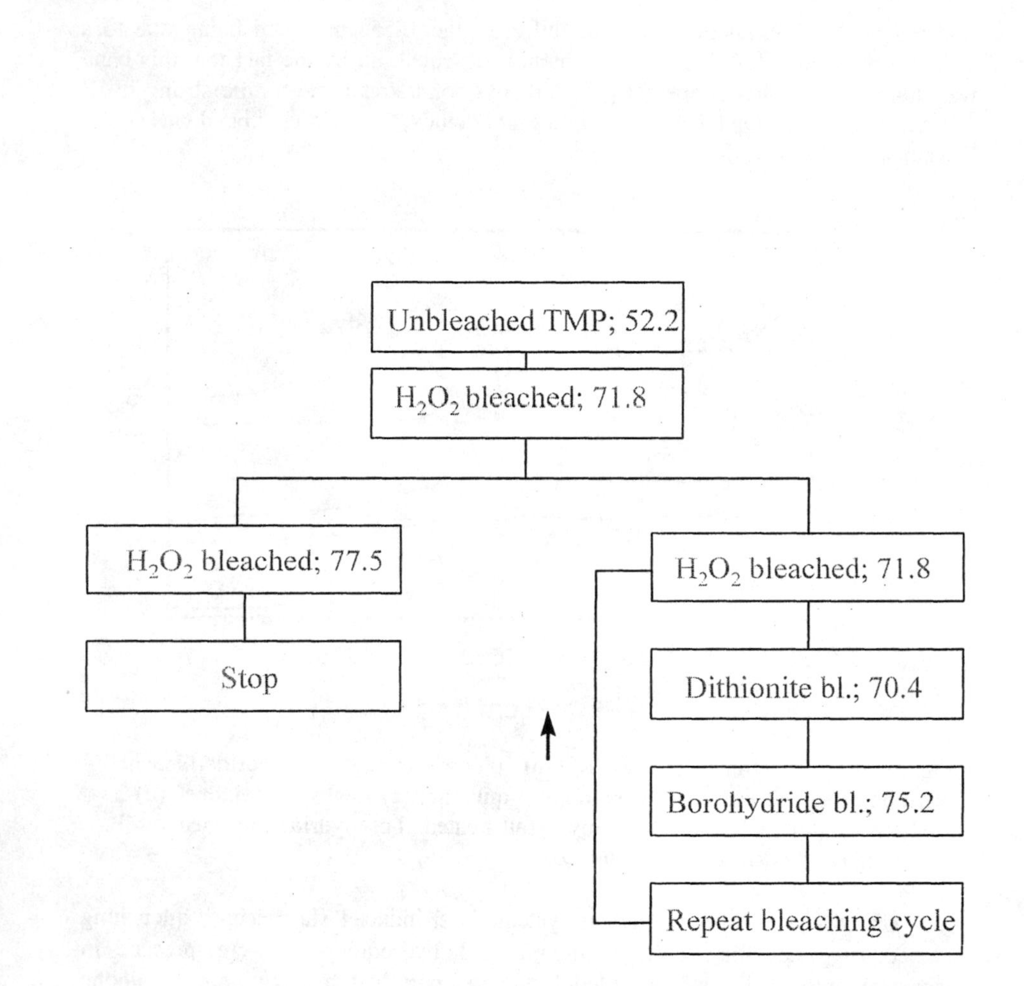

Figure 10. Flow diagram showing sequence of bleaching for thermomechanical pulp. Brightness values of bleached pulps are also given. Brightness data for pulps bleached the second time (using the repeat cycle) are not shown. These values were 78.8, 79.0, and 79.9, respectively, for H_2O_2, dithionite, and borohydride bleached TMPs.

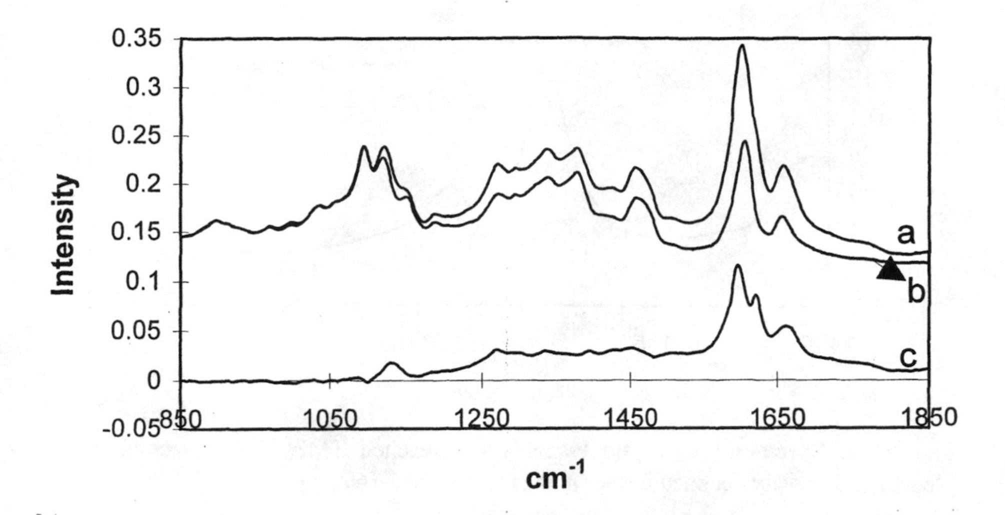

Figure 11. FT Raman spectra of unbleached (a, brightness 52.5) and H_2O_2 bleached (b, brightness 71.8) TMPs. Plot "c" is the difference of unbleached and bleached pulp spectra.

As pointed out previously [*5*], when TMP is bleached the decline in intensity at 1135, 1595, and 1622 cm^{-1} can be explained by the fact that coniferaldehyde groups are being removed. Moreover, the presence of a peak at 1663 cm^{-1} (in the difference spectrum) can be interpreted in terms of removal of the contributions of both coniferaldehyde and *p*-quinone groups. The C=O stretch in coniferaldehyde has a Raman band at or near 1654 cm^{-1} whereas the *p*-quinones are expected to be in the 1665-1690 cm^{-1} region. Some variation in the vibrational frequencies of coniferaldehyde and *p*-quinones, especially for the latter, is expected because of the dependence of the frequency on molecular structure and intermolecular interactions. This is supported by plot "c" in Fig. 11 where instead of a narrow band a wider contribution in the 1665-1690 cm^{-1} region is detected. Because *p*-quinones entities in TMP are likely to have different molecular structures and different intermolecular interactions, the broadness of the Raman feature can be explained.

After the initial bleaching by peroxide, bleaching with either peroxide or dithionite resulted in only small changes in the spectrum (at 1600 cm^{-1} and in the 1665-1690 cm^{-1} region, Fig. 12 and 13). When spectra of pulps treated with any of the remaining bleaching steps (sequence in Fig. 10) were compared, either no changes or only small changes were detected. This was interpreted in terms of *p*-quinones being oxidized and reduced depending upon the nature of the bleaching agent.

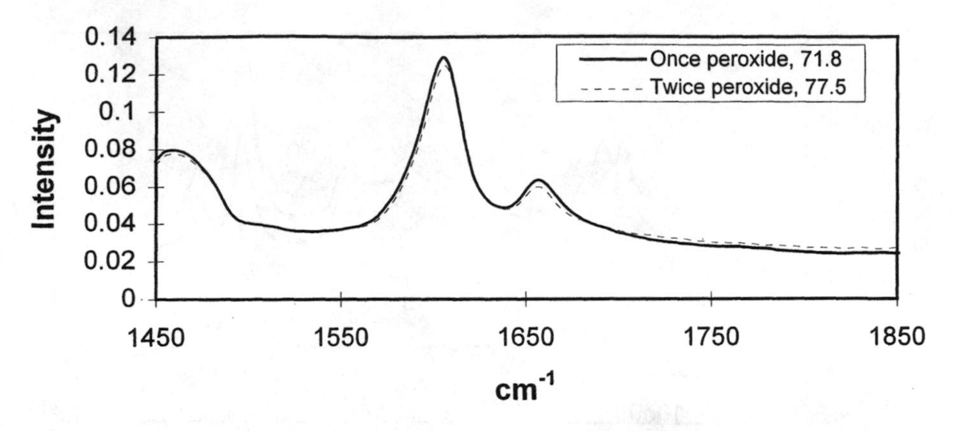

Figure 12. Spectra of once and twice H_2O_2 bleached TMPs. Upon second time bleaching, intensity declined further in the region 1645-1690 cm^{-1}.

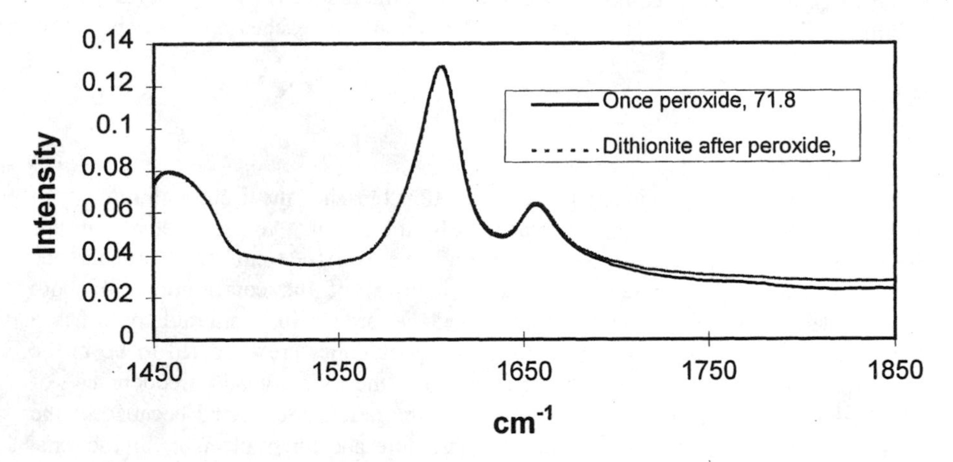

Figure 13. Spectra of once H_2O_2 bleached (dark line) and once dithionite bleached after-once-H_2O_2-bleached (less dark line) TMPs. Upon dithionite bleaching no significant change occurred in the pulp's spectrum.

If *p*-quinones play a major role in determining pulp brightness, as suggested by the Raman analysis, one should be able to correlate brightness with the degree to which these structures are removed upon bleaching. To determine if this occurs pulp brightness was plotted against decline in Raman intensity (in the 1665-1690 cm^{-1} region). As shown in Fig. 14, brightness was found to be positively correlated (linear regression correlation coefficient 0.89) with the extent of *p*-quinones removed, thus indicating that for TMP, brightness is very likely to be related to the degree to which *p*-quinones groups are present.

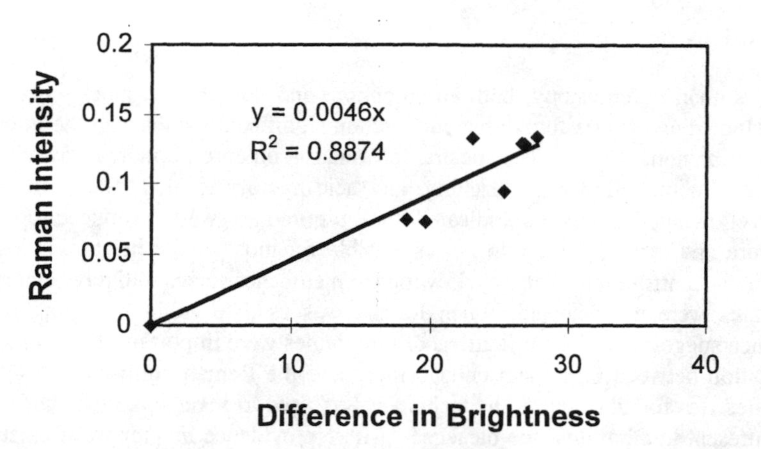

Figure 14. Linear regression between increase in brightness (upon bleaching) and decline in the 1665-1690 cm^{-1} region Raman intensity. The contribution in the 1665-1690 cm^{-1} region is primarily due to *p*-quinones.

Acid Hydrolyzed TMP. It has been reported that peroxide bleached mechanical pulp contain stilbenes [*14*]. This finding was based on HPLC analysis of the acid hydrolysate of peroxide bleached pulp. To investigate whether stilbenes existed in borohydride bleached TMP, the pulp was acid hydrolyzed (using the same method as used in [*14*]) and the pulp and its hydrolysate were analyzed using Raman spectroscopy. Raman was used because its sensitivity to detect stilbenes is extremely good (Fig. 2). If present in any reasonable amount, the contribution of stilbenes should be detected near 1635 cm^{-1} (Table II and Fig. 5). However, Raman spectra of the hydrolyzed pulp and its extract (Fig. 15, plots a and b, respectively) showed no band at or near 1635 cm^{-1}. This indicated that significant quantities of stilbenes were not present in the borohydride bleached pulp.

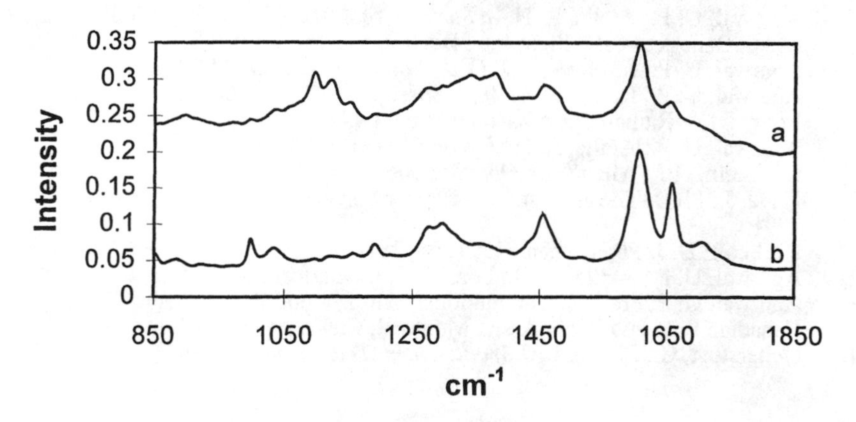

Figure 15. Raman spectra of acid hydrolyzed TMP (a) and extracted material (b). No Raman band at 1635 cm^{-1} was present in either spectrum.

Conclusions

Using Raman spectroscopy, both chromophore and non-chromophore lignin models were studied and it was found that conjugation significantly enhanced the intensity of certain vibrational modes. The spectral information on chromophores was used in the studies of photoyellowing, bleaching, and acid-hydrolysis of TMP. In the area of photoyellowing, the results indicated that *p*-quinones were produced upon light exposure and they are likely to be responsible for most of the brightness loss. Any significant contribution to photoyellowing from stilbenes seems unlikely because these structures were not detected. When Raman was used to study bleaching of TMP, evidence suggested that contributions of *p*-quinones were important. Reasonably good correlation between brightness enhancement and the Raman contributions due to *p*-quinones was found. Finally, the technique was used to investigate whether stilbenes were present in a borohydride bleached-TMP. No evidence in support of existence of stilbenes in pulps or its acid hydrolysate was obtained.

Acknowledgments

The authors thank Jim McSweeny for assistance with experiments involving bleaching of pulps. Help received from Sally Ralph in carrying out acid hydrolysis of bleached TMP and in drawing model structures is very much appreciated.

Literature Cited

1. Agarwal, U. P.; Atalla, R. H. Proc. 8th International Symp. Wood Pulp. Chem., Vol. III, **1995,** Gummerus Kirjapainooy, Jyvaskyla, pp. 67.
2. Agarwal, U. P. Fifth Chemical Congress of North America, Cancun, Mexico, **1997,** Book Abstract, Agrochemistry, Abs. # 503.
3. Agarwal, U. P.; Atalla, R. H. *Planta* **1995,** *169,* 325.
4. Agarwal, U. P.; Atalla, R. H. In *Surface Analysis of Paper*; Eds. T.E. Conners and S. Banerjee; CRC Press Inc.: Boch Raton, FL, 1995; Chap. 8.
5. Agarwal, U. P.; McSweeny, J. D. *J. Wood Chem. Tech.* **1997,** *17,* 1.
6. Agarwal, U. P.; Ralph, S. A. *Appl. Spectro.* **1997,** *51,* 1648.
7. Long, D. A. Raman Spectroscopy, McGraw Hill: London, 1977.
8. Agarwal, U. P.; Atalla, R. H. *J. Wood Chem. Tech.* **1994,** *14,* 227.
9. Bernstein, H.J.; Allen G. *J. Opt. Soc. Am.* **1955,** 45, 237.
10. Bond, J.; Ph.D. Thesis, Institute of Paper Science and Technology, Atlanta, GA, 1991.
11. Becker, E.D. J. Phys. Chem. **1991,** 95, 2818.
12. Agarwal, U. P.; Atalla, R. H., Forsskahl, I. Holzforschung **1995,** *49,* 300.
13. Agarwal, U. P. Proc. 9th International Symp. Wood Pulp. Chem., **1997,** Canadian Pulp and Paper Assn., Montreal, Canada, paper K4-1.
14. Gellerstedt, G.; Zhang, L. *J. Wood Chem. Tech.* **1992,** *12,* 387.

Chapter 12

Limiting Molecular Weight of Lignin from Autocatalyzed Organosolv Pulping of Hardwood

H. L. Hergert[1], G. C. Goyal[2], and J. H. Lora

Repap Technologies Inc., 2650 Eisenhower Avenue, P.O. Box 766, Valley Forge, PA 19482

Approximately 60 - 80% of the lignin of various hardwoods (poplar, maple, birch, etc.) was found to have a Mw of 1600 - 2000 and a Mn of 900 when isolated from pulping liquors of the Alcell® process. The remaining 20 - 40% consists of low molecular weight lignin fragments remaining soluble during the precipitation procedure. Extending the cooking time up to 16 hours does not materially reduce the molecular weight or change the chemical structure. Molecular weight data on lignin isolated from thousands of semi-commercial cooks shows the same molecular weight phenomena suggestive of a carbon-carbon linked "core" of about seven to ten units which remains after cleavage of all alpha and beta ether linked monomers. The implication of these findings is that biosynthesis of hardwood lignin is under genetic control and is not the totally random process advocated by Freudenberg and others.

Commercial chemical pulping of woody materials primarily involves depolymerization or derivatization of lignin to products soluble in the pulping medium. Lignin can then be separated from the pulp by washing or extraction. In the sulfite process, for example, lignin is converted to salts of sulfonic acids which are water soluble. In the kraft process lignin is cleaved to phenols which are soluble in aqueous alkali. Each of these processes has environmental problems associated with the use of sulfur during pulping. Much research has been conducted during the past two decades to find processes which produce pulp with good papermaking properties but do not emit malodorous compounds or discharge

[1]Current address: 901 Burdan Drive, Pottstown, PA 19464.
[2]Current address: Potlatch Corporation, East End Avenue B, Cloquet, MN 55720.

toxic effluents. This has lead to extensive investigation of pulping with organic solvents, i.e. organosolv pulping [for a comprehensive review, see reference (*1*)].

One of the simplest of these methods is heating woodchips under pressure in the presence of a mixture of an organic solvent and water. Acetyl groups in the wood are released and provide a mildly acidic environment which permits hydrolytic cleavage of lignin to products soluble in the solvent. This approach has been used in the pre-commercial Alcell process. It has been extensively investigated in the laboratory using batch pressure vessels or a small pilot plant and in a semi-commercial plant at Newcastle, New Brunswick, Canada, capable of pulping about 10 tonnes of wood per batch. This plant, started-up in 1989, uses a mixture of hardwood chips (50% maple, 35% birch and 15% poplar) which is heated to 195°C in the presence of 50% ethanol in a pressure reactor. Pulping takes place in three displacement stages to remove solubilized lignin. Washed pulp is bleached and used for the manufacture of coated paper. Dissolved lignin is recovered through aqueous precipitation. Since lignin is a major co-product along with pulp, recovery and sale is important for good process economics.

Many hundreds of tonnes of Alcell process lignin have been produced and sold during the last eight years. One of the major distinguishing characteristics of this lignin compared to all other commercial lignins is the freedom from sulfur, metal ions, etc. normally encountered in kraft or sulfite process-derived lignins. Another key property is a relatively low glass transition temperature and molecular weight. In this paper we present our studies on molecular weight distribution of Alcell process lignins and the effect of extended cooking time. Unexpectedly this work showed that the lignin molecular weight decreased to a limiting size. As will be seen, this finding has significant implications for the probable structure of lignin as it exists in hardwood trees.

Experimental

Wood samples were procured from forests in the vicinity of Newcastle, New Brunswick. Logs were chipped at the Repap pulpmill or hand-chipped in the laboratory. Pulping was carried out: (a) as 40 gm samples with 6:1 ratio of solvent in Parr bombs in a rocking autoclave; (b) as 6 kg samples in a laboratory pilot plant; or (c) as 10 tonne samples in a semi-commercial installation. Lignin was recovered from (a) by filtration of black liquor followed by washing of chips with alcohol, acetone and water, vacuum evaporation, and precipitation into water. Further purification involved redissolving lignin in dioxane followed by precipitation into diethyl ether to remove occluded wood resins. In (b) and (c) above, black liquor was flashed to reduce organic solvent content and precipitated in 90% water - 10% solvent, washed with water, and dried in a proprietary system.

Elemental analyses were performed commercially. Infrared spectra were measured as KBr wafers in a Digilab FTS-40 FTIR spectrometer. Proton and C-13 NMR spectra were measured in deuteroacetone or as acetate derivatives in deuterochloroform in a Bruker 300 MHz instrument at Bryn Mawr University.

Molecular weight was measured by HPSEC in a Spectra Physics SP8810 liquid chromatography apparatus controlled by an Epson Computer. Samples were dissolved in tetrahydrofuran, filtered, and submitted to HPLC on a series of Phenomenex Phenogel 500, 100 and 50 Angstrom columns. Detection was by U.V. at 254 and 280 nm or refractive index. Columns were calibrated with polystyrene standards. Correct molecular weight were given for a series of phenolic model compounds up to a molecular weight of 900 ± 10% daltons. Further experimental details can be found in references 9 and 10.

Molecular Properties of Commercial Alcell Lignin

As part of the Alcell lignin quality control program, production samples of lignin were collected at the demonstration plant and forwarded to the RTI research laboratory for measurement of molecular properties. A portion of the data for a typical year's production is presented in Table I.

Table I. Properties of Alcell Lignin for the Year 1994

	Avg.	Std. Dev.
Weight Avg. Mol. Wt.	1618	167
Number Avg. Mol. Wt.	759	53
Tg, deg. C (atm.)	82.7	3.7
Carbon, %	66.5	0.37
Hydrogen, %	6.1	0.06
Methoxyl, %	17.3	0.39
Syringyl per C9	0.63	0.07
Guaiacyl per C9	0.44	0.05

The lignin analyzed in Table I was produced during a combination of normal operating conditions and experimental runs. While there was no reason to believe that the measured variability in weight and number average molecular weight had any significant impact on the intended enduse of the lignin (primarily adhesive resins), good industrial practice dictated identification of the sources of variability.

Some of the possible sources for molecular weight variability considered were (a) wood species, (b) cooking time and temperature, (c) alcohol concentration, and (d) time that the solubilized lignin in the black liquor remained at temperature in the accumulator. Factors (a) to (c) were brought under good control, but the time of black liquor holdup [factor (d)] could vary as much as three or four hours depending upon interruptions in the lignin separation and washing part of the overall process sequence. The time at temperature was thought to impact molecular weight either through cleavage to lower molecular products or condensation to higher molecular weight. Either of these possibilities would, in turn, impact elemental analyses, ratio of syringyl to guaiacyl groups, and polymer glass transition temperature.

Molecular Weight of Alcell Lignin Fractions

Our technique for molecular weight measurements was VPO (vapor pressure osmometry), which yields only a number average, or HPSEC which gives both number and weight averages through computerized analyses. A series of columns with substantial resolving power (Phenomenex Phenogel, 500, 250 and 50 Angstroms in series) gives molecular weight distribution curves with substantial fine structure. Examination of the curves of the various Alcell lignin samples showed peaks at 169, 480, 1040 and 2235 which could correspond to a mixture of "low" molecular weight, i.e. one to three monomeric units, and "higher" molecular weight, i.e. ten to twelve units.

In order to verify the nature of the mixture, a sample of lignin was fractionated by extraction of dry lignin in a Sohxlet extractor with solvents of increasing polarity as shown in Table II.

TABLE II. Fractionation of Alcell Lignin (Sample ADI900622) by Extraction with Solvents of Increasing Polarity

Percent of Original	Solvent	Mn	Mw	Mz	Principle Peaks
1.0	Pet. Ether	-	-	-	167, 180, 350
13.8	Ethyl Ether	326	436	562	165, 180, 430, 650
15.2	Ethyl Acetate	763	1196	1759	440, 1010
58.3	Acetone	1029	2109	4157	1220, 2600
10.1	Methanol	1184	2442	4613	2650
1.7	Residues*	3637	8035	11910	10,500
1.0	Insoluble	-	-	-	-
101.1 = Total Recovered					
Original Sample		752	1702	3856	169, 480, 1040, 2235

*Precipitates from acetone and methanol extraction.

This procedure was adapted from that of Morck, et.al. (*2*) who used successive extraction of birch kraft lignin with methylene chloride and methanol to give two solubilized fractions and a residue in roughly equal portions as shown in Table III.

Table III. Fractionation of Birch Kraft Lignin by Extraction with Solvents of Increasing Polarity

Percent of Original	Solvent	Mn	Mw
32	Dichloromethane	650	910
38	Methanol	1320	2110
30	Residue	3470	9760

More recently, Thring, et.al. (*3*) slightly varied this procedure on Alcell lignin to give an ether-soluble fraction (27%, 48 Mn, 720 Mw), a methanol-soluble fraction (53%, 1040 Mn, 2410 Mw) and an insoluble fraction (18%, 2400 Mn, 6950 Mw). We used a more comprehensive system but our results were basically the same as Thring et.al. except the petroleum ether extraction removed waxes and fats derived from the wood. This indicates that the CH_2 and CH_3 groups in their NMR spectra are mostly of non-lignin origin. Furthermore, acetone is a better solvent for the lignin than methanol, so we were able to solubilize virtually all of the original lignin. Schuerch (*4*) much earlier had noted that lower molecular weight lignin fractions are much more soluble in solvents such as ether, ethyl acetate or acetone which have a weaker hydrogen bonding capacity, i.e. lower Hildebrand solubility parameter, than ethanol or dioxane.

One of the peculiarities of autocatalyzed organosolv pulping as employed in the Alcell process is that best results are obtained with a 50% mixture of alcohol and water. This solvent mixture is not particularly good for solubilization of dried, isolated Alcell lignin at low temperatures. Acetone or hot 70% ethanol, as noted by vanHeiningen (*5-6*), are better solvents and can be used to remove additional quantities of lignin from the pulp. SEC showed that this "residual" lignin had a significantly higher molecular weight than normal Alcell lignin (*7*), but Mw and Mn were not calculated from the distribution curves. Thus, variations in pulp washing could result in poor removal or reprecipitation of this high molecular weight fraction. Consequently, the amount in the isolated lignin would vary and affect Mw. There also could be variation in the amount of the lowest mol.wt. fraction.

For economic operation of the Alcell process, lignin precipitation at the demonstration plant is carried out following flashing off part of the alcohol from the black liquor and subsequent dilution with water. The lowest molecular weight portions of the solubilized lignin remain in the filtrate. It can be recovered following removal of the alcohol in a stripper and subsequent concentration in an evaporator. This material has interesting potential applications (*8*) and was reported to have a Mw and Mn of 765 and 435, respectively. Obviously isolated Alcell lignin could be fractionally precipitated by graded dilution of an alcohol solution with water. Mechanical problems were encountered when this was attempted, so the fractionation with increasingly polar solvents, as reported here, was more convenient.

Limiting Molecular Weight With Extended Cooking

Pilot Plant Experiment. To establish the effect of extended time at temperature upon Alcell lignin, typical hardwood furnish was pulped in the RTI pilot plant at 193°C. The pulping liquor was composed of 55/45 ethanol water mixture and was circulated continuously through the chip mixture at a 7.3 liquid to wood ratio. When the cooking liquor temperature reached temperature (25 minutes), a sample of the pulping liquor was withdrawn. Samples were withdrawn every 20 minutes during the 2 hour pulping time. The spent liquor was then pumped into an accumulator and kept for an additional 12 hours at 193°C with sampling being done each hour. The lignin from each sample was precipitated with acidified water at a dilution ratio of 3 to 1. After washing with water, the lignins were dried and analyzed by a variety of spectrometric techniques. Acetylated lignins were examined by H-NMR so that the ratio of syringyl to guaiacyl groups could be estimated. Abstracted results are shown in Table IV.

TABLE IV. Effect of Extended Time at Temperature on Molecular Weight of Organosolv Lignin from Hardwood

Time, Minutes	Mw	Tg, °C	Ratio Syr./ Guaiacyl
Start, (193°C)	1758	116	1.39
40	1692	107	1.16
100	1692	106	1.39
220	1653	100	1.50
460	1615	97	1.38
820	1560	90	1.19

It will be noted that the initial molecular weight quickly leveled off and then declined very slightly during the following fourteen hours. Although some condensation on side chains may have occurred through the splitting out of water (carbon content increased 3 percentage points, hydrogen content decreased 2% and methoxyl remained constant from the initial 64.1% carbon, 6.15% hydrogen and 19.1% methoxyl), there was no evidence of shortening or lengthening of lignin chains. Spectra and molecular weight of the six hour samples resembled a typical Alcell lignin. Based on this detailed experiment, we would not expect the commercial Alcell lignin to vary significantly in molecular weight or chemical properties resulting from normal variation of lignin solution held at temperature.

We have already noted that our molecular weight data was derived by HPSEC using polystyrene standards. Thus, our reported molecular weights are, technically, "apparent" molecular weights. Since we were able to achieve very good results with phenolic compounds including glucosides and lignans up to a molecular weight of a thousand daltons, we believe that our data are useful in the relatively low molecular weight ranges that we are observing.

Laboratory Experiments with Varying Time and Solvent Composition. A series of birch organosolv lignin samples remaining from a previous study (*9*) were submitted to HPSEC. In this case the wood was cooked with 50% ethanol in a batch reactor for varying times. Precipitation procedure was the same as in the preceding section. Results are shown in Table V.

Table V. Molecular Weight of Lignin from Organosolv Pulping at Various Times at Temperature (195°C)

Time at Temperature (minutes)	Mn	Mw	Polydispersity Index
25	799	1814	2.27
30	777	1745	2.24
45	811	1770	2.18
60	798	1629	2.04
90	778	1573	2.02
120	782	1608	2.05

While the molecular weight did not change much, phenolic hydroxyl groups nearly doubled while aliphatic hydroxyl groups decreased by almost one-half. Since the lowest molecular weight cleavage products are excluded from the lignin isolated by this method, it was considered to be of interest to change the lignin isolation procedure to more completely represent the total solubilized fraction.

Aspen chips were preextracted in a Sohxlet extractor to remove resins, etc., which might subsequently interfere with spectral studies on the solubilized lignin. Chips were then pulped with 50% ethanol in a rocking autoclave at 195°C. The first experiment was terminated as soon as the cooking temperature was reached (approximately 25 minutes) by quenching the autoclave in ice water. The contents were filtered and the pulp washed with hot 90% ethanol until the filtrate was colorless. The combined filtrate and washings were reduced to near dryness in a vacuum rotary evaporator. The solids were dispersed in hot water, filtered, and washed twice to give a light brown lignin product which was vacuum dried. The experiment was repeated for 30 and 90 minutes cooking time. Pulp yield, viscosity (a measure of cellulose chain length), degree of delignification (based on a 20.8% combined Klason and acid-soluble lignin content of the extractive-free wood), and amount of recovered lignin are shown in Table VI.

TABLE VI. Organosolv Pulping of Aspen with 50% Ethanol at 195°C

Cooking Time, Minutes at Temperature	0	30	90
Pulp Yield, %	82	65	56
Pulp Kappa Number	53	44	14
Pulp Viscosity	34	41	20
Delignification, %	67	78	94
Solubilized Lignin Recovered, % of Original	25	68	77
Solubilized Lignin Not Recovered, % of Original	41	10	17

The lignin recovery in the zero time experiment was not very good because it did not cleanly precipitate, probably because of the peptizing effect of hemicelluloses present in the spent liquor. Most of the lignin not recovered in the 30 and 90 minute cooks was present as water-soluble or dispersible very low molecular weight products. They could be recovered by liquid-liquid extraction with a water immiscible solvent such as ether or ethyl acetate. GLC, TLC, and GPC showed that this fraction contained typical lignin monomeric degradation products such as vanillin, syringaldehyde, acetosyringone, sinapyl alcohol, etc., and a number of compounds presumably dimers and trimers based on GPC retention times.

Molecular properties were measured, including quantitative carbon 13 NMR of the acetate derivatives, and compared with those of an aspen milled wood lignin sample (see Table VII).

TABLE VII. Comparison of Organosolv Lignins from Aspen Pulped for Varying Times with 50% Solvent:Water

Lignin	Milled Wood	Zero Minutes	90 Minutes	Dioxane-Water (30 minutes)
Composition Calculated to a C9 Basis				
Carbon	9	9	9	9
Hydrogen	8.50	8.51	8.34	-
Oxygen	2.90	2.64	2.63	-
Methoxyl	1.43	1.38	1.36	-
Aromatic Protons	2.54	2.16	2.08	1.76
Benzylic Protons	0.53	0.28	0.21	0.38
Molecular Weight				
Number Average	2601	833	998	833
Weight Average	4530	3192	1949	1488
Polydispersity	1.74	3.83	1.96	1.77

The molecular weight distribution curve of the zero and thirty minute cooking time lignin samples is shown in Figure 1. The presence of peaks at 180, 360, 600, 1200, 3600, and 10,500 daltons is particularly noticeable. The latter two peaks largely disappear in the 90 minute curve (not shown) leaving a strong peak at 1150 and small amounts of a series of peaks at 180 to 560.

Experiments were conducted at 60:40 solvent to water ratio to compare the efficacy of dioxane or diethylene glycol dimethyl ether ("diglyme") with ethanol pulping for 30 minutes. Results are shown in Table VIII.

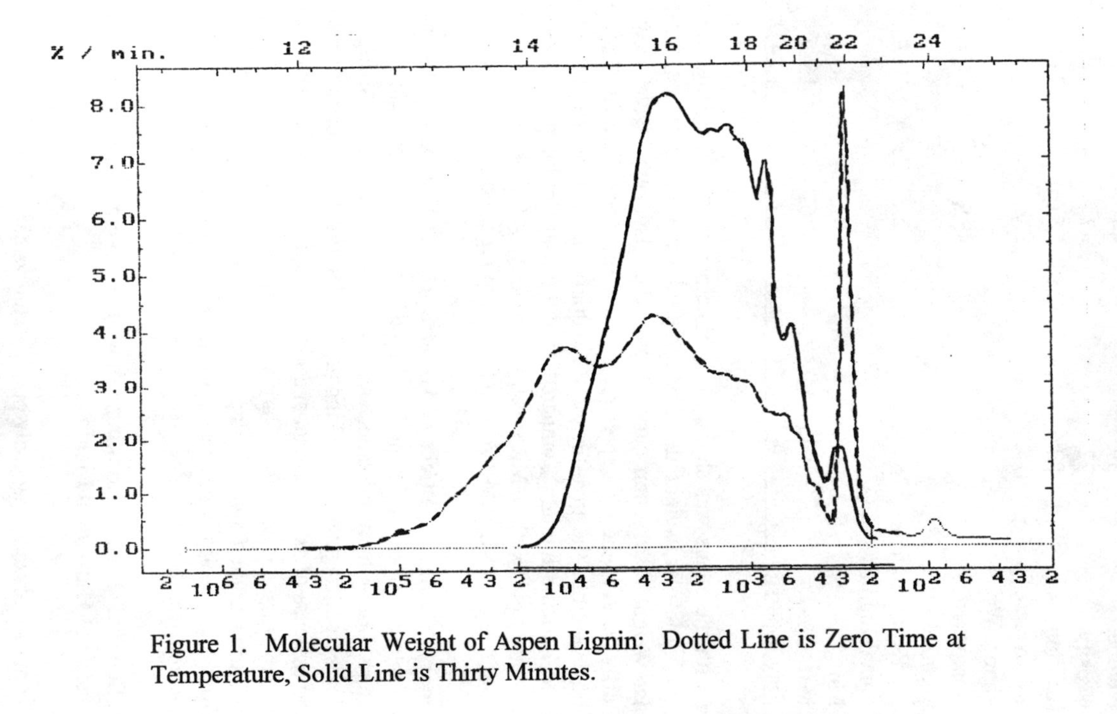

Figure 1. Molecular Weight of Aspen Lignin: Dotted Line is Zero Time at Temperature, Solid Line is Thirty Minutes.

TABLE VIII. Role of Organic Solvent in 60:40 Solvent:Water of Aspen for Thirty Minutes at 195°C

Solvent	Ethanol	Dioxane	Diglyme
Pulp Yield, %	58	71	66
Kappa Number	39	44	50
Pulp Viscosity	22	47	27
Delignification, %	82.6	76.0	74.6
Lignin Recovered, % of Original Wood	15.8	9.0	13.4
Lignin Recovered, % of Original Lignin	76.0	43.3	64.4
Lignin not Recovered, % of Original Lignin	6.6	32.7	10.2
Lignin Mn	998	1033	864
Lignin Mw	1949	3262	1587
Lignin Polydispersity	1.96	3.16	1.84

Pulping with the ethers was slightly less effective in terms of delignification but more low molecular weight degradation products seemed to be formed (see molecular weight distribution curve, Figure 2). This may have accounted for greater difficulties in the lignin precipitation and recovery. The main advantage in using dioxane from the investigator's standpoint is that no ethoxyl groups are introduced into the lignin. Compared to organosolv pulping with ethanol, the dioxane-water lignin C-13 NMR spectra complexity is thereby reduced. A portion of the details of the C-13 NMR spectra have been reported elsewhere (*10*).

Preliminary Studies on Effect of Wood Species. Comparison of organosolv lignins isolated from different hardwood species under comparable time, temperature, liquor to wood ratio, etc. showed different ratios of high and low molecular weight fragments. Cooking at extended times did not further reduce the size as in the case of aspen, maple, and birch already noted. The exact size of the major lignin fragment seemed to vary with different species, however. This is a very intriguing finding since it suggests that lignin structure is not precisely the same from one species to the next. Considerably more careful experiments are needed to confirm this preliminary observation.

Implications of Lignin Limiting Molecular Weight Upon Structure and Biosynthesis of Hardwood Lignin

In all the autocatalyzed organosolv pulping studies carried out in the laboratory at 185 - 195°C, we see strong evidence that the lignin is cleaved into 20 - 40% of the original wood lignin as very low molecular weight fragments (1 to 4 D.P.) and the remainder as a higher molecular weight polymer that cannot be reduced in size beyond 900 - 1300 daltons upon extended cooking time. The only cleavage mechanism known to be operative in the type of organosolv pulping discussed herein is the splitting of alpha and beta aryl ethers (so called α-0-4 and β-0-4 ethers) and hydrolysis of esters, such as para-hydroxy benzoic acid present in

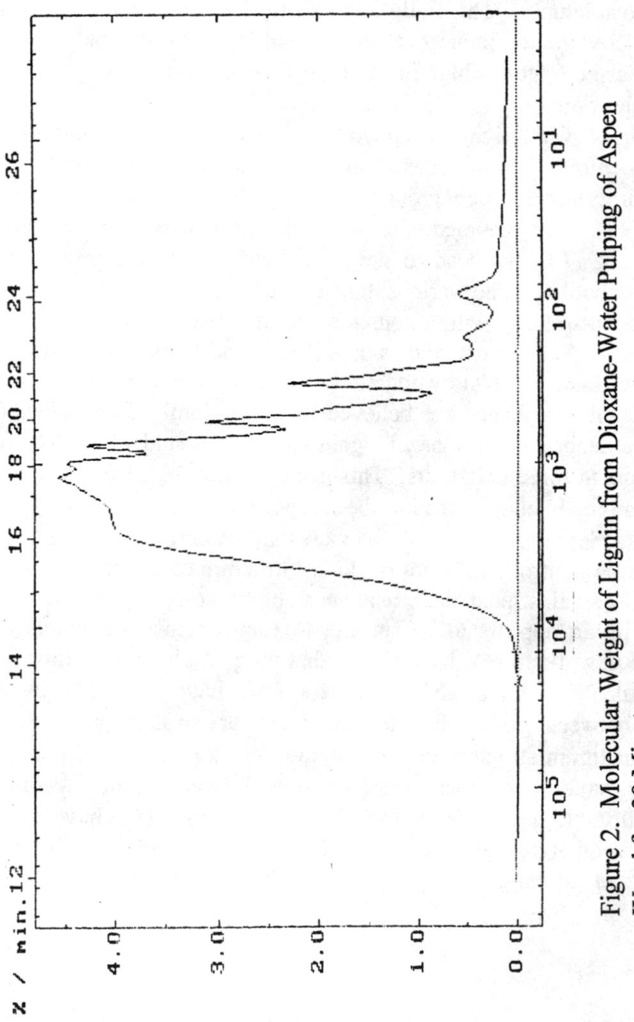

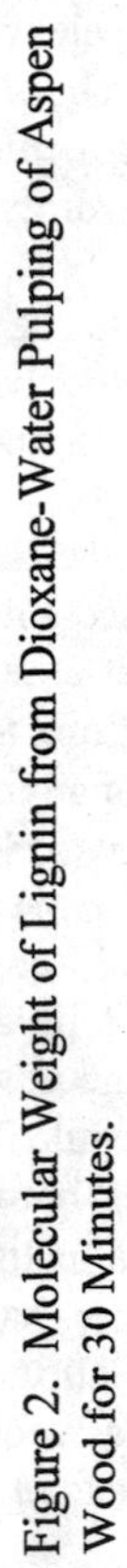

Figure 2. Molecular Weight of Lignin from Dioxane-Water Pulping of Aspen Wood for 30 Minutes.

certain hardwoods, i.e. poplar. The clear implication is that the hardwood lignins contain two types of lignin: one of these (20 - 40%) is relatively easily cleavable to monomeric or dimeric entities. It must, therefore, be a polymer based on guaiacyl or syringyl glycerol joined through alpha or beta ether linakges. The other (60 - 80%) consists of units about 6 to 10 D.P. joined to each other by non-cleavable linkages, i.e. carbon-carbon bonds. These units must be joined to each other by cleavable units. This finding of structural heterogeneity is consistent with the trimethylsilyl iodide-induced cleavage studies of Meshitsuka and coworkers (*11*) who describe fractions high in carbon-carbon linakges (phenylcoumaran type) resistant to cleavage.

Much of what lignin chemists know (or think they know) about the chemical structure of lignin is based on identification of monomeric, dimeric and trimeric hydrolytic cleavage products directly from wood by Nimz (*12*) and Sakakibara (*13*). This, coupled with the *in vitro* synthesis of lignin-like polymers by Freudenberg (*14*), has lead to proposed lignin structures presented in wood chemistry textbooks. The various linkages in lignin are said to be randomly distributed in accordance with a non-enzymic directed mechanism first advocated by Freudenberg. Subsequent studies on milled-wood lignin have indicated that up to 65% of the linkages in hardwood are of the easily cleavable alpha or beta ether type. Since these linkages are believed to be randomly distributed, one could expect that extended conditions of lignin cleavage would ultimately result in a mixture of one to three D.P. units. This is clearly not the case in our work nor in the work of anybody else that has yet been reported.

Our work is consistent with the view that the entire process of lignification is under genetic control and is not random. Structure of the carbon-carbon linked larger-sized "domains" and their relative amounts compared to easily cleavable portions of lignin appears to be specific to each angiosperm genus or species. Monties and coworkers (*15*) have shown that monomeric composition varies with species, while Terashima and coworkers (*16*) have pointed out significant differences between middle lamella and secondary wall lignin. Some of the difficulties in lignin structure determination may be related to the fact that most investigators have based their work on milled wood lignin assuming that it represents all the lignin in the wood. Maurer and Fengel (*17*) have clearly shown that milled wood lignin only represents that in the secondary wall. Our studies using extended cooking time represent nearly all the lignin including middle lamella.

Conclusions

Based on the findings in this report, we conclude:

1. Cleavage of lignin during autocatalyzed organosolv pulping of hardwoods results in 20 - 40% low molecular weight (1 - 3 D.P.) fragments and 60 - 80% higher molecular weight (7 - 10 D.P.) polymers.

2. Prolonged pulping does not reduce nor increase the size of the higher molecular weight fraction once it reaches a limiting size.

3. Hardwood lignin in the tree appears to be constructed of domains of

alpha and beta aryl ether linked monomers and modules of carbon-carbon linked monomers of Mn 900 - 1200.

4. Generic differences in hardwood lignin are implied. These appear to be under enzymic control.

Acknowledgments

This work was done on behalf of Alcell® Technologies Inc. and was financially supported in part by the Government of Canada through their Industry Science and Technology Office.

Literature Cited

1. Hergert, H.L.; In *Environmentally Friendly Technologies for the Pulp and Paper Industry*; Young, R.A.; Akhtar, M., Eds.; John Wiley and Sons, Inc.: New York, NY, 1998; pp 8-69.
2. Morck, R.; Reimann, A.; Kringstad, K.P., *Holzforschung* **1988**, *42*, pp. 111-116.
3. Thring, R.W.; Vanderlaan, M.N.; Griffin, S.L., *J. Wood Chem. Tech.* **1996**, *16*(*2*), pp. 139-154.
4. Schuerch, C., *J. Am. Chem. Soc.* **1952**, *74*, pp. 5061-5067.
5. Ni, Y.; vanHeiningen, A.R.P., *Tappi* **1996**, *79*(*3*), p. 255.
6. Girard, R.D.; vanHeiningen, A.R.P., *Preprints 82nd Annual Mtg., Tech. Sect. CPPA*, Feb. 1996; p. B297.
7. Lora, J.H.; Goyal, G.C.; Raskin, M., *Proc. 7th Intl. Symp. Wood Pulping Chem.*, Beijing, China, 1993; *1*, pp. 327-336.
8. Creamer, A.W.; Blackner, B.A.; Lora, J.H.; *Proc. 9th Intl. Symp. Wood Pulping Chem.*, Montreal, Canada, 1997, pp. 21(1) - 21(4).
9. Goyal, G.C.; Lora, J.H.; Pye, E.K., *Tappi*, **1992**; *75*(*2*), pp. 110-116.
10. Gallagher, D.K.; Hergert, H.L.; Cronlund, M.; Landucci, L.L., *Proc. 5th Intl. Symp. Wood Pulping Chem.*, 1989; pp. 709-717.
11. Nakahata, A.; Kim, Y.S.; Meshitsuka, G., *Proc. 8th Intl. Symp. Wood Pulping Chem.*, Helsinki, Finland; *1*, pp. 123-129.
12. Nimz, H.H., *Proc. 8th Intl. Symp. Wood Pulping Chem.*, Helsinki, Finland; *1*, pp. 1-31.
13. Sakakibara, A., *Wood Sci. Tech.*, **1980**, *14*, pp. 89-100.
14. Freudenberg, K., In *Lignin Structure and Reactions*; J. Marton, Ed.; Advances in Chemistry Series 59, Am. Chem. Soc., Washington, DC, 1966; pp. 1-21.
15. Tollier, M.-T.; Monties, B.; Lapierre, C., *Holzforschung*, **1980**, *40*(*suppl.*), pp. 75-79.
16. Terashima, N.; Fukushima, K.; Imai, T., *Holzforschung*, **1992**, *46*(*4*), pp. 271-275.
17. Maurer, A., Fengel, D., *Holzforschung*, **1992**, *46*(*5*), pp. 417-423, *46*(*6*), pp. 471-475.

Chapter 13

Structural Investigation of Dehydrogenation Polymer (models of lignin) Films at the Air–Water Interface by Neutron Reflectivity

B. Cathala, N. Puff, V. Aguié-Béghin, R. Douillard, and B. Monties

INRA, Equipe de biochemie des macromolécules végétales, UPBP, CRA, 2 Esp. R. Garros, 51686 Reims Cedex 2, France (cathala@reims.inra.fr)

Two DHPs (Dehydrogenation Polymers, models of lignins) were spread at the air/water interface at several surface concentrations. The structure of the interfacial layer was investigated by neutron reflectivity. At low surface concentrations DHPs form an uniform layer containing a high percentage of water (70-80%). The thickness of this layer increases with the amount of DHP deposited. At high surface concentrations a two-layer structure is observed with a water content of 50-60% on the air side and 80-90% on the bulk side. These films exhibit long relaxation times after compression (determined by dynamic surface pressure measurements) indicating that the DHP molecules are probably involved in a network.

Plant cell walls are heterogeneous structures mainly composed of three types of polymers: cellulose, hemicelluloses and lignins. Cellulose and hemicelluloses are polysaccharides and they are hydrophilic (1). On the contrary lignins are thought to be more hydrophobic because they are composed of phenylpropane units containing fewer hydroxyls groups (2). In the plant cell walls these polymers are associated with each others and the contact zone (the interface) between such hydrophilic and hydrophobic polymers should be very important for the molecular architecture of the cell walls. In this study, we investigate the structure and the properties of dehydrogenation polymers, models of lignins, at the air/water interface modeling an hydrophobic / hydrophilic interface in order to evaluate some features of the macromolecular organization and properties of DHPs in an heterogeneous medium.

Interfacial adsorption layers have a thickness which is of the order of magnitude of the size of the adsorbed macromolecule (3). Thus they provide a means to study, at the molecular level, the properties and the organization of the

DHPs in an anisotropic environment. These studies are complementary to those made in a solution which is a homogeneous and isotropic medium. It is usually accepted that lignins are amorphous and have a low degree of organization, however some general orders can be detected. For example, Agarwall and Atalla (4) using Raman spectroscopy have demonstrated that the aromatic ring of the phenyl propane units are parallel to the plane of the plant cell wall surface. This was also demonstrated by molecular modeling by the same authors and also by Jurasek (5) who moreover pointed out that lignins can be porous. The same conclusion was also drawn by studying air/water interfacial films of lignin by ellipsometry (6) and in our team by neutron reflectivity (7). Luner and Kempf (8) have also found that lignin at the air/water interface can form a gel-like structure indicating that lignin is involved in a network.

DHPs are not lignins, however they can mimic lignin properties and structures (9). They can be obtained in a reproducible way and free of any type of contamination and particularly free of polysaccharides which can have a significant effect on the surface properties. They allow also large and easy changes in the chemical composition. As a consequence DHPs are useful to support hypotheses on structure-properties relations in lignins. In the present work, we have used neutron reflectivity to determine, at several concentrations, the structure of DHP interfacial layers and their surface concentrations which are relevant parameters for the interfacial properties. These structural results are compared with the compressibility behavior of the films which is determined by dynamic surface pressure measurements.

Experimental

DHP synthesis. DHPs were prepared according to Higuchi's procedure (10) from coniferyl alcohol (11). All the parameters of the synthesis were identical for the Zutropfverfahren DHP (ZT) and Zulaufverfahren DHP (ZL) except the addition rate of the precursors. In the ZT, the precursors (coniferyl alcohol and hydrogen peroxide) were added in a slow and continuous way on a solution of peroxidase, whereas for the ZL, all the reagents were added at the same time. Some characteristics of DHPs are listed in Table I, and they are published in details elsewhere (12). They were dissolved and used in a 2 g/L 9/1 dioxan/water solution (v/v). ZT DHPs have a higher content of β-O-4 linkages and a lower content of cinnamyl alcohol end groups compared to ZL DHPs. In the latter case, at the beginning of the reaction, the concentration of monomeric radicals is high (due to the fast addition) leading to the formation of dimeric products. These dimers polymerize rapidly to larger molecules. On the contrary, in the ZT method, the concentration of radicals is low (due to the slow addition) and as a consequence the coupling between two monomers is less frequent. Here, the monomeric radicals react with polymeric radicals leading mainly to linkages of the β-O-4 type involving a decrease of phenolic hydroxyl groups. This way of addition results also in higher molecular weight products.

Table I : Main characteristics of the ***ZT*** and ***ZL DHPs***

	ZT DHP	***ZL DHP***
Elemental composition	%C : 58.9 %H : 5.64 %O : 35.5	%C : 63.8 %H : 5.85 %O : 30.3
GPC determination [a]		
Mw	1800	1400
Mn	1400	1050
polydispersity	1.3	1.312
Thioacidolysis yield [b] (µmol/g of DHP)	630	527
ratio phenolic OH groups versus aliphatic OH groups [c]	0.26	0.32
Scattering length density Nb (10^{-6} Å^{-2})[d]	2.75	2.8

a. Weight average molecular weight (M_w) = $\Sigma_{i=1\ to\ n}\ C_i/\Sigma_{i=1\ to\ n}(C_i/M_i)$; Number average molecular weight (M_n) = $\Sigma_{i=1\ to\ n}\ (C_iM_i)/\Sigma_{i=1\ to\ n}\ C_i$ (C = concentration; M = molecular weight from calibration curve), polydispersity = Mw/Mn. GPC analysis in THF, flow rate :1ml/mn, T° :40°C, injection volume :100µl, detection UV 280nm. (12)
b. Thioacidolysis is an acid catalysed degradation in dioxan-ethanethiol with boron trifluoride which results in the cleavage of aryl-glycerol βaryl ether (21) These values reflect the content of βO4 linkages which is the most abundant intermonomeric linkage in lignins and in DHP
c. Based on the intensity of carbonyl infrared bands of acetylated DHPs according to Faix's procedure (22). The contents of phenolic and aliphatic OH change according to the distribution of intermonomeric linkages. ZL DHPs have a higher content of phenolic OH than ZT DHPs (9)
d. Calculated for 1.5 OH group per C9 unit (13) and a density of 1.41 g/cm^3 (8) (see text)

Neutron reflectivity.

Theoretical background. Neutrons interact with matter in the same manner as light, when their electromagnetic properties are concerned and so neutron reflectivity is linked with the neutron index of refraction, n, of the medium crossed by the neutron beam. The index of refraction of the medium is : $n = 1-\lambda^2 Nb/2\pi$ where λ is the neutron wavelength, N is the atomic number density, and b is the coherent length density, thus, the product N b is the scattering length density. The b parameter is characteristic of the nuclei of atoms and is very different when considering hydrogen or deuterium. Thus in order to increase the contrast between the layers and the subphase, all the neutron reflectivity experiments were performed using D_2O instead of hydrogenated water. In these cases the replacement of the hydroxyl hydrogen by deuterium has to be considered for DHP Nb calculation. It can be assumed that the average number of OH groups per C9 unit is 1.5 for ZL and ZT DHP based on the results obtained by Saake (13). The DHP density has also to be taken in account for the Nb calculation. The density value used in this study is 1.41 g/cm^3 according to the value obtained on Milled Wood Lignin (MWL) of spruce by Luner et al (8). The use of these two approximations is not crucial in the context of this work. It may only change the absolute values but not the main tendencies observed here (variation on DHP volume fraction should be less than +/- 0.1).

Experimental. The reflectivity spectra were determined with a polychromatic beam of neutrons at a fixed incident angle θ_0 using the « time of flight » method. Reflectivity is the ratio of the intensity of the specularly reflected beam to the intensity of the incident beam. When an adsorption layer occurs at the interface, the value of n(z) in the direction perpendicular to the interface plane is not constant. The deviation of the reflectivity from the Fresnel reflectivity (n(z) constant) provides information on the variation of the refractive index n(z). Because the refractive index is a function of the scattering length density of the atoms, one can therefore deduce the solvent and DHP composition of the interfacial layer. The method used to calculate the variation of refractive index is based on the replacement of the continuous variation by a series of discrete homogeneous layers and the application of the standard optical methods. In each layer, the DHP volume fraction Φ is linked to the scattering length density of the DHP (Nb_{DHP}) and of water (Nb_{water}) by (14) :

$$N\,b(z) = \Phi(z)\, Nb_{DHP} + [1 - \Phi(z)]\, Nb_{water}$$

Experiments have been performed at the « time of flight » reflectometer DESIR (15) in the Orphée reactor (Léon Brillouin Laboratory, Saclay) at the grazing angle θ_0 of 1.265°. The useful neutron wavelength ranges from 3 to 15.5 Å. The reflectivity measurements were performed using a teflon trough 6.5 x 13.5 x 0.3 cm in a gas tight cell thermostated at 20°C enclosed in a second cell which helps to maintain constant temperature of the air surrounding the first cell. The trough was filled up with 9 mL of water forming a meniscus 2-3 mm over its edge. DHPs were spread on the air / water interface from a 9/1 dioxan/D_2O solution (2g/L). The reflectivity spectrum was recorded during eight hours and a flat background was subtracted before reflectivity calculation

Dynamic surface pressure experiments : DHP layers were formed by droplet deposition of a 2 g/L (dioxan/water, 9/1, v/v) solution on the surface of ultra pure water filling the trough (LB 105, Atemeta, Paris; maximum area 286 cm^2). After deposition, the DHP layer was allowed to settle during 30 min, then the barrier was moved with a rate proportional to the area : $dA/dt = -A/\tau$ where A is the area of the film and τ is a constant whose value is obtained by integration : $\tau = -\Delta t / \ln (Ae/Ao)$, where Ao and Ae are the values of A at the beginning and at the end of the compression (Ao-Ae =10 cm here) and Δt its duration (presently 30 min).

When the compression was completed, relaxation was allowed to take place during 180 min.
The limit concentration was determined by deposition of low amounts of DHP. It is the surface concentration where the film starts to develop a significant pressure. The results presented here are the average of four independent determinations.

Results and discussion :

Neutron reflectivity. Figure 1 shows the experimental and calculated reflectivity curves of DHPs spread on deuterium oxide (24 mg/m^2 deposited). Spreading DHPs on deuterium oxide yields a favorable situation for neutron reflectivity measurements because the Nb of DHPs (2.8 10^{-6}Å^{-2}) is practically half-way between those of D_2O (6.37 10^{-6} Å^{-2}) and air (≈ 0 Å^{-2}). Experimental neutron reflectivity spectra were fitted with calculated models and the best mathematical result was retained. Tables II and III report the volume fractions and the thicknesses found for ZL and ZT DHPs at several deposited surface concentrations ranging from 3 to 48 mg/m^2. For the low concentration regime (less than 12 mg/m^2 deposited for ZT and 18 mg/m^2 deposited for ZL) a good fit is obtained using a one layer model. Two layer or more complicated models do not improve the χ^2 of the fits. At higher concentrations (high concentration regime), an improvement of the results is achieved with a two layer model.

Volume fraction. The layer formed at low concentration has a constant DHP volume fraction of around 0.2-0.25 for all the concentrations and for ZT and ZL DHP (Tables II and III). A similar stability of the volume fraction is observed in the case of high concentration deposits: the air side layer has a DHP volume fraction of around 0.4 and the water side layer has a lower content of DHP (ϕDHP=0.15, ϕwater=0.85). These results show that all these layers have a high content of water implying that the interfacial organization of the DHPs results in a large free volume. These data are in good agreement with those of Constantino et al (6) obtained by ellipsometric measurements on extracted lignins from Pine wood. Wheat straw lignins studied by neutron reflectivity exhibit the same general behavior with roughly 60-70% water in the interfacial layer (7). Even though DHPs are usually thought to be hydrophobic, strong associations between water and DHP are not very surprising because it is very well known that lignins (and DHPs) require a mixture of organic solvent and water to be solubilized. It was already demonstrated by Schuerch (16) that lignin can act as donor and acceptor of hydrogen bonds and that solvents for lignin solubilization need both of these characteritics. In the case of the interface, one can suppose that interactions between DHPs themselves can exhibit the two behaviors in complement with water association to form a gel-like phase. Association with water was also proposed in solution by Gilardi (17) who reported that the relaxation time of water protons measured by NMR is decreased in the presence of lignins. Since this parameter is sensitive to molecular motions,

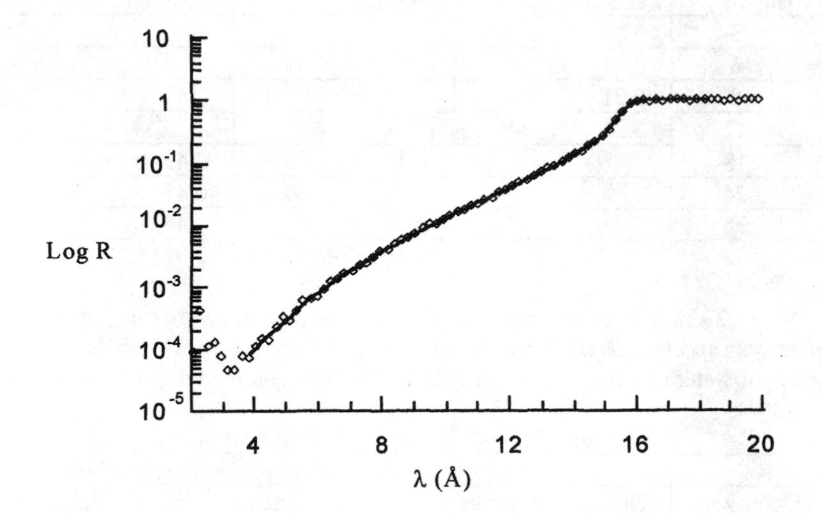

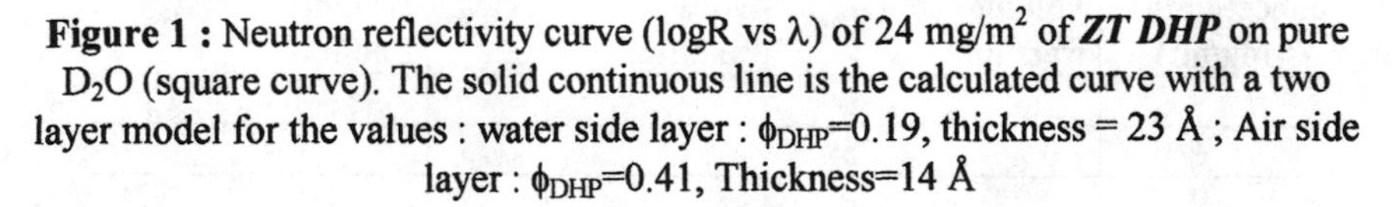

Figure 1 : Neutron reflectivity curve (logR vs λ) of 24 mg/m^2 of ***ZT DHP*** on pure D_2O (square curve). The solid continuous line is the calculated curve with a two layer model for the values : water side layer : ϕ_{DHP}=0.19, thickness = 23 Å ; Air side layer : ϕ_{DHP}=0.41, Thickness=14 Å

these results can be interpreted as the « trapping » of the water molecules in the lignin aggregates.

Table II : Modeling of the ***ZT DHP*** interfacial layers from neutron reflectivity spectra. ***ZT DHP*** volume fraction and film thickness were determined at the air/water interface a one layer or a two layer model.

	One layer model		***Two layer model***			
			Water side layer		**Air side layer**	
Deposited concentration (mg/m^2)	DHP volume fraction	Thickness (Å)	DHP volume fraction	Thickness (Å)	DHP volume fraction	Thickness (Å)
3	0.21	15.				
6	0.23	24				
9	0.21	35				
12	0.25	49				
18			0.19	23	0.41	14
24			0.16	41	0.48	26
36			0.10	64	0.43	35
48			0.14	60	0.47	36

Table III : Modeling of the ***ZL DHP*** interfacial layers from neutron reflectivity spectra. ***ZL DHP*** volume fraction and film thickness were determined at the air/water interface using a one layer or a two layer model. (nf= not fitable)

	One layer model		***Two layer model***			
			Water side layer		**Air side layer**	
Deposited concentration (mg/m^2)	DHP volume fraction	Thickness (Å)	DHP volume fraction	Thickness (Å)	DHP volume fraction	Thickness (Å)
3	nf	nf.				
6	0.24	25				
9	0.28	30				
12	0.27	47				
18	0.22	46				
24			0.16	41	0.48	26
36			0.16	42	0.45	29
48			0.15	56	0.43	38

The water content also changes with the DHP surface concentration. At low concentrations, a uniform film is formed whereas at higher concentration DHPs form a more « dry » association on the air side and a more hydrated phase on the water side. The structure of the adsorption layer is not completly described by a two homogenous layer profile and variations in the water content are probably more or less continuous perpendicularly to the surface. The « transition » between the two regimes occurs for the ZT DHP at a lower deposited concentration than for the ZL (between 12 and 18 mg/m2 for the ZT and between 18 and 24 mg/m^2 for the ZL), however in both cases the transition occurs at the same measured surface concentration (around 1.5 mg/m^2). These differences are probably related to structural differences existing between the type of DHP. This « transition » could be understood as a change in the type of associations. At low concentration, the DHPs are mostly associated with water. The increase of the concentration allows the formation of more frequent DHP-DHP associations (instead of DHP-water associations) resulting in the formation of a more dense phase. The ZT DHPs form a more dense phase at lower deposited concentration. This could indicate that this type of DHP has more affinity for the surface and so ZT can be consider as more hydrophobic than the ZL type which has a higher content of phenolic hydroxyl groups and lower molecular weight. These two parameters can have a large impact on the hydrophilicity (and the solubility) of the molecules. It is clear from these results that increasing the DHP concentration induces the formation of layers with higher density and the resulting association could be related to the molecular structure of the DHP, even if the general behavior is the same. This behavior seems to be a typical feature of lignin and « lignin like » structures and the formation of denser structures is reminiscent of the observations made by Takabe et al (18) concerning the decrease of the water content of the cell wall during lignification

Thickness of the adsorbed layers. The thickness of the layers increases with the amount deposited starting from 15 Å to 96 Å for ZT DHP (Table II). In the case of the ZL DHPs, no fit was obtained for 3 mg/m^2 deposited due to the small difference between the Fresnel reflectivity (spectrum corresponding to pure D_2O) and the DHP spectrum. The first data available are for the 6 mg/m^2 deposited. The thickness increases from 25 to 96 Å (Table III). These data are in general in good agreement with the thickness observed by previous ellipsometry measurements which give a thickness of 60 Å per layer for a multi layer system built by the Langmuir-Blodgett method (6). For both DHPs increasing the amount deposited results in an increase of the thickness of the layers whereas the DHP volume fraction remains constant. It can therefore be concluded that DHPs have a three dimensional organization at the air/water interface leading to somewhat of a stoichiometric ratio with water.

Surface concentration. Determination of the DHP volume fraction and layer thickness allows the calculation of the concentration of DHP in the interfacial layer. Table IV reports these concentrations for each deposit. The surface concentration increases with the amount deposited meaning that the saturation of the interface is not reached. The other important result coming from these data is that around 10% of the DHPs remain at the interface. Determination of the surface concentration of wheat straw lignin by neutron reflectivity (7) has led to the same conclusions (roughly 20-30% remains at the interface). This ratio decreases slowly when the amount deposited increases ranging from 15% to 7%. Thus it is clear that the surface concentration has to be determined for interfacial studies. Otherwise, large errors can be made on surface concentration evaluations from deposited amounts. However the low concentration of DHP remaining at the interface is not completely surprising in view of the small but effective solubility of DHPs in water and of the small amount deposited with respect to the trough volume. Thus, these data can reflect the efficiency of the spreading procedure, the solubility of DHP in water, or the precipitation of DHP aggregates in the subphase. However since the DHP deposition is very reproducible (19) it can be supposed that an equilibrium between the subphase and the interface is responsible of the low measured surface concentration. Therefore, if there is some possibility for DHPs to move from the surface to the bulk (or the reverse), chemical heterogeneities of DHPs and lignins have to be taken into account. DHPs (like lignins) differ essentially from most of the other polymers by variations in the type and frequency of the intermonomeric linkages leading to structurally heterogeneous polymers. Among them, some may have more affinity for the interface than others. So it will be essential in future studies to examine the chemical structure of the molecules adsorbed at the interface.

Table IV : Surface concentrations calculated from neutron reflectivity data for ***ZL and ZT DHP***. R is the ratio :surface concentration measured by neutron reflectivity / concentration deposited and surface concentration=ϕDHP*layer thickness*$DHP_{density}$

Concentration deposited (mg/m^2)	3	6	9	12	18	24	36	48
surface concentration ***ZT DHP*** (mg/m^2)	0.45	0.81	1.06	1.77	1.46	2.68	3.04	3.51
R	0.15	0.13	0.12	0.14	0.08	0.11	0.08	0.07
surface concentration ***ZL DHP*** (mg/m^2)	-	0.84	1.21	1.82	1.43	2.73	2.84	3.49
R	-	0.14	0.13	0.15	0.08	0.11	0.08	0.07

Dynamic surface pressure experiments

Limit area. The limit area is the area (or concentration) where the film starts to develop a significant pressure and it allows the calculation of the area occupied by one molecule of polymer (or by one monomer) when the molecules start to interact. The deposited concentrations were corrected by the data of neutron reflectivity experiment assuming that below 3 mg/m^2 (lowest concentration deposited for neutron reflectivity experiment) the general behavior described at higher concentrations is still observed (DHP volume fraction remains constant). This calculation is based on Mn values which correspond to the average number molecular weight obtained from the GPC determination (12). The limit areas per monomer or molecule calculated for ZT and ZL DHPs are significantly different (Table V). ZT is more expanded than ZL (Area/per monomer : ZT=124 Å^2; ZL=95 Å^2). Sarkanen (20) has suggested that ZT DHPs are more linear than the ZL DHPs. In this case it will be easier for the ZT DHPs to be more expanded than the more « compact » ZL DHPs. The areas reported here are larger than those reported by Luner and Kempf (8) (17.1 Å^2 per monomer for spruce MWL). This comes from the fact that the values we report are corrected by neutron data. The uncorrected data (area per monomer 19.5 Å^2 for ZT DHP and 12 Å^2 for ZL DHP) are in good agreement with those reported previously. Nevertheless, the area calculated is significantly larger than the area of a monomer deduced from a molecular model, evaluated to be close to 57 Å^2 (8). This discrepancy should very likely be interpreted as a « swelling » by water of the DHP network formed at the limit concentration. In other words, the interfacial layer is not a DHP layer but a mixed DHP-water layer.

Table V: Limit area of ***ZT and ZL DHP*** measured on Langmuir trough and corrected using neutron reflectivity data

	ZT	ZL
Limit concentration deposited (mg/m^2)	1.7	2.75
Limit concentration neutron corrected (mg/m^2)	0.25	0.33
Area per molecules (Å^2)	1172	704
Area per monomer (Å^2)	124	95

Based on the neutron data, we can also assume that the thickness of the layer at the limit area is somewhat lower than 15Å. This value is a little higher than the height of one phenyl propane unit (10 Å) but much less than the height of one DHP molecule containing at least 2 or 3 monomeric units and as a consequence the DHPs are certainly laying flat in the interfacial layer, a fact which is consistent with the large limit area obtained. Moreover, this hypothesis is reminiscent of the molecular modeling made by Jurasek showing that lignin deposition in a virtual secondary wall occurs with a preferred orientation parallel to the carbohydrate fibrils (5).

Relaxation Behavior. DHP films spread on a Langmuir trough were compressed and their relaxation behavior was monitored. In all cases, the compression led to a large increase of the surface pressure as compared to values obtained at equilibrium (19). On relaxation, the surface pressure decreased slowly during several hours (Fig.2). This behavior indicates that the DHPs are

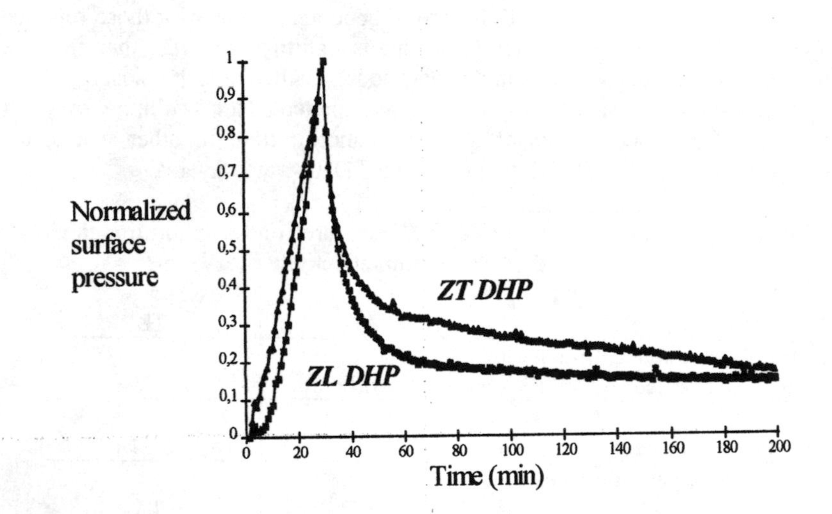

Figure 2 : relaxation behaviour of ***ZT*** and ***ZL DHP***.
DHP were deposited on the trough (6 mg/m^2) and compressed at 75% of the original area in a period of 30 min. The normalized surface pressure is (Πt-Πo)/(Πmax-Πo), where Πt is the surface pressure at t time, Πo the initial surface pressure and Πmax the maximum surface pressure during the compression process. Πmax-Πo is roughly 15mN for both experiment.

involved in a network and that the molecules interact with their neighbors. Layers of ZT DHPs relax more slowly than layers formed of ZL DHPs. This would indicate that interactions between DHPs are stronger in the former case. This relaxation behavior is probably a key point to understand DHP and lignin association behavior and it is presently under investigation. Several parameters can affect this behavior, and these include the molecular weight and chemical composition. ZT DHPs have a higher molecular weight and as a consequence they can have more interaction sites leading to more interlinked structures. Regarding the chemical composition, the phenols are often considered to be important for the aggregation processes (17). However, for the DHPs the ZL have more phenolic groups (13) and relax more easily. It can be concluded that the relaxation behavior is a complex phenomenon which involves probably several parameters and characteristics of the DHP structure.

Conclusion

Several conclusions can be drawn from the results presented here. The first one is that DHPs at the air water interface form a gel-like phase containing a large amount of water and where the molecules form a physical network. The increase of DHP amount at the interface results in the increase of the film thickness and in the formation of a denser layer on the air side. In this study the structure of the adsorption layer of two DHPs has been investigated. These model compounds show the same general behavior although some differences can be observed. This general trend is also in agreement with previous studies performed on lignin samples, confirming that DHPs can mimic lignin's properties at the air/water interface and as a consequence the use of different DHPs with several monomeric composition and structures in parallel with studies on natural lignins will lead to a better understanding of the macromolecular organization of lignins in heteregeneous media and will help in the understanding of some critical problem in the lignin field like the aggregation process, the reactivity and also the general scheme of lignification.

Acknoledgements

Dr L. T. Lee's (laboratoire Léon Brillouin, CEA-CNRS, Saclay, France) help for the neutron reflectivity experiments is gratefully acknowledged. Helpful discussions and advice on the plant cell wall organization was also provided by Dr B. Chabbert.

Litterature cited :

1 Houtman, C.J. ; Atalla, R.H. *Plant Physiol.*, 1995, **107**, 977-984.
2 Shigematsu M. ; Morita M ; Sakata I. *Macromol. Chem. Phys.* 1994, **195**, 2827-2837.
3 Guiselin O. ; Lee L. T. ; Farnoux B. ; Lapp A. ; *J. Chem. Phys.* 1991, **95**, 6, 32-42 ; Wiener M. C. ; King G. I. ; White S. H. ; *Biophys. J.* 1991, **60**, 568-576.

4 Agarwal, U.P. ; Atalla, R.H. *Planta*, 1986, **169**, 325-332.
5 Jurasek, L. *J. Pulp Paper Sci.*, 1996, **22**, 376-380.
6 Constantino, C.J.L. ; Juliani, L.P. ; Botaro,V.R. ; Balogh, D.T. ; Pereira, M.R. ; Ticianelli, E.A. ; Curvelo, A.A.S. *Thin solid films*, 1996, **284-285**, 191-194
7 Aguié-Béghin, V. ; Baumberger, S. ; Lapierre, C. ; Monties, B. ; Douillard, R., Groupe français des polymères symposium, 1997, Lyon, France.
8 Luner, P. ; Kempf, U. *Tappi*, 1976, **53**, 2069-2076.
9 Luner P. ; Roseman G. ; *Holzforschung*, 1986, **40**, 61-66
10 Tanahashi, M. ; Aoki, T. ; Higuchi, T. *Holzforschung*, 1982, **36**, 117-122.
11 Quideau, S. ; Ralph, J. *J. Agric. Food. Chem.*, 1992, **40**, 1108-1111.
12 Cathala, B. ; Saake, B. ; Faix, O. ; Monties, B. ; *Polymer degradation and stability*, 1998, **98**, 65-69.
13 Saake, B. ; Argyropoulos, D.S. ; Beinhoff, O. ; Faix, O. *Phytochem.*, 1996, **43-2**, 499-507.
14 Penfold, J., Thomas, R.K. *J. Phys :Condens Matter 2*, 1990, 1369-1412.
15 Lee, L.T. ; Mann, E.K. ; Langevin, D ; Farnoux, B. *Langmuir*, 1991, **7**, 3076-3080.
16 Schuerch, C. *J. Am. Chem. Soc.*, 1952, **74**, 5061-5067.
17 Gilardi, G. ; Cass, A.E.G. *Langmuir*, 1993, **9**, 1721-1726.
18 Takabe, N. ; Fugita, M. ; Haide, H. ; Saiki, K. *Mokuzai Gakkaishi*, 1981, **27**, 813-820.
19 Cathala, B. ; Aguié-Béghin, V. ; Douillard, R. ; Monties, B. *Polymer degradation and stability*, 1998, **59**, 77-80.
20 Sarkanen, K.V., In *Lignins : occurrence, formation, structure and reactions*, Sarkanen, K.V. ; Ludwig, C.H., ed, Wiley Interscience, New York, 1971, 95-155.
21 Lapierre, C.; Monties, B.; Rolando, C. *Holzforschung*, 1986, **40**, 47-50.
22 Faix, O.; Grunwald, G.; Beinhoff, O. *Holzforschung*, 1992, **46**, 425-430.

Chapter 14

Comparative Analysis of the Lignins of Cork from *Quercus suber L.* and Wood from *Eucalyptus globulus L.* by Dry Hydrogen Iodide Cleavage

L. G. Akim[1,4], N. Cordeiro[2], C. Pascoal Neto[3], and A. Gandini[1]

[1]Ecole Française de Papeterie et des Industries Graphiques, St. Martin d'Hères, France
[2]Department of Chemistry, University of Madeira, Funchal, Portugal
[3]Department of Chemistry, University of Aveiro, Aveiro, Portugal

Lignin from cork (*Quercus suber L.*) was isolated by two procedures: organosolv extraction and dioxane-water (9:1) extraction in presence of HCl. These lignins were characterized using a mild hydrogen iodide-cleavage method followed by ^{1}H NMR and GPC analysis. The results were compared with those for eucalyptus lignins (*Eucalyptus globulus*) isolated by the same procedures. The method used provided syringyl/guaiacyl ratios for the linear parts of the macromolecules and the degrees of crosslinking. The prevalence of guaiacyl units was demonstrated for cork lignin. Syringyl units were found to be minor components and present mainly in the linear parts of macromolecules. *p*-Hydroxyphenyl units were mainly condensed. Cork lignin was found to be significantly more cross-linked than eucalyptus lignin.

Lignin represents about 20% of cork, the outer bark of *Quercus suber* L. (1). Unfortunately, both the determination of lignin content and the analysis of its structure are hindered by the presence of suberin and extractive compounds. This is a general problem in bark lignin analysis (2). Unlike suberin, the major cork component responsible for the characteristic mechanical properties of cork, lignin has attracted little attention (3, 4). The chemical structure of cork lignin is still under discussion being chemically close and difficult to distinguish from suberin. Lignin, however, plays an important role in cork-adhesive interactions during industrial bonding processes and therefore the search for new polymeric compositions requires a sound knowledge of the structure of cork lignin.

This work is a continuation of previously published studies (5, 6) in which cork lignin was isolated and characterized using conventional procedures, such as

[4]Present address: Pulp and Paper Research Institute, McGill University, 3420 University Street, Montreal, Quebec H3A 2A7, Canada.

quantitative ^{13}C NMR, FTIR, nitrobenzene and permanganate oxidation. In our previous papers, it was shown that cork lignin posses some peculiar structural features which differ from those of traditional lignins. In this paper, we characterize cork lignin using a recently developed method of selective mild cleavage with dry hydrogen iodide. A characterization of a hardwood lignin from *Eucalyptus globulus* isolated using the same procedure is included for comparison.

The method of HI-cleaving (depolymerization with dry hydrogen iodide) was developed a few years ago (7, 8) and has proven to be a powerful tool in the study of lignin structure. This is a mild analytical method in which samples are subjected to a short treatment with dry hydrogen iodide that leads to a mixture of polymeric, oligomeric and monomeric products. The monomeric products can be characterized by NMR spectroscopy (9) and the mixture of oligolignols by GPC and ^{13}C NMR methods (8, 10). The data thus obtained gives information about the degree of lignin cross-linking, i.e. the ratio between linear parts and branching points of the macromolecular structure. In this study, we applied this method also to evaluate the guaiacyl/syringyl ratio in the linear chains of lignin.

Materials and Methods

Isolation of lignins. High quality reproduction cork from *Quercus suber L.* was milled in a cross-beater mill and sieved to 20 mesh. The cork powder was then sequentially soxhlet-extracted with methylene chloride, ethanol and water (8 hrs. for each solvent) and dried at 40°C. Eucalyptus (*Eucalyptus globulus* L.) wood was milled to 40-60 mesh, extracted with ethanol-toluene (1:2, v/v) in a soxhlet extractor for 4 hours, water for 4 hours, and then was air dried.

The cork organosolv lignin (COL) was isolated with a yield of 2.6% (o.d. cork) from solvent extracted cork powder by an acid catalyzed ethanol-water (1:1, v/v) organosolv method (170°C, 4 hrs.) as previously described (5). The same procedure was applied to solvent extracted eucalyptus sawdust, to obtain eucalyptus organosolv lignin (EOL) with a yield ranging from 10 to 12 % (o.d. wood).

The cork dioxane lignin (CDL) was prepared according to the procedure described in (11). Cork powder, extracted as described above, was suspended in dioxane-water (9:1) containing 0.2 N HCl. The mixture was refluxed for 1 hour, cooled, and filtered. The filtrate was vacuum evaporated to a syrup, which was then poured into water, precipitating the lignin. The precipitated lignin was centrifuged, dried under vacuum, and purified by dissolution in a 1,2-dichloroethanol:ethanol (2:1,v/v) mixture followed by precipitation in ethyl ether. The purified precipitated lignin was centrifuged, washed with ethyl ether, and dried yielding 1. % of o.d. cork. The eucalyptus dioxane lignin (EDL) was isolated using the same procedure, except for a 4 hours reflux time. The yield of EDL was 9.3% (o.d. wood).

Analytical methodology. Elemental microanalyses were performed at the laboratories of the Service Central d'Analyse of CNRS (Vernaison, France). The methoxy group content was determined by the modified Zeisel method (12).

The hydrogen iodide treatment was conducted by suspending 10-mg samples in 0.5 ml $CDCl_3$ in 0.5-mm (O.D.) NMR tubes. Dry HI was bubbled in an argon flow through the suspension for 1.5 hours. The ^1H NMR spectra were recorded using a Bruker AC-300 instrument at ambient temperature; spectra of the reaction mixture were recorded every 30 minutes. Before recording the last spectrum, a coaxial capillary containing a solution of known amount of 1,1,2,2-tetrachloroethane in $CDCl_3$ was inserted in the tube as an external standard for the quantitative measurements. The amount of diiodides was determined based on the intensities of the NMR signals to the 1,1,2,2-tetrachloroethane standard. The signals for OCH_3, H_α and aromatic protons could be distinguished and integrated. Under the chosen conditions, only monomeric products remained in solution; the other minor products formed a colloidal suspension giving no signal in the NMR spectrum. The G/S ratio was determined from the relative intensities of the corresponding OCH_3, H_a and aromatic proton signals of 1,3-diiodo-1-(4-hydroxy-3-methoxyphenyl)propane (guaiacyl product) and 1,3-diiodo-1-(3,5-dimethoxy-4-hydroxyphenyl)propane (syringyl product). After recording of the last ^1H NMR spectrum, the suspension was filtered and the residue washed five times with $CHCl_3$, dried *in vacuo* and dissolved in DMSO before GPC analysis. Gel permeation chromatography of the lignins was performed using an 840x10 mm Sephadex LH-20 column with. DMSO + 0.03M H_3PO_4 + 0.03M LiBr as the eluent. Chromatograms were monitored by UV light at a wavelength of 280 nm.

^1H NMR spectrum of 1,3-diiodo-1-(4-hydroxy-3-methoxyphenyl)propane **I** (d, ppm; J, Hz): 2.45 (H_β, m), 2.78 ($H_{\beta'}$, m), 3.16 (H_γ, t, $J_{\gamma-\beta}$ 6.7), 3.92 (OCH_3, s), 5.25 (H_α, t, $J_{\alpha-\beta}$ 7.5), 6.83 (H_5, d, J_{5-6} 8.1), 6.90 (H_2, d, J_{2-6} 2.1), 6.95 (H_6, dd, J_{5-6} 8.1, J_{6-2} 2.1).

^1H NMR spectrum of 1,3-diiodo-1-(3,5-dimethoxy-4-hydroxyphenyl)propane **II** (d, ppm; J, Hz): 2.47 (H_β, m), 2.74 ($H_{\beta'}$, m), 3.13 (H_γ, d, $J_{\gamma-\beta}$ 6.7), 3.88 (OCH_3, s), 5.20 (H_α, $J_{\alpha-\beta}$ 7.6), 6.63 ($H_{2,6}$, s).

Results and Discussion

The method used to isolate lignin from lignocellulosics while preserving its chemical structure, involves milling the material in a vibrating or rotary ball mill, followed by solvent extraction of the released lignin (13, 14). When these methods are applied to cork, some problems may arise because of its particular chemical composition and mechanical properties. The Poisson ratio of cork is extremely low due to its cellular microstructure and the high elasticity of its cell walls (15). Significant recovery of linear dimensions takes place even after strong compression (16). Therefore, cork is a material that is not easy milled, and, consequently, the yields of the "milled cork lignin" are very low (3). In addition, the isolated lignin-like polymer is highly contaminated with aliphatic suberin fragments and carbohydrates (3). This occurs because lignin is a minor component of cork and is strongly associated with other cork components. This drawback may be overcome if cork is previously desuberized (17). However, desuberization by strong alkaline solutions may induce structural changes in the lignin macromolecule. These are the reasons for choosing other methods of lignin isolation that avoid the mechanical treatment stage and provide higher selectivity and increased yield and purity of cork lignin.

The first of these methods is mild organosolv extraction in ethanol-water solution (1:1) catalyzed by 0.1M acetic acid (5). The yield of lignin (COL) thus obtained, 2.6%, is higher than that from milled cork lignin (3). This lignin is not contaminated with carbohydrates and contains only small amounts of aliphatic structures, which were assumed to be covalently bound to the phenylpropane units of lignin. Analysis of the permanganate oxidation products of intact cork and COL (6) suggested that this lignin was representative of the structure and composition of the "in situ" lignin of cork. The acidolysis method was chosen because it was reported to be a convenient method for the isolation of lignin from barks (18, 19). Cork dioxane lignin (CDL) was obtained in 1.6% yield.

Table I. Elemental Analyses and Molecular Formulae of Cork and Eucalyptus Lignins.

Sample	Elemental Analysis, %				Molecular Formula
	C	H	O	OCH_3	
COL	61.9	5.9	32.2	12.0	$C_9H_{8.91}O_{3.07}(OCH_3)_{0.73}$
CDL	62.1	6.4	31.5	6.8	$C_9H_{10.40}O_{3.18}(OCH_3)_{0.40}$
EOL	59.6	6.0	34.4	21.3	$C_9H_{8.25}O_{3.08}(OCH_3)_{1.44}$
EDL	59.3	5.9	34.8	18.2	$C_9H_{8.52}O_{3.28}(OCH_3)_{1.21}$

The results of elemental and methoxyl content analyses of the lignins are given in Table I. Empirical molecular formulas calculated based on this data are also presented. The main difference between cork and eucalyptus wood lignins is the methoxyl content which is much lower for both cork lignins compared to eucalyptus. This obviously results from a low content of syringyl structures in cork lignins. Permanganate oxidation of intact cork and COL has previously shown that cork lignin is a guaiacyl-rich HGS-type lignin with H:G:S molar ratios of 2:93:5 for OCL and 3:93:4 for the "in situ" lignin of cork (6). The methoxyl content of CDL is particularly low. This may by attributed to the fact that this lignin, contrary to COL, is highly contaminated with suberinic aliphatic chains, as evidenced by its FTIR spectrum (not shown) which exhibits strong absorption bands at 2854 and 2930 cm^{-1}. The eucalyptus wood lignins (EOL and EDL) gave high methoxyl contents in agreement with their high syringyl contents. Previous results of permanganate oxidation of *Eucalyptus globulus* dioxane lignin had shown that it is a syringyl-rich lignin with H:G:S ratios of 3:36:67 (20). Both cork and eucalyptus organosolv lignins (COL and EOL) displayed methoxyl contents higher than those in the corresponding dioxane lignins (CDL and EDL). This may be explained by the formation of ethoxy structures during the organosolv treatment in ethanol-water system. The methoxyl group calculations have been adjusted accordingly.

Hydrogen iodide cleavage. The treatment of the lignin suspension in chloroform with dry hydrogen iodide led to extensive destruction of the polymer. Because of the low solubility of the polymeric and oligomeric products in solvents such as chloroform, only monomeric products can be registered in the ^1H NMR spectrum of the resulting solution.

Previously, it was demonstrated that only one type of monomeric product is released after HI-cleavage of guaiacyl-type lignins and lignin model polymers, namely 1,3-diiodo-1-(4-hydroxy-3-methoxyphenyl)propane **I** (9). Due to the specific mechanism of its formation (Fig. 1), the yield of this compound corresponds to the amount of structural units in the linear parts of the lignin macromolecule and reflects the degree of cross-linkage of the polymer. On the other hand, α-ethers could provide branching, but the abundance of non-cyclic α-O-4 ethers is very low (21).

The treatment of eucalyptus syringyl-rich lignins released the second monomeric product, i.e. the syringyl analog of **I**, 1,3-diiodo-1-(3,5-dimethoxy-4-hydroxyphenyl)propane **II** (Figures 2 and 3). Compound **II** was also found in the products of the cork lignin treatment. Monomers **I** and **II** come from coniferyl/sinapyl alcohol end-groups and linear end-wise parts of the polymer (9). The amount of diiodides formed corresponds to the degree of cross-linking in the polymer. Obviously, the guaiacyl/syringyl (G/S) monomer ratio must correspond to the G/S ratio in the linear parts of the macromolecules. No monomeric compound resulting from *p*-hydroxyphenyl structural units was identified in the spectra of the decomposition products.

The data on G/S ratio for the lignins is presented in Table II. The ratios for the linear parts of both the organosolv and dioxane lignins were equivalent indicating that these are similar lignins. Cork lignins showed a high G/S ratio of 4:1 as opposed to eucalyptus lignins that displayed a reverse trend of 1:5.

Table II. Guaiacyl/Syringyl (G/S) Ratios in the Monomeric Products of HI-Cleavage

Sample	G/S ratio
COL	4:1
CDL	4:1
EOL	1:5
EDL	1:5

The presence of products **I** and **II** in the spectra of the decomposition products of cork lignins in a 4:1 ratio supports the idea that cork lignin is guaiacyl rich lignin which confirms our previous results (5, 6). Although cork comes from a hardwood tree, the G/S value of cork lignin is similar to a G/S value for a softwood lignin. Generally, many bark lignins display higher G/S ratios in comparison with wood lignins (2).

The G/S values for organosolv cork lignins were found to be quite different from the 19:1 ratio obtained by the permanganate oxidation (6) and 56:1 ratio obtained by nitrobenzene oxidation. Even when the inaccuracies caused in the three techniques by the low syringyl contents are taken into account, the general tendency is that the G:S ratio in the products of HI cleavage is lower than in the products of the permanganate or nitrobenzene oxidation. This suggests that the linear end-wise fragments of cork lignin were enriched with syringyl units. The same tendency was

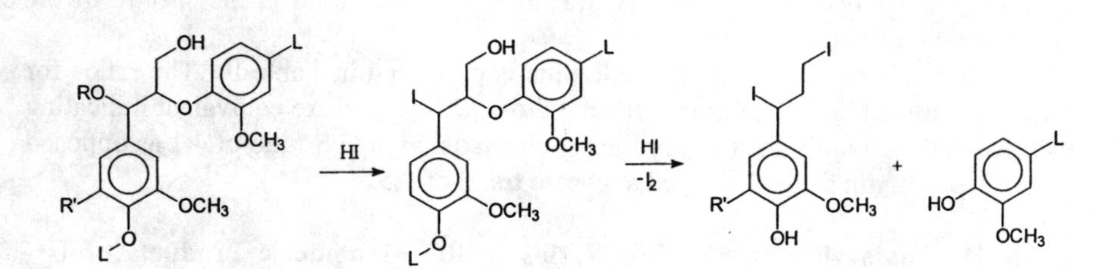

Figure 1. Chemical reactions involved in the HI-cleavage of an β-O-4 linkage in lignin.

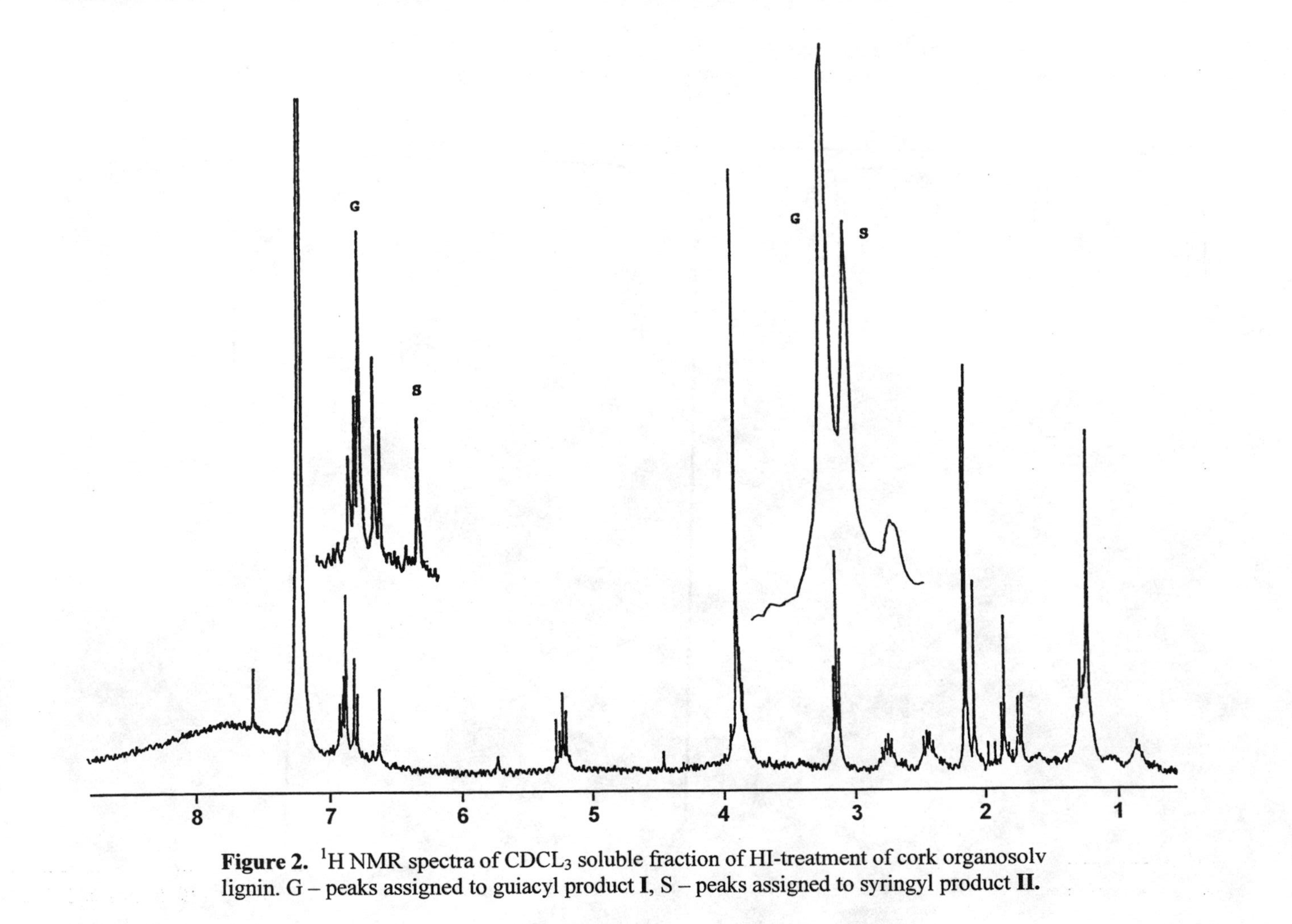

Figure 2. ^{1}H NMR spectra of $CDCL_3$ soluble fraction of HI-treatment of cork organosolv lignin. G – peaks assigned to guiacyl product **I**, S – peaks assigned to syringyl product **II**.

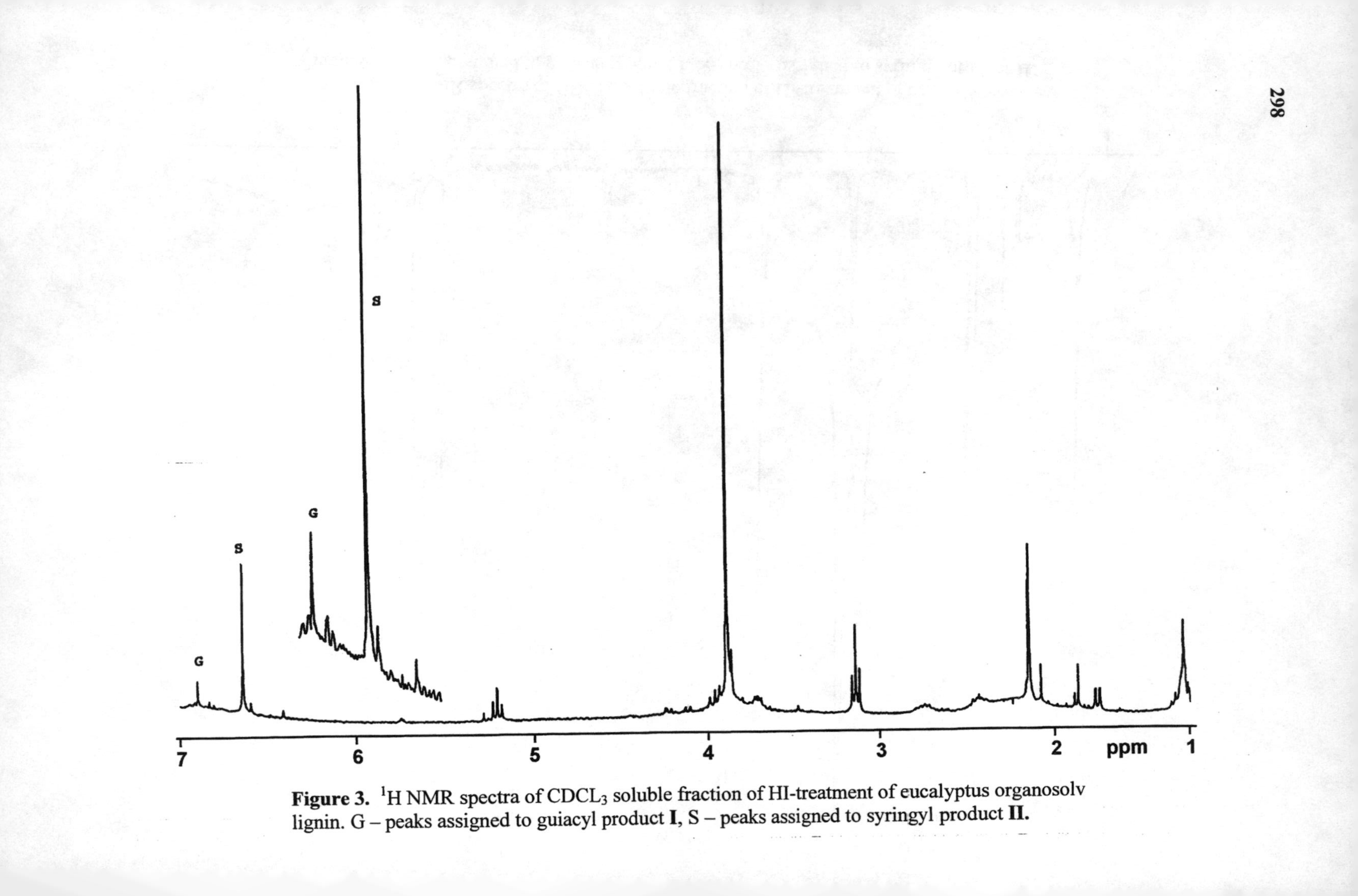

Figure 3. ^{1}H NMR spectra of $CDCL_3$ soluble fraction of HI-treatment of eucalyptus organosolv lignin. G – peaks assigned to guiacyl product **I**, S – peaks assigned to syringyl product **II.**

observed for the eucalyptus wood lignin. The G/S value obtained for eucalyptus dioxane lignin from permanganate oxidation was approximately 1:2 (20), whereas that obtained from the HI cleavage was 1:5. We believe, therefore, that the linear end-wise fragments in eucalyptus lignin consist mainly of syringyl units. The guaiacyl moieties are mainly present in branching sites of the lignin macromolecule due to their ability to form β-5 and 5-5' bonds.

Decomposition products from cork and eucalyptus wood lignins arising from *p*-hydroxyphenylpropane units were not detected. Since *p*-hydroxyphenyl units were previously observed in both types of lignin by other techniques (20, 6), it can be concluded that almost all the *p*-hydroxyphenyl units are condensed and incorporated into the branching points of the macromolecule instead of the linear portions. This would preclude the formation of the corresponding diiodides. This conclusion is in agreement with our previous results on the characterization of COL by ^{13}C NMR (5). It should also be added that unpublished results on some grass lignins (*p*-hydroxyphenyl type of lignin) confirm this conclusion.

Polymeric/oligomeric/monomeric distribution of lignin cleavage products. Data on the distribution of different types of decomposition products after HI-splitting is presented in Table III. Because cork organosolv lignin (COL) was found to be representative of the "in situ" cork lignin and mostly free of carbohydrates and suberin, it was chosen for the experiments on the distribution of depolymerization products as was an equivalent sample of eucalyptus organosolv lignin. The polymer/oligomer ratio was determined from the area measurements of gel-permeation chromatograms (Fig. 4). The amount of monomeric fraction was calculated from ^{1}H NMR spectra.

Table III. Ratios of Polymers/Oligomers/Monomers in Products of Lignin HI-Cleavage

Sample	Polymeric Fraction (%)	Oligomeric Fraction (%)	Monomeric Fraction (%)
EOL	15	65	20
COL	25	65	10

In both samples the amounts of oligomeric structures were almost equal and the main difference rested in the ratio of polymeric and monomeric fractions. The yield of monomeric species in eucalyptus lignin was twice as high as that of cork lignin. This suggests that cork lignin contains significantly less linear fragments or that these fragments are much shorter. The increasing amount of polymeric species indicates that the number of hydrogen iodide unsplittable carbon-carbon and Ar-O-Ar bonds was much higher in cork lignin and that the polymeric matrix was more cross-linked. The high degree of condensation of cork lignin was already proposed based

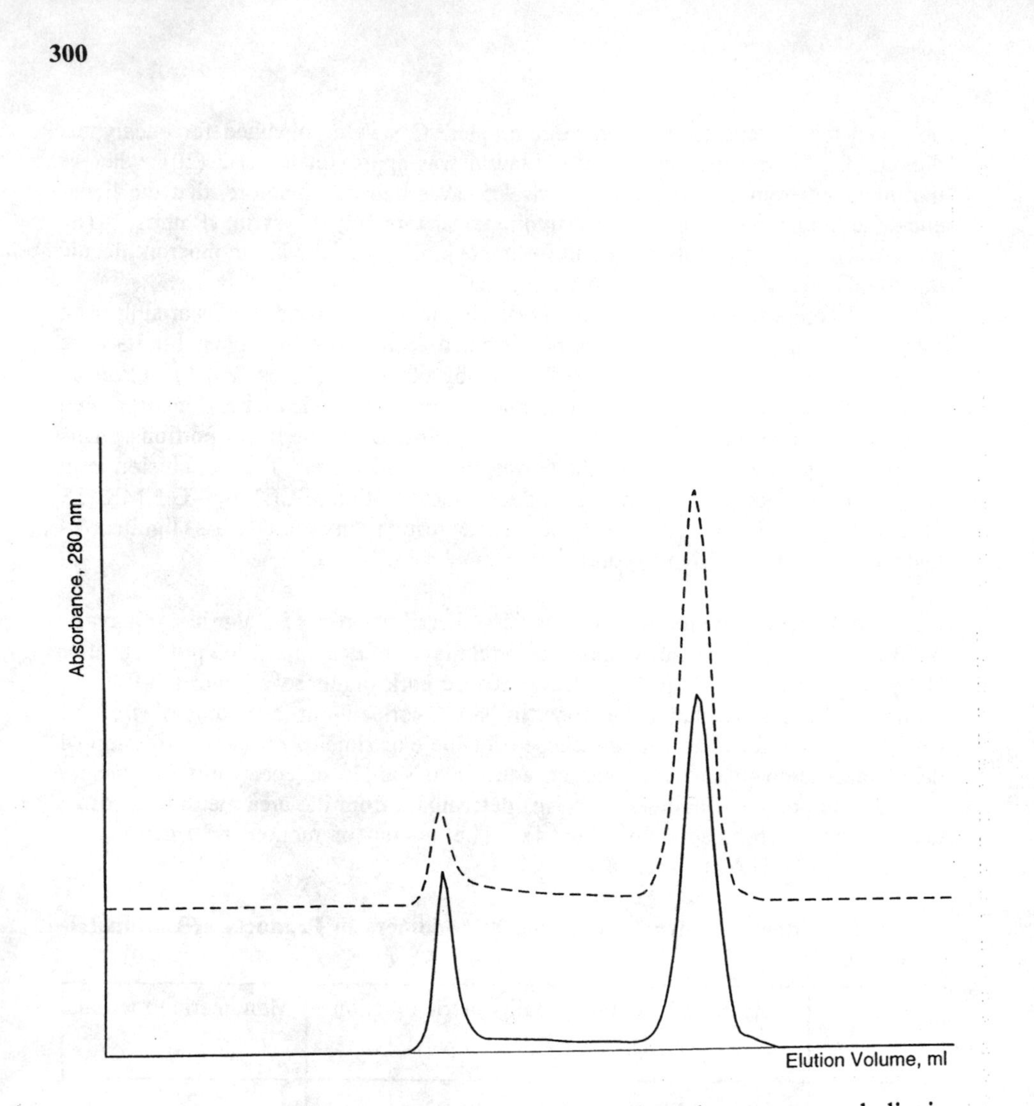

Figure 4. Gel-permeation chromatograms of cork and eucalyptus organosolv lignins depolymerized with dry hydrogen iodide in $CDCl_3$: α - COL; β - EOL.

on the results of ^{13}C NMR (5) and may be one of the reasons for the low yield of the isolated cork lignins.

The low yield of the monomeric fraction in eucalyptus lignin compared to that observed for birch MWL (up to 40%) can be explained by the differences in the isolation methods. The conditions of the organosolv delignification procedure are more severe than those found in the Bjorkman procedure. The organosolv treatment can lead to acid catalyzed condensation of lignin structural units.

The molecular-mass (MM) distribution of the decomposition products from cork lignin (Fig. 4, curve **a**) was found to be very similar to that of corresponding eucalyptus lignin (curve **b**) and to the MM distribution of the products from other types of wood lignins (8, 9).

Conclusions

The method of mild cleavage with dry hydrogen iodide provides an opportunity to evaluate the degree of cross-linking of lignins and the G/S ratio of the linear parts of macromolecule.

Cork lignin was found to posses predominantly guaiacyl type moieties although significant amounts of syringyl units were also present in the linear chains. Only traces of a *p*-phenylpropane monomeric product were found after HI-treatment of cork lignin. The *p*-hydroxyphenyl units were mainly condensed. The eucalyptus lignin was found to be a syringyl-rich structure, with linear end-groups composed mainly of syringyl units.

The HI-cleavage suggested a higher degree of cross-linking of cork lignin compared with that of eucalyptus lignin. Together with the low content of lignin in cork, the high cross-linking determines, at least partially, the low yield of cork lignin during isolation procedures.

Literature cited.

1. Pereira, H. *Wood Sci. Technol.* **1988**, *22*, 211-218.
2. Laks, P.E. In *Wood and Cellulosic Chemistry.* Hon, D.N.-S., Shiraishi N., Eds.; Marcel Dekker: New York, NY, 1991, Basel.
3. Marques, A.V.; Pereira, H.; Meier, D.; and Faix, O. *Holzforschung* **1994**, *48 (Suppl.)*, 43-50.
4. Zimmerman, W.; Nimz, H.; Seemüller, E. *Holzforschung* **1985**, *39,* 45-49.
5. Pascoal Neto, C.; Cordeiro, N.; Seca, A.; Dominiques, F.; Gandini, A.; Robert, D. *Holzforschung* **1996**, *50*, 563-568.
6. Lopes, M., Pascoal Neto, C.; Evtuguin, D.; Silvestre, A. J. D.; Gil, A.; Cordeiro, N., Gandini, A.. *Holzforschung* **1997**, *51*, in press.
7. Akim, L.G.; Shevchenko, S.M.; Zarubin, M.Ya. *Khim. Tekhnol. Volokn. Polufabr*. **1990**, 69-73.
8. Shevchenko, S.M.; Akim, L.G.; Pranovich, A.V; Zarubin, M.Ya. *TAPPI J.* **1991**, *74*, 257-262.

9. Shevchenko, S.M.; Akim, L.G.; Tanahashi, M.; Higuchi, T. *J. Wood Chem. Technol.* **1995**, *15*, 163-178.
10. Akim, L.G.; Shevchenko, S.M; Zarubin, M.Ya. *Wood Sci. Technol.* **1993**, *27*, 241-248.
11. Browning, B. L. *Methods in Wood Chemistry;* John Wiley and Sons: New York, NY, 1967, Vol II, p. 732.
12. Chen, C-L. In *Methods in Lignin Chemistry;* Lin, S. Y., Dence, C.W., Eds.; New York, NY, 1992; pp.465-471 .
13. Björkman, A. Svensk Paperstidn. **1956**, *59*, 477-482.
14. Brownell, H.H. *Tappi J.* **1965**, *48*, 513-518.
15. Gibson, L.J.; Easterling, K.E.; Ashby, M.F. *Proc. R. Soc.* **1981**, *A377*, 99-117.
16. Rosa, M.E.; Fortes, M.A. *J. Mater. Sci.* **1988**, *23*, 879-885.
17. Marques, A.V.; Pereira, H.; Meier, D., Faix, O. *Holzforschung* 1996, *50*, 393-400.
18. Solar, R.; Melcer, I. *Cellulose Chem.Technol.* **1985**, *19*, 159-166.
19. Swan, E.P. *Pulp Pap. Mag. Con.* **1966**, *67*, 456-459.
20. Pascoal Neto, C.; D. Evtuguin and P. Pedrosa Paulino. In: Proceedings of the 9th International Symposium on Wood and Pulping Chemistry. Tappi. Montreal, Canada, **1997**, Vol. 2, 78.1-78.4.
21. Ede, R.M. and Kipelainen, I. In: Proceedings of 8th International Symposium on Wood and Pulping Chemistry. Helsinki, Finland, **1995**,Vol. 1, 487-494.

MODIFICATION AND UTILIZATION

Chapter 15

Commercial Use of Lignin-Based Materials

J. D. Gargulak and S. E. Lebo

LignoTech USA, 100 Highway 51 South, Rothschild, WI 54474

The usefulness of industrial lignins has been demonstrated by the profitability of the lignin chemicals business operated worldwide. The total sales in 1980 amounted to 180 million dollars. By 1996 that number had grown to 600 million dollars. This growth is fueled by demand in the traditional industries: oil well drilling additives, concrete additives, dyestuff dispersants, agricultural chemicals, animal feed and other industrial binders. Within the past few years new potential uses for modified commercial lignins have appeared that will continue to fuel this growth. Specialty polymers for the paper industry, enzyme protection, neutralization of biocides, precious metal recovery aids and wood preservation are some examples of these advanced uses.

Historically, lignosulfonates and other lignin based products have been viewed as waste materials of limited industrial usefulness. Advances in lignin technology, however, have led to the development of specialty products that can compete with higher cost synthetics. The improved performance of these products together with the fact that they are essentially non-toxic and derived from a renewable resource make them versatile, cost-effective chemicals for today's environmentally conscious consumer. In this chapter, we will attempt to summarize the mature markets for lignin materials, as well as look at some promising future technologies.

Definitions

Commercial lignosulfonates are complex anionic polymers obtained as co-products of wood pulping. They are obtained from spent sulfite pulping liquors (Figure 1) or from postsulfonation of kraft (i.e. sulfate) lignins (Figure 2). Lignosulfonates derived from spent pulping liquors are actually mixtures of sulfonated lignins and other wood

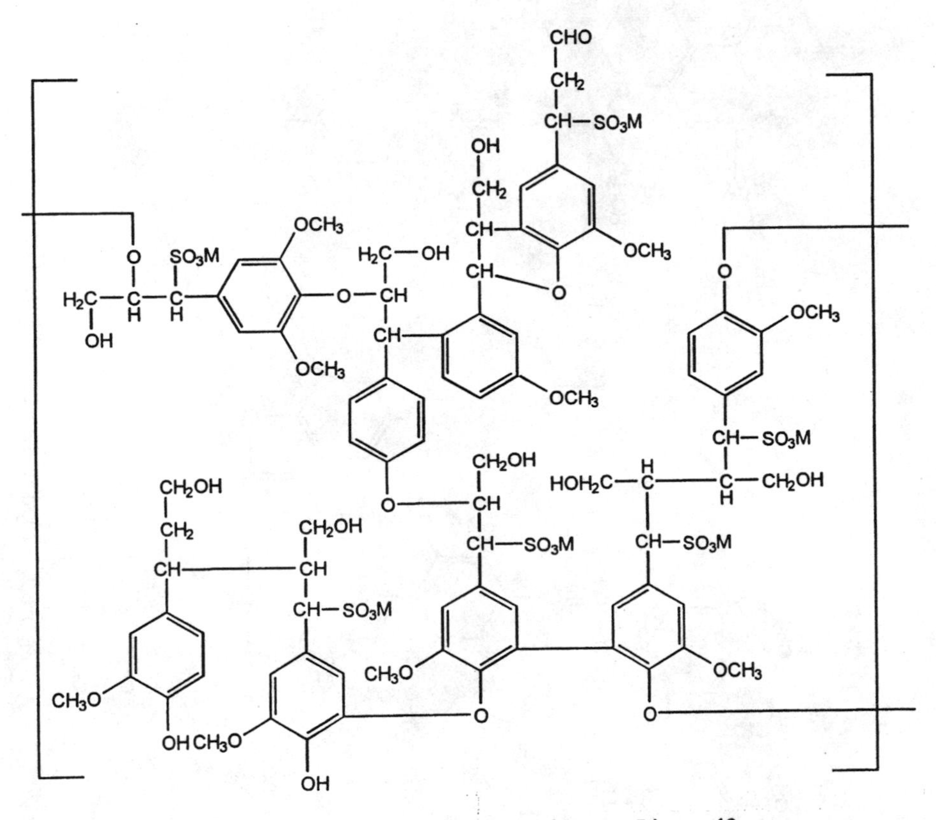

Figure 1. Structural representation of Spruce Lignosulfonate.

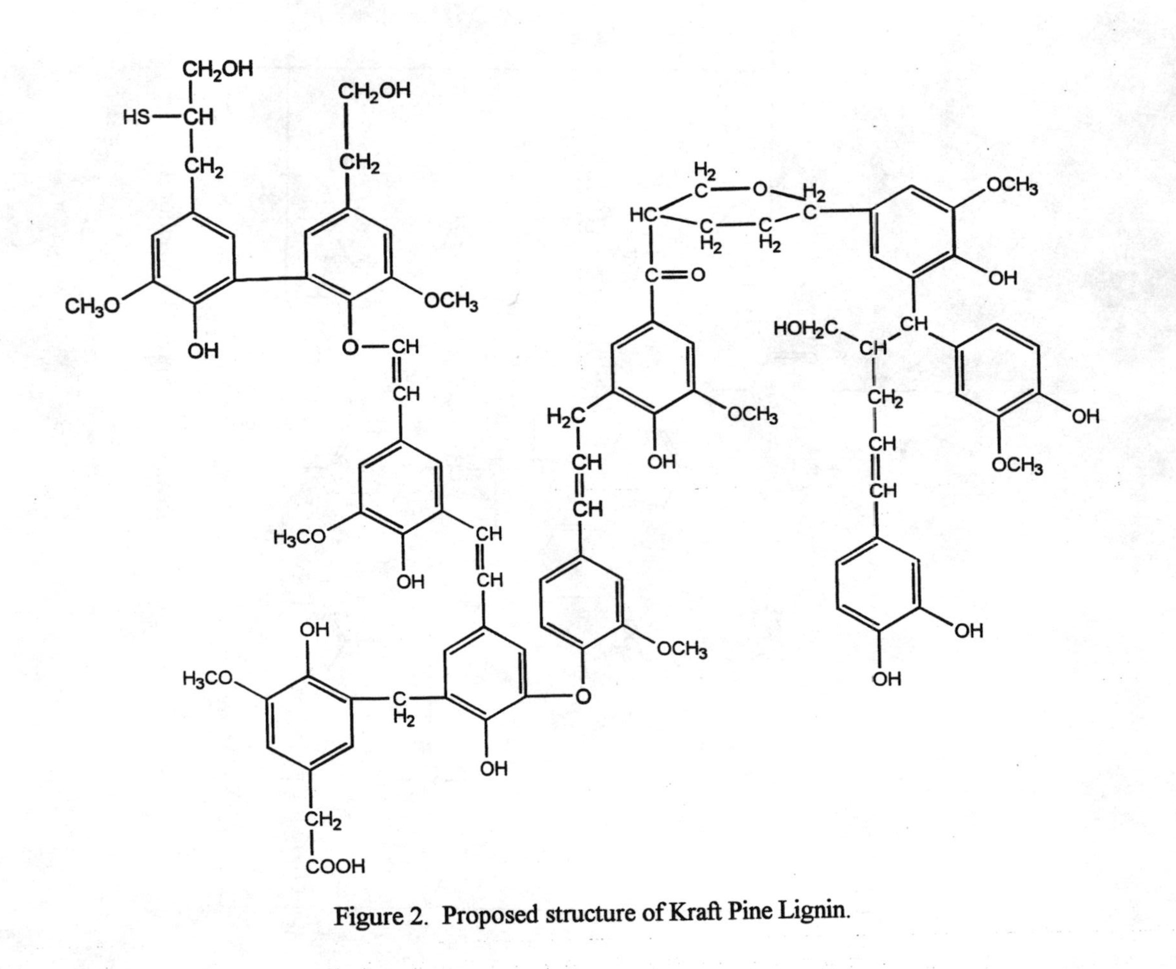

Figure 2. Proposed structure of Kraft Pine Lignin.

derivatives that may be used with little or no modification, or they may be modified to produce specialty materials with chemical and physical properties specific to certain end-use markets. A number of specialty uses also exist for kraft lignins isolated from sulfate pulping liquors.

Producers

Excluding the former Soviet Union, lignosulfonate capacity worldwide is estimated to be about 975M solid tons/year (Table 1). Production is concentrated among a few well-established companies and is dominated by lignosulfonate derived from spent sulfite liquor. Only Westvaco and LignoTech Sweden produce lignosulfonates by postsulfonation of kraft lignin.

Not listed on Table 1 is the recently announced lignosulfonate producer LignoTech South Africa. Slated to start up in 1999, production at this facility will add 200,000 solid tons/yr capacity by 2003.

Table I: Major Lignosulfonate Producers

Producer	*Country*	*Annual Capacity (solid tons/year)*
Borregaard LignoTech	Norway	160,000
LignoTech Sweden	Sweden	60,000
Borregaard Germany	Germany	50,000
LignoTech Iberica	Spain	30,000
LignoTech Finland	Finland	20,000
LignoTech USA	United States	60,000
Georgia Pacicfic	United States	200,000
Westvaco	United States	35,000
Flambeau Paper	United States	60,000
Tembec	Canada	20,000
Avebene	France	40,000
Tolmezzo	Italy	30,000
Sanyo Kokusaka	Japan	50,000
Others		150,000
		975,000

Production Chemistries

For some applications, spent sulfite liquors containing crude lignosulfonate are used without further modification. For most specialty applications, however, impurities negatively impact performance and purification and/or modification is required. Purification methods include sugar removal by fermentation or chemical destruction. Ultrafiltration and chemical precipitation are also used commercially to produce high purity lignosulfonates. Chemical modifications include: sulfonation, sulfoalkylation, desulfonation, formulation, oxidation, carboxylation, amination, crosslinking, depolymerization, graft polymerization, and combinations of the above. Chemical modifications are generally performed to enhance the dispersing, complexing or binding properties of the finished product.

Chemical And Physical Properties

As stated earlier, lignosulfonates are complex polymers. As such, detailed characterization of their chemical and physical properties is often difficult. Analytical techniques for determining gross properties such as purity, degree of sulfonation and degree of carboxylation are fairly well established (1). Analysis of minor functional groups is more difficult and often yields ambiguous results. Even more difficult is the analysis of physical properties, particularly molecular weight.

From an industrial standpoint, it is not always critical that properties of a given lignosulfonate are known in detail. While not belittling the advances lignosulfonate producers have made in this area, semi-quantitative characterizations are often sufficient to provide information for determining the usefulness of a given product in a particular application. In fact, it is there wide variety of physical properties that make industrial lignosulfonates useful in a wide variety of applications.

Toxicological Properties

Extensive testing has shown that lignosulfonates are non-toxic at typical use levels (Table II). Their use in a wide variety of food and food packaging applications has been approved under the following U.S. Food and Drug Administration regulations (2).

Markets

Several, more detailed, reviews have been done on this subject (2,3). However, a breif mention of the commercial markets is appropriate for this tutorial. The markets for lignosulfonates can, in general, be divided into two classes, commodity and specialty. Many, but not all, of the large volume markets use lignosulfonates from spent pulping liquors without further processing. Most, but not all, of the specialty markets use modified or highly modified lignosulfonates or lignosulfonates derived by postsulfonation of kraft lignin.

Table II. Toxicity of Lignosulfonates.

Property	*Calcium Lignosulfonate*	*Sodium Lignosulfonate*	*Ammonium Lignosulfonate*
Acute Toxicity[a]	> 2,000	> 10,000	>10,000
Eye Irritant	No	No	No
Skin Irritant	No	No	No
Fish Toxicity[b]	> 1,000	> 1,000	> 1,000
Bacteria Toxicity[c]	5,000	5,000	343

[a](LD_{50} - mg/kg). [b](LC_{50} - mg/l), [c](EC_{10} - mg/l)

Commodity Markets

Concrete Admixtures. Approximately 50% of all lignosulfonates produced worldwide are used in concrete admixtures. Their functions in concrete include water reduction (i.e. dispersion), quicker strength development, and improved workability. They can also act synergistically with air entraining agents, thereby reducing the amount of such agents required (4). Lignosulfonates retard the setting, or hydration, of concrete. This is the primary reason why they are not used as superplastcizers. Typical dosages in concrete are 0.1-0.3% by weight on cement. Typical salts are calcium and sodium.

Animal Feed. In animal feeds, lignosulfonates function primarily as binders. As such, they improve pellet durability and increase abrasion resistance. As a result, less fines are generated in handling and feed conversion is improved. A second function of lignosulfonates in animal feeds is as a lubricant. In this function, they reduce energy costs, reduce downtime and lower extruder wear (5). A maximum dosage of 4% is possible in finished pellets. Typical salts are calcium and ammonium.

A related use of lignosufonates in animal feed is as an additive for molasses. Their principle function in this application is to lower viscosity. The desire for lower viscosity stems primarily from ease of pumping, but it can also contribute to increased resistance to fermentation. Inclusion rates of up to 11% are allowed by the U.S. FDA.

Oil Well Drilling Muds. Lignosulfonates, primarily chrome and ferrochrome salts, are well established products in the drilling well mud market. They function as mud thinners (i.e. dispersants), clay conditioners, viscosity control agents and fluid loss additives. Mud systems conditioned with lignosulfonates perform well at high pressures and at temperatures up to 175°C. Typical dosages are 0.2-0.5%.

More recently lignosulfonate-acrylate graft copolymers have been used in applications where chrome is unacceptable (6). These products perform similarly to chrome and ferrochrome lignosulfonates but are considered more environmentally friendly. Condensates of lignosulfonates and commercial phenol-formaldehyde resins have also been developed for use as fluid lose control additives (7).

Dust Control. Dust control applications such as road and mineral ore dedusting are one of the largest single end uses for lignosulfonates. Concentrated spent sulfite liquors containing crude lignosulfonates are sprayed directly onto substrates. Typical application rates in road dust control range from 1-2 gallons of 25% solids solution per square yard while mineral ore application rates are 1-2 gallons of 45% solids solution per ton.

A more recent advance in dust control is to blend concentrated spent sulfite liquors containing crude lignosulfonates with surface active agents. The resulting blends are then applied as foams giving excellent dust control at low application levels.

Specialty Markets

Vanillin Production. Only one company worldwide, Borregaard LignoTech in Norway, still produces vanillin from softwood lignosulfonates. It is used in vanilla flavors and as a feedstock for specialty chemical derivatives.

Pesticides. Lignosulfonates are used in the formulation of pesticides (8,9). In wettable powders, suspension concentrates and water dispersible granules, products modified for enhanced performance are used as dispersant to prevent sedimentation. In the production of granular pesticides, less modified products are used primarily as binders or stickers. In water soluble formulations, highly modified lignosulfonates are used as complexing agents. Typical usage rates in wettable powders, suspension concentrates and water dispersible granules range from 2-10%. Usage rates in water soluble formulations range from 0.1-0.5%.

A recent advance in the formulation of pesticides has been the move towards paste extruded granules. Such formulations require lignosulfonate dispersants that have been modified for reduced binding and improved wetting. Usage rates in paste extruded granules range from 1-5%. Another recent advance in this area has been the development of lignosulfonate-based UV protectants. Methods based on simple additives (10), complexation (11) and microencapsulation (12) have all been reported.

Oil Well Cement Retarders. Lignosulfonates are widely used as retarders in oil well cements. They maintain fluidity over extended periods of time allowing proper placement within the targeted well zone (13). Depending on how they are formulated or modified, they are effective over a wide temperature range. Calcium and sodium salts are used at dosages that vary with temperature.

Gypsum Board. Lignosulfonates are used to disperse stucco in the manufacture of gypsum board. They reduce the amount of water required to fluidize the slurries fed to the forming rolls of gypsum board machines. This in turn means faster dryer through-put and lower fuel costs. Finished board quality is also improved as a result of more uniform drying (14). Calcium salts are most typically used in this application at dosages in the 0.1-0.3% range.

Carbon Black, Inks and Pigments. Sodium lignosulfonates are used as dispersants in water-based paints and inks. Addition inhibits settling of particles and increases solids at a given viscosity. Addition also improves the density and reduces the water permeability of dried films (15).

Water Treatment. In water treatment, lignosulfonates are used as dispersants and scale-deposit inhibitors in the treatment of waters used in boilers and cooling towers. They are also used as sludge conditioners and scale inhibitors in steam generators and desalinization units. Modified spent sulfite liquors containing lignosulfonates and sugar acid salts are used as complexing agents for zinc in cooling water corrosion control applications. Spent sulfite liquors containing lignosulfonates and sugar acid salts are also used to control slime in the paper making process (16). Their action in these applications is attributed to the complexing properties of such products. Typical dosages in the preceding applications are 100-500 ppm and sodium salts are used exclusively.

A recent advance in this area is the use of lignosulfonate-acrylate copolymers as scale-deposit inhibitors/descalants (17). Such products are claimed to have significantly improved performance compared to conventional lignosulfonates.

Industrial Cleaners. Lignosulfonates are used in both alkaline and acid cleaning formulations. Usually these products are highly modified for compatibility with other cleaning formulation components such as caustic, silicates, carbonates, phosphates and nonionic and anionic wetting agents. They function as dispersants for dirt particles, as mild surfactants for improved rinsing and as complexing agents for metal ions. They have been used in soak tank, spray, electrolytic and maintenance cleaners and in paint strippers.

Emulsion Stabilizers. Lignosulfonates are well known dispersants. They also can act as emulsions stabilizers. The most important property of any emulsion system is it's stability. Stability is defined as the ability of that system to maintain the state of subdivision of the discontinuous phase. Some advantages to using lignosulfonates are stability to salt contamination, mechanical stress and temperature variations. They are also very cost effective. Lignin products will not act as emulsifiers. Often emulsion formulations are mixtures of hydrophobe, water, emulsifier and lignosulfonate that acts as a stabilizer. Lignosulfonates are used in a variety of wax/water and oil/water emulsions for applications varying from paper sizing to asphalt stabilization.

For asphalt, cationic and anionic emulsion stabilizers have been developed. The first cationic products where lignin amines (19) prepared by the reaction of glycidylamine with Kraft lignin. Subsequent materials have been prepared by Mannich condensation of various primary and secondary amines with Kraft lignin (20).

Micronutrients. Complexes of lignosulfonates and metal ions such as calcium, iron, copper, manganese, zinc and magnesium provide essential micronutrients to plants. When applied as floiar sprays, such conplexes can be readily absorbed by plants without undesired leaf burn. Such complexes are also useful in soil treatment although to a lesser extent. In general, sodium lignosulfonates are used in this application with dosages varying with metal ion dosage.

Dyestuffs. In disperse dye systems lignosulfonates act as primary dispersants, extenders, protective colloids, and grinding aids. Materials produced by the reaction of lignosulfonates with benzyl alcohol have low azo dye reduction properties, low fiber-staining properties, high dispersion efficiency, good grinding aid qualities, and increased heat stability (18). The superior performance of lignosulfonates, combined with their low cost allow them to dominate the dye market.

Lead Acid Batteries. Modified lignosulfonates are an essential component in lead acid batteries. The role of lignin is to act as an expander, or surface modification additive, on the negative plate. Few studies have been devoted to understanding the mechanism by which lignin optimizes the dimensional structure of the negative plate. It is believed to act as a crystal growth inhibitor, causing the negative plate to maintain a spongy character, thus optimizing surface area and current discharge. A relatively small amount of lignosulfonate, roughly a few grams per battery, will be the difference between a short or long battery life. Two recent publications discuss the role of lignosulfonate in batteries (21,22).

Bricks, Refractories and Ceramics. Lignosulfonates are widely used in the brick industry. They are generally used on plastic raw materials which require little help with increased green strength. Increased green strength translates into high yield to unbroken bricks during manufacture. Lignin based products have also been developed to increase dry strength and lubricate the die to increase extrusion rate and decrease equipment wear, and to decrease the migration of salts that cause scumming. Recent developments in this industry include products targeted to solve specific tasks. A lignosulfonate-acrylate co-polymer has been developed that is designed to improve slip homogeneity and provide greater dry strength for ceramics, at very low dose (23).

Select Potential New Markets

Wood Preservation. The treatment and preservation of wood is another industry which lignin technology has advanced over the past 10 years. Lignin based wood preservatives offer an advantage over current preservatives in that they offer a system that is less hazardous to the environment.

Two major destructive forces that act on wood are fire and decay. While fire is often caused by human factors, decay is due to biological attack by microorganisms, such as fungi and bacteria. The most common method of preserving wood against fungal attack is to impregnate it with a wood preservative. Chromated copper arsenate (CCA) is the current, most widely used, lumber preservative in the U.S (24). About 95% of treated lumber, 4 billion board feet in 1984, is treated with this water-borne preservative. This material gives the best available protection against decay and insects. Unfortunately, two main constituents, chromium and arsenate, are considered to constitute a significant risk to human health and the environment. Selected groups within the wood preservation industry have taken note of the possible effects of CCA, and have partnered with different technology companies to overcome this problem.

A lignin based wood preservation system has been patented (25). In this system the wood is impregnated, using the same equipment as CCA, with a pressure treatment of a preservative composed of an aqueous solution of an ammonium salt of lignin and $Cu(NH_3)_4^{++}$. The impregnated wood is either air dried at ambient temperature or oven dried to let ammonia escape from the interior wood, resulting in the insolubilization and fixation of the lignin and metal salt in the wood. In this process ammonia is recovered and reused. The treated wood shows good retention and penetration of preservative. Leachability of the active material, Copper, is low. Treatment of the wood with $Cu(NH_3)_4^{++}$ alone results in copper leaching from the wood at a high rate, depleting the active biocide.

Figure 3 displays a graph of some preliminary data for this preservation system (Ziobro, R. J. Osmose. 980 Ellicott Street, Buffalo, NY 14209, unpublished data). The Fahlstrom test stake data was collected in Gainsville, Florida per AWPA Standard E7-93. To summarize, the test is an accelerated biological decay study done on small dimension stakes made of Southern Yellow Pine. The control sample in this environment has completely decayed after 2 years. Intermediate performance is observed by Lignin/Cu (L/C) at a 0.3/0.1 pounds per cubic foot (pcf) dose. Performance of the (L/C) system at 0.7/0.2 pcf is on par with the current industry standard of CCA at 0.4 pcf.

Efforts will continue to commercialize this technology. However, several hurtles exist. The most prominent is the cost advantage of CCA. Assuming a similar efficacy for CCA at 0.4 pcf and L/C at 0.8 pcf, the L/C system is approximately 30% more expensive from a material standpoint. Capital requirements also need to be considered. While the same pressure treating facilities used for CCA may be used for L/C, recovery systems for ammonia would have to be installed, adding additional capital. Given these economic parameters, one should take into account the environmental issue caused with the use of arsenic. In Europe several countries have taken the environmental initiative and have banned CCA. If a ban of CCA were to occur in the US, this system would earn obvious consideration as a possible replacement technology.

Enzyme Stabilization. The storage stability of enzymatic compositions is greatly improved by the inclusion of a lignosulfonate (26). A wide variety of lignosulfonates are effective in a wide variety of enzyme compostions. Practical dosages range from 0.01-0.1%.

Pharmaceutical Applications. Medicinally, lignosulfonates have been perported to have value as an antiviral agent (27,28). Fractionated lignosulfonates have also been purported to have value as antithrombotics (29).

Mining Aids. The use of lignosulfonates as ore floatation aids is well known. One new application for lignin derived materials in the mining industry that has been recently disclosed involves increased yields of precious metals from the extraction of sulfided and carbonaceous ores (30). This technology describes the use of modified lignosulfonates as primary catalysts for the oxidation of sulfided ores to metal ores which are readily available for conventional gold/metal recovery techniques.

Their are several classifications of precious metal ores. These classifications are dependent on the physical makup of the ore, and the techniques required to recover the desired metals. The easiest to recover, and previously most abundant, are known as oxide ores. Due to their relative ease of metal recovery via leaching, oxide ores have historically been the most rapidly depleted. Two other broad classifications for metal bearing ores are sulfide and carbonaceous. Metals in these systems are unavailable to traditional leaching techniques. Larger quantities of sulfide and carbonaceous ores have traditionally been passed by during mining opperations because more expensive recovery systems were required, for example bio or pressure oxidation. Subsequently, lesser quality ores have often been left behind because of an inherent requirement for low capital processing. If one could find a way to rapidly free up the metals for conventional recovery, it would be considered a great improvement over the existing technology.

The structural feature of lignin believed to be responsibe for catalytic activity is catechol / orthoquinone. In conjuction with an oxidant, usually oxygen, the system oxidizes sulfide and carbon to liberate locked gold. Air is used to regenerate reduced oxidant. The catalytic cycle is depicted in the Figure 4. This is a very simplistic view of the system, since order to covert the sulfide sulfur completely to sulfate, multiple cycles through the system are required (31).

The system has been demonstrated to be catalytic. In a similar system, oxidation of hydrogen sulfide by quinone to sulfur and peroxide has been demonstrated (32). The technology is compatible with conventional cyanidation recovery, so, zero capital is required for implementation of this lignin based oxidation system. Lignosulfonate materials that perform with a high degree of activity are

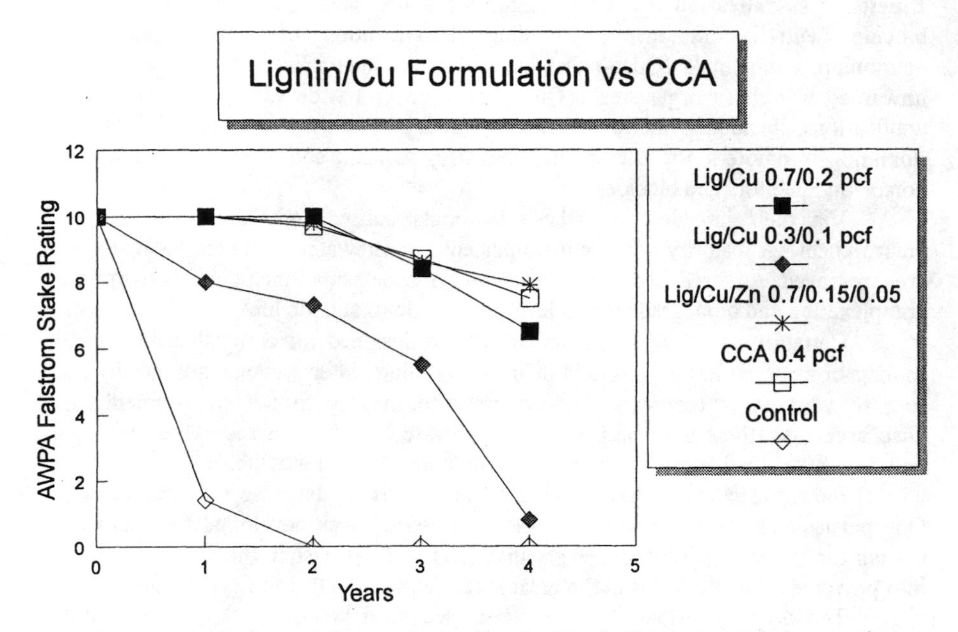

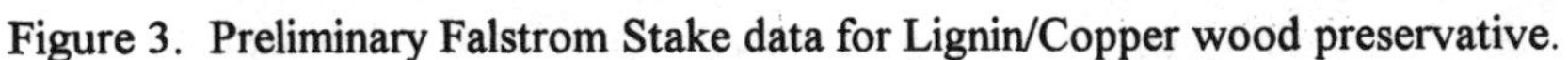
Figure 3. Preliminary Falstrom Stake data for Lignin/Copper wood preservative.

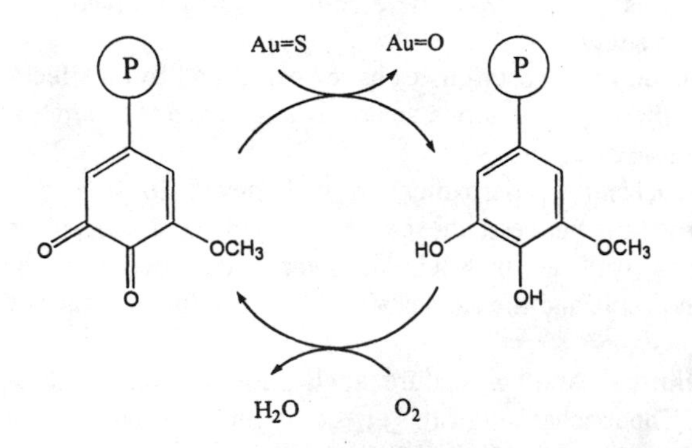

Figure 4. Proposed catalytic cycle for sulfide oxidation.

hardwood lignosulfonates which have been demethoxylated. Products tailored for this process have been manufactured via proprietary processes by LignoTech USA, and commercialization of this system is in progress.

Biocide Neutralization. Modified lignosulfonates have found unique utility as biocide neutralization agents, including detoxification of certain quaternary ammonium compounds (Quats) that have been commercialized for the control of unwanted waterborn organisms. Quats are used in a wide variety of commercial applications, including fabric softeners, laundry detergents, anti static sprays, floatation promoters for the mining industry, asphalt and petroleum additives, corrosion inhibitors and biocides.

Via their intended use, these materials subsequently enter the aquatic environment. A majority of these materials enter wastewater treatment facilities and are removed or reduced by numerous mechanisms including adsorption, complexation and biodegradation to levels non toxic to aquatic life.

Certain uses of Quats, particularly those designed for controlling biological fouling or as corrosion inhibitors in industrial cooling water systems, are usually not sent to wastewater treatment, but are released directly to the environment. Discharge from these systems may contain elevated residual Quats. These releases may cause undesirable sort-term or long term impacts to aquatic life.

Industrial chemists have sought out methods to decrease effluent toxicity. One particular system that works to decrease toxiciy was developed for a biocide system for the control of nuisance shellfish (33). In this system the quat is metered into power station inlet pipes in the great lakes basis to control the Zebra Muscle.

The original system that was developed used bentonite clay to neutralize toxicity. The mechanism of activity is thought to be complexation of the quat to negative sites on the clay. However clay, being solid, is difficult to pump and meter into aquatic systems. Because of these solid handling problems, an alternative to bentonite clay was sought.

Highly modified lignosulfonates have been found to be effective materials for detoxification of these quanternary amines. Studies are under way to determine how best to employ these materials.

The mechanism of detoxification is believed to be a combination of electrostatic attraction between the cationic biocide and the sulfonate groups on lignosulfonate, as well as hydrophobic interactions between the bulk of the lignosulfonate molecule and the alkyl chain or benzylic functionality of the Quat.

Specialty Chelants. Another mature application for technical lignoslufonate is micronutrients. The mechanism of this effect is sequestration through the suphonate group. Unfortunatly, only limited amounts of metal can be sequestered by lignosulfonates. This makes technical grade materials poor candidates for applications where higher levels of chelation are required, for example water treatment or industrial cleaning processes. Recent technology has been developed that has greatly increased the ability of lignin derived materials to sequester metals

(Raskin, M. H., U. S. patent applied for). Earth Sciences, a Boston based company, has been able to use oxidation chemistry to introduce a large portion of salicilate functionality to hydrolysis lignin. This modified hydrolysis lignin is a polydisperse low molecular weight polymer with a complexibility for iron of 2-3 mM per gram.

One industry where this material could stake a position in a market currently occupied by synthetics is pulp production. The use of synthetic chelants is becoming more prevalent in the manufacture of pulp. Consumption of chelants in the pulping industry was estimated to be 40 million lbs. in 1998. This is directly linked to the use of hydrogen peroxide as a bleaching chemical for Kraft and TMP pulps, which has increased substantially during recent years, due to the development of ECF and TCF bleaching (34).

Much effort has been directed to the removal of transition metal ions from pulp, particularly iron, manganese and copper, that decompose hydrogen peroxide and reduce its effect in bleaching. Currently the removal of these metals prior to peroxide bleaching is facilitated with chelating agents such as EDTA or DTPA. In addition to being high priced, traditional cheating agents are synthetic substances with poor biodegradablity characteristics. These chelating agents accumulate in the soil and water in considerable quantities. Therefore, research and investigation of new biodegradable chelation agents, particularly those produced from renewable resources, is very attractive.

This modified hydrolysis lignin was evaluated for its ability to remove transition metals in the Q stage. The metal removal capacity under different conditions was studied with Kraft pulp. A significant improvement in metal removal was obtained from pulp treated with oxidized hydrolysis lignin. This improvement translated, after subsequent peroxide bleaching, into significant increased brightness (M. Raskin, unpublished internal report).

In lab trials, Kraft and TMP pulps wer pretreated with the experimental chelant and hydrogen peroxide (HP) bleached. Table III illustrates work done on TMP pulp, Table 2 on Kraft pulp. As one can see, performance is not equivalent to the control chelant, EDTA. However, the results are interesting enough to indicate that this material should warrant further study. Both trials show a trend for higher brightness when chelant levels are *decreased*. This has been explained by an affinity of the lignin product for cellulose. Since the oxidized hydrolysis lignin is readily available to react with HP, it is believed to consume peroxide in a competitive reaction. In these examples the lignin derived chemical does not have performance equal to EDTA, however if superior biodegradability is seen for this material, and it's metal complexes, it may find a high level of commercial interest.

Retention Aids. Due to it's competitive nature, the paper industry has been driven to improve sheet quality, increase paper machine productivity and control rising pulp costs. During the manufacture of paper improvements in retention, drainage and formation properties are most desirable. One can effect each of these through thoughtful use of wet end chemistry. Unfortunately, improvements in one area often lead to deficiencies in others. For example, polymeric chemicals used for drainage

Table III. Effect of Complexone on Bleaching of TMP pulp.

Experiment	*Chelant*	*Dose, % on pulp*	*Brightness (ISO)*
384	Blank	0	43.7
276	Complexone C75R	0.2	61
280	Complexone C75R	0.4	53.6
284	Complexone C75R	1	55.7
388	EDTA	0.1	61.3

Chelation: T = 60-70 °C, pH = 5-6 t = 30 min. Washing: 2 stages with pH 10-11. Bleaching: 2.5% HP, T = 70 °C, t = 4 hr. TMP pulp was obtained from Weyerhaeuser Corporation.

and retention (D/R) aids to adjust the stock freeness have been in use for quite some time. Increased machine productivity and improved retention of various additives, particularly fines and fillers, has been the result. Unfortunately, these benefits in D/R come at the expense of sheet formation, which is also a desirable property in paper making. However, when D/R aids are used at low doses significant improvements in drainage can be obtained with minimal loss in formation (35).

Additional potential for drainage and retention of filler exists if the problem of D/R and sheet formation could be overcome simultaneously. One way to improve D/R with minimal loss of formation is to find some means of associating filler, fines, and other additives with the fibers without having the fibers associate with each other. Microparticle systems have been employed for years to solve this problem, but they have certain drawbacks, not the least of which is poorer that obtainable formation. A system is being commercialized using a synergistic combination of two water-soluble products: A modified lignosulfonate and a high molecular weight cationic synthetic polymer. This system works with both acid and alkali furnish. By varying dose, polymer ratios, and feed points The inventors were able to tailor this system for most any paper machine. Comparisons between this new two polymer system and conventional microparticle and monopolymer systems have shown that superior D/R/F can be obtained. Good results were obtained with both virgin and recycled fibers (36). High molecular weight lignosulfonate products have been tailored to give optimal performance to this system.

Enhanced Oil Recovery and Profile Control. In enhanced oil recovery, surfactant floods are use to mobilize hard to displace oil. Adsorption of surfactants onto resevoir rocks, however, affects the economics of this process. By using lignosulfonates as sacrificial agents (i.e. by preflushing with lignosulfonates prior to surfactant flooding), surfactant loss can be reduced significantly, and overall economics can be improved (37).

Lignosulfonates have also been used in oil well drilling as profile control agents (38). In such applications, lignosulfonates (primarily ammonium lignosulfonates) are pumped downhole where temperature and pressure cause them to gel. The formation of such gels improves steam sweep in the surrounding areas and increases overall oil recovery.

Conclusions

It is safe to say that the lignosulfonate industry has come a long way from it's humble beginings. Many lignosulfonate producers are now international companies with multiple manufacturing sites capable of producing both unmodofied and modified product lines. They have modern analytical and technical services laboratories staffed by well trained scientists to serve their customer's needs.

Modern lignosulfonate producers also employ a variety of chemistries for modifying base feed stocks. Using these chemistries, they produce specialty chemicals for use in numerous applications. Through research and development, they strive towards the future by developing new and improved products.

In short, lignosulfonate producers have become true specialty chemical companies with bright futures.

Liturature Cited

1. Lin, S. Y.; Dence, C. W. *Methods in Lignin Chemistry;* Springer-Verlag: Berlin, 1994
2. Marton, J In *Lignins*: Sarkanen, K. V.; Ludwig, C. H., Eds.; Wiley-Interscience: New York, 1971; p. 698
3. Ramachandran, V. S. *Concrete Admixtures Handbook;* Noyes Publications: Park Ridge, 1994
4. Knodt, C. B. U. S. Patent 3 035 920, 1963
5. Kelley, J. R. U. S. Patent. 4 374 738, 1983.
6. Your Natural Partner for More Natural Drilling Fluids, Metsä-Serla Group, Tampere, Finland, 1990.
7. The Pesticide Manual, LignoTech USA Inc., 721 Route 202/206, Bridgewater, NJ 08807
8. Surfactant Systems for Pesticides, Westvaco Corporation Polychemicals Department, Charleston Heights, South Carolina, 29415.
9. Hobbs, D.G. International Patent Application WO 95/22253, Feb. 13, 1995.
10. Simmons, R. W. U. S. Patent 3 447 889, 1969
11. Lebo, S. E. U. S. Patent 5 529 772, 1996.
12. Detroit, W. J.; Sanford M. E. U. S. Patent 4 846 888, 1989
13. Kirk, G. B. U. S. Patent 2 856 304, 1958
14. Hale, N; Xu, M. U. S. Patent 5 640 180, 1996
15. Slime and Deposit Control in the Paper Industry, Petromontan, Obersteinergasse, Germany.

16. Lin, S. Y.; Bushar, L. L U. S. Patent 4 891 415, 1990
17. Lin, S. Y; In *Progress in Biomass Conversion;* Vol. 4 Academic Press, Inc, Orlando, 1983, pp 31-78
18. Falkehag, I; Dilling, P. U. S. Patent 3 718 639, 1973
19. Schilling, P.; Brown, P. E. U. S. Patent 4 775 744, 1988
20. Szava, G. J. *Journal of Power Sources.* **1988**, pp 119 -124
21. Szava, G. *J. Journal of Power Sources.* **1989**, pp 149-153
22. Additive-A 373 Bullitin. LignoTech USA Inc., 721 Route 202/206, Bridgewater, NJ 08807
23. *Forest Industries.* **1985,** 11, pp 36-38
24. Lin, S. Y. U. S. Patent 5 246 739, 1993
25. Van Dijk, W. R.; Ouwendijk, M.; Hall, P.J.; International Patent Application WO 97/00932, 1997
26. Toda, S. Japan Patent 02 262 524
27. Sakagami, H.; Kawazoe, Y; Konnom K. Japan Patent 03,206,043
28. Clough, T. J. U. S. Patent 5 575 334, 1996
29. Clough, T. J. U. S. Patent 5 344 625, 1996
30. Clapp, P. A. Evans, D. F. U. S. Patent 5 545 391, 1996
31. Petrille, J. C.; Vasconcellos, S. R;. Werner, M. W. U. S. Patent 5 486 296, 1996
32. Dence, C.; Reeve, C. *Pulp Bleaching. Principals and Practice*; Tappi Press, Atlanta, GA, 1996
33. Elliot, D. L.; Falcione, R. J.; Hunter, W. E. U. S. Patent 5 501 773, 1996
34. Vaughan, C. W. *Tappi J.* **1996**, 79(7) 103-9
35. Hong, S. A., Bae, J. H. and Lewis G. R.; An Evaluation of Lignosulfonate as a Sacrificial Absorbate in Surfactant Flooding, SPE/DOE Fourth Symposium on Enhanced Oil Recovery, Tulsa, OK, April 15-18, 1984.
36. Hong, S.A. and Bae, J.H.; *SPE Reservoir Engineering*, **1990**, 11.

Chapter 16

Synthesis of ^{14}C-Labelled 1-Ethenylbenzene Using Wittig Reaction and Synthesis of Graft Copolymer of Lignin and ^{14}C-Labelled 1-Ethenylbenzene

Meng-Jiu Chen[1] and John J. Meister

Department of Chemistry, New Mexico Institute of Mining and Technology, Socorro, NM 87801

Laboratory-scale methods have been optimized to convert radioactive 1-phenylmethanal (benzaldehyde) to 1-ethenylbenzene (styrene, S). The Wittig reaction of an aldehyde and methyltriphenylphosphonium iodide converts 69.3% and 89.2% of the original 1-phenylmethanal to 1-ethenylbenzene. Separation of 1-ethenylbenzene from benzene was successfully done with recovery of 1-ethenylbenzene being 66.9% and 88.5%, and with the final styrene concentration in benzene being 78.3% and 90.7%. Tetrahydrofuran contaminated, ^{14}C- (1-phenylmethanal) was used directly in the reactions without further treatment, and appeared to have no major effect on the Wittig reaction or analysis of 1-ethenylbenzene by gas chromatography. Small-scale copolymerizations of lignin and ^{14}C-labelled 1-ethenylbenzene were successfully run. Based on yield data, we now have a radiotraced graft copolymer of lignin and 1-ethenylbenzene with which to conduct biodegradation tests. The copolymer contains about 7830 counts / second / mg radioactivity and about 52% poly(1-phenylethylene) in it.

The monomer 1-ethenylbenzene is a common and cheap compound which has been produced by a dehydrogenation of 1-phenylethane on a large scale for years (*1*). Most 1-ethenylbenzene is used as an ethylene monomer to prepare crystalline poly(1-phenylethylene) (PS), high impact poly(1-phenylethylene) (HIPS) and copolymers (ABS resin, SBR rubber, SAN rubber, unsaturated polyester, etc.). Numerous studies in polymer chemistry were previously done on radiotraced poly(1-phenylethylene) but from the late 1960's onward, commercial synthesis of radiotraced molecules created compounds for biochemistry rather than polymer research (*2*). With no traced monomers available from commercial suppliers in 1994, making monomers requires that archaic synthesis routes be updated to more modern methods and reagents. This paper discloses a synthesis method for radiotraced 1-ethenylbenzene monomer and a

[1]Current address: Research Division, Cardolite Corp., 500 Doremus Avenue, Newark, NJ 07105–4805.

synthesis and recovery process by which to create multigram graft copolymer samples for biodegradation testing. Analytical accuracy, the rate of decompostion of orgainc matter by fungi, and the process of recovery of CO_2 from fungal decay dictate that the synthesis must produce a product with a radioactivity of at least 800 counts per second per milligram.

In recent years, there has been a growing interest in the research and development of biodegradable polymers, including poly(1-phenylethylene) and its copolymers. Social attitudes about environmental protection and public perceptions that plastics are a source of litter or waste make it very important to develop biodegradable polymers and study the process and mechanism of their biodegradation and metabolism. Because of the slow rates of degradation, the complex mixture of products produced, and the multiple chemical pathways followed during degradation, radiotracer analysis is a preferred method for documenting biodegradation. The need to create biodegradable plastics and document the environmental decay rate of these materials is one of many needs that makes practical, multigram synthetic methods to prepare radiotraced polymers critical.

Our previous work strongly supported the hypothesis that white rot fungi (basidiomycete) were able to biodegrade graft copolymer of lignin and 1-ethenylbenzene (*3*). White rot fungi degraded the plastic samples at a rate that increased with increasing lignin content in the copolymer sample. Both poly(1-phenylethylene) and lignin components of the copolymer were readily degraded. The degradation of lignin graft copolymers by white rot fungi was followed by determining weight loss, observing with scanning electron microscopy, and monitoring aromatic absorption bands by FTIR and NMR. The results show loss of mass, loss of functional groups, and reduction of aromatic ring resonances from the copolymer after 50 days of incubation with the white rot fungi (*4*). However, in order to guarantee that the observed losses were due to decay instead of solubilization, envelopment, or fragmentation of the sample, ^{14}C-labelled 1-ethenylbenzene was made and used to prepare a graft copolymer of lignin and 1-ethenylbenzene. The synthesis method for radiotraced copolymer consists of a Wittig reaction on ^{14}C-labelled 1-phenylmethanal, quantification of 1-ethenylbenzene yield by gas chromatography (GC), fractional distillation to recover concentrated monomer, and free radical graft copolymerization onto lignin. Because the raw material, ^{14}C-labelled 1-phenylmethanal, was a tetrahydrofuran (THF) solution, the effect of the tetrahydrofuran on the results of the Wittig synthesis and gas chromatography analysis was studied.

After enough experience was acquired in running the Wittig reaction to synthesize trace amounts of 1-ethenylbenzene, separating the mixture of 1-ethenylbenzene and benzene from the reaction system, and concentrating 1-ethenylbenzene from its benzene solution, two real reactions were run with radiotraced 1-phenylmethanal. Before the copolymerization of lignin and ^{14}C-labelled 1-ethenylbenzene was conducted, the copolymerization was optimized by conducting several small scale reactions.

Experimental

Materials. [Ring-U- ^{14}C] 1-phenylmethanal (3 mCi) was supplied in 1.5 mL of THF by Amersham International plc, Amersham Place, Little Chalfont, Buckinghamshire HP7 9NA, England. Its specific activity determined by mass spectrometry was 1.89 GBq/mmol, 51 mCi/mmol (4). It was used directly in the reaction without further treatment. Carbon-14 is a low energy, beta emitter with a half life of 5730 years. It should never be ingested or inhaled. Petrochemical 1-phenylmethanal was from J. T. Baker Chemical Co., Phillipsburg, NJ 08865. Before the reaction, it was vacuum-distilled and the central cut was collected. Methyltriphenylphosphonium Iodide

($(C_6H_5)_3PCH_3I$) [2065-66-9] was purchased from Aldrich Chemical Company, Inc., Milwaukee WIS 53233. Its purity was 97%. All percents are weight percent unless otherwise specified. Benzene was from Fisher Scientific Company, Fair Lawn, NJ 07410. It was of reagent grade. Sodium hydroxide was of reagent grade. It was dissolved in distilled water to form a 5M solution. Diluent 1-ethenylbenzene was from Fisher Scientific Co., Fairlawn, NJ 07410. Before being used, it was vacuum distilled and the central cut was collected and kept in a -4°C refrigerator.

The lignin used in the copolymerization was a kraft pine lignin prepared in "free acid" form with a number-average molecular weight of 9,600, a weight-average molecular weight of 22,000, and a polydispersity index of 2.29. The "free acid" form of lignin has virtually all of the metal cations replaced with hydrogen ions. The radiotraced 1-ethenylbenzene (I-1D and I-2D) was derived from our synthesis. Hydrogen peroxide was its 30% water solution. Anhydrous calcium chloride and dimethylsulfoxide (DMSO) were commercial products of reagent grade. Nitrogen was commercial grade bulk gas purified over copper filaments at 500°C.

Equipment. A reflux-reaction system mounted on a magnetic stirrer was used to run the Wittig reaction. A vacuum-distillation system with at least two cold traps in dry ice-(2-propanone) baths was used to do vacuum distillation. During the Wittig reaction and distillation, water baths were used to control the temperature. Gas chromatography on a Hewlett-Packard 89531A equiped with a flame ionization detector was used to determine the concentration of 1-ethenylbenzene in benzene.

Synthesis. Wittig Reaction Method. The monomer 1-ethenylbenzene can be synthesized by dehydrogenation of ethylbenzene, dehydration of 1-phenylethanol and dehydrochlorination of alkyl chlorination products of ethylbenzene (*5*). The Wittig reaction of 1-phenylmethanal and methyltriphenylphosphonium iodide can also be used to prepare a small amount of 1-ethenylbenzene for research (*5*). We selected the Wittig reaction to run the synthesis because it is one step, high yield, and offers comparatively easy purification of the product. Equations showing the Wittig reaction are:

$$(C_6H_5)_3P^+CH_3\ I^- + NaOH \Longleftrightarrow (C_6H_5)_3P{=}CH_2 + NaI + H_2O$$

$$(C_6H_5)_3P{=}CH_2 + C_6H_5CH{=}O \Longleftrightarrow C_6H_5CH{=}CH_2 + (C_6H_5)_3P{=}O$$

Benzene was placed in a round-bottom flask equipped with a water condenser, a thermometer, and a stirrer, and the flask mounted in a water bath. Sodium Hydroxide (aqueous, 5M), methyltriphenyl-phosphonium iodide, and distilled 1-phenylmethanal or radiotraced 1-phenylmethanal in THF were added to the flask in that order. The reaction mixture was stirred at 40°C for 48 hours, and separated into a benzene layer and a base layer. The mixture of 1-ethenylbenzene and benzene was separated from the mixture of $(C_6H_5)_3P{=}CH_2$ and $(C_6H_5)_3P{=}O$ by controlled vacuum distillation which was done under nitrogen atmosphere at 250 mm Hg pressure at room temperature (23°C) for 10 minutes, then at 50°C for 15 minutes, and finally at 90°C for 15 minutes. This was the first separation. During this process, the receiving flask was kept in a dry ice / 2-propanone bath. The condensed liquid in the receiving flask and cold traps, and the unevaporated residual solid in the distillation flask were recovered and weighed respectively. The recovery of the condensed liquid and the residual solid were calculated. The 1-ethenylbenzene in its benzene mixture was further concentrated by controlled vacuum distillation. This was the second separation. This second separation was manditory since dilute solutions of 1-ethenylbenzene in benzene gave poor copolymerization yields. Before vacuum distillation, common 1-ethenylbenzene was added to the above 1-ethenylbenzene-benzene mixture to obtain mixed 1-ethenylbenzene of fixed radioactivity. The mixture was vacuum distilled under nitrogen atmosphere at 250 mm Hg pressure at 50°C for 45 minutes. Two fractions were obtained: the fraction recovered in the receiving bottle with high benzene concentration in it, and the fraction remaining in the distillation flask with a high concentration of 1-ethenylbenzene in it.

Graft Copolymer. Lignin, calcium chloride, and dimethylsulfoxide were mixed in a 30-mL glass sample bottle and stirred with a magnetic stirrer until a uniform solution was formed. The lignin solution was bubbled with nitrogen for 30 minutes, then hydrogen peroxide was dripped into the solution, and this mixture was further bubbled with nitrogen for 20 minutes. In the meantime, the radiotraced 1-ethenylbenzene was bubbled with nitrogen in a 10-mL glass sample bottle for 5 minutes and added to the above lignin solution with a glass delivery tube. The one-phase reaction mixture was bubbled with nitrogen for 10 minutes. Finally, the reaction bottle was sealed and placed in a 30°C water bath. During a 48-hour reaction period, the reaction mixture gradually changed its color, from black to brown to yellow brown. The reaction was terminated by adding 10 mL 2N HCl and 30 mL water to it. The solid precipitate was separated from the liquid phase by centrifugation and washed with water at least 4 times until the PH value of the wash was 7. The solid was dried in a 45°C oven for 72 hours to obtain the final lignin-^{14}C 1-ethenylbenzene graft copolymer.

Analysis by Gas chromatography. Gas chromatography was used to determine the concentration of 1-ethenylbenzene in its benzene mixture, thus determining the conversion of 1-phenylmethanal. Fraction of peak area (A%) gives the mole fraction of a material in the mixture (mol%). Standard solutions of 1-ethenylbenzene in benzene were used to calibrate peak area of 1-ethenylbenzene to mole fraction of 1-ethenylbenzene. The 1-ethenylbenzene mole fractions in the standard solutions were chosen to be close to the 1-ethenylbenzene mole fractions in the reaction products (1-ethenylbenzene-benzene mixtures): 3.14 mol% to 3.93 mol%. Therefore, standard solutions of 1, 2, 3, 4, and 5 mol% of 1-ethenylbenzene in benzene were prepared, and a standard curve and a regression equation of peak area of 1-ethenylbenzene, Y, vs. 1-ethenylbenzene molar concentration in benzene, X, were obtained:

$$Y\ (A\%) = a + bX\ (mol\%).$$

For each assay, Y is the percentage of total peak area produced by 1-ethenylbenzene and X is the percentage of 1-ethenylbenzene mole fraction in a standard benzene mixture. The unknown samples were analysed for peak area of 1-ethenylbenzene at the same time the standards were run. The 1-ethenylbenzene mole fraction in the product was obtained from the regression equation for that analysis run,

$$X\ (mol\%) = (Y\ (A\%) - a)/b.$$

Because X (mol% of 1-ethenylbenzene) =

$$\frac{(\text{weight of 1-ethenylbenzene} / 104.16)}{[(\text{weight of 1-ethenylbenzene} / 104.16) + (\text{weight of benzene} / 78.12)]},$$

and weight of 1-ethenylbenzene + weight of benzene = weight of the mixture of 1-ethenylbenzene and benzene, the weight percent 1-ethenylbenzene in its benzene mixture can be calculated. Yield was based on the theoretical conversion of one mole of 1-phenylmethanal to one mol of 1-ethenylbenzene.

All the gas chromatography separations were done under the following conditions: Oven temperature, detector temperature and injector temperature were 80, 120, and 100°C, respectively. Column head pressure was 142 kPa. Split flow rate and purge flow rate were 107.1 and 0.60 mL/min, respectively. Sample size was 2 μL.

Results and Discussion

Wittig Reaction. Several factors affect the Wittig reaction. These factors are: addition order of the reactants, molar ratio of $(C_6H_5)_3PCH_3I$ to C_6H_5CHO, reaction temperature, reaction time, agitation speed, and solvent type. The highest yields were obtained by using the following reaction conditions.

An addition order of (1) C_6H_6, (2) $(NaOH+H_2O)$, (3) $(C_6H_5)_3PCH_3I$, and (4) C_6H_5CHO. This addition order avoided the decomposition of $(C_6H_5)_3PCH_3I$ as evidenced by a color change from white to yellow and the formation of solid lumps or an emulsion layer in the reaction mixture. The best molar ratio of $(C_6H_5)_3PCH_3I$ to C_6H_5CHO was 2 to 1. The best reaction temperature was 20 to 40°C. The preferred reaction time was 48 to 75 hours. Two solvents, pentane and benzene, were tested. The results show that benzene is much better than pentane for this reaction. In reactions in pentene, (1) $(C_6H_5)_3PCH_3I$ could not be completely dissolved, (2) the final reaction mixture contained a white sticky emulsion layer, (3) conversion of 1-phenylmethanal or yield of 1-ethenylbenzene was 11% or less, and (4) pentane evaporated quickly, causing an error in the calculation of conversion of 1-phenylmethanal or yield of 1-ethenylbenzene. Rapid agitation was very important in this two-phase reaction since low agitation caused low yield. After the reaction, about 80% of benzene and benzene-soluble materials were recovered. The formulations of the reactions are listed in Table I.

Radiotraced bezaldehyde was supplied as a tetrahydrofuran (THF) solution in a borosilicate glass ampoule sealed under nitrogen. About 6.24mg (listed as 3 mCi) of ^{14}C 1-phenylmethanal were dissolved in 1.5 mL THF and it could not be easily separated from THF. To determine the effect of THF on the Wittig reaction, two reactions contaminated with 0.7 mL of THF were run. The results show that THF did not effect the reaction. All the reaction phenomena and results were nomal. Gas chromatography results showed that the peak of THF did not interfere with the 1-ethenylbenzene peak but did overlap the benzene peak. Since the molecular weights of benzene (78.12) and THF (72.11) are close to each other and the amount of THF in the 1-ethenylbenzene-benzene mixture was small (about 12 %), the estimated relative gas chromatography error caused by this amount of THF was less than 1.0%.

Separation of 1-Ethenylbenzene from the Reaction Product System by a Controlled Vacuum Distillation. This process consists of two separations: (1) separation of the mixture of 1-ethenylbenzene and benzene from the mixture of $(C_6H_5)_3P=CH_2$ and $(C_6H_5)_3P=O$, and (2) separation of 1-ethenylbenzene from benzene. For the first separation, a GC analysis of standard benzene/1-ethenylbenzene mixtures with 0.5 to 5.0 mol% concentration of 1-ethenylbenzene in benzene gave a standardization curve of

$$Y \text{ (1-ethenylbenzene area\%)} = -0.26778 + 0.37294X$$

Here, X equals 1-ethenylbenzene concentration in its benzene solution, expressed as mole percent, and R for the fitting curve is 0.987. The recovered aromatic mixture from each synthesis was diluted with common 1-ethenylbenzene and vacuum

Table I. Formulations of the Reactions

Reaction:	C_6H_5CHO +	$(C_6H_5)_3PCH_3I$ +	$(NaOH+H_2O)$ +	C_6H_6 +	C_6H_5CHO (radiotraced, in THF) $\Leftrightarrow$	$C_6H_5CH{=}CH_2$ +	$(C_6H_5)_3P{=}O$
Experiment I-1							
mmol	3.0	6.0	90	67.32	5.88mCi	3.0	6.0
Grams	0.32	2.43	20.71 (5N)	5.26 (6mL)	0.81	0.3125	1.67
Experiment I-2							
mmol	3.0	6.0	90	67.32	5.88mCi	3.0	6.0
Grams	0.32	2.43	20.66 (5N)	5.27 (6mL)	0.79	0.3125	1.67
Addition Order	(4)	(3)	(2)	(1)	(5)		

Reaction conditions: 40°C, 48 hours, with agitation.

distilled to remove benzene. For this second separation, a GC analysis of standard benzene/1-ethenylbenzene mixtures with 1.0 to 9.0 mol% concentration of benzene in 1-ethenylbenzene gave a standardization curve of

$$Y^{*}(\text{benzene area\%}) = 0.37441 + 1.3255X^{*}.$$

Here, X^{*} equals benzene concentration in its 1-ethenylbenzene solution, expressed as mole percent, and R for the fitting curve is 0.988. Separation results are given in Table II and show that the overall recovery of radioactive 1-ethenylbenzene from two reactions is 59.7% and 61.3% with concentration of 1-ethenylbenzene in benzene of 90.7% and 78.3%.

Retention of Radioactivity in Samples. The data on retention of radioactivity in samples are summarized in Table III. It shows that 2.37g and 2.42g 1-ethenylbenzene with radioactivity of 8.75×10^{6} and 8.97×10^{6} count / second, respectively, were obtained. The next step was to synthesize graft copolymer of lignin with the radiotraced 1-ethenylbenzene.

Synthesis of Graft Copolymer. Graft copolymers of lignin and 1-ethenylbenzene were synthesized by our research group several years ago (*6*). Quantified extractions to to recover purer homopolymer, graft, and lignin phases were performed and the results published (*7*). Because of the limited amount of radiotraced 1-ethenylbenzene available and the need to make an approximately 50% 1-ethenylbenzene in its lignin copolymer to insure rapid biodegradation, the copolymerization of lignin and ^{14}C-labelled 1-ethenylbenzene was run on a scale 90% smaller than that previously published (*6*). The whole reaction mixture was less than 13.0 grams. To insure that the synthesis with the radiotraced 1-ethenylbenzene would be successful; model, small-scale reactions were run. The reactions were conducted at 30°C for 48 hours with a lignin content in the polymerizable materials (lignin plus 1-ethenylbenzene) of 46%. The conversion of 1-ethenylbenzene ranged from 67.1% to 88.9%, the yield of the product ranged from 82.3% to 94.0%, and the lignin content in the product ranged from 49.3% to 56.3%. The results are given in Table IV. The results show that there was no problem in running small-scale copolymerization of lignin and 1-ethenylbenzene. To obtain 50% lignin in the final reaction products, however, the lignin content in the polymerizable materials should be reduced from 46% to 40%. Results for copolymerization reactions containing 40% lignin were: a conversion of 1-ethenylbenzene of 66.0% and 74.5%, a yield of product of 79.6% and 84.7%, and a lignin content in the product of 50.2% and 47.2%, respectively. After successful small-scale copolymerizations of lignin and common 1-ethenylbenzene, two real copolymerization reactions of lignin and concentrated radiotraced 1-ethenylbenzene were conducted. Data from these syntheses are contained in Table V.

Conclusions

Wittig reactions were successfully run with conversion of 1-phenylmethanal or yield of 1-ethenylbenzene being 69.3% and 89.2%. Separation of 1-ethenylbenzene from benzene was successful done with recovery of 1-ethenylbenzene being 66.9% and 88.5%, and with the final 1-ethenylbenzene concentration in benzene being 78.3% and 90.7%. Tetrahydrofuran (THF) contaminated, ^{14}C 1-phenylmethanal was used directly in the reactions without further treatment, and appeared to have no major effect on the Wittig reaction or analysis of 1-ethenylbenzene by gas chromatography. Small-scale

Table II. Separation of ^{14}C 1-Ethenylbenzene from Reaction Product.

Sample No.	I-1	I-2
The First Separation:		
Recovered[1] (g)	5.93	4.25[2]
Recovery of 1. (%)	92.9	66.7
^{14}C 1-ethenylbenzene yield, (g)	0.279	0.217
Recovery (%)	89.2	69.3
The Second Separation:		
Common 1-ethenylbenzene added (g)	3.27	2.53
Liquid remaining in flask (g)	2.61[3]	3.10[3]
Calculated 1-ethenylbenzene available for later reaction		
weight (g)	2.368	2.427
1-ethenylbenzene concen. (wt. %)	90.7	78.3
recovery (%)	66.9	88.5

1. benzene+^{14}C 1-ethenylbenzene. 2. During distillation, some liquid was lost because of a sudden pressure change. 3. Labeled product D.

Table III. Synthesis Data on Retention of Radioactivity.

Experiment. No.	I-1D	I-2D
Original 1-phenylmethanal		
radiotracer content (mCi)	5.88	5.88
radioactivity (counts/s)	$5.88x10^7$	$5.88x10^7$
^{14}C 1-ethenylbenzene in benzene mixture after the first separation		
grams	0.279	0.217
recovery	89.2%	69.3%
radioactivity (counts/s)	$5.25x10^7$	$4.08x10^7$
1-ethenylbenzene in condensed solution after the second separation		
^{14}C 1-ethenylbenzene used (g)	0.273	0.211
1-ethenylbenzene added[1] (g)	3.27	2.53
recovered 1-ethenylbenzene (g)	2.368	2.427
recovery (%)	66.9	88.5
radioactivity (counts/s)	$3.43x10^6$	$3.52x10^6$

1. Common, non-radiodactive 1-ethenylbenzene.

copolymerizations of lignin and ^{14}C-labelled 1-ethenylbenzene were successfully run. Based on yield data, we now have final graft copolymer of lignin and 1-ethenylbenzene with which to conduct biodegradation tests. The copolymer contains about 2,000 count / second / mg radioactivity and about 52% poly(1-phenylethylene) in it.

Acknowledgment

Portions of this work were supported by the U. S. Department of Agriculture under Grant No. 89-34158-4230, Agreement No. 71-2242B, Grant No. 90-34158-5004, Agreement No. 61-4053A and Grant No. 93-34158-8384, Agreement No. 61-4108A.

Table IV. Scaled Copolymerization of Lignin and 1-Ethenylbenzene.

Total weight of reaction (g)	6.0
Lignin / (lignin+1-ethenylbenzene) (%)	46
Reaction vessel	20 mL sample bottles
Reaction temperature (°C)	30
Reaction time (hours)	48

Experiment Number:	IV-1	IV-2	IV-3
1-ethenylbenzene (g)	0.78	0.78	0.78
Lignin (g)	0.67	0.67	0.67
$CaCl_2$ (g)	0.50	0.50	0.50
DMSO (g)	3.34	3.34	3.34
H_2O_2 (35%) (g)	0.73	0.73	0.73
Polymerization product (g)	1.19	1.36	1.19
Conversion of monomer (%)	67.1	88.9	67.1
Yield of product (%)	82.3	94.0	82.3
Lignin in product (%)	56.3	49.3	56.3

Table V. Copolymerization of Lignin and ^{14}C-(1-Ethenylbenzene).

Formulation		
Experiment No.	V-1	V-2
Radiotraced monomer from	I-1D	I-2D
total weight, distillate (g)	2.21	2.56
monomer concentration (%)	90.7	78.3
Radiotraced 1-ethenylbenzene used		
weight (g)	2.00	2.00
radioactivity (count / s)	2.97×10^7	2.98×10^7
Lignin (g)	1.33	1.33
$CaCl_2$ (g)	1.00	1.00
DMSO (g)	6.66	6.66
H_2O_2 (35%) (g)	1.46	1.46
Total weight (g)	12.6	13.0
Results		
Appearance of product:	yellow-brown powder	
Product weight (g)	2.83	2.79
Conversion of monomer (%)	75.0	73.0
Yield of product (%)	85.0	83.8
Lignin in product (%)	47.0	47.7
Radioactivity of the product (count / s)	2.23×10^7	2.18×10^7
Radioactivity per mg of product (count / s / mg)	7868	7801

Literature Cited

1. Lewis, P. J.; Hogapran, C.; Koch, P.; in *Kirt-Othmer Encyclopedia of Chemical Technology*, Grayson, M., Ed.; Third Edition, Vol. **21**, p772, John Wiley & Sons, Inc., New York, NY, USA, 1983.
2. *New England Nuclear Catalog*, 1960 and 1995, New England Nuclear, Inc., 549 Albany St., Boston, MA.
3. Milstein, O.; Gersonde, R.; Huttermann, A.; Chen, M. J.; Meister, J.J.; "Fungal Biodegradation of Lignopolystyrene Graft Copolymers", *Applied and Environmental Microbiology* **1992**,*58* (#10), pp. 3225-3232.
4. Milstein, O.; Gersonde, R.; Huttermann, A.; Frund, R.; Feine, H. J.; Chen, M. J.; Meister, J.J.; "Infrared and Nuclear Magnetic Resonance Evidence of Degradation in Thermoplastics Based on Forest Products", *J. of Environmental Polymer Degradation* , **1994**, *2* (#2), pp. 137-152.

Bq = Disintegrations/sec.   Ci = Curie = 3.7×10^{10} Bq.
5. Tagaki, W.; Inoue, I,; Yano, Y.; Okonogi, T.; "Wittig Reaction in Benzene-Aqueous Alkaline Solution", *Tetrahedron Letters* **1974**, *30* pp.2587-2590.
6. Meister, J.J.; Chen, M. J.; "Graft 1-Phenylethylene Copolymers of Lignin. 1. Synthesis and Proof of Copolymerization", *Macromolecules* **1991**, *24* (#26), pp. 6843-6848.
7. Chen, M. J.; Meister, J.J.; Gunnells, D. W.; Gardner, D. J.; " Alteration of the Surface Energy of Wood Using Thermoplastic Graft Copolymers. ", *Jour. Wood Chem. Tech.,* **1995** 15 (2), 287-302.

Chapter 17

Blends of Biodegradable Thermoplastics with Lignin Esters

Indrajit Ghosh, Rajesh K. Jain, and Wolfgang G. Glasser

Biobased Materials and Recycling Center, Department of Wood Science and Forest Products, Virginia Polytechnic Institute and State University, Blacksburg, VA 24061–0324

Thermoplastic blends of several biodegradable polymers with lignin and lignin esters were prepared by solvent casting and melt processing. Among the biodegradable thermoplastics were cellulose acetate butyrate (CAB), a starch-caprolactone copolymer/blend (SCC), and poly(hydroxybutyrate) (PHB). Lignin esters included the acetate, butyrate, hexanoate, and laurate of organosolv lignin, LA, LB, LH, and LL, respectively. Blend properties were analyzed by thermal, mechanical, and optical (transmission electron microscopy, TEM) analysis. The results indicate widely different levels of interaction between the two polymer constituents. Blends of LA and LB with CAB exhibited a high level of compatibility that was lost when the acyl substituent increased in size. The addition of unmodified lignin to PHB greatly retarded crystallization and produced blends with lower melting points. The same was true for SCC blends, which were found to crystallize and melt at lower temperatures if lignin was present. However, a significantly increased modulus at room temperature resulted with the addition of lignin, and this was attributed to increased crystallinity in the presence of lignin.

Biodegradable thermoplastics are making inroads into the polymer and materials market on account of their ease of disposal [1-4]. In general, biodegradable thermoplastics include starch and starch derivative-based formulations [5,6]; fermentation-produced polymers such as poly(hydroxy butyrate) (PHB), poly (hydroxy butyrate-co-valerate) (PHB/V), and polyactic acid (PLA) [7-9] and cellulose esters [10,11]. The engineering of specific target properties of polymeric materials has long taken advantage of the ability to mix polymers with different properties in solution-cast or melt-processed blend systems [12]. Additives of a wide variety are available for engineering the material properties of man-made polymers. However, fully biodegradable materials, consisting of biodegradable thermoplastics and biodegradable additives, have a much smaller variety of potential additives at their disposal. Qualified biodegradable additives include (again) fermentation-produced additives or those derived from the chemical modification of natural polymers [1]. Among the latter is lignin, the high modulus amorphous binder of woody plants [13]. Thermoplastic lignin and lignin derivatives are commercially or semi-commercially available [14], and they are candidates for biodegradable additives [15, 16].

The qualifications of a polymeric additive to influence blend properties depend primarily on its ability to associate or develop a molecular interaction with the thermoplast in question. Secondary association-driven interactions between blend components are necessary to establish some degree of molecular compatibility [17]. While complete miscibility is virtually unachievable, phase compatibility on the nano-level has been observed in many systems. The degree of phase compatibility as opposed to macroscopic separation is usually established using thermal analysis or transmission electron microscopic techniques [18].

The objectives of this report deal with the determination of the effect of lignin and selected biodegradable lignin derivatives on the phase morphology and selected properties of blends of three biodegradable thermoplastics, cellulose acetate butyrate (CAB), poly(hydroxybutyrate) (PHB), and starch/ caprolactone copolymer/blend (SCC). Additional detail regarding blends of CAB with lignin derivatives have been given elsewhere [19].

Materials and Methods

Materials : CAB (CAB 381-20) was obtained from Eastman Chemical Company, Kingsport, Tennessee; PHB from Biopol (Marlborough Biopolymers), Cleveland, UK; and starch-polycaprolactone blend (SCC) (tradename 'Envar') from Biomaterials Research Center, Michigan State University [20]. The approximate starch content of SCC is 30% by wt. and the polycaprolactone component has a M_n of 148,000 and a M_w of 473,000. Organosolv Lignin (L) was obtained from Aldrich Chemical Company, WI, USA (Catalog #: 37,101-7). Lignin was esterified with acetic, butyric and hexanoic anhydrides, and lauryl chloride to yield the corresponding esters by adopting the esterification methodology described elsewhere [15]. Selected physical properties and molecular weight parameters of these polymers are summarized in Table I.

Methods: Blends were prepared by either casting appropriate mixtures of the polymers from solutions or by melt-processing. For solvent casting, the blend solutions were prepared (~5 wt.%) using chloroform as solvent. The solutions were stirred for approximately 6 hours before they were cast into Teflon molds. The castings were kept at ambient temperature for 48 hours, allowing slow evaporation of the solvent by partially covering the molds. Extruded blends were prepared with a

Table I : Some Physical Properties of Biodegradable Polymers and Lignin Additives

Preparation	**Synthesis Method**	**Physical Form**	**T_g (°C)**	**M_n x 10^{-3}**	**M_w x 10^{-3}**
CAB	Anhydride	Powder	134	69.6	127.6
PHB	Fermentation	Powder	0.8	-	-
SCC	Extrusion	Pellets	-60	148.0	473.0
Lignin (L)	Precipitation	Powder	107	0.82	3.14
LA	AA[1]; PPT-W[2]	Powder	92	1.55	5.89
LB	AA[1]; PPT-BC[3]	Powder	52	2.31	7.73
LH	AA[1]; PPT-BC[3]	Powder	30	2.65	9.44
LL	AC[4]; PPT-E[5]	Tar	2	13.4	33.2

[1] Acid anhydride, AA
[2] Precipitation in water, PPT-W
[3] Precipitation in bicarbonate solution, PPT-BC
[4] Acid Chloride, AC
[5] Extraction with chloroform/acetone 3/2, E

Custom Scientific Instruments Mini-Max injection molder with the extrusion temperatures in the range of 200-220°C for CAB blends, 165-190°C for PHB blends and 175-190°C for SCC blends, and a residence time of no more than 2 mins to avoid degradation. The extruded samples for testing included dog-bone and rectangular specimens. All samples were dried under vacuum at 30°C for 4 hours and stored in air-tight plastic bags in a vacuum desiccator till tests were performed.

Analytical methods included thermal analysis by differential scanning calorimetry (DSC) using a Perkin-Elmer Model DSC-4 interfaced to a thermal analysis data station; and dynamic mechanical thermal analysis (DMTA) with a Polymer Laboratories Ltd. dynamic mechanical thermal analyzer in the single or dual cantilever bending mode, or in the shear mode depending on sample geometry and sample consistency. The spectra for DMTA were collected at a heating rate of 4°C min^{-1} at a frequency of 1 Hz. Transmission electron microscopy (TEM) was performed using a JEOL JEM-100CX-II electron microscope operated at an accelerating voltage of 80 kV. The mechanical properties (modulus, strength and ultimate strain) of the blends were determined on a Miniature Materials Tester (Minimat model # SM9-06) by Polymer Laboratories, Loughborough, England. Tests were conducted at room temperature with a 1000 N load beam using strain rates of 5 mm/min.

Results and Discussion

1. Cellulose Acetate Butyrate (CAB) Blends :

Blends of CAB with lignin esters were examined by DSC, DMTA and TEM. Results by DSC revealed single glass transition temperatures for all polymer blends with LA (Fig.1) and LB (not shown). Glass transitions of CAB shifted progressively to lower temperatures as LA and LB content rose. Single glass transitions were also observed by DMTA (Fig.2), where single and sharp tan δ-transitions were observed at increasingly lower temperatures following the addition of LA and LB. When LH was blended with CAB instead of LA or LB, however, a low temperature shoulder appeared in the DMTA-thermograph (tan δ-transition) at >20% LH-content that was indicative of macroscopic phase separation. This shoulder was even more apparent in the loss modulus (E'' and G'') thermogram of the CAB/LH blends, and this clearly suggests the separation of polymeric phases on a scale of 0.5 to 1 μm. The onset of blend softening on a temperature scale, however, revealed a significant dependence on lignin content for all blends with lignin esters.

By examining the impact of lignin esters on the shift in glass transition temperature using the well-known Fox-equation [21],

$$\frac{1}{T_g} = \frac{w_1}{T_{g1}} + \frac{w_2}{T_{g2}} \qquad (1)$$

where T_g is the glass transition temperature in degrees K for polymer 1, 2, of the blend; and w is the weight fraction of component 1 and 2 in the blend, a quantitative measure of compatibility can be established (Fig. 3a). Whereas LA and LB are in apparently full compliance with copolymer behavior as required by equation (1), CAB/LH blends reveal incompatibility at lignin derivative contents in excess of 10% (Fig. 3b). This suggests that, whereas LA and LB form compatible blends over the entire mixing range, LH is incompatible with CAB on a scale in excess of ca. 0.1 μm [18].

A comparison of transmission electron micrographs with resolution on the nano-scale reveals phase separation between the cellulose and lignin ester components on different levels (Fig. 4). Solvent-cast blends exhibit included (discontinuous) phases that vary with lignin ester content (Figs. 4a and 4b), and these are always larger than those observed by the corresponding melt blending technique (Fig. 4c). This is expected because phase dimensions are a consequence of chemical intermolecular affinity (i.e. thermodynamic effects) as well as the time available for different polymers to separate from each other (i.e., kinetic effect). Since polymers have a virtually unlimited amount of time to phase separate during solvent casting, and they are near their equilibrium condition at all times, the included polymer phases are always larger during solvent-casting than during melt blending (Fig. 4a vs. 4c). A comparison between 20% CAB/LA and LH blends (Fig. 4d vs. 4c) reveals dimensional differences favoring CAB/LA blends. Since the latter display an included phase on the scale of 10 to 30 nm (Fig. 4d), whereas they are 30 to 100 nm for the corresponding LH-blends (Fig. 4c), the results are seen to agree with the thermal analysis experiments.

The superior phase compatibility of the LA and LB blends of CAB as compared to LH and LL, find further support in the mechanical stress vs. strain experiments. Whereas all blends reveal a decrease in tensile strength, strain and modulus with lignin ester content rising to >20% lignin derivative content, CAB blends with LA and LB reveal increases in tensile properties and in modulus with

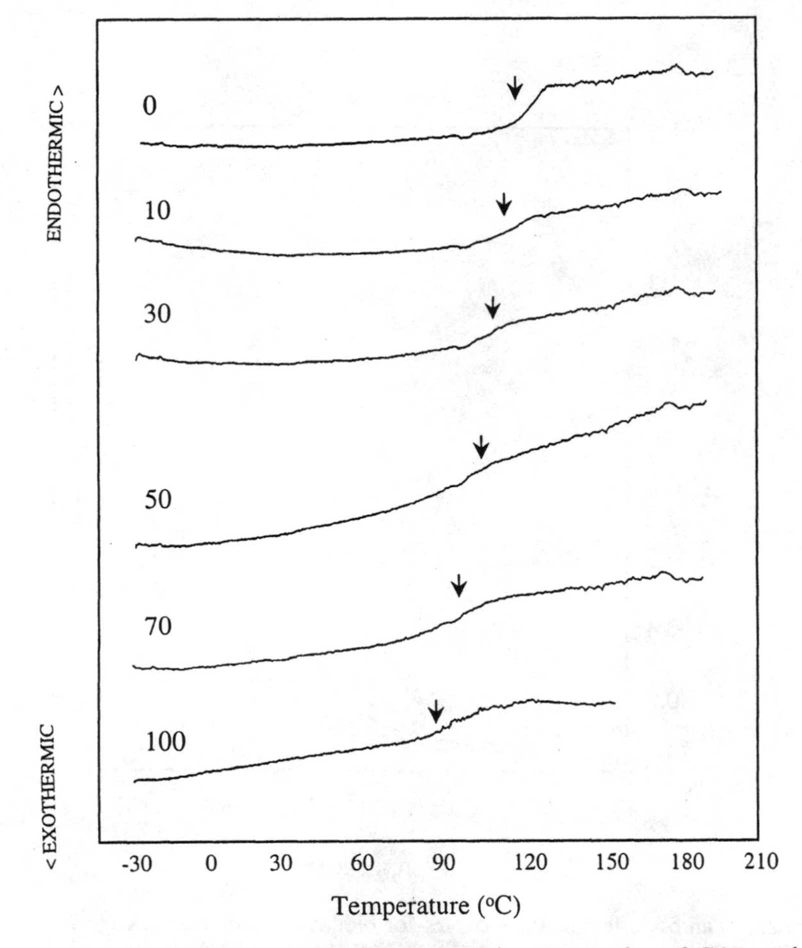

Figure 1. DSC thermograms of solvent ($CHCl_3$) cast samples of CAB and LA. Numbers on each curve denote LA content (wt.%) in the blend. These traces are from the second heating scan (after quenching from melt at a rate of 300°C/min). (Reproduced with permission from ref.19, copyright 1998 by John Wiley & Sons, Inc.)

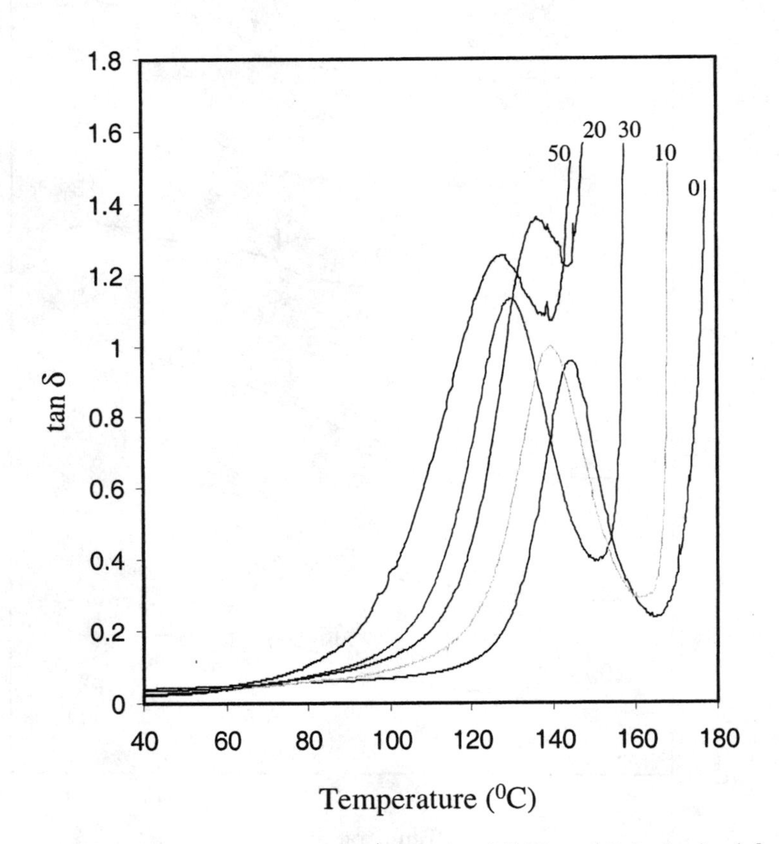

Figure 2. **Tan δ vs. temperature curves for blends of CAB and LA obtained from DMTA experiments at 1 Hz. (Adopted from ref.19)**

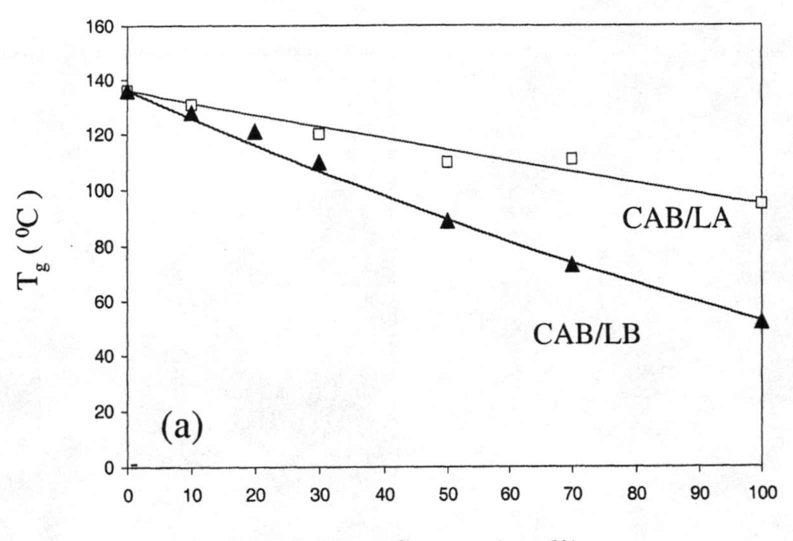

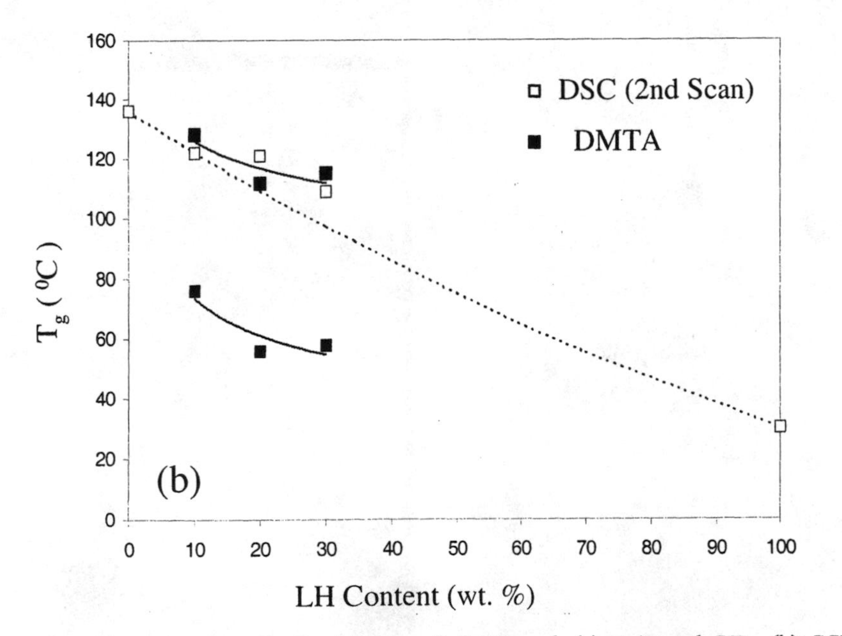

Figure 3. Fox equation fit for blends of CAB and (a) LA and LB; (b) LH. (Adopted from ref.19)

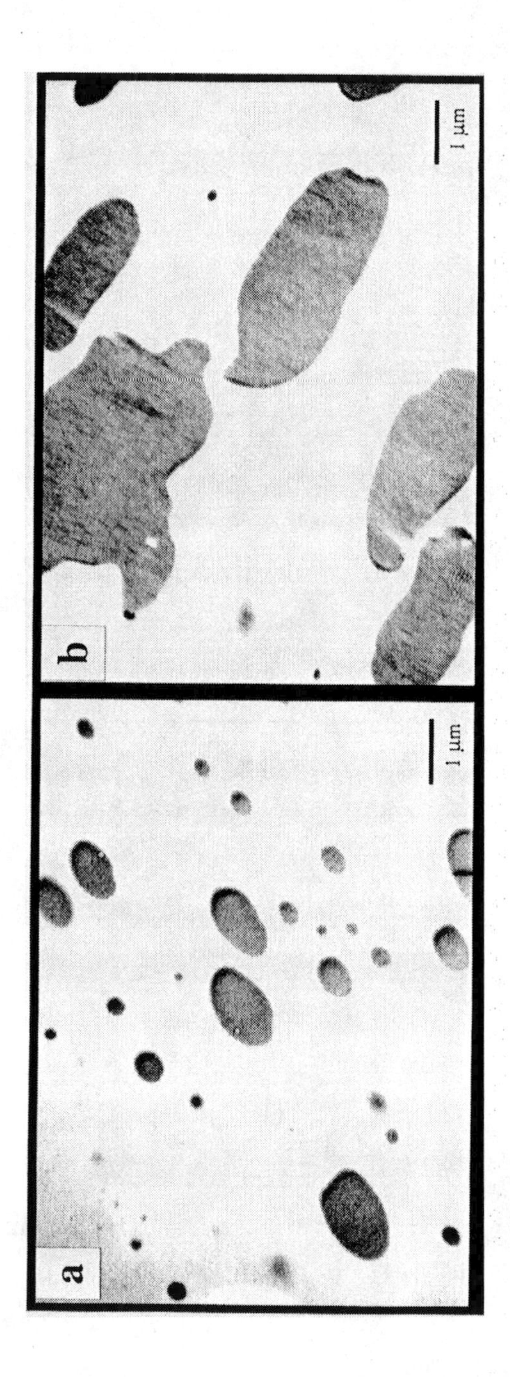
a
1 μm
b
1 μm

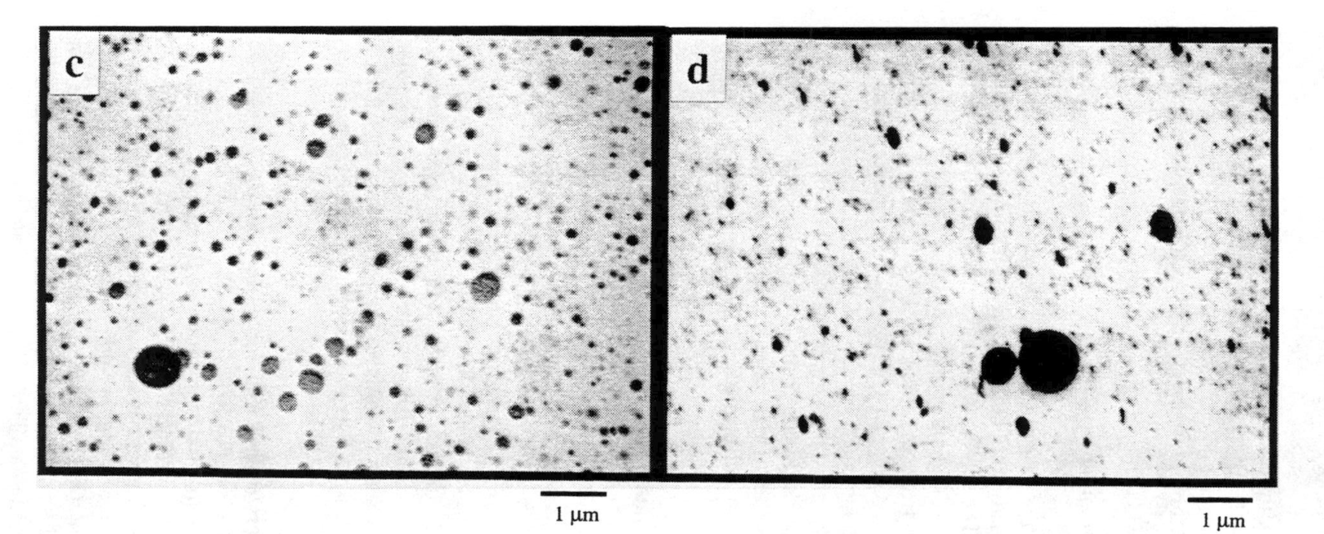

Figure 4. Transmission electron micrographs of CAB/L ester blends with (a) solvent-cast, 20% LH; (b) solvent-cast, 50% LF; (c) melt-blended, 20% LH; (d) melt-blended, 20% LA by weight (magnification at 10,000 X). (Reproduced with permission from ref.19, copyright 1998 by John Wiley & Sons, Inc.)

<20% lignin (Fig. 5). This increase is not observed with LH or LB blends, which reveal an almost monotonous decline in all strength properties with lignin derivative content rising.

In summary, CAB blends with LA and LB are compatible at all blend compositions on the 10 to 50 nm-level. By contrast, CAB blends with LH and LL become incompatible when lignin ester content rises to >20%. Although lignin esters are effective plasticizers of CAB by lowering T_g, they contribute to an increase in modulus at lignin contents below 20%.

2. Poly(hydroxybutyrate) (PHB) Blends:

PHB is a thermoplastic polymer with significant propensity for crystallization [9]. One of its biggest handicaps constraining commercial utilization is its rapid loss of stability when heated above T_m. The effect of lignin on PHB must, therefore, be examined in terms of the impact on melting and crystallization behavior.

When neat PHB is cooled from the melt at 10°C min^{-1}, a significant exothermic crystallization peak is observed at 81°C (Fig. 6). When heated again, the corresponding endotherm at 175°C indicates melting. Cooling the sample for a second time (at 10°C min^{-1}), a broader crystallization event is observed compared to the first time reflecting partial thermal degradation and loss of molecular uniformity. Reheating and recooling the pre-melted PHB at different heating/cooling rates produces T_c and T_m-events that vary in uniformity and location on the temperature scale (Fig. 6).

The presence of (unmodified) lignin in a solvent-cast PHB/L blend inhibits or retards crystallization at any cooling rate and at lignin contents of 10 and 20% (Fig. 7). Glass transition temperatures are observed at around 5°C for PHB, and this undergoes spontaneous crystallization when heated to >50°C at any heating rate and any lignin content (Fig. 7). Similar results were observed for DSC curves of PHB and LB blends (not shown). The observed T_g values for the PHB phase reveal shifts towards the glass transitions of L (or LB) indicating some interaction between PHB and L or LB (Fig.8a). The normalized ΔH_m values for PHB/LB are higher than those for PHB/L blends (Fig.8b). This might be an indication of greater interaction between PHB and L as compared to PHB and LB. On the other hand, the T_m values for PHB/LB blends are observed to shift towards lower temperatures (Table II). These results suggest that PHB and L (or LB) have a high degree of compatibility, and that lignin inhibits or retards PHB-crystallization.

3. Starch-Caprolactone Copolymers/Blends (SCC) :

Melt processable SCC undergoes thermal transitions and softening events that are dictated by the presence of the polycaprolactone component. The introduction of unmodified lignin or lignin ester derivative by solvent-casting or melt-processing methodology causes the crystallization and melting points of the PCL component to decline in relation to lignin content (Fig. 9). Melting behavior thereby mirrors crystallization behavior (Fig. 9a and 9b). When analyzing the overall energetics of the crystallization and melting phenomena (Table III), it is apparent that different

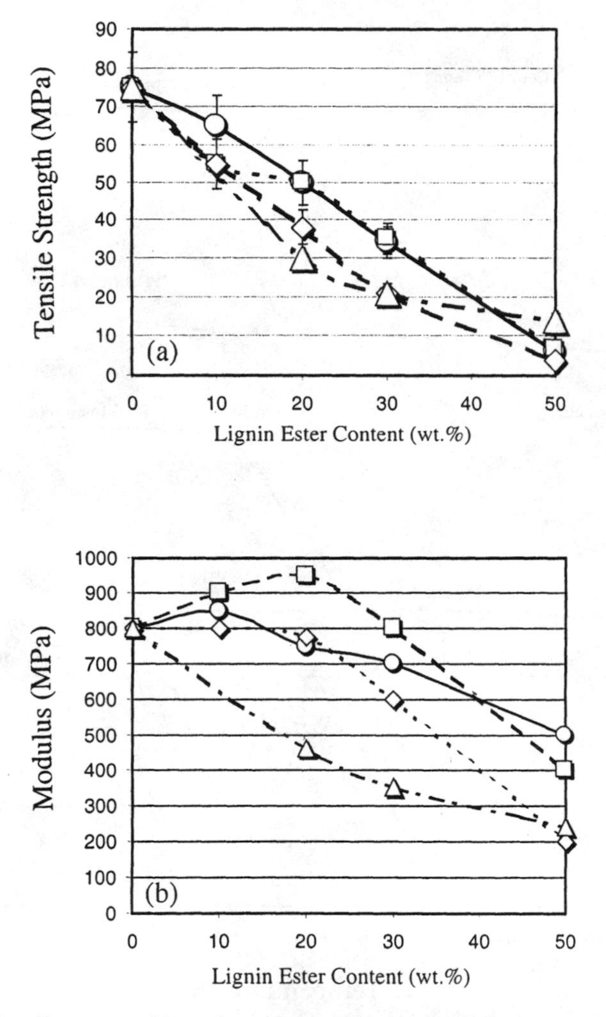

Figure 5. Tensile stress (a) and modulus (b) vs. lignin ester content for melt-blended CAB/lignin ester blends: (○) CAB/LA, (□) CAB/LB, (◇) CAB/LH, (Δ) CAB/LL. (Adopted from ref.19)

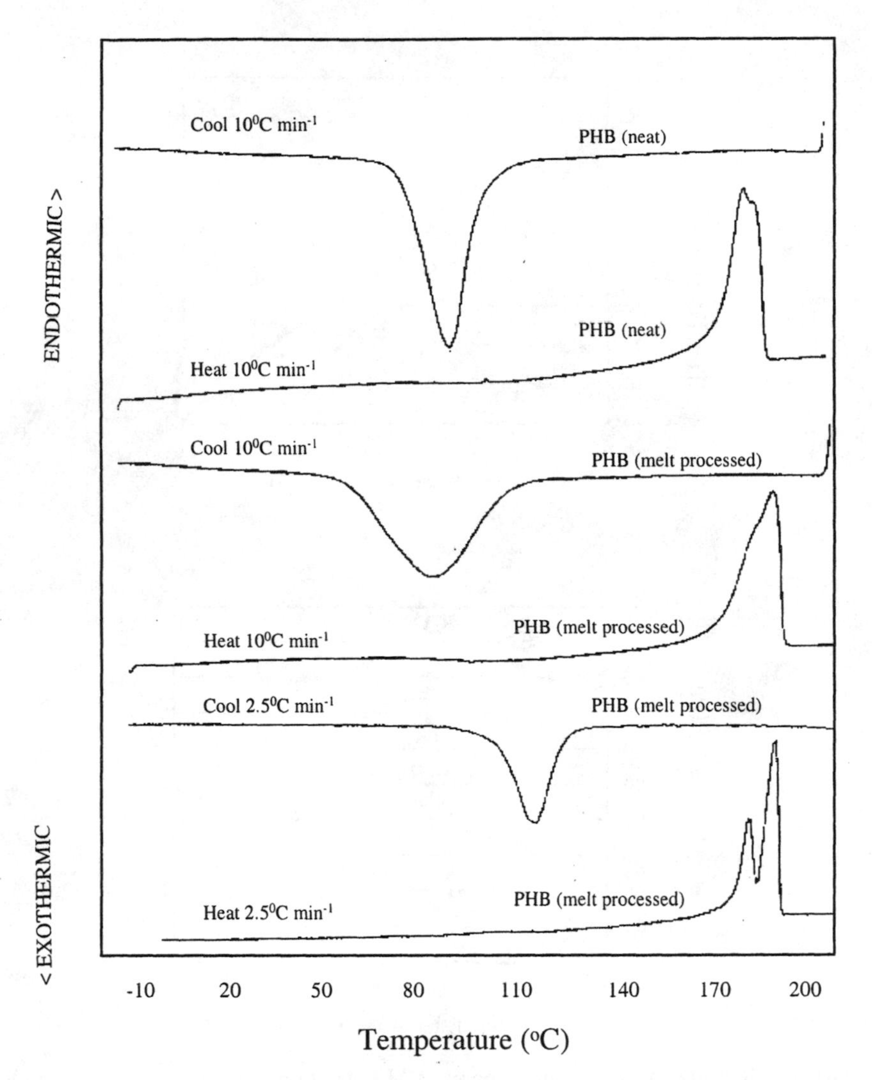

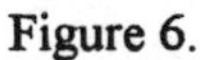
Figure 6. Effect of processing and crystallization times on PHB. The cooling traces are from the first cooling scans from melt and the heating traces are from the second heating scan. (Some curves have been expanded on the y-axis for greater clarity; y-axis has arbitrary scale).

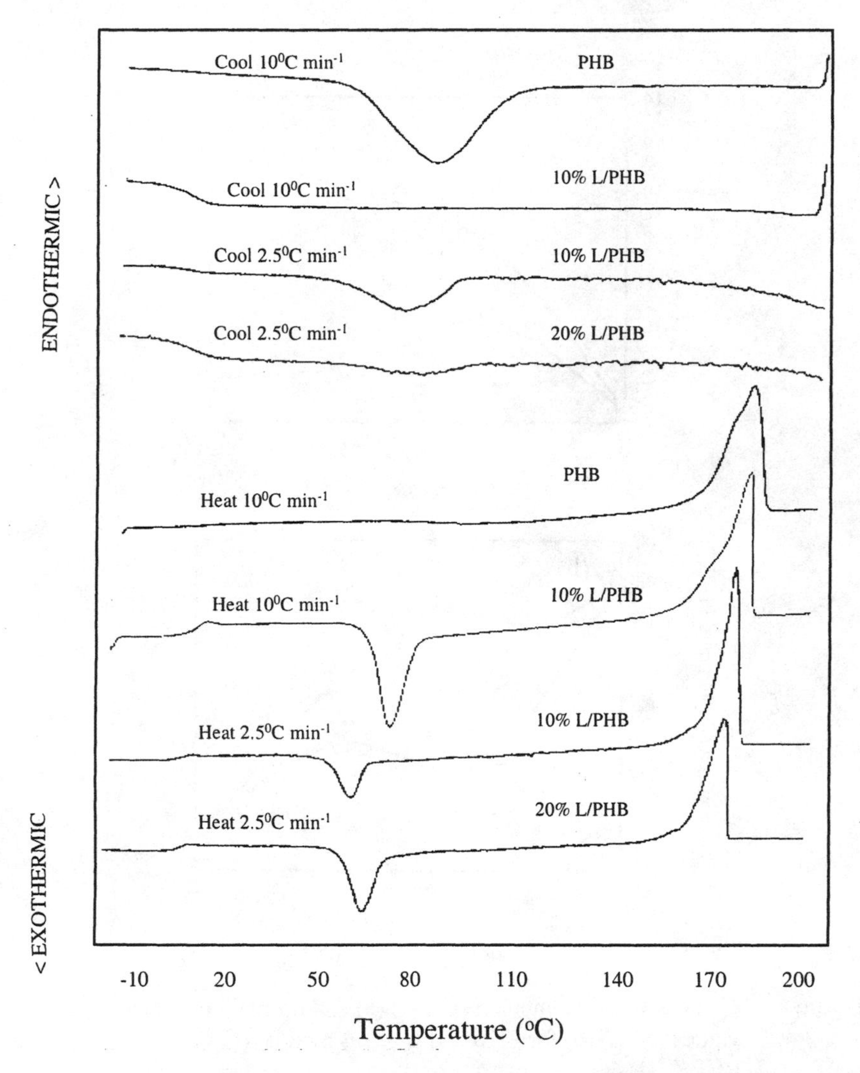

Figure 7. **DSC thermograms of melt blended samples of PHB and L. The cooling traces are from the first cooling scans from melt and the heating traces are from the second heating scan.**

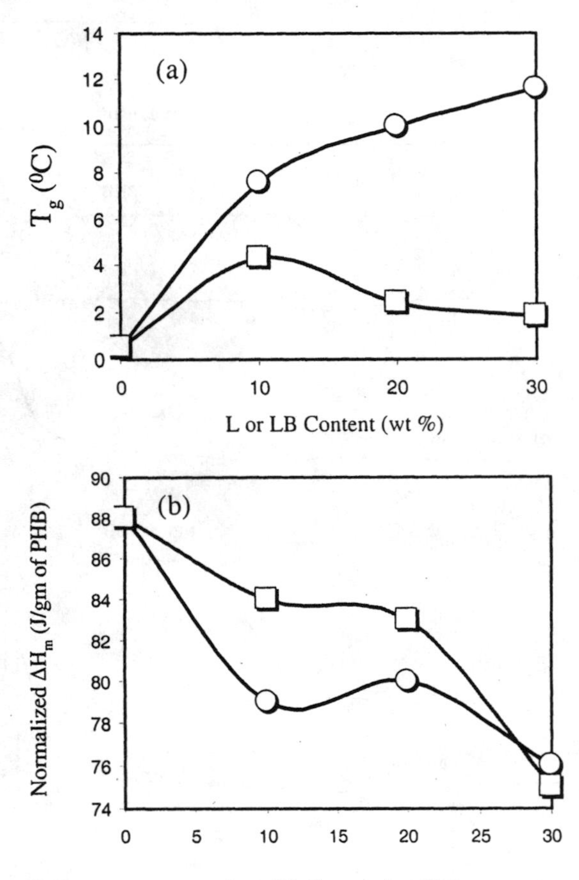

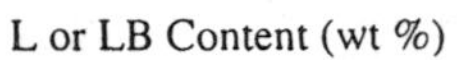

Figure 8. **Glass transition temperatures (T_g) (a) and normalized heat-of-fusion values (ΔH_m) (b) for PHB and L or LB blends: (○) PHB/L, (□) PHB/LB.**

Table II : Thermal Characteristics of Blends of PHB.

Preparation	T_g (°C)	T_m (°C)	T_c* (°C)	ΔH_c* J/gm	ΔH_m J/gm	Normalized ΔH_c J/gm of PHB	Normalized ΔH_m J/gm of PHB
PHB (native)	0.5	173.1	81.5	-61.3	87.6	-61.3	87.6
PHB (melt processed)	-	177.8	78.1	-63.3	92.8	-63.3	92.8
PHB/L Blends							
90/10	7.6	177	69*	-70.3*	71.1	-78.1	79.0
80/20	10.0	175	74*	-62.7*	64.2	-78.4	80.3
70/30	11.6	175	81*	-51.6*	53.1	-73.7	75.9
PHB/LB Blends							
90/10	4.3	170	63*	-73.7*	75.4	-81.9	83.8
80/20	2.4	161	75*	-64.6*	66.7	-80.8	83.4
70/30	1.8	156	80*	-50.8*	52.7	-72.6	75.3

* The reported crystallization temperatures (T_c) and heat of crystallization (ΔH_c) values for the blends are the peak temperatures of the crystallization endotherms observed in the second heating scans (usually reported as cold crystallization) at a scanning rate of 10°C/min. No crystallization was observed during the cooling scans at a scanning rate of 10°C/min for any of the blend samples. The T_cs reported for PHB (without lignin component) are from the crystallization endotherms observed during the first cooling scans from melt at a cooling rate of 10°C/min.

lignin components vary in their effect on PCL morphology. Whereas the presence of unmodified lignin in melt-processed blends enhances the crystallization of PCL, LB appears to have the opposite effect. The latter is more pronounced when lignin is introduced by solvent-casting than by melt-processing (Table III). These results are consistent with earlier observations [16] that suggest that high-T_g unmodified lignin may serve as nucleating agent for PCL, thereby enhancing crystallization and consequently also melting, and low-T_g lignin esters show a significant compatibility with PCL enhancing an amorphous component and reducing crystallization and fusion.

The impact of lignin and lignin ester content on the stress vs. strain properties of SCC blends mirrors the changes in morphology (Fig. 10). Dramatic increases in tensile stress, by 300%, following the addition of lignin at the 10%-level suggest greatly enhanced nucleation of PCL. Presumably smaller crystallites and greater overall crystallinity (at 10% L-content) contribute to significant strength gains that are not realized if (a) greater amounts of L or (b) lower-T_g lignin derivatives (LB) are used for blending.

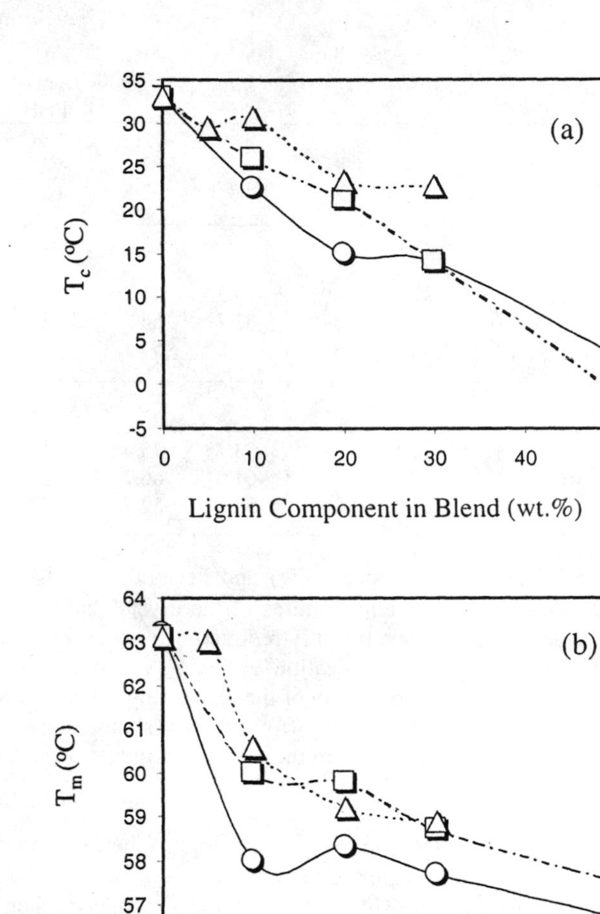

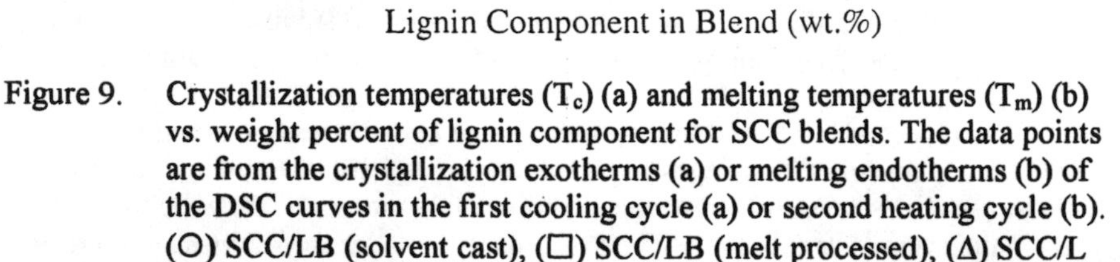

Figure 9. Crystallization temperatures (T_c) (a) and melting temperatures (T_m) (b) vs. weight percent of lignin component for SCC blends. The data points are from the crystallization exotherms (a) or melting endotherms (b) of the DSC curves in the first cooling cycle (a) or second heating cycle (b). (O) SCC/LB (solvent cast), (□) SCC/LB (melt processed), (Δ) SCC/L (melt processed).

Table III: Crystallization and Melting Data of SCC-Blends.

Preparation	ΔH_c (J/gm)	ΔH_f (J/gm)	Normalized* ΔH_c [(1)] (J/gm PCL)	Normalized* ΔH_f [(1)] (J/gm PCL)	Normalized* X_c (%)
SCC (original)	-49	58	-69	81	60
SCC ($CHCl_3$ cast)	-49	57	-69	81	60
SCC (Melt Extruded)	-48	54	-69	78	58
SCC/L Blends					
95/5	-49	54	-74	81	60
90/10	-48	56	-77	89	66
80/20	-40	43	-71	77	57
70/30	-36	38	-73	78	58
SCC/LB					
(Melt Blended)	-43	48	-68	76	56
90/10	-40	42	-71	75	56
80/20	-33	40	-68	81	60
70/30	-6	26	-16	76	56
SCC/LB					
($CHCl_3$ Cast)	-39	43	-62	69	50
90/10	-38	42	-68	75	56
80/20	-31	38	-63	77	57
70/30	-9	26	-26	74	55

* ΔH_c, ΔH_f and X_c represent the heat of crystallization, heat of fusion and degree of crystallinity in the sample, respectively. The crystallization data are from the first cooling scans from the melt, by DSC; and the melting data are from the subsequent (second) heating scans.

(1) ΔH_c or ΔH_f multiplied by (content of PCL in blend)$^{-1}$.

(2) Calculation of X_c is based on $\Delta H^o f \approx 135$ J/gm for pure crystalline PCL.

This suggests that, while clearly immiscible, L in particular has much to contribute to SCC by virtue of interacting with the PCL component. L and LB have opposite effects on crystallization: whereas L enhances the crystallization of PCL (at 10% L-content), LB depresses T_c, T_m and ΔH_c probably by virtue of favorably interacting with PCL. Gains in tensile stress of 80-300% are observed when 10-20% of L or LB are added to SCC.

Conclusions

Lignin and lignin esters show a significant degree of interaction with all thermoplastic biodegradable polymers studied. The degree of compatibility varies

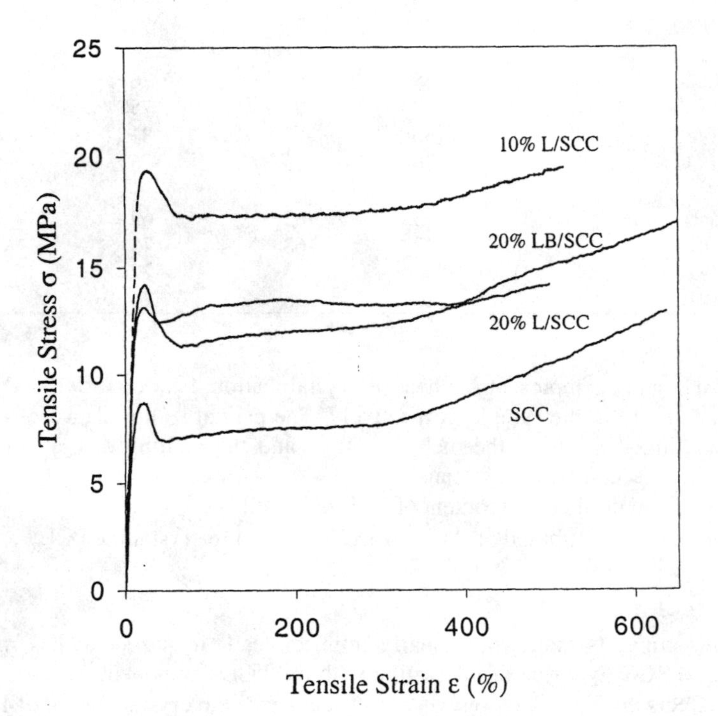

Figure 10. Tensile stress vs. strain curves for SCL/L and SCL/LB blends.

with lignin ester type; with lignin and lignin derivative content; with thermoplastic polymer type; and with method of blending (i.e., solvent casting or melt processing). The degree of compatibility is consistently revealed by thermal analysis, including DSC and DMTA. Degree of phase compatibility is also revealed quantitatively by TEM.

Acknowledgment

The authors wish to express gratitude to the Eastman Chemical Company for supplying CAB, and to Professor Ramani Narayan, Michigan State University, for providing the thermoplastic starch derivative (Envar). Thanks is also given to Professor H. Marand, Department of Chemistry, Dr. Mike McLeod, Department of Chemical Engineering, and Professor Charles E. Frazier, Department of Wood Science and Forest Products, all Virginia Tech, for valuable advice and assistance as part of this study.

This publication represents Part 16 of the series entitled "Multiphase Materials with Lignin." Earlier publications have appeared in the Journal of Wood Chemistry and Technology 14, 119 (1994), Polymer 35, 1977 (1994), Journal of Applied Polymer Science 51, 563 (1994), Macromolecules 27, 5 (1994), and elsewhere.

Literature References

1. Kaplan, D. L.; Mayer, J. M.; Ball, D.; McCassie, J.; Allen, A. L.; and Stenhouse, P. Chapter in "Biodegradable Polymers and Packaging," Ching, C.; Kaplan, D. L. and Thomas, E. L., eds. Technomic Publishers, Lancaster, **1993,** pg. 1-42.
2. Huang, S. J., and Edelman, P. G., Chapter in Degradable Polymers : Principles and Applications, Scott, G., and Gilead, D., eds., Chapman & Hall, **1995**, pg. 18-28.
3. Swift, G., Chapter in Agricultural and Synthetic Polymers- Biodegradability and Utilization, Glass, J. E. and Swift, G. , eds. American Chemical Society Symposium Series No. 433, **1990**, pg. 2-12 (1990).
4. Narayan, R., and Bloemberger, Polym. Prepr., **1991**, 32(2), 119-120.
5. Griffin, G. J. L., "Degradable Plastic Films," Proceedings of Symposium on Degradable Plastics, Washington, D. C.: The Society of Plastics Industry, Inc., **1987**, pg. 47-49.
6. Narayan, R., Chapter in Assessment of Biobased Materials, H. L. Chum, ed., Solar Energy Research Institute, Colorado, SERI/TR-234-3610, **1989**, pg. 7.1-7.25.
7. Doi, Y., Microbial Polyesters, VCH Publishers, New York, **1990**.
8. Gilmore, D. F., Fuller, R. C., Lenz, R., Chapter in Degradable Materials: Perspectives, Issues and Opportunities," Barengerg, S. A.; Brash, J. L.; Narayan, R., and Redpath, A. E., eds., CRC Boston, **1990,** pg. 481-514.
9. Organ, S. J., and Barham, P. J., *Journal of Material Science*, **1991**, 26, 1368-1374.

10. Buchanan, C. M.; Gardner, R. M.; Komarek, R. J., Gedon, S. C.; and White, A. W., Chapter in Biodegradable Polymers and Packaging, Ching, C.; Kaplan, D. L. and Thomas, E. L., eds., Technomic Publishers, Lancaster, **1993**, pg. 133-140.
11. Sealey, J. E.; Samaranayake, G.; Todd, J. G.; and Glasser, W. G.; *Journal of Polymer Science: Part B: Polymer Physics*, **1996**, 34, 1613-1620.
12. Walsh, D. J., Higgins, J. S., Maconnachie, A., eds. Polymer Blends and Mixtures, Martinus Nijhoff Publishers, Dordrecht/Boston/Lancaster, **1985**, 469 pg.
13. Glasser, W. G., and Kelley, S. S., Lignin, *Encyclopedia of Polymer Science and Engineering*, Vol. 8, 2nd edition, John Wiley & Sons, Inc., **1987**, 795-852.
14. Aldrich Catalog, product # 37101-7, 37102-5, 37106-8, 37104-1, 37103-3, Aldrich Chemical Company, Inc. Milwaukee, WI, USA.
15. Glasser, W. G., and Jain, R. K., *Holzforschung*, **1993**, 47 (3), 225-233.
16. Oliveira, W. de, and Glasser, W. G.; *Macromolecules*, 27, 5-11 (1994).
17. Paul, D. R., Chapter in Polymer Blends, vol. 1, D. R. Paul and S. Newman, eds., Academic Press, New York, **1978**, pg. 1-14.
18. Kaplan, D. S., *Journal of Applied Polymer Science*, **1976**, 20, 2615-2629 (1976).
19. Ghosh, I., Jain, R. K., and Glasser, W. G.; *Journal of Applied Polymer Science*, in press.
20. Narayan, R., and Krishnan, M., *PMSE Proceedings*, *American Chemical Society*, **1995**, 72, 186.
21. Fox, T. G., *Bull. Am. Phys. Soc.*, **1956**, 2(2), 123.

Chapter 18

Thermoplastics with Very High Lignin Contents

Yan Li and Simo Sarkanen

Department of Wood and Paper Science, University of Minnesota, St. Paul, MN 55108–6128

Annually huge quantities of kraft lignins are generated as byproducts from the chemical conversion of wood chips into pulp for making paper. Traditionally these industrial byproducts have been employed as fuel, but the imperative to maximize output can create a surplus of kraft lignin which is difficult to use profitably. An important key to the effective utilization of kraft lignins may be discerned in their colligative behavior, which is dominated in a unique way by pronounced noncovalent interactions between the constituent molecular species; the resulting associated complexes appear to be assembled from the individual components *in a specific order*. The ensuing macromolecular domains in the solid state would be expected to contribute substantially to cohesiveness in lignin-based polymeric materials. Previously it had been very difficult to exceed an effective limit of 25-40% (w/w) for incorporating lignin derivatives into polymeric materials, but four or five years ago it proved possible to fabricate 85% kraft lignin containing thermoplastics that possess tensile strengths and Young's moduli which vary monotonically with the degree of intermolecular association. These could be the first authentic lignin-based plastics ever to have been made. More recently, alkylated 100% kraft lignin-based polymeric materials have been produced with tensile properties that are virtually identical to those of polystyrene. This may be the first time that substantial tensile strengths have been achieved with materials composed only of a simple lignin derivative. Efficacious plasticizers for alkylated kraft lignins could thus serve as a vital means for realizing large-scale lignin utilization.

The utilization of lignin derivatives as thermoplastics has challenged the ingenuity of lignin chemists for more than 20 years. Despite the extensive efforts that have been invested in trying to incorporate lignin derivatives into polymeric materials, only limited success has been achieved for the broad variety of formulations examined (*1*). These attempts have been prompted by the potentially enormous quantities of industrial lignin derivatives that could be used in this way if only such uses could be reliably and profitably established. For example, the chemical pulping industry each year produces a huge amount of byproduct lignin from the conventional kraft process. It has been estimated that more than 25 million tons of kraft lignins are generated annually in the United States. However, less than 0.1% of these byproducts are typically employed in applications other than as a relatively low value fuel in the recovery furnaces of pulp mills (*2*). Attempts to maximize pulp production in certain cases can result in a surplus of kraft lignin that can no longer be burned because the capacity of the recovery furnace has been exceeded (*2*). On the other hand, insistent demands to reduce pollution have fostered an atmosphere that is conducive to the development of environmentally friendly pulping methods; perhaps steam explosion and Organosolv pulping could have matured to the point of economic viability in this regard. The success of such new pulping methods relies heavily on the creation and marketing of high value-added lignin-based products (*3*). Moreover, there has been growing interest in producing bioethanol (*4-6*) and chemicals (*7, 8*) from sugars; to the extent that these feedstocks are derived from lignocellulosic plant materials, such processes will generate lignin as a residual byproduct.

Prospects for Lignin-Based Plastics

Among all the conceivable uses to which byproduct lignins could be put, the development of lignin-based plastics with a wide range of predictable mechanical properties seems to be the most compelling. This is one of the few foreseeable ways of elevating lignin derivatives to the level of high volume commodity products. An advantage of lignin-based polymeric materials is that they are, if suitably formulated, biodegradable by white-rot fungi (*9*). However, it is difficult to use lignin derivatives as polymeric materials with reproducible properties when the structures of lignin components themselves have been imperfectly elucidated: polymers with even very small variations in their chemical configurations can exhibit pronounced differences in mechanical behavior.

The inexactness that has typified depictions of lignin structure has arisen from a particular assumption about the configuration of the native biopolymer to which most workers in the field have subscribed. For several decades, it was thought that lignins in wood are three-dimensional network polymers constituted from (*p*-hydroxyphenyl)propane units by random distributions of about ten different linkages (*10*). In the past, most work devoted to analyzing lignin structures has only provided partial characterization in terms of the relative ratios of the three monomer residues, namely the *p*-hydroxyphenyl, guaiacyl and syringyl units, and the frequencies of different interunit linkages determined by degradative methods (*11*). Analytical protocols have long been in place to estimate the contents of different functional groups in lignins (*11*), which are related to their reactivity when they are subjected to chemical modification before, for example, their incorporation into polymeric materials.

The main functionality in lignins is actually the hydroxyl group. Accordingly most of the attempts to use byproduct lignins as polymeric materials have depended on their phenolic nature or simply their hydroxyl functional group frequencies. On this basis byproduct lignins have been employed as polymeric components in phenol-formaldehyde resins, polyurethanes, epoxies and acrylics (*1*). Alternatively they can be used as grafting backbones for the attachment of other synthetic polymer chains by means of free radical polymerization reactions (*12*). On the other hand, a contrasting approach has advocated the degradation of lignins into various monomeric and oligomeric components which would then be polymerized or copolymerized into a variety of different polymeric materials (*13*). However, the degradation reactions are usually very complicated and expensive, resulting in mixtures of many products, and so it has been difficult to justify the feasibility of such approaches.

Thus no serious attempts have been made to use lignins as polymeric materials in their own right. This is due, at least in part, to the fact that the structural characteristics and the physicochemical properties of lignins are not fully understood. Owing to the lack of adequate analytical degradative methods, the primary structures of lignins, namely the sequences of interunit linkages along the macromolecular lignin chains, are still unknown. However, certain aspects of the physicochemical properties of lignins have been found to be inconsistent with what would be expected for a random three-dimensional network polymer (*14*). Among them, the associative behavior of kraft lignin components has been found to be remarkably specific in character and, as such, has been interpreted in terms of vestiges of native macromolecular configurations nonrandomly disposed along the original polymer chains (*15-21*). Recent developments in the field of lignin biosynthesis have affirmed that the dehydrogenative polymerization of lignin precursors and the assembly of lignins in plant cell walls are under highly regulated biochemical control. Indeed, a mechanism for the formation and replication of macromolecular lignin primary structures has been proposed (*22-24*). If these ideas are correct, native lignins should possess nonrandom primary structures and well-defined configurations just like all other biopolymers. This could be critically important as far as the development of lignin-based polymeric materials is concerned.

Thus a particular consequence of native macromolecular structure, namely the evident specificity in the intermolecular associative interactions between kraft lignin components (*15-21*), can be expected to favor the formation of cohesive domains in kraft lignin-based polymeric materials. Indeed, the idea of producing plastics with very high lignin contents was first enunciated in 1990 (*17*) on the basis of these intrinsically powerful intermolecular forces. About four years ago their impact on the mechanical properties of a series of 85% kraft lignin-based thermoplastics was revealed for the first time (*25*). At that point the feasibility of creating plastics with very high lignin contents was established. Thereafter thermoplastics based entirely on simple lignin derivatives, such as alkylated kraft lignins, have been in the process of being developed at the University of Minnesota.

Physicochemical Properties of Kraft Lignins

There is no doubt that the structures of lignin components, the nature of the intermolecular interactions between them, and the resulting associated complexes play very key roles in determining the physical properties, such as chain-segmental mobility and response to deformation, of lignin-based polymeric materials. Therefore, it has become vitally important to elucidate lignin structures explicitly in order to establish meaningful structure-property relationships which could facilitate the systematic modification of lignin-based polymeric materials.

Although the actual configurations and conformations of lignin components in solution and in the solid state have not been clearly established, investigations of the physicochemical properties of lignin derivatives have proven to be both informative and suggestive (*15-21*). Especially relevant is the finding that the associated macromolecular complexes, which are the predominant species in kraft lignin preparations, are not randomly assembled entities at all (*16, 19*). In solution, association between kraft lignin components takes place spontaneously and reversibly. Under aqueous alkaline conditions, dissociation of kraft lignin species occurs in dilute solutions (~0.5 gL^{-1}), while a marked tendency to associate prevails at higher concentrations (~100 gL^{-1}). The modes of these associative/dissociative processes were delineated through plots of $\overline{M}_w$ *vs.* $1/\overline{M}_n$, which are characterized by slopes of $-2\langle m_i m_j \rangle$ where $\langle m_i m_j \rangle$ is the number-average product of the molecular weights for the interacting species at any point during the process. From the observed invariance of $\langle m_i m_j \rangle$, it was deduced that association between kraft lignin species is quite specific: not only do the higher molecular weight complexes interact selectively with lower molecular weight individual components, but each complex seems to possess a locus which is complementary to only one of the components present. Intermolecular nonbonded orbital interactions, dominated by those of the *HOMO-LUMO* type between the aromatic rings, were presumed to govern the underlying mechanism of the associative processes (*16, 19*). These findings clearly imply that the configurations of the kraft lignin components are nonrandom; moreover, the individual molecular species must be able to adopt, at least transiently, more extended conformations that are compatible with macromolecular associated complex formation.

The hydrodynamic behavior of the associated kraft lignin complexes was found to be in accord with a flexible lamellar configuration for these species (*26*), while the discrete kraft lignin components seemed to adopt appreciably expanded random coil conformations in aqueous alkaline solutions (*20*). Through an analysis that has become almost classical in its importance (*26*), attention was drawn to the fact that the thicknesses of monomolecular films of various lignin derivative fractions disposed on a water subphase appear to maintain constant values of about 2 nm despite significant differences in sample molecular weight (*27*). This observation is consistent with the working hypothesis that lignin derivatives consist of disk-like macromolecular fragments that have been cleaved from lamellar parent structures which are about 2 nm thick. Therefore it seems that lignin derivatives are predominantly composed of lamellar macromolecular complexes in which the individual molecular components are strongly held together by multiple noncovalent intermolecular interactions.

Consequently, lignins may represent a class of macromolecules that are quite different from most synthetic polymers. Conventional methods for characterizing polymers cannot disclose the properties of the individual lignin components without the intrusion of the associated complexes. For example, size exclusion chromatography reveals the apparent molecular weight distributions of both associated complexes and individual components (*15-21*). Differential scanning calorimetry (DSC) of kraft lignin fractions (isolated by successive organic solvent extractions from a parent preparation) were profoundly influenced by the effects of intermolecular noncovalent interactions: the variation of T_g with relative log $\overline{M}_n$ exhibited a marked discontinuity (*17, 28*) unlike anything encountered with a conventional homologous polymeric series. The very low values of the Kuhn-Mark-Houwink-Sakurada exponents characterizing the variation of intrinsic viscosity with the molecular weight of kraft lignin (*29*) could arise from these associative phenomena and thus need to be reevaluated.

Indeed, in neutral aqueous solution and in organic solvents where association is favored, the apparent average molecular weights of lignin preparations are often very large (above 10^7). Moreover, the apparent molecular weight distributions of acetylated methylated kraft lignin samples in DMF exhibit well-defined multimodal patterns of species encompassing these very high molecular weights, which indicates that the supramacromolecular kraft lignin complexes are entities with well-defined configurations (*17*). The sizes of the species in electron micrographs, which are representative of acetylated methylated kraft lignin samples recovered from DMF, extend to apparent dimensions comparable to those of 20 million molecular weight polystyrene. In marked contrast, the profiles described by eluting kraft lignin samples from Sephadex G100 with 0.10 *M* NaOH are monomodal in shape and reflect effective molecular weight distributions which approach those of the discrete components very closely: the apparent weight-average molecular weights of kraft lignin preparations in aqueous 0.10 *M* NaOH seldom exceeds 5000 (*16, 17, 19, 21*). It should be borne in mind that the solubilities of lignin samples in aqueous alkaline solution and polar organic solvents such as DMF and DMSO could indicate that lignins are not highly cross-linked. In fact it has been estimated that the cross-linking density of native lignin macromolecules is only 0.052 or 1 in 19 units (*29*).

Lignin-Containing Polymeric Materials

Despite considerable effort, it had in the past been remarkably difficult to improve upon an effective incorporation limit of 25-40% (w/w) for lignin derivatives in polymeric materials without severely compromising their mechanical integrity (*31*). The major problem has been the brittleness caused by high levels of lignin components in such formulations. Blends of lignin derivatives with other synthetic polymers also have been investigated (*32*), but in most cases these resulted in multiphase morphologies and the corresponding lignin contents were typically quite low.

Use of lignin derivatives as polyols in polyurethanes has been the most intensively investigated application (*3, 31, 33-35*). This is not surprising since synthetic polyurethanes can be strong and tough, soft and flexible, or very elastomeric, depending upon the components used and the cross-linking density. The effects of cross-linking density, the molecular weight of the incorporated lignin derivative, and the lignin content on lignin-containing polyurethanes were all investigated through a series of materials based on a kraft lignin-polyether triol-poly(methylene diphenyldiisocyanate) system (*33, 34*). It was found that, at low and intermediate NCO/OH ratios, the tensile strengths and Young's moduli of the polyurethanes derived from a particular kraft lignin fraction isolated for the purpose generally increased with lignin content up to 25-33%. In contrast, at high NCO/OH ratios, the tensile strengths and Young's moduli attained their respective maximum values of ~45 MPa and ~1.2 GPa when the content of the kraft lignin fraction fell between 5 and 10%. When the contents exceeded 30-35%, however, the polyurethanes obtained were always hard and brittle regardless of the NCO/OH ratios employed (*33*).

A series of four fractions were obtained from a parent kraft lignin preparation by solvent extraction for the purpose of investigating the effects of their molecular weights upon the properties of the kraft lignin-polyether triol-polymeric MDI polyurethanes (*34*). At a fixed NCO/OH ratio of 0.9, the tensile strengths and Young's moduli of the polyurethanes generally (with one exception) increased with kraft lignin content up to 30%. Maximum values of ~45 MPa and ~1.5 GPa for tensile strength and Young's modulus, respectively, were obtained with particular fractions when the kraft lignin content was about 30%. On the other hand, the ultimate strains of the polyurethanes prepared from the two higher molecular weight fractions decreased monotonically with kraft lignin content, while those of the polyurethanes produced from the two lower molecular weight fractions exhibited maxima of about 180-250% at kraft lignin contents around 10%, whereafter they decreased rapidly as the lignin content increased. Overall, the tensile behavior of the polyurethanes did not exhibit any uniformly systematic relationship with respect to the molecular weight of the kraft lignin fractions. As before, when the lignin contents exceeded 30%, the polyurethanes obtained were glassy and brittle regardless of the molecular weights of the fractions employed (*34*).

Polyurethanes derived from hydroxypropyl kraft lignin-diisocyanate with and without poly(ethylene glycol) (*31*), and Organosolv (Alcell) lignin-poly(ethylene glycol)-polymeric MDI formulations (*3*) exhibited roughly comparable trends. Interestingly, lignin-derived polyurethanes were capable of exhibiting tensile strengths between 50 and 100 MPa when the lignin contents were 25-40%. However, the ultimate elongations fell to quite low values as the lignin contents approached 40%. Chain-extended hydroxypropyl lignin-diisocyanate polyurethanes showed ~50-100% ultimate elongations with lignin contents in the 20-30% range, but these fell to low values when lignin contents exceeded 30% (*31*).

The hydroxyl functionality also has been employed to incorporate lignins or lignin derivatives into epoxy resins (*37-40*). Hydroxyalkyl lignin derivatives were epoxidized by reaction with epichlorohydrin in aqueous alkali, and the resulting epoxy resins were cured with a suitable diamine (*37, 38*). The network polymers thus produced exhibited behavior leading to ductile failure with tensile strengths and ultimate strains of ~55 MPa and ~8%, respectively, at lignin contents around 40%. In one case, a 57% lignin-containing thermoset material was produced with a tensile strength of 42 MPa and an ultimate strain of 6%. The tensile modulus of these cured epoxy resins increased linearly with lignin content, while the tensile strength increased with lignin content initially, but leveled off beyond 40% (*38*).

Kraft lignins have also been employed as fillers or extenders in a diglycidyl ether of bisphenol A (DGEBA)-polyamine based epoxy adhesive resin (*39*). When the resins had been thermally cured, partial chemical bonding between the kraft lignin and the curing agent was supposed to have taken place. The cured lignin-containing epoxy resins were found to be homogeneous with lignin contents up to 20%, at which point the shear strength also attained its maximum value. However, the latter was drastically reduced in the blend specimens having lignin contents in excess of 35% (*39*), a problem that probably resulted from phase separation. Another successful epoxy resin system was developed by mixing an alkaline solution of kraft lignin with the water-soluble epoxy compound, poly(ethylene glycol) diglycidyl ether, and a curing reagent, triethylene tetramine or hexamethylene diamine (*40*). It was found that gelation took place over a broad range of lignin contents between 20 and 80%, indicating that the functional groups of the kraft lignin reacted with the epoxide. From dynamic mechanical measurements, the peak temperature for the loss modulus, which roughly can be considered to correspond to the glass transition temperature (T_g), increased uniformly from ~-18 to ~85°C as the lignin content varied between 0 and 60%. Thus the glass transition temperature could be modified over a wide range by appropriately selecting the epoxy compound and the curing reagent (*40*).

The First 85% Lignin-Based Thermoplastics

The macromolecular domains resulting in the solid state from the strong noncovalent interactions between the individual molecular components (*15*, *16*, *18-21*) would be expected to contribute substantially to cohesiveness in lignin-based polymeric materials. The successful fabrication of plastics with very high lignin contents should thus be attainable (*17*). The feasibility of doing so was demonstrated for the first time more than four years ago with blends composed of 85% underivatized softwood kraft lignin with poly(vinyl acetate) and two plasticizers (*41*). The tensile behavior of these polymeric materials depended directly upon the degree of intermolecular association between the intrinsic kraft lignin components (Figure 1). While the polymeric material based on the most associated kraft lignin preparation was not very deformable, the ultimate strain extended to over 65% in the case of the plastic containing the most dissociated kraft lignin preparation (*25*). The latter result affirms that true lignin-based materials can exhibit plastic behavior when the individual components are dissociated, *i.e.* separated sufficiently to allow the macromolecular chains to be drafted past one another in response to mechanical stress. In other words, it demonstrated that lignin-based materials are not inherently brittle because of their chemical structures, and they can indeed exhibit substantial flexibility. However, it is the extent to which the individual molecular species interact with one another that determines the mechanical properties of such materials.

In respectively extending to 25 MPa and 1.5 GPa, both the tensile strength (σ_{max}) and Young's modulus (E) of the 85% lignin containing thermoplastics increased linearly with the degree of intermolecular association (Figure 2). Thus such polymeric materials are truly lignin-based in a very fundamental way. Moreover, these new thermoplastics are homogeneous and possess T_g's near room temperature. Melt-flow index measurements have furthermore suggested that the 85% kraft lignin-based polymeric materials can be successfully extrusion-molded (*25*).

Thus it is possible to produce plastics with mechanical properties that are based directly on the molecular structures of lignin derivatives. Now alkylated 100% kraft lignin-based polymeric materials have been created with very encouraging mechanical properties.

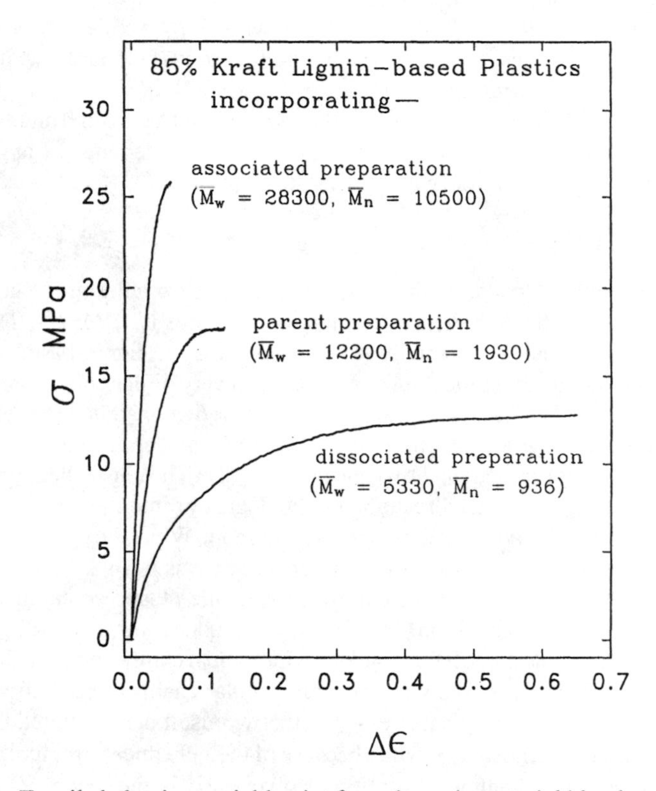

Figure 1. Tensile behavior to yield point for polymeric material blends containing 85% (w/w) kraft lignin preparations that differ only in degree of intermolecular association (stress-strain σ-ϵ curves delineated by Instron Model 4026 Test System employing 0.25 mm min^{-1} crosshead speed with 9 mm specimen gauge lengths). (Data from ref. *25.*)

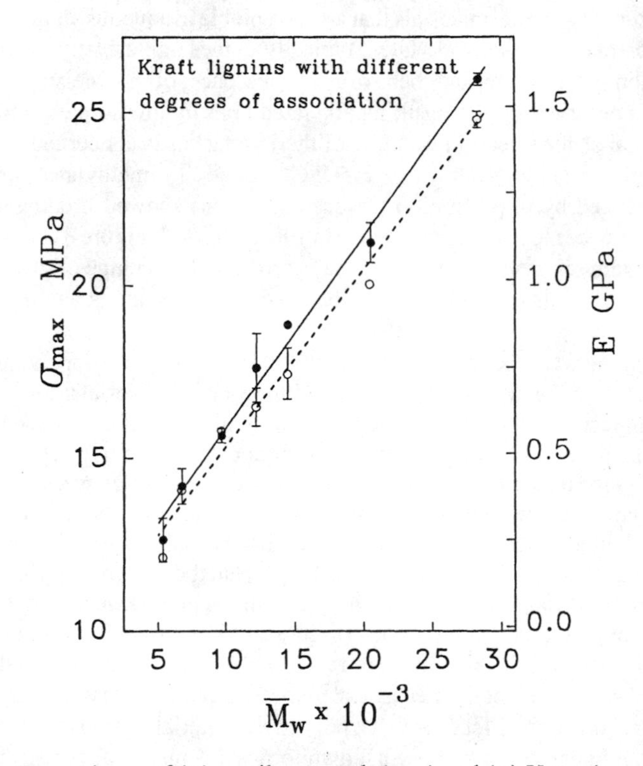

Figure 2. Dependence of (●) tensile strength (σ_{max}) and (○) Young's modulus (E, determined from initial slope of stress-strain curve) upon weight-average molecular weight ($\overline{M}_w$) of kraft lignin preparations incorporated at 85% (w/w) levels in polymeric material blends with same composition (averages from two series of measurements). (Data from ref. *25*.)

Alkylated 100% Kraft Lignin-Based Polymeric Materials

Kraft lignins themselves do not exhibit melting behavior, and even their glass transition temperatures are difficult to detect clearly. Probably the melting points are so high that kraft lignins would undergo pronounced chemical changes before the onset of melting. The high melting point may be of structural origin, in that it could reflect a highly organized supramacromolecular structure characteristic of the materials in the solid state. In an attempt to obtain lignin preparations with more desirable thermal properties (and in the process generate materials that are not soluble in aqueous alkaline solution), kraft lignin preparations were alkylated, typically in the simplest way by methylation and ethylation. Quasi-melting behavior (coalescence of powdery particles into globules) was observed for the resulting alkylated kraft lignin samples, which is very important as far as the injection-molding of thermoplastics is concerned.

Differential scanning calorimetric (DSC) studies of methylated kraft lignin fractions produced by ultrafiltration (before alkylation) showed broad endothermic transitions that were most prominent between 100 and 150°C (Figure 3). This indicates the onset of chain segmental motion and may correspond to a range of glass transition temperatures that could overlap with the observed quasi-melting behavior of the kraft lignin derivatives.

To determine what the effects of the low molecular weight components might be on the strengths of the alkylated 100% kraft lignin-based materials, the parent kraft lignin was subjected to single-step fractionation by ultrafiltration in aqueous 0.10 *M* NaOH through a 10,000 nominal molecular weight cutoff membrane (Figure 4). The parent kraft lignin preparation and higher molecular weight fraction were alkylated with the corresponding dialkyl sulfates in aqueous 60% dioxane solution at pH 11-12 followed by complete methylation with diazomethane in chloroform. It was found that solvent-casting from DMSO of the ethylated methylated derivatives yielded cohesive materials. These alkylated 100% kraft lignin-containing polymeric materials exhibited very encouraging mechanical properties (Figure 5). The Young's moduli (~1.9 GPa) and the tensile strengths (~37 MPa) are remarkably high in view of what had been previously accomplished for the mechanical properties of materials with extremely high lignin contents: this is the first time that a polymeric material composed exclusively of any simple lignin derivative has been shown to possess measurable tensile strength. Therefore these alkylated kraft lignin preparations are promising thermoplastic materials in their own right, even though their tensile energy absorption to fracture and maximum strain are still fairly low. Interestingly, the ethylated methylated higher molecular weight kraft lignin sample exhibited better tensile behavior than the parent preparation, as would be expected since polymers with higher molecular weight generally produce stronger materials.

Earlier results showed that, under these casting conditions, variations in the degree of association between the individual molecular components before derivatization seem to have no systematic effect upon the Young's moduli of the corresponding alkylated kraft lignin-based polymeric materials (*41*). This indicated that the high concentrations in DMSO during solvent casting promote the formation of highly associated species from the alkylated kraft lignin components regardless of the degree of association of the lignin samples before derivatization. Thus, while intermolecular noncovalent

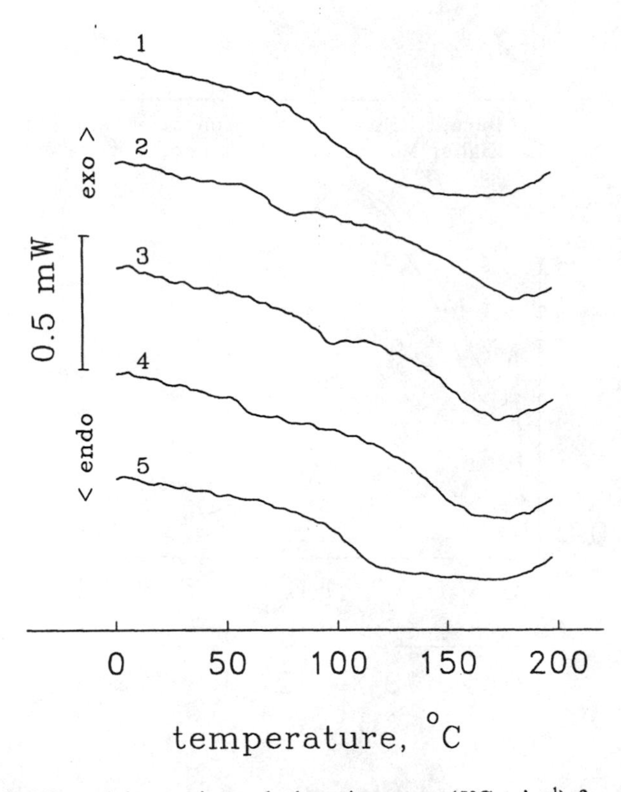

Figure 3. Differential scanning calorimetric curves (5°C min^{-1}) for methylated kraft lignin samples: (1) parent kraft lignin and (2-5) kraft lignin fractions isolated by consecutive ultrafiltration of the parent preparation through a series of membranes with nominal molecular weight cutoffs of (2) 10,000, (3) 5,000, (4) 3,000 and (5) 1,000.

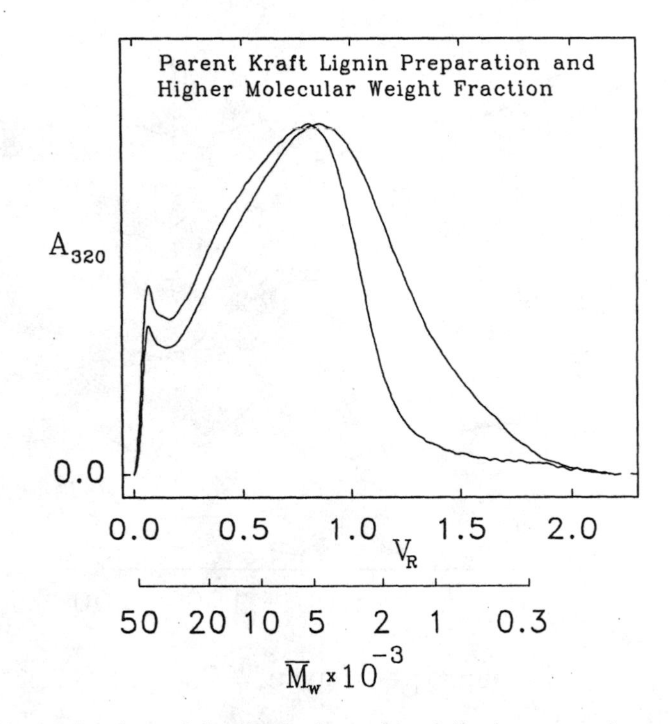

Figure 4. Molecular weight distributions of kraft lignin preparation and higher molecular weight fraction (Sephadex G100/aqueous 0.10 *M* NaOH elution profiles plotted in terms of relative retention volume).

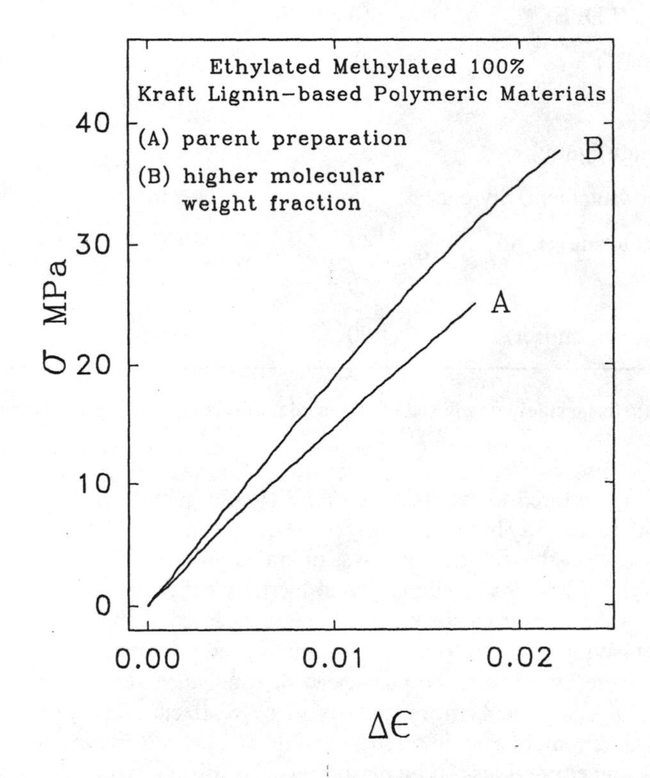

Figure 5. Tensile behavior to failure for polymeric materials composed entirely of ethylated methylated kraft lignin: (A) parent preparation and (B) higher molecular weight fraction (stress-strain σ-ϵ curves delineated by Instron Model 4026 Test System employing 0.05 mm min^{-1} crosshead speed with 9 mm specimen gauge lengths).

Table I. Some Common Polymeric Materials:
Tensile Strength, Young's Modulus and Elongation to Failure (*42*)

Polymeric Material	Tensile Strength MPa	Young's Modulus GPa	Elongation to Failure %
Polyethylene (LDPE)	14	0.22	400
Polystyrene (HI)	28	2.1	2
Polypropylene	35	1.4	400
Alkylated kraft lignin [a]	37	1.9	2
Acrylonitrile-Butadiene-Styrene	38	2.0	4
Poly(vinyl chloride) (rigid)	47	1.6	60
Epoxy cast	59	2.4	5
Polyurethane (thermoset)	90	0.41	1000

[a] Polymeric material solely composed of ethylated methylated kraft lignin fraction.

interactions are required to maintain the essential cohesiveness of the materials, it is presumably the unusually high propensity for extensive intermolecular association that is inherently responsible for the brittleness of kraft lignin-based polymeric materials.

The alkylated 100% kraft lignin-based thermoplastics are virtually identical to polystyrene as far as their tensile properties are concerned (Table I). The problem is that both polystyrene and the 100% kraft lignin-based polymeric materials are quite brittle; they therefore need to be plasticized or toughened if the goal of injection-molding useful components from them is to be realized. The difference is that polystyrene is extremely easy to plasticize while 100% kraft lignin-based polymeric materials had never previously been plasticized. The difficulty in achieving adequate plasticization for the latter presumably resides in the strong noncovalent forces that prevail between the individual kraft lignin components. Therefore suitable plasticizers must be able to overcome, but not totally destroy, the strong intermolecular interactions between kraft lignin components in facilitating their chain segmental mobility while preserving the integrity of the material under externally imposed tensile stresses. In this connection, it should be borne in mind that kraft lignin-based materials certainly do not have to be brittle, as was established by the tensile behavior of the thermoplastics based on highly dissociated kraft lignin preparations (*25*).

Thus major efforts are under way at the University of Minnesota in the search for plasticizers that act effectively with alkylated 100% kraft lignin-based polymeric materials to achieve sufficient degrees of plasticization for producing tough components by injection-molding techniques.

Conclusions

Levels of incorporation for lignin derivatives in polymeric materials have been limited either by brittleness resulting from association between the lignin components, or by phase separation arising from incompatibilities within the incipient blends. However, it has now been shown that thermoplastic materials with very high lignin contents can be created. Simple kraft lignin derivatives have been found to be very promising in this regard. Alkylated 100% kraft lignin-based polymeric materials have been successfully produced with tensile properties that are virtually identical to those of polystyrene. With the development of effective plasticizers (*43*), entirely new types of thermoplastics possessing a broad range of mechanical properties, while embodying very high contents of simple lignin derivatives, can be anticipated in the not too distant future.

Acknowledgments

Acknowledgment for support of this work is made to the United States Environmental Protection Agency through the National Center for Clean Industrial and Treatment Technologies (although it does not necessarily reflect the views of the Agency or the Center, so no official endorsement should be inferred) and to the Minnesota Agricultural Experiment Station.

Paper No. 984436801 of the Scientific Journal Series of the Minnesota Agricultural Experiment Station, funded through Minnesota Agricultural Experiment Station Project No. 43-68, supported by Hatch Funds.

Literature Cited

1. *Lignin—Properties and Materials*; Glasser, W. G., Sarkanen, S., Eds.; American Chemical Society: Washington, D. C.; *ACS Symp. Ser.* **1989**, *397*.
2. Kirkman, A. G.; Gratzl, J. S.; Edwards, L. L. *Tappi J.* **1986**, *69*, 110-114.
3. Vanderlaan, M. N.; Thring, R. W. *Biomass and Bioenergy* **1998**, *14*, 525-531.
4. Gregg, D.; Saddler, J. N. *Appl. Biochem. Biotechnol.* **1996**, *57/58*, 711-727.
5. Cao, N. J.; Krishnan, M. S.; Du, J. X.; Gong, C. S.; Ho, N. W. Y.; Chen, Z. D.; Tsao, G. T. *Biotechnol. Letters* **1996**, *18*, 1013-1018.
6. Vinzant, T. B.; Ehrman, C. I.; Adney, W. S.; Thomas, S. R.; Himmel, M. E. *Appl. Biochem. Biotechnol.* **1997**, *62*, 99-104.
7. Dorsch, R.; Nghiem, N. *Appl. Biochem. Biotechnol.* **1998**, *70-72*, 843-844.
8. Donnelly, M. I.; Millard, C. S.; Clark, D. P.; Chen, M. J.; Rathke, J. W. *Appl. Biochem. Biotechnol.* **1998**, *70-72*, 187-198.
9. Milstein, O.; Gersonde, R.; Hüttermann, A.; Chen, M.-J.; Meister, J. J. *Appl. Environ. Microbiol.* **1992**, *58*, 3225-3232.
10. Sakakibara, A. *Wood Sci. Technol.* **1980**, *14*, 89-100.
11. Adler, E. *Wood Sci. Technol.* **1977**, *11*, 169-218.
12. Meister, J. J. In *Renewable-Resource Materials—New Polymer Sources*; Carraher, C. E., Jr., Sperling, L. H., Eds.; Plenum Publishing Corp.: New York, 1986; pp 305-322.
13. Lindberg, J. J.; Kuusela, T. A.; Levon, K. *ACS Symp. Ser.* **1989**, *397*, 190-204.
14. Goring, D. A. I. *ACS Symp. Ser.* **1989**, *397*, 2-10.

15. Garver, T. M., Jr.; Sarkanen, S. In *Renewable-Resource Materials—New Polymer Sources*; Carraher, C. E., Jr., Sperling, L. H., Eds.; Plenum Publishing Corp.: New York, 1986; pp 287-303.
16. Dutta, S.; Garver, T. M., Jr.; Sarkanen, S. *ACS Symp. Ser.* **1989**, *397*, 155-176.
17. Dutta, S.; Sarkanen, S. *Mat. Res. Soc. Symp. Proc.* **1990**, *197*, 31-39.
18. Himmel, M. E.; Mlynár, J.; Sarkanen, S. In *Handbook of Size Exclusion Chromatography*; Wu, C.-s., Ed.; Chromatographic Science Series 69; Marcel Dekker: New York, 1995; pp 353-379.
19. Sarkanen, S.; Teller, D. C.; Stevens, C. R.; McCarthy, J. L. *Macromolecules* **1984**, *17*, 2588-2597.
20. Sarkanen, S.; Teller, D. C.; Abramowski, E.; McCarthy, J. L. *Macromolecules* **1982**, *15*, 1098-1104.
21. Connors, W. J.; Sarkanen, S.; McCarthy, J. L. *Holzforschung* **1980**, *34*, 80-85.
22. Guan, S.-y.; Mlynár, J.; Sarkanen, S. *Phytochemistry* **1997**, *45*, 911-918.
23. Lewis, N. G.; Davin, L. B.; Sarkanen, S. *ACS Symp. Ser.* **1998**, *697*, 1-27.
24. Sarkanen, S. *ACS Symp. Ser.* **1998**, *697*, 194-208.
25. Li, Y.; Mlynár, J.; Sarkanen, S. *J. Polym. Sci. B: Polym. Phys.* **1997**, *35*, 1899-1910.
26. Goring, D. A. I. *ACS Symp. Ser.* **1977**, *48*, 273-277.
27. Luner, P.; Kempf, U. *Tappi J.* **1970**, *53*, 2069-2076.
28. Yoshida, H.; Mörck, R.; Kringstad, K. P.; Hatakeyama, H. *Holzforschung* **1987**, *41*, 171-176.
29. Dong, D.; Fricke, A. L. *Polymer* **1995**, *36*, 2075-2078.
30. Yan, J. F.; Pla, F.; Kondo, R.; Dolk, M.; McCarthy, J. L. *Macromolecules* **1984**, *17*, 2137-2142.
31. Kelley, S. S.; Glasser, W. G.; Ward, T. C. *ACS Symp. Ser.* **1989**, *397*, 402-413.
32. Ciemniecki, S. L.; Glasser, W. G. *ACS Symp. Ser.* **1989**, *397*, 452-463.
33. Yoshida, H.; Mörck, R.; Kringstad, K. P.; Hatakeyama, H. *J. Appl. Polym. Sci.* **1987**, *34*, 1187-1198.
34. Yoshida, H.; Mörck, R.; Kringstad, K. P.; Hatakeyama, H. *J. Appl. Polym. Sci.* **1990**, *40*, 1819-1832.
35. Rials, T. G.; Glasser, W. G. *Holzforschung* **1984**, *38*, 191-199.
36. Rials, T. G.; Glasser, W. G. *Holzforschung* **1986**, *40*, 353-360.
37. Nieh, W. L.-s.; Glasser, W. G. *ACS Symp. Ser.* **1989**, *397*, 506-514.
38. Hofmann, K.; Glasser, W. G. *J. Wood Chem. Technol.* **1993**, *13*, 73-95.
39. Feldman, D.; Banu, D.; Natansohn, A.; Wang, J. *J. Appl. Polym. Sci.* **1991**, *42*, 1537-1550.
40. Nonaka, Y.; Tomita, B.; Hatano, Y. *Holzforschung* **1997**, *51*, 183-187.
41. Li, Y.; Sarkanen, S. *Proc. 9th Internat. Symp. Wood Pulp. Chem.* **1997**, *63*, 1-6.
42. *Handbook of Plastic Materials and Technology*; Rubin, I. I., Ed.; Wiley: New York, 1990.
43. Li, Y.; Sarkanen, S. In *Proceedings of the Fourth Biomass Conference of the Americas*; Elsevier Science: Oxford, England, 1999; in press.

Chapter 19

Plasma Modification of Lignin

G. Toriz, F. Denes, and R. A. Young

Department of Forestry and Engineering Research Center for Plasma-Aided Manufacturing, University of Wisconsin, Madison, WI 53706

Plasma-chemistry technologies offer an alternative and efficient way for functionalization of lignin. Plasma modification approaches have a number of advantages which include environmental benefits because the technique involves dry chemistry, the active species only penetrate about 10 nm deep so it does not alter the base structure and even the most inert surface can be functionalized. In this paper the plasma-enhanced functionalization of lignin is demonstrated with oxygen, argon, and silicon chloride plasma gases using three different cold-plasma installations: A parallel-plate, diode configuration 30 kHz RF-reactor; a rotating 13.56 MHZ RF-plasma reactor and a dense medium (liquid) plasma (DMP) reactor. Silicon chloride plasma functionalized lignin has been successfully grafted with polydimethlsiloxane in a post-plasma procedure.

Lignin is a renewable raw material and it represents the third largest source of organic matter in the plant kingdom. About 95%, of about 60 million tons of annually produced lignin worldwide by processing biomass into cellulose and pulp, is used as an energy source or disposed as waste material (1). It is estimated that only a couple of percents of available lignin of pulping origin are recovered and marketed in the USA in processes other than fuel technologies.

The present scientific and technological efforts for lignin modification are oriented towards converting it into high-performance lignin-based polymeric materials. The successful syntheses of lignin-based polymers, engineering plastics and lignin-based copolymers with special properties were recently reported from various laboratories. Two main directions have been developed in connection with the utilization of lignin for polymeric materials: a) direct use of unfractionated lignin and lignocellulosic materials for manufacturing of low cost lignin-polymer blends, and b) modification of lower molecular weight degradation products of lignin by graft-copolymerization reactions.

Blending various polymers is a less expensive, alternative way for the creation of desired polymeric characteristics. The properties of blends are controlled by compatibility of the components which often have distinctly different character. Glasser and coworkers (2,3) studied extensively lignin-based polymeric blends. Graft copolymerization reactions have also been evaluated to create modified lignin and lignin-based composites (4,5). Some of these approaches have been successful for creation of high-performance novel materials.

Despite the fact that lignin is composed of a phenolic-type polymeric network very little of this natural polymer is used in phenolic resins. Competition of aminoplast- and polyphenolic-type adhesives and the moderate prices of crude oil (the source of phenol) represent the main technical and economical limitations for the application of lignin based adhesives. The synthesis of epoxy resins from lignin and lignin-derivatives have been studied as well (6-8).

Most of the previous research has not resulted in the development of commercial technologies due to the following reasons:

1. The lignin modification approaches involve reactions that are based on wet chemistry involving large amounts of organic solvents.
2. The lignin derivatives produced by wet chemistry require sophisticated procedures for separating the desired reactants from a complex reaction mixture.
3. These synthetic- and separation-routes are based on high energy consumption processes and are not cost effective.
4. Some of the larger volume reactions can create serious environmental problems.

Plasma-chemistry technologies offer an alternative and efficient way for functionalization of lignin. The plasma state denotes a partly or completely ionized gas (9). This gaseous complex may be composed of electrons, ions of either polarity, gaseous atoms and molecules in the ground or any higher state, and light quanta. There are two plasma categories: thermal and non-thermal. A cold or non-thermal plasma occurs when gas atoms are at ambient temperatures while electrons are maintained at highly elevated temperatures. Only 10^{-1}-$5x10^{-3}$ percent of the species are ionized. Glow discharge experiments causing surface modifications such as deposition or plasma polymerization require this state because the hot electrons have enough energy to bombard and rupture bonds at the surface, while the cooler ions do not excessively heat the substrate. The cold type of plasma is used for all surface modifications of organic polymeric materials. The use of plasmas for the modification of lignin are desirable due to its unique characteristics:

1. The particle-energies of plasma states are high enough (e.g. average electron energy: 0.2-5 eV) to alter all chemical bonds of organic structures.
2. The active species of the discharge interact only with the very top layers of the exposed substrates (around 10 nm deep surface chemistry).
3. Most of the plasma technologies involve dry chemistry.
4. Even the most inert polymeric surfaces (e.g. polyolefins, Teflon) can be efficiently functionalized under proper plasma environments (10-12).

In this paper the plasma-enhanced functionalization of lignin will be demonstrated with O2, Ar, and SiCl4-plasma gases using three different cold-plasma installations: A parallel-plate, diode configuration 30 kHz RF-reactor; a rotating 13.56 MHZ RF-plasma installation and a Dense Medium Plasma (DMP) reactor.

Experimental.

Powdery kraft lignin (Indulin AT) from Westvaco Corporation, Charleston, SC, was dried in oven at 60°C for at least one week prior to use. The following reactants were purchased from Aldrich Chemical Co.: silicon tetrachloride, dichlorodimethylsilane (DDS), polydimethylsiloxane (PDMS) (viscosity 100,000 centistokes), sulfuric acids (99%, A.C.S. regent), HPLC grade ethyl ether, acetone and absolute ethanol.

Both survey and high resolution (HR) ESCA multiplex spectra were taken with a Perkin Elmer PHI 5400 Spectrometer (Mg source; 15 kV and 300 W; angle: 45 degrees). A Mattson Galaxy Series 7000 FT-IR Spectrometer was used for collecting KBr-IR signatures in the 700-4000 cm-1 wavenumber region (resolution 4 cm^{-1}), of the lignin samples. The solid state ^{29}Si NMR spectra were obtained on a Varian Unity 300 Fourier transform NMR spectrometer.

Plasma Reactors.

The *Parallel Plate Reactor* has a stainless steel diode configuration reaction chamber, provided with two disc-shaped electrodes, and it is operated by using a 70 kHz (0-1000 W) RF power supply. This reactor has been described in our earlier work (13).

In a typical experiment a thin layer of vacuum-dried powdery lignin is placed on the lower, grounded electrode, the reactor is closed and the base-pressure is established. Then the preselected reaction pressure and flow rate of the plasma gas or vapor is created and the plasma is ignited and sustained under the desired RF power and treatment time conditions. At the end of the plasma exposure, the base pressure was established in the reactor, then the system is re-pressurized with argon. The lignin powder is then removed from the reactor and stored under vacuum for later analysis. The following experimental conditions were employed for surface functionalization of lignin in the parallel-plate reactor:

Plasma gases, vapors: $SiCl_4$, O_2, and Ar;
RF-power: 100-250 W;
Base pressure: 30-50 mT;
Pressure in the absence of plasma: 200 mT;
Gas flow rate: 5-7 standard cubic cm (oxygen units);
Treatment time: 0.5, 1, 3, 5, 7, and 10 minutes.

The *Rotating Plasma Reactor* can be used to create novel, advanced composites from dissimilar polymeric materials (e.g. renewable, natural polymers and synthetic polymers) through surface functionalization. Cold plasma environments are the best approaches for creating compatible polymeric surfaces. High efficiency can only be achieved if the composite components have high specific surface areas (e.g. powders).

To plasma-process powdery materials an original 13.56 MHZ-RF-plasma reactor was designed and developed (Figure 1). The reactor is composed of a Pyrex glass chamber [11] provided with connecting rubber and stainless steel rings [9, 13] on both ends. The vacuum-tight connection of the reactor to the monomer- and gas-supply system and to the vacuum line is assured with the aid of two ferrofluidic feed-throughs [8, 14]. The hollow shaft of the stainless steel chambers [6, 15] are made of special magnetic material and they are a part of the ferrofluidic sealing system. The RF power is transferred to the reactor through two semi-cylindrical copper electrodes [10] located outside from a 1000 W, 13.56 MHZ RF power supply and matching network assembly [12]. A large cross- section gate valve [17] separates the reactor from the vacuum line and allows the control of the out-flow of the plasma gases. The vapor and gas flow into the reactor is controlled through individual flow controllers [4, 5]. The rotation of the reactor at various angular velocities is assured by a digitally speed-controlled electric engine-transmission system [18, 19].

In a typical experiment vacuum-dried lignin powder is introduced into the reactor, the system is closed and the base pressure is created. The rotation of the reactor is started at the selected speed and the system is kept under these conditions for 30 minutes in order to complete the gas- and moisture-desorption from the extremely large powder surface. In the next step the selected gas-flow and pressure conditions are established, the plasma is ignited and sustained for the desired treatment time. At the end of the reaction the system is vacuumed to base pressure level, re-pressurized with argon , and the sample is removed and stored in vacuum desiccator until analytical procedures and/or second stage reactions are initiated.

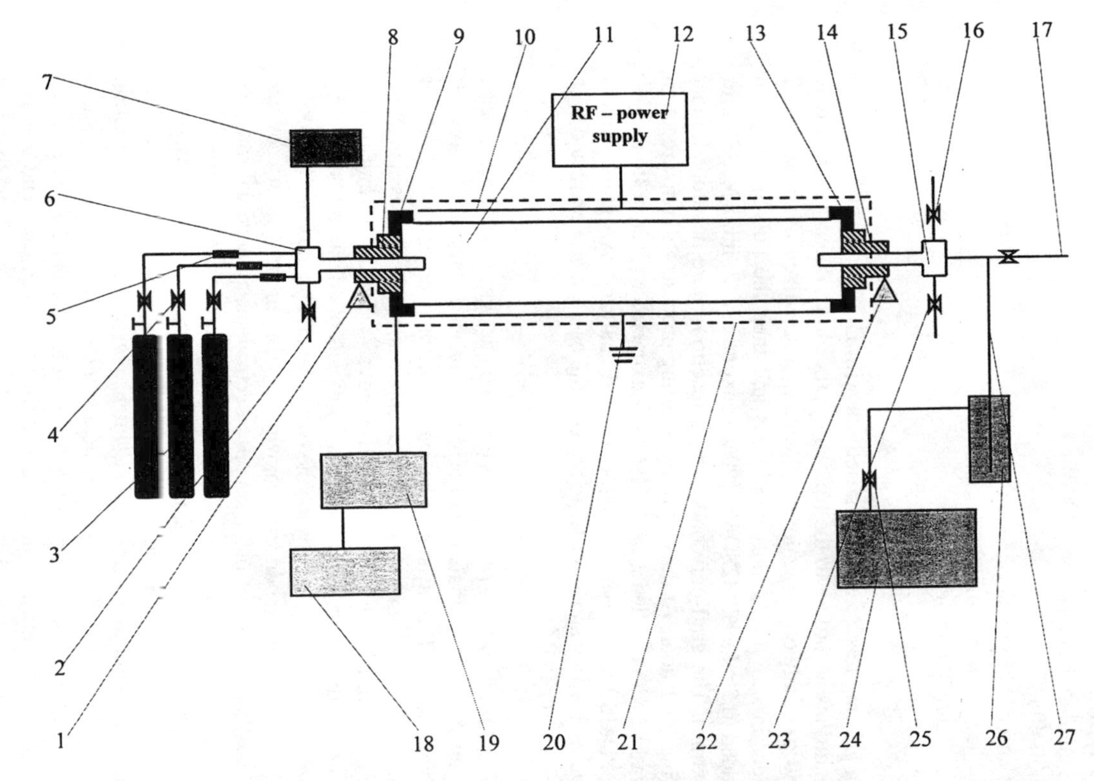

Figure 1. Schematic diagram of rotating RF-plasma reactor: 1, 22. Metallic rigid support; 2. Vent valve; 3. Monomer and gas reservoirs; 4. Needle valves; 5. Gas and vapor flow controllers; 6, 15. Stainless steel gas-mixing chambers; 7. Pressure gauge-MKS Baratron; 8, 14. Ferrofluidic sealings; 9, 13. Stainless steel connecting rings; 10. Semi cylindrical copper electrodes; 11. 1 m long and 0.1 m ID Pyrex glass reaction chamber; 12. 13.56 MHz RF power supply; 16, 23. Additional gas connecting lines; 17; Gate valve; 18. Angular speed digital controller; 19. Electric engine; 20. Ground connection; 21. Faraday cage; 24. Vacuum pump; 25. Large cross section valve; 26. Liquid nitrogen trap; 27. 1 inch diameter stainless steel connection.

The following experimental conditions were used for the plasma treatments in the rotary reactor:

Substrates: lignin (Indulin AT)
Grafting monomer : dimethyldichlorosilane (DDS);
Plasma gases: Ar, $SiCl_4$;
RF power dissipated to the electrodes: 300 W
Base pressure: 30-40 mT;
Pressure in the absence of plasma: 250 mT;
Gas flow rate: 7 sccm (oxygen units);
Treatment time: 5 and 10 minutes;
Angular speed: 30 rpm;

The post plasma grafting reactions were carried out with DDS under both acid and base catalyst conditions according to the following procedure: Equal amounts (e.g. 20/20 or 250/250 g) of silicon tetrachloride-plasma functionalized lignin and DDS were mixed under in-situ or ex-situ conditions and then the polycondensation catalysts were added (0.5 ml 99% H2SO4 for 20/20 mixture and 20 ml, 1 N, KOH for the 250/250 mixture). The graft-polycondensation reactions were developed under continuous stirring for 1 hour. At the end of the grafting processes the raw reaction products were washed with distilled water (pH; 6.5-7), extracted with ether (to remove the non-grafted PDMS homopolymer), then dried and analyzed. All grafted lignin samples were washed with water in order to remove the catalyst and extracted with ether (a good solvent for PDMS).

The *Dense Medium Plasma (DMP) reactor* was especially designed to plasma-process liquids and suspensions of materials at atmospheric pressure. Figures 2 show the schematic diagrams of the DMP system. The DMP reactor is composed of tubular shaped Pyrex glass (length-150 mm; ID-60 mm), double walled (Pyrex glass jacket for thermostating purposes), a reaction vessel, with removable Teflon top and bottom caps. The caps are tightly fastened to the reactor by means of silicon rubber O-ring. The electrical discharge is sustained between the two vertically positioned electrodes. Both electrodes have a hollow-cylindrical form, with disc-shaped smooth surfaces, and a triangular vertical plane cross section. The upper electrode is connected through an elastic coupling [3] to a rotating system, which allows the controlled rotation of the upper electrode between the limits of 0-5,000 rpm. Three symmetrically positioned (120 degree) holes located in the middle range of the upper electrode and the axial inner channel assure the centripetal-force-driven intense recirculation of the reaction media. The lower electrode has a fixed position during the discharge, however, it can be vertically translated by a metric thread system for selecting the desired distance between the electrodes. The inner channel permits the feeding of inert and reactive gases before and/or during the plasma processes, both for degassing and gas-mediated reaction purposes. In and out thermostat connections allow the re-circulation of cooling agents (e.g. alcohol, water, etc.). Teflon-tubing connections [5] (ID-5 mm), mediate

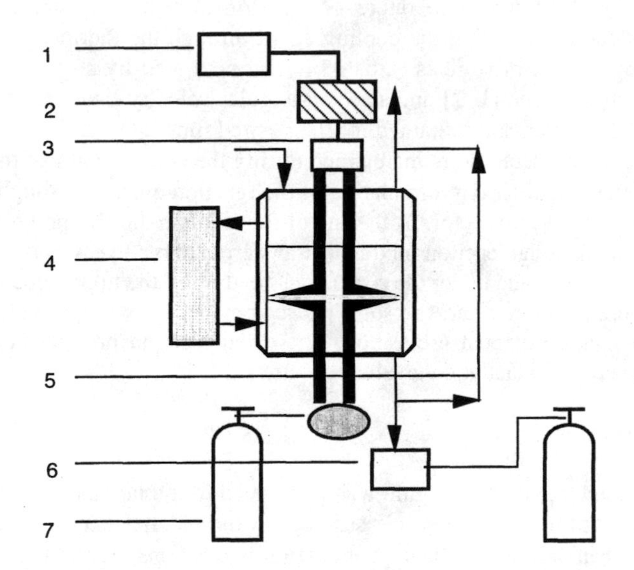

1 Motor speed controller; **2** Electric motor; **3** Elastic coupling; **4** Thermostat; **5** Teflon tubing outer loop for recirculation of reaction media; **6** Pyrex glass vessel for reaction product; **7** Gas cylinder.

the centripetal-force driven re-circulation of the reaction media from the bottom to the upper part of the reactor. Stainless steel tubing (ID-5 mm), assures the removal of the final reaction products and the evacuation and trapping of plasma generated gaseous components.

A typical reaction is performed in the following steps: 3.0g dried lignin was added under an argon-purged-glove-box environment, and under vigorous stirring, to 250 ml silicon tetrachloride to form the lignin/SiCl4 suspension. After selecting the required distance between the electrodes, the suspension was transferred through a stainless steel connection into the reactor. It is noteworthy that the level of the liquid suspension should cover the re-circulating holes of the shaft in order to assure vigorous stirring of the system. During these operations argon is passed continuously through the reaction medium, and the temperature of reaction system is maintained at the pre-selected value by recycling the cooling agent through the thermostat [4]. The stirring of the reaction mixture is initiated in the next step by starting the speed controlled electric engine [1, 2] and selected angular velocity established. The DC discharge is then ignited and sustained for the desired time interval. The high speed rotation of the upper electrode is maintained during the entire length of the plasma reaction to sustain an intense re-circulation of the reaction mixture through both the inner and outer loop of the reactor. At the end of the reaction the DC-power supply is disconnected and the final reaction mixture is transferred into a Pyrex glass vessel [6] under an argon blanket, and under the continuous rotation of the upper electrode. The reaction mixture is filtered and the solid phase (powder) is washed with absolute ethanol in an argon-protected glove box. The dried reaction product is stored in vacuum dessicator until analytical evaluations are started.

Plasma Reactions with Lignin.

$SiCl_4$ (ST)-plasma treatment of lignin was performed to implant extremely reactive SiClx groups onto the lignin particle surfaces in the parallel plate reactor. These functionalities can initiate graft-polycondensation reactions, for instance, in the presence specific bifunctional compounds (e.g. dimethyldichlorosilane, diacids, diamines, etc.) and can lead, in the presence of moisture, to very hydrophilic Si(OH)x groups.

Comparative survey and high resolution (HR) Electron Spectroscopy for Chemical Analysis (ESCA) data collected from unmodified and ST-plasma treated lignin indicated that Si is present in the plasma exposed samples. Besides the characteristic C1s binding energy peaks of lignin (C-C, C=C 284.7 eV; C-O 286.3 eV; CO=O 288.6 eV), C-Si (284.2 eV) and a high binding energy area O-CO-O (291 eV) peaks can also be identified in the HR ESCA spectrum of plasma exposed lignin (Figure 3). The relative surface atomic composition of untreated and ST-plasma processed samples at various treatment times, show that the higher the plasma exposure period the higher the Si and oxygen relative surface concentrations, and that long treatment times and

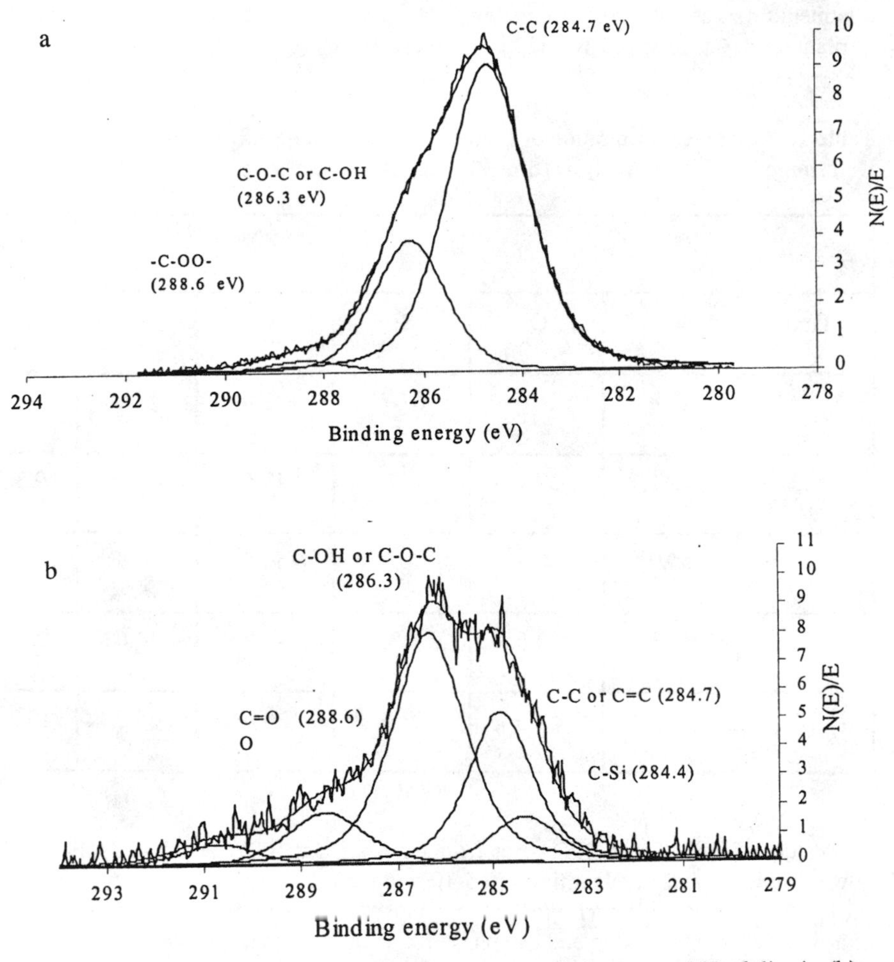

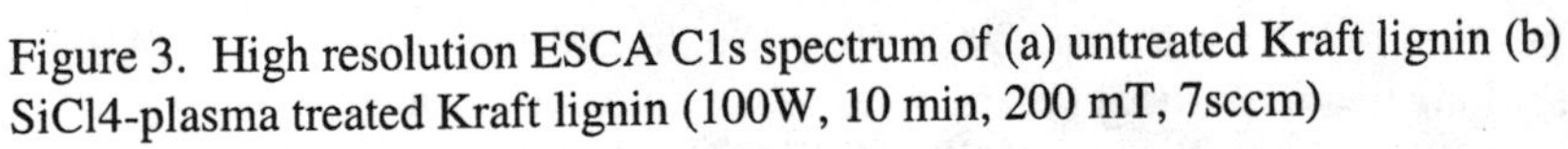

Figure 3. High resolution ESCA C1s spectrum of (a) untreated Kraft lignin (b) $SiCl_4$-plasma treated Kraft lignin (100W, 10 min, 200 mT, 7sccm)

higher RF-power conditions result in lower Si atom concentrations (Table 1). This phenomenon might be explained by the presence of a more intense fragmentation of ST at higher RF powers resulting in significantly increased amounts of very high volatility $SiCl_x$ species ($x<3$), which would preferentially leave the system; and by the development of a more intense plasma-induced crosslinking mechanism of lignin causing the generation of a relatively inert lignin surface. It should also be noted that treatment times as low as 0.5 minutes are high enough to result in significant Si atom implantation (6.9 % at 100 W and, respectively, 9.2 % at 200 W).

Table 1. Atomic Concentration of Lignin Samples After $SiCl_4$-Plasma Treatment Determined by ESCA Analysis (time in minutes, atomic concentration as %).

	100W			200W		
Time	C	O	Si	C	O	Si
0	72.8	26.6	0.07	72.8	26.6	0.07
0.5	52.1	37.7	6.9	48.5	44.2	9.3
2	55.9	38.4	5.7	50.9	40.8	8.2
5	42.8	43.6	8.8	41.9	44.6	10.9
10	24.5	57.3	17.9	38.0	45.6	10.2

The values of the O/C and Si/C ratios from treatments at 100W and 200W RF powers and various treatment times (0.5-10 minutes) shown in Figures 4-6 substantiate the possible occurrence at long plasma-exposure times and high power of a significant crosslinking and gas-phase fragmentation mechanism (Figure 4). The O/C ratios from 100 and 200W RF power treatment are comparable except at very long treatment times (10 minutes) where the value of the ratio at 100W is twice as high in comparison to the 200 W C/O value. A less significant incorporation of oxygen atoms at higher RF powers can only be explained by the presence of a intense crosslinking mechanism. Certainly in this case a free radical mediated, ex-situ , post plasma oxidation process is less probable.

The implantation of Si atoms onto the lignin substrate surfaces follows a similar pattern. Significant differences between 100 and 200 W plasma-exposures were only observed at long treatment times (Figure 5). The low Si/C values at 200W are probably associated with the presence of a more inert surface structure and/or by the existence of intense plasma-induced ST fragmentation reactions.

These findings are significant because they indicate that both plasma induced surface and gas-phase chemistries play a important role during the surface functionalization processes, and that different competitive processes (crosslinking, fragmentation, etc.) will control the surface functionalization mechanisms.

Oxygen and argon-plasma functionalization of lignin also indicate intense surface oxidation of lignin particles. The relative surface atomic concentrations evaluated from ESCA measurements show in both cases (100W RF power) an increased oxygen atom incorporation with increased treatment times (Tables 2 and 3). However, while at seven minutes O2-plasma exposure the oxygen atomic concentration doubles (53.3%) relative to the untreated sample (24.7%), argon-plasma modifications generate a moderate oxygen up-take (28.5%). Higher RF powers (e.g. 250 W) result in lower relative oxygen atomic concentrations, regardless of the treatment times, similarly to the ST-plasma exposures. Surface crosslinking mechanisms could also be responsible for this behavior.

Table 2. Atomic Concentration of Lignin Samples After O_2-Plasma Treatment determined by ESCA Analysis (time in minutes, atomic concentration as %).

	100W			250W		
Time	C	O	S	C	O	S
0	75.0	24.7	0.24	75.0	24.7	0.24
0.5	53.1	43.8	3.2	--	--	--
1	52.9	44.1	2.9	58.2	39.4	2.42
3	49.0	46.3	4.7	--	--	--
5	46.5	47.6	5.9	--	--	--
7	39.8	53.3	6.9	50.2	44.6	5.2

The variation of reactive surface sulfur concentration in the lignin was significantly different after oxygen versus argon plasma treatments. In both cases low RF power and longer treatment conditions result in a substantially increased sulfur concentration. It should also be noticed that oxygen-plasma environments generate much higher sulfur

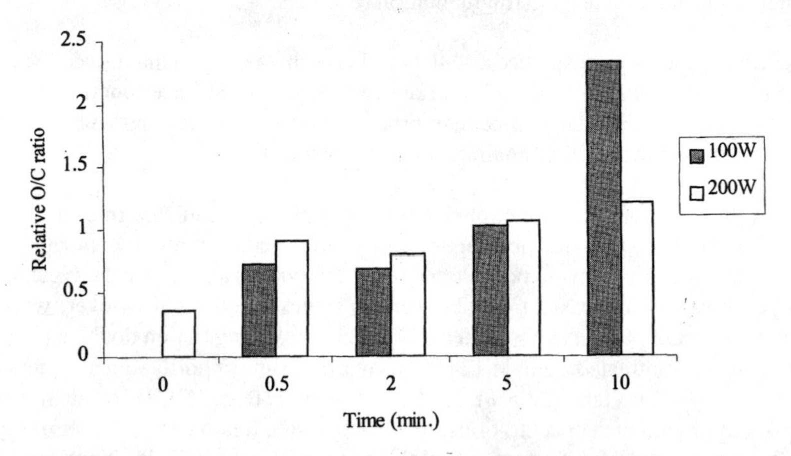

Figure 4. Treatment time dependent relative O/C concentration ratios of $SiCl_4$-plasma treated kraft lignin at 100W and 200W.

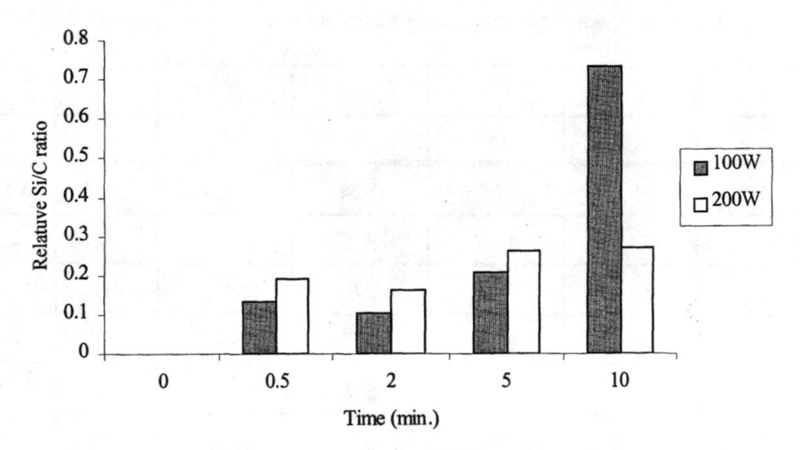

Figure 5. Treatment time dependent relative Si/C concentration ratios of $SiCl_4$-plasma treated kraft lignin at 100W and 200W.

concentrations (e.g. 6.9%; 7 minutes) in comparison to argon plasmas (1.2%; 7 minutes). Etching mechanisms are probably responsible for these phenomena. Oxygen plasmas induce more intense etching in comparison to the argon plasmas resulting in higher relative quantities of process-origin, non-volatile,SO_x species at the lignin surface.

It should also be noted that higher powers diminish the relative surface atomic concentrations under oxygen-plasma conditions, or reduce it completely, in the case of argon, as a result of crosslinking and sputtering processes. The relative ratios of O/C and S/C as a function of treatment time and RF power dissipated to the electrodes show the effect of the plasma gas and RF power-dependent surface chemistry (Figures 6 and 7). While oxygen-plasma environments lead to intense surface oxidation and etching processes, argon-discharges induce mainly surface crosslinking mechanisms.

Table 3. Atomic Concentration of Lignin Samples (%) after Ar-Plasma Treatment Determined by ESCA Analysis (time in minutes).

	100W			250W		
Time	C	O	S	C	O	S
0	75.0	24.7	0.24	75.0	24.7	0.24
0.5	73.8	25.8	0.43	--	--	--
1	71.85	27.4	0.72	78.9	21.1	0.00
3	75.0	24.6	0.37	--	--	--
5	67.4	31.1	1.46	--	--	--
7	70.2	28.6	1.20	77.9	22.1	0.00

Surface Functionalization of Lignin in the Rotating-RF-plasma Reactor.

Survey ESCA data indicated that lignin treated in a SiCl4 plasma in the rotary reactor resulted in comparable relative surface silicon atom concentrations (7-7.8 %; 300 W) to that from the parallel plate installation (10.2 %; 200 W). The small difference might be due to the "electrodeless" nature of the rotating system. High resolution ESCA data revealed the different C1s functionalities from the surface of the modified lignin particles and the presence of a fairly intense Si-C binding energy peak at 284.4 eV indicated that SiClx groups had been attached to the lignin particle surfaces.

Post-plasma grafting reactions were carried out with DDS under both acid and base catalyst conditions as described in the experimental section. The silicon content of the final reaction products synthesized in the presence of H2SO4 and KOH indicated that

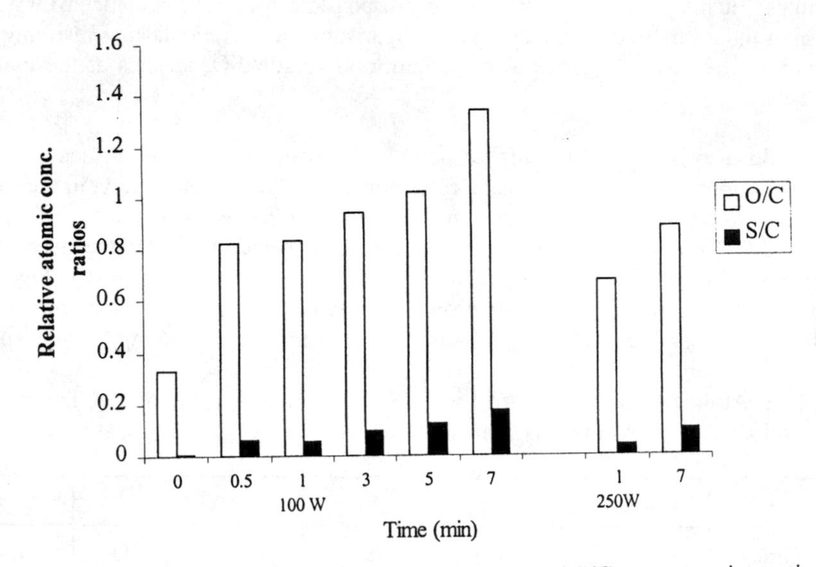

Figure 6. Treatment time dependent relative O/C and S/C concentration ratios of O_2-plasma treated Kraft lignin (100W and 250 W).

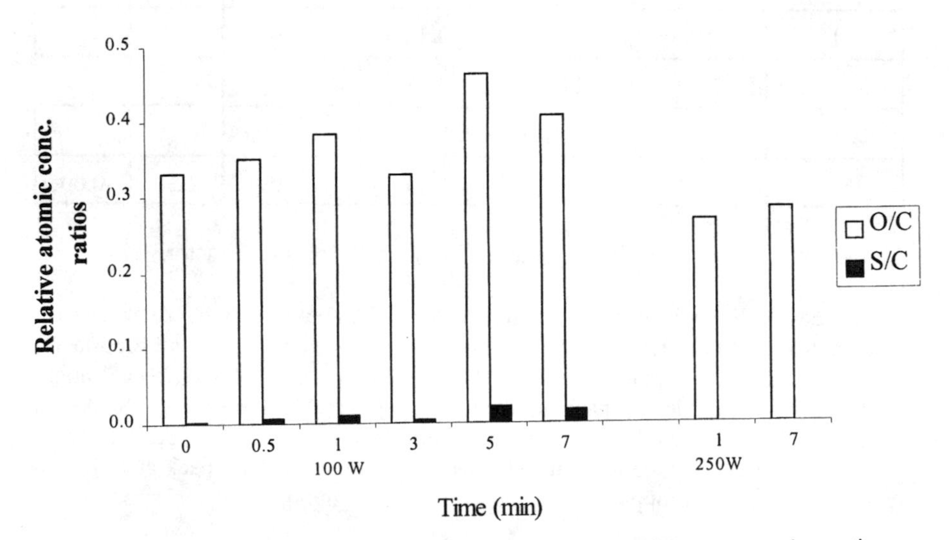

Figure 7. Treatment time dependent relative O/C and S/C concentration ratios of Ar-plasma treated Kraft lignin (100W and 250 W).

the graft-polycondensation reactions proceeded efficiently from plasma functionalized lignin surfaces (Table 4). The HR ESCA diagram of ST-plasma functionalized and polydimethylsiloxane (PDMS)-grafted lignin exhibited a sole, symmetric peak corresponding to Si-C bond at 284.2 eV.

Table 4.Atomic Concentration of Lignin Samples (untreated, $SiCl_4$-plasma treated in the rotary reactor and grafted either with H_2SO_4 or KOH as catalyst in %, time in minutes).

Lignin Sample	Element concentration (%)			
	C	O	Si	Cl
Untreated	72.0	27.4	--	--
$SiCl_4$-plasma treated L	55.5	30.5	7.8	6.1
Grafted with H_2SO_4 catalyst	52.2	24.6	23.2	--
Grafted with KOH catalyst	51.4	28.2	20.4	--
PDMS	32.4	21.6	37.8	

Comparative ATR-FTIR analysis of lignin, ST-plasma functionalized lignin, plasma-functionalized and PDMS-grafted lignin, and PDMS (Figures 8-11, respectively) indicate that substantial surface modification of the lignin was induced during the plasma exposure, and that the presence of a polydimethylsiloxane (PDMS) layer on the lignin particle surfaces can be identified. As a result of ST-plasma exposure the band at 1414 cm^{-1} was significantly reduced possibly due to opening of lignin aromatic rings and a much higher OH (3100-3600 cm-1)/CH (2820-2990 cm^{-1}) ratio characterized the spectrum (Figure 9). These modification can be related to the interaction of SiClx-plasma generated species with C-OH groups of the lignin and by the formation of Si-OH functionalities as a result of hydrolysis of SiClx groups under open laboratory conditions. Differential ATR-FTIR spectrum of ST-plasma functionalized and DDS-grafted lignin (Figure 10) shows a strong similarity to the ATR-FTIR spectrum of DDS (Figure 11). In both spectra the basic PDMS IR absorption can be identified: -Si(CH3)2- stretching: 800-850 cm^{-1}, and bending: 1250-1260 cm^{-1}; Si-O-Si-O, and Si-O-C stretching: 1020-1090 cm^{-1}; C-H bending: 1400-1420 cm^{-1}. These findings substantiate the earlier suggestion that the grafting reactions have successfully been developed from ST-plasma functionalized lignin surfaces.

Argon plasma exposure of lignin in the rotating reactor (250 mT; 100 W; 5 minutes; 5 sccm; 30 rpm) resulted in similar relative surface atomic concentrations (C: 71.2%; O: 28.8%) in comparison to the oxygen-mediated parallel plate reactions (C: 67.4%; O: 31.1 %; S: 1.46%). This indicates that electrode and electrode-less RF plasma

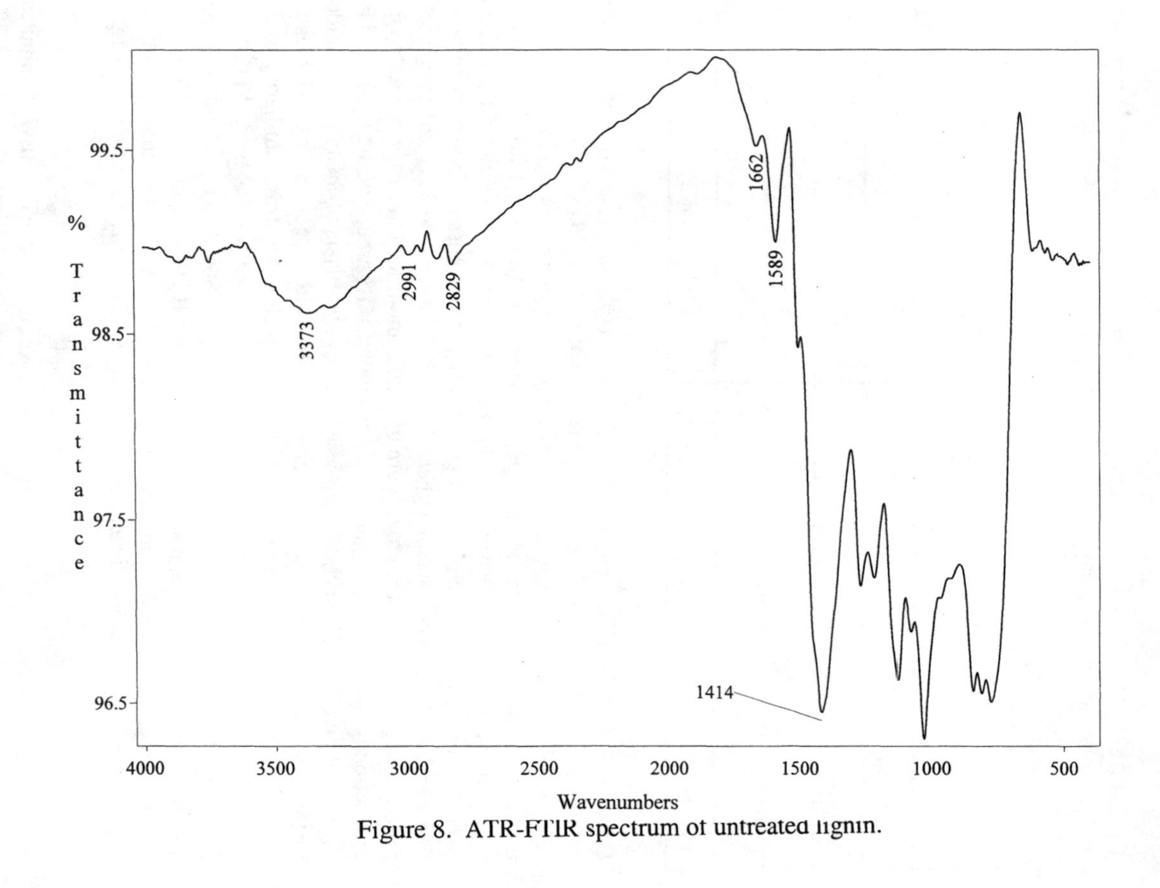

Figure 8. ATR-FTIR spectrum of untreated lignin.

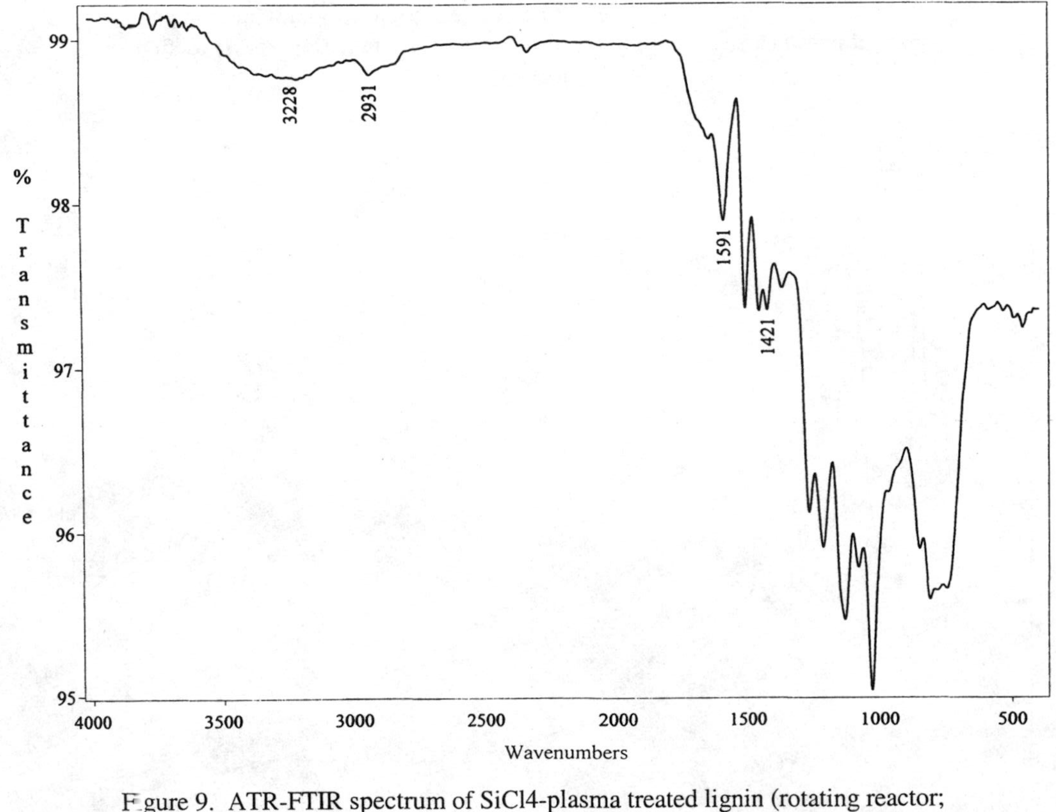

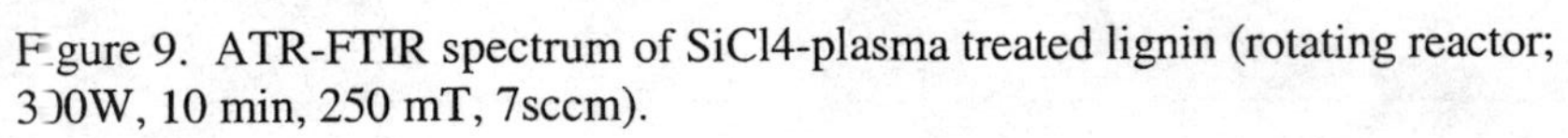
Figure 9. ATR-FTIR spectrum of $SiCl_4$-plasma treated lignin (rotating reactor; 300W, 10 min, 250 mT, 7sccm).

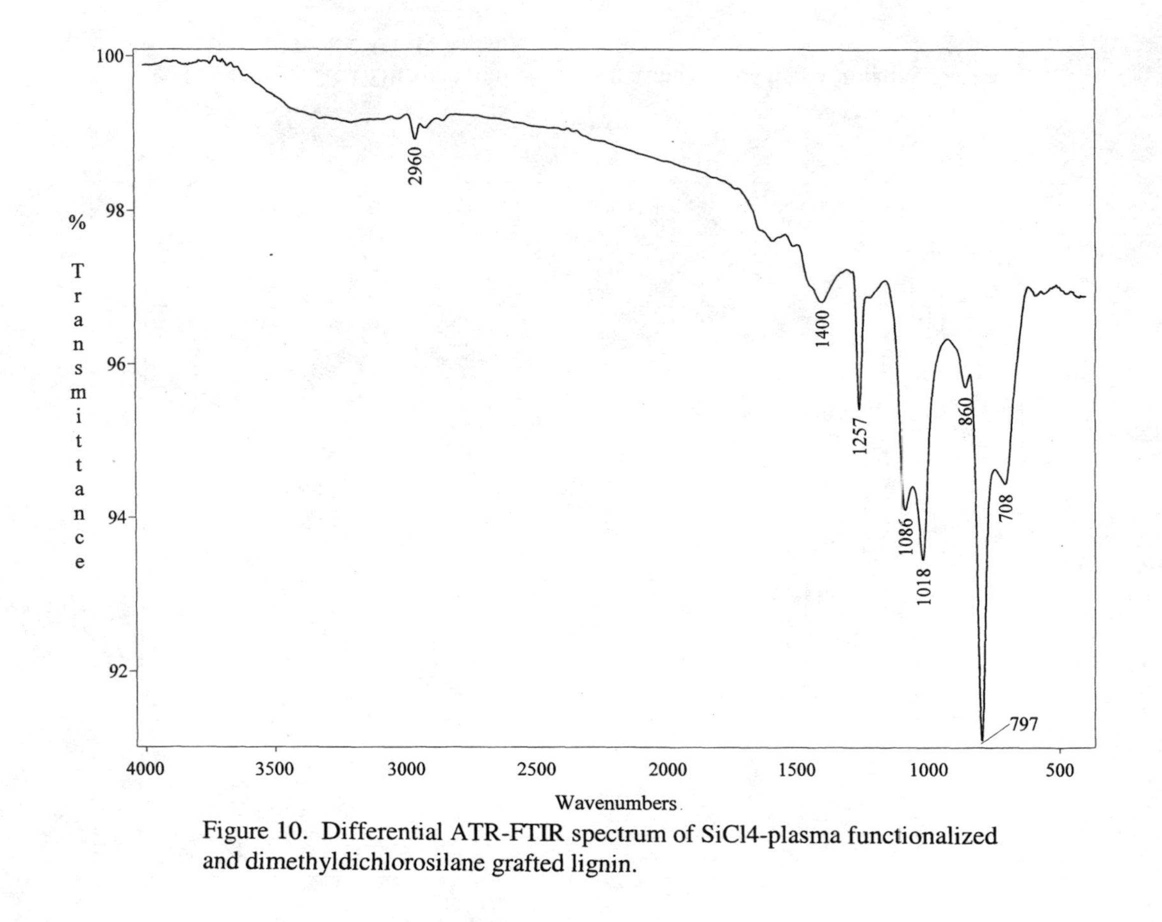

Figure 10. Differential ATR-FTIR spectrum of SiCl4-plasma functionalized and dimethyldichlorosilane grafted lignin.

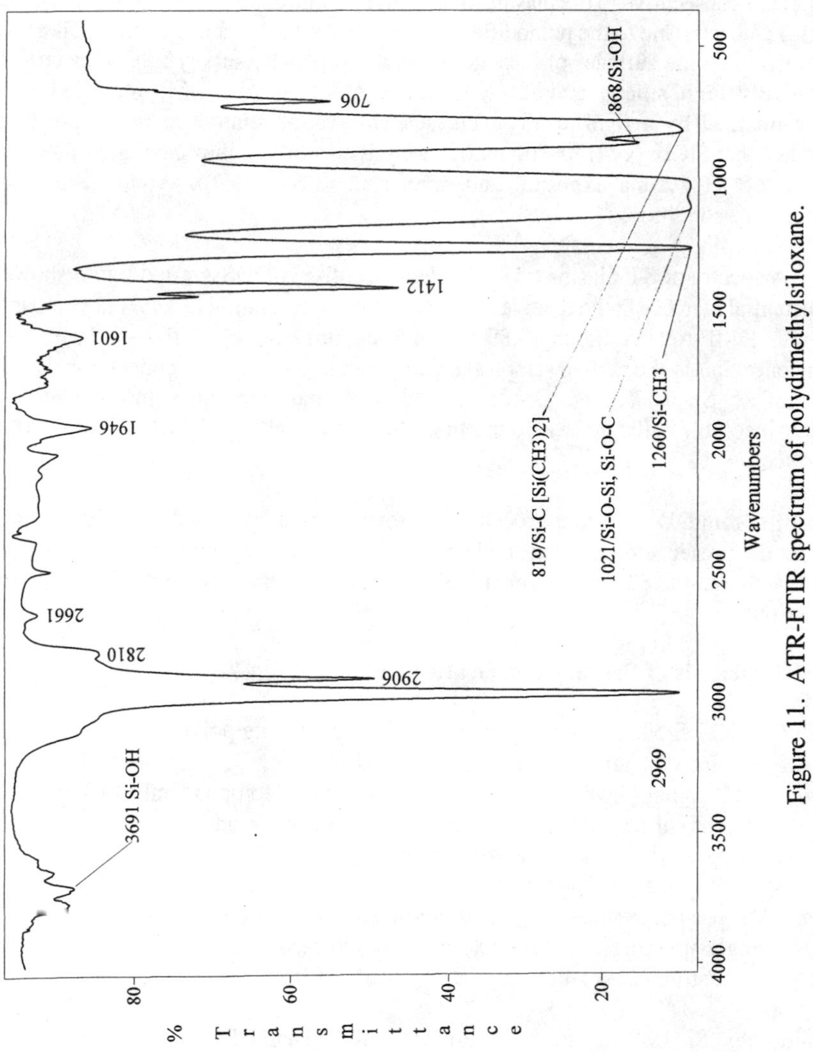

Figure 11. ATR-FTIR spectrum of polydimethylsiloxane.

reactors, operating even in significantly different frequency ranges (30 KHz and 13.56 MHz) induce similar surface chemistries.

Surface Functionalization of Lignin under DMP Conditions.

Survey and high resolution ESCA spectra of the ST-plasma treated lignin were collected consecutive to the plasma treatment. In addition to carbon (C1s) and oxygen (O1s) characteristic to the unmodified lignin, silicon (Si2p) and chlorine (Cl2p) were identified on the surfaces of plasma exposed samples by survey ESCA spectra. The relatively high silicon and low chlorine contents of the DMP-modified lignin, accompanied by a significantly increased O/C ratio, relative to the virgin lignin, suggest that SiClx ($x<4$) functionalities were incorporated into the lignin matrix as a result of ST-plasma exposure and converted into Si(OH)x groups under open laboratory conditions.

ESCA spectra of ST-plasma treated and consecutively DDS-grafted lignin showed a substantial increase in the relative silicon atomic concentration (18.7%) in comparison to the DMP-treated lignin (9.8%); and the comparable relative surface atomic concentration of the grafted samples in comparison to that of the conventional PDMS (C: 50%; Si: 25%, and O: 25%) indicated that additional (non-volatile and ether-insoluble) silicon based structures have been attached to the lignin-particle surfaces.

Results from DDS-grafting-procedure performed on unmodified kraft lignin did not show the presence of any detectable amounts of Si on the lignin particle which also indicated that the ST-plasma functionalized lignin mediates the initiation of grafting reactions.

NMR Analysis of Polysiloxane-Grafted Lignin.

Solid state 29Si NMR has been widely used for analyzing polysiloxane structures. However, since the relative intensity of the signal from ^{29}Si is significantly weaker than the 1H signal (the natural abundance of the 29Si isotope is only 4.67 %), the application of solid state ^{29}Si NMR analysis for surface-modified polymers is significantly limited (low signal-to-noise ratio).

The DMP-plasma treated and grafted lignin was examined by cross-polarization and magic-angle-spinning (CP-MAS) solid state ^{29}Si NMR. This solid state ^{29}Si NMR technique distinguishes silicon atoms involved in different chemical bonds.

Solid state ^{29}Si NMR spectra of DMP-plasma-treated and DDS-grafted powdery lignin were recorded at a spinning speed of 2175 Hz. Two resonance peaks were identified at -68 ppm and -23 ppm values. The NMR spectrum collected from an identical sample at a lower spinning speed (1544 Hz) resulted in resonance peaks at identical

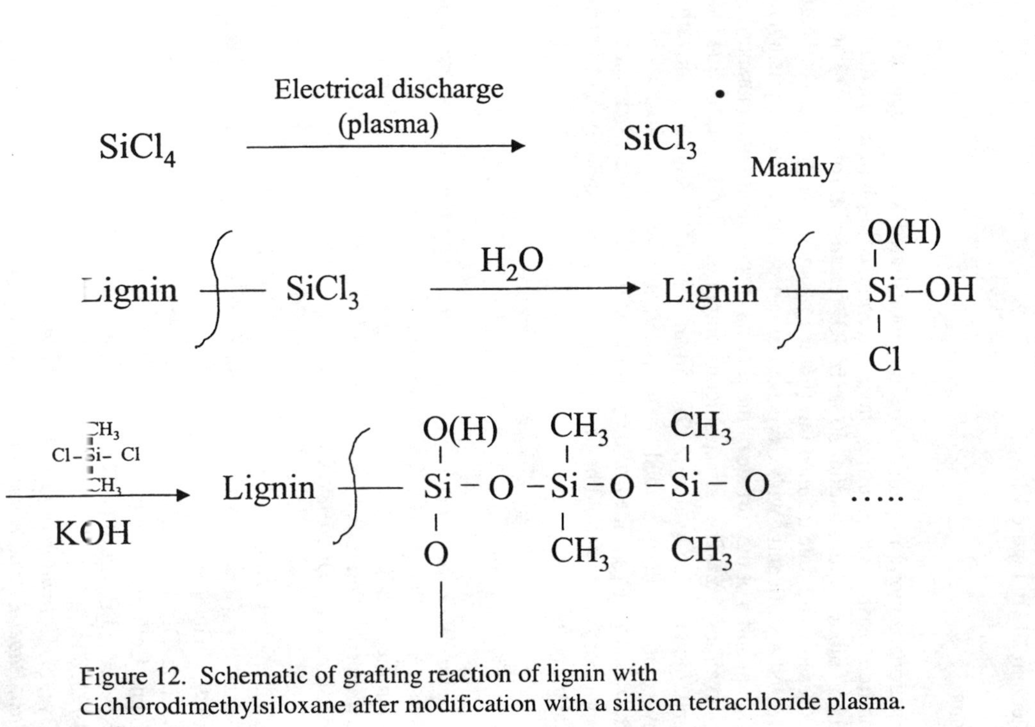

Figure 12. Schematic of grafting reaction of lignin with dichlorodimethylsiloxane after modification with a silicon tetrachloride plasma.

chemical shifts (-23 and -68 ppm). Based on solid state NMR PDMS data, the origin of these resonance peaks were related to the presence of two distinctive silicon-based structures, namely , -O-SiO-O- and -O-Si(CH3)2-O-. A silicon signal could not be detected from lignin samples which were not plasma treated and grafted. These data indicate that the grafting reactions were successful resulting in a possible structure shown schematically in Figure 12.

Conclusions

Plasma functionalization of lignin has been demonstrated with oxygen, argon, and silicon chloride plasma gases using three different cold-plasma installations: A parallel-plate, diode configuration 30 kHz RF-reactor; a rotating 13.56 MHz RF-plasma reactor and a dense medium (liquid) plasma (DMP) reactor. The O/C ratios increased significantly with oxygen plasma treatment, and surprisingly, the concentration of sulfur at the surface of the plasma treated lignin was enhanced from about 0.2% to almost 7%. The surface silicon content of silicon tetrachloride plasma treated lignin increased up to almost 19%. In air the surface silicon is converted to silicon hydroxide with provided a convenient site to graft lignin with polydimethlsiloxane. This is the first demonstration of plasma modification of lignin in a rotating reactor which could be utilized to modify large quantities of lignin for commercial purposes.

References

1. Glasser, W.G.; Sarkanen, S., Eds., *Lignin: Properties and Materials*, ACS Symposium Series, No. 397, Washington DC., 1989.
2. Ciemniecki, S.L.; Glasser, W.G. In Lignin Properties and Materials, ACS Symposium Series 397, Eds. Glasser, W.G.; Sarkanen, S., ACS, Washington, DC, 1989, Chapter 35.
3. Rials T.G.;Glasser, W.G. ,Phase Morphology of Hydroxypropylcellulose Blendswith Lignin. In *Lignin Properties and Materials*, ACS Symposium Series 397, Glasser, W.G.; Sarkanen, S., Eds., ACS, Washington, DC, 1989, Chapter 36.
4. Valeria, D.; Bozena, K., Lignin Utilization in Polyolefin Blends, In Ligno-Cellulosics-Science, Technology, Development and Use, Eds. Kennedy, J.F.,. Phillips, G.O; Williams, P.A. , Ellis Horwood, New York, 1992, 827.
5. Bye, F., *Utilization of Lignins and Lignin Derivatives, Bibl. Ser. No. 292, Inst. Paper Chem., Appleton (Inst. Paper Science and Technology, Atlanta), USA, 1984-1985.*
6. Wu, C.-F.L.; Glasser, W.G.,*J. Appl. Polym. Sci.*, **1984**, *29*, 1111.
7. Lee, D.-S.; Perlin, A.S., *Carbohyd. Res.*, **1982**,*106, 1.*
8. Tai, S.; Nagata, M.; Nakano, J.; Migita, N., *J. Jap. Wood Res. Soc.*, **1967**, *13*, 102.

9. Chapman, B. *Glow Discharge Processes: Sputtering and Plasma Etching*. John Wiley & Sons, Inc., NY, 1980.
10. Denes, F.,*TRIP,* **1997**, *5 (1)*, 23.
11. d'Agostino, R., *Plasma Deposition, Treatment and Etching of Polymers*, Academic Press, New York, 1990.
12. Mittal, K.L., Ed., *Polymer Surface Modification: Relevance to Adhesion, Part 1, Plasma Surface Modification Techniques, VSP-International* Science Publishers, Amsterdam, 1995.
13. Denes, F.; Hua, Z.Q.; Barrios, E.; Evans, J.; Young,R.A. Journal of Macromolecular Science-Pure and Applied Chemistry, **1995**, *A32 (8&9),* 1405.

PULPING AND BLEACHING

Chapter 20

Chemistry of Pulping: Lignin Reactions

Josef S. Gratzl and Chen-Loung Chen

Department of Wood and Paper Science, North Carolina State University, Raleigh, NC 27695–8005

The chemical reactions of lignins in both acidic and alkaline pulping processes are reviewed. The acidic pulping processes include the sulfite process in the pH range of 2-4, and the bisulfite process in the pH range of 5-6. The alkaline pulping processes include soda, kraft, polysulfide, alkaline sulfite, neutral sulfite, and AQ/AHQ processes. The reactions of lignins with the active chemical species in these pulping processes are discussed in terms of lignin solubilization, degradation, condensation, the redox cycle of AQ/AHQ processes, and stabilization of carbohydrates. Furthermore, the formation of volatile compounds is discussed.

Among the industrial chemo-technical processes, the manufacture of pulp and paper is rather unique. The industries producing basic and specialty chemicals and polymers, which are based on non-renewable resources, such as coal, oil and natural gas, have been developed exclusively employing chemical and physico-chemical principles. The pulp and paper industry, on the other hand, relying on renewable raw materials, evolved empirically, mainly by trial and error employing state of the art mechanical and process engineering.

The objective of chemical pulping is the selective removal of lignin without extensive degradation of the polysaccharides, cellulose and hemicelluloses, as well as the removal of extractives and colored structural moieties (chromophores) in residual lignin or other wood components, originally present or generated during pulping. The pulping processes are conducted either in acidic or alkaline aqueous solutions. In the solvent pulping processes under consideration, organic solvents (alcohols) are used exclusively, or are added to the pulping liquor to enhance the penetration of the chips by the pulping chemicals, as well as the removal of lignin and lignin degradation products.

The conditions required for the degradation of lignins and the removal of their degradation products are drastic. The high temperatures and rather high charges of acids or alkali lead not only to the desired fragmentation but also to undesired condensations of lignins and their fragments that are produced during the pulping. In addition, substantial degradation and removal of polysaccharides occur, resulting in losses in yield and fiber strength. However, the degradation and removal of lignin and retention of polysaccharides can be enhanced by process modification and use of pulping additives.

Lignins

Lignins are three-dimensional polymers consisting of phenylpropane or C_9 units. There are three different types occurring in a wide range of proportions in the woody tissues of different plant species. These three C_9 units are *p*-hydroxyphenylpropane, guaiacylpropane and syringylpropane units (Figure 1). The proportion of these three C_9 units in lignin occurring in the woody tissues of plants depends on the nature of plant species. In general, gymnosperm (softwood) lignins consist mostly of guaiacylpropane units. Angiosperm (hardwood) lignins are mainly composed of guaiacyl- and syringylpropane units with varying ratios depending on the wood species. By contrast, grass (greamineae) lignins are comprised of p-hydroxyphenyl-, guaiacyl- and syringylpropane units and consist of lignin cores and peripheral units. The lignin cores are essentially lignins of the guaiacyl-syringyl type, while the peripheral units consist of p-hydroxy-cinnamic acid and ferulic acid groups that are bonded by ester linkages mostly to hydroxyl groups at C-γ in the C_9-units of the lignin cores.

The phenylpropane units are joined together by ether (C-O-C) and carbon-carbon (-C-C-) linkages. The type of linkages connecting the phenylpropane units (substructures) and their estimated frequencies are given in Figure 2 (*1*). Except for the diarylether type linkages, interunit ether linkages readily undergo both acid- and base-induced hydrolysis under the given conditions. Side-chains may be cleaved depending on the type of substructures, particularly under alkaline conditions.

As a result of the structural complexity of lignins and of the drastic reaction conditions applied, an understanding of the principal chemical reactions has been established by experiments with model compounds featuring lignin substructures, and employing state-of-the-art analytical methodologies. In the following discussion, we would like to concentrate on the very basic, underlying chemical reactions rather than on topochemical phenomena involved in the removal of lignin.

Acid Processes

The degradation and removal of lignin is accomplished with aqueous solutions of sulfur dioxide (SO_2) in presence of bases, such as magnesium, sodium, potassium and ammonium hydroxide. Replacement of the originally used calcium oxide by the aforementioned bases allowed not only their recovery, but also provided flexibility in the selection of raw materials and pulping conditions for the production of a wide

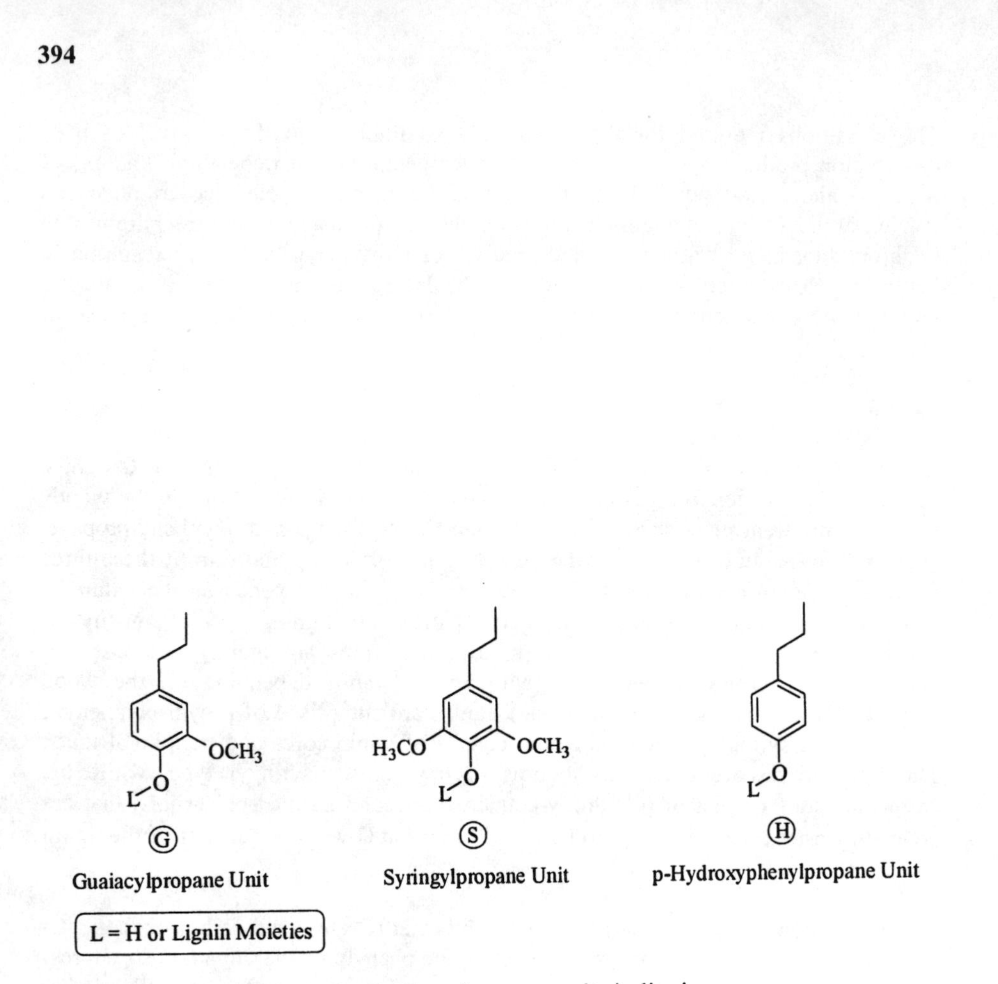

Figure 1. Phenylpropane units in lignins.

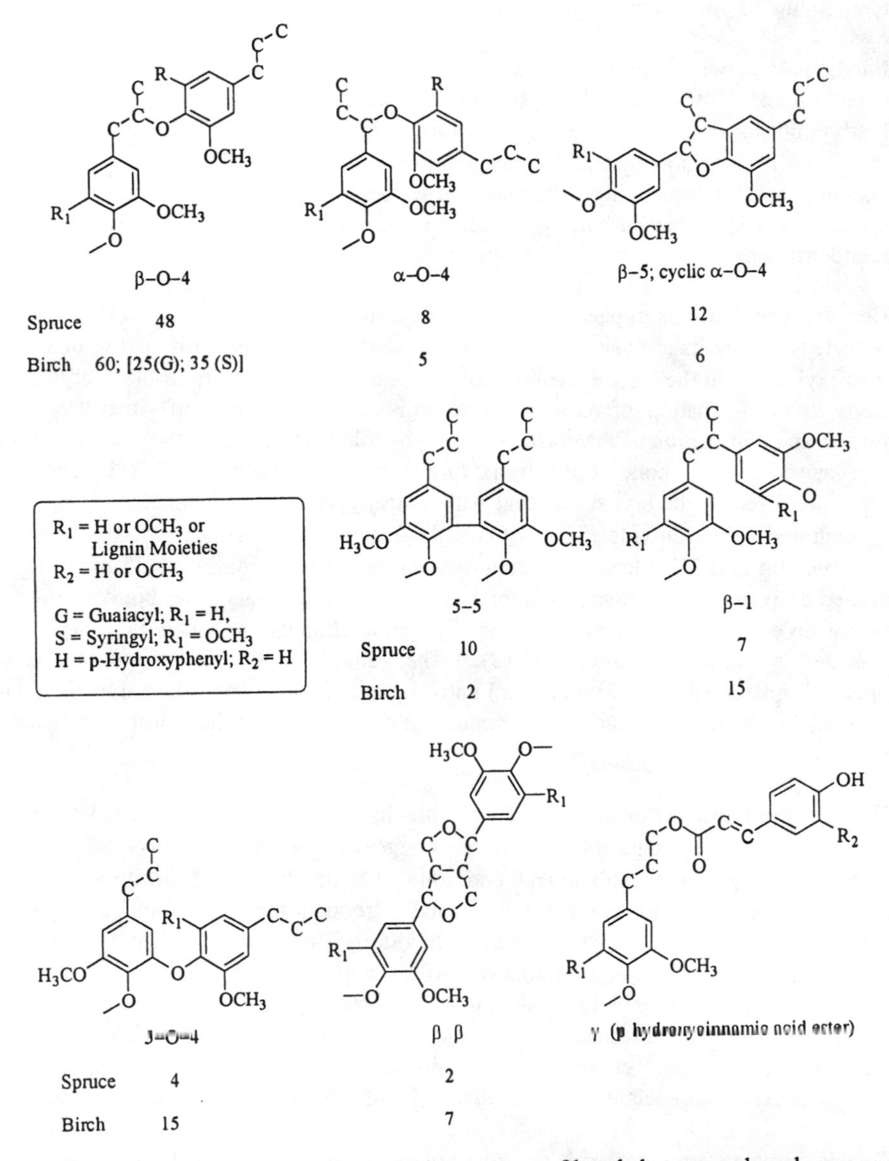

Figure 2. Estimated frequencies of different types of bonds between phenylpropane (C_9) units of milled wood lignins (MWLs) per 100-C_9-units. Adapted with permission from reference 1. Copyright 1977, *Wood Science and Technology*.

variety of pulp types. Figure 3 shows the equilibria of the components in aqueous solutions of sulfur dioxide. The relative concentrations are functions of pH and temperature. Pulping temperatures are in the range of 130-170°C.

Interactions between lignin substructures and the acid (hydrated proton; H_3O^+) with concomitant dehydrations lead to the formation of electron-deficient centers (carbonium ions). These centers react with electron-rich centers of chemical species (nucleophiles), such as hydrated sulfur dioxide ($SO_2 \bullet H_2O$; H_2SO_3) and corresponding chemical species, producing sulfonates. Alternatively, the centers react with certain electron-rich aromatic carbons generating carbon-carbon (-C-C-) interunit linkages (condensations).

Benzyl alcohol groups in phenolic and nonphenolic subunits, and in the corresponding α-arylether subunits are very reactive towards acid sulfonation. Protonation of the α-hydroxyl group in the benzyl alcohol moieties and the oxygen atom in the α-arylethers leads to the formation of oxonium intermediates. Elimination of water from the former and elimination of the corresponding phenols from the latter, with concomitant cleavage of the ether bond, results in the formation of corresponding benzyl carbonium ions, stabilized by the aryl substituent with hydroxyl and alkoxyl functions. Addition of hydrated sulfur dioxide ($SO_2 \bullet H_2O$; H_2SO_3) or related chemical species, such as bisulfite ion (HSO_3^-), leads to the formation of corresponding α-sulfonic acid or related derivatives, as shown in Figure 4. Cleavage of the α-arylether bonds is mainly responsible for fragmentation of lignins. The more abundant nonphenolic β-aryl ether linkages are relatively unreactive (*2*). The same overall mechanism applies to phenylcoumaran (β-5) and pinoresinol substructures (β-β), where the furan rings are cleaved followed by sulfonation or condensation of the benzylium ion intermediate (*3*).

Figure 5 shows the sulfonation of coniferyl aldehyde type subunits. Protonation at the carbonyl of the α,β-unsaturated aldehyde endgroup leads to formation of a benzyl carbonium intermediate with an enol endgroup. The addition of a bisulfite ion on C-α of the carbonium intermediate results in the introduction of a sulfonic acid group. Furthermore, addition of hydrated sulfur dioxide (H_2SO_3) to the aldehyde endgroup may occur, resulting in the formation of corresponding γ-hydroxysulfonic acid. Thus, sulfonic acid groups can be introduced in both α- and γ-positions of side-chains in the coniferyl aldehyde substructures. Although the 4-O-alklyated coniferyl aldehyde substructure is a minor component in lignin, the introduction of α- and γ-sulfonate groups increases appreciably the solubility of lignin fragments in the pulping process.

Guaiacylpropane moieties with an α-carbonyl group are sulfonated at the terminal group via the corresponding γ-carbonium ion intermediates (Figure 6) (*3*). The intermediates are produced via keto-enol tautomerism of the α-carbonyl group to the corresponding 4-O-alkylconiferyl alcohol intermediates, protonation of the terminal hydroxymethyl group and subsequent dehydration. Thus, 4-O-alkylconiferyl alcohol substructures are produced as intermediates in pulping. As shown in Figure 7, a similar mechanism is operating in the reactions of substructures with conjugated

$$SO_2 + H_2O \rightleftharpoons H_2SO_3\ (SO_2{\bullet}H_2O) \rightleftharpoons H^+ + HSO_3^- \rightleftharpoons 2\,H^+ + SO_3^{2-}$$

pKa ~2 ~7

Strongly Acidic	Acid Sulfite	Bisulfite
pH < 2	pH 2~4	pH 5-6
H_3O^+	H_3O^+	H_3O^+
SO_2	$SO_2{\bullet}H_2O$	H_2SO_3
$SO_2{\bullet}H_2O$	H_2SO_3	HSO_3^-
H_2SO_3	HSO_3^-	

pH-Adjustment with hydroxides (oxides) of Na^+, K^+, NH_4^{2+} Mg^+ and Ca^{2+}

Figure 3. Reactive chemical species in acid pulping processes.

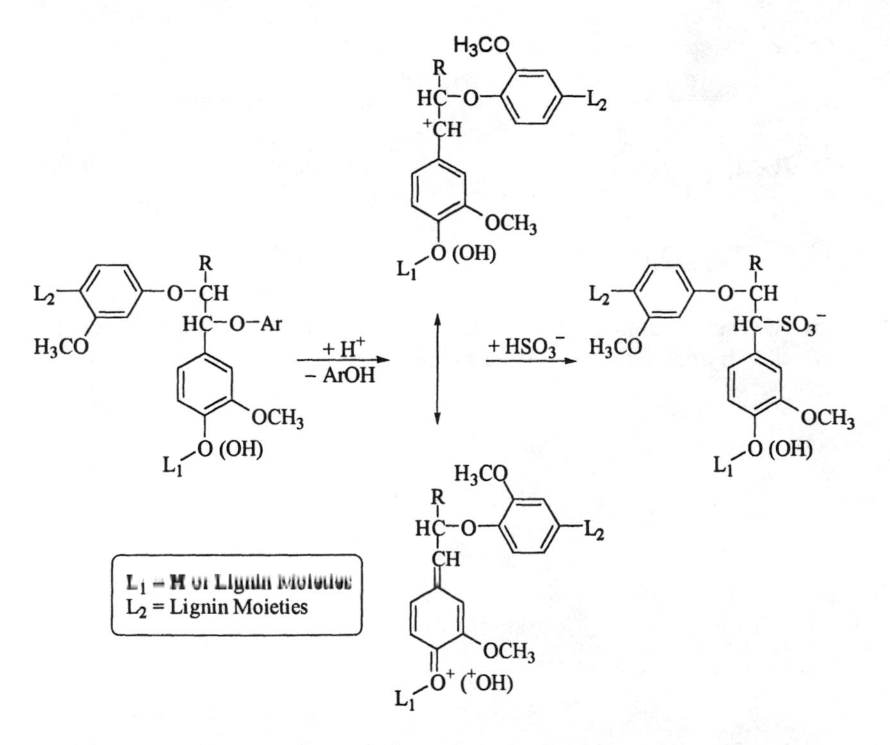

Figure 4. Cleavage of α-arylether substructures followed by sulfonation.

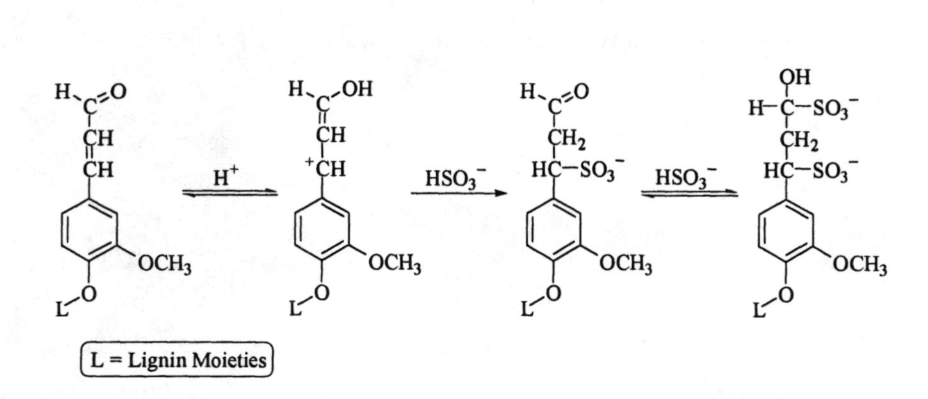

Figure 5. Acid sulfonation of coniferyl aldehyde end groups. Adapted with permission from reference 2. Copyright 1971, Arbor Publishing.

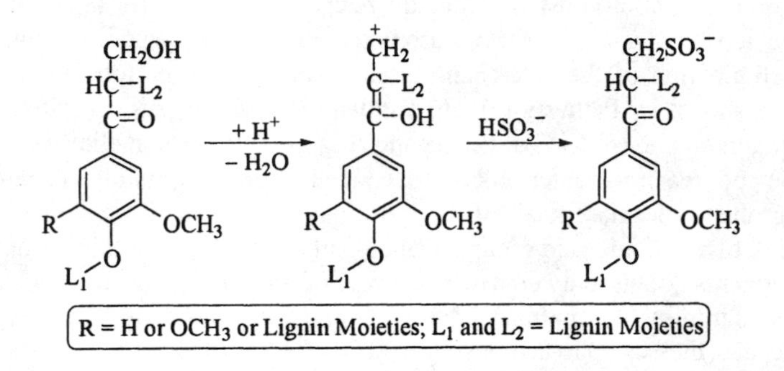

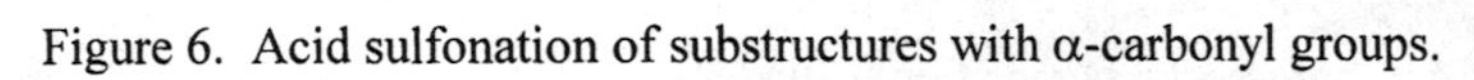

Figure 6. Acid sulfonation of substructures with α-carbonyl groups.

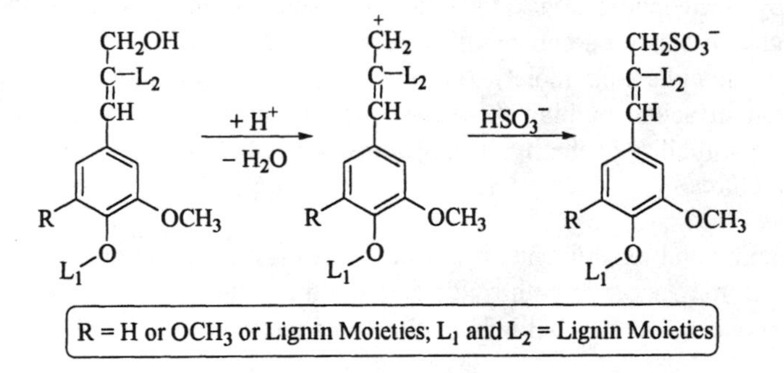

Figure 7. Acid sulfonation of substructures withconjugated double bonds.

double bonds to aromatic moieties, such as 4-O-alkylated coniferyl alcohol substructures.

Competing with sulfonation of carbonium ion intermediates are reactions with aromatic structures (weak nucleophiles) to form carbon-carbon (-C-C-) interunit linkages. The frequency of such condensation reactions increases with the acidity of the pulping liquor, and decreases with the concentration of bisulfite ions (hydrated sulfur dioxide species). Figure 8 shows two examples of condensation reactions resulting in the formation of diphenylmethane substructures involving C-1 C-6 and C-5 (weakly nucleophilic sites) of aromatic nuclei. Electrophilic addition of benzyl carbonium ion on C-1 of the second aromatic ring with a benzylcarbinol side-chain leads to elimination of the side-chain as aldehyde or of a proton via reverse Prins reaction, as shown by Pathway (a). In Pathway (b), electrophilic addition of benzyl carbonium ion on C-6 (or C-5) of the second ring involves proton elimination only (*4*). This type of reaction, intermolecular condensation, is mainly responsible for increasing the molecular mass of lignosulfonates. It should be mentioned that introduction of a sulfonic acid group enhances substantially the solubility of lignin and lignin fragments. This outweighs the impact of increasing the molecular mass on removal of lignin by acid pulping liquors. Lignosulfonates have been determined to have molecular masses up to and beyond 100,000 Daltons (*5*).

Pinoresinol substructures in lignin undergo "internal" condensation forming a cyclohexene ring. This results from an acid-catalyzed cleavage of the two furan rings followed by nucleophilic attack by a sulfonic acid group on C-α in the phenolic moiety, and acid-catalyzed condensation between the C-α of the nonphenolic moiety and C-6 of the phenolic moiety (Figure 9) (*3*). In spite of the acid-catalyzed condensation in acidic pulping processes, the introduction of sulfonic acid groups increases the solubility of the lignin fragments derived from mono-4-O-alkylated β-β type substructures.

In model compound experiments, it has been demonstrated that aryl migration along the side-chain may occur in both acidic and alkaline conditions. In acidic media aryl migration between C-α and C-β in both directions was observed (*6*).

Spent sulfite liquors contain substantial amounts of low molecular mass sulfonated lignin degradation products including monomeric and dimeric compounds. Considerable efforts have been made in isolating and characterizing these compounds (*7,8*). A number of monomeric and dimeric model compounds of the guaiacyl- and veratryl-type sulfonates featuring several side-chain configurations of phenolic and nonphenolic lignin substructures, were converted to the corresponding acetylated sulfonic acid methylester, which allowed characterization by application of ^{1}H NMR-spectrosccopy (*9,10*). Employing this method of derivatisation to lignosulfonates obtained from pulping of western hemlock wood and its milled wood lignin (MWL) preparations, allowed a quantitative assessment of hydroxyl and sulfonate groups in the various fractions (*11*). A substantial portion of the low molecular mass

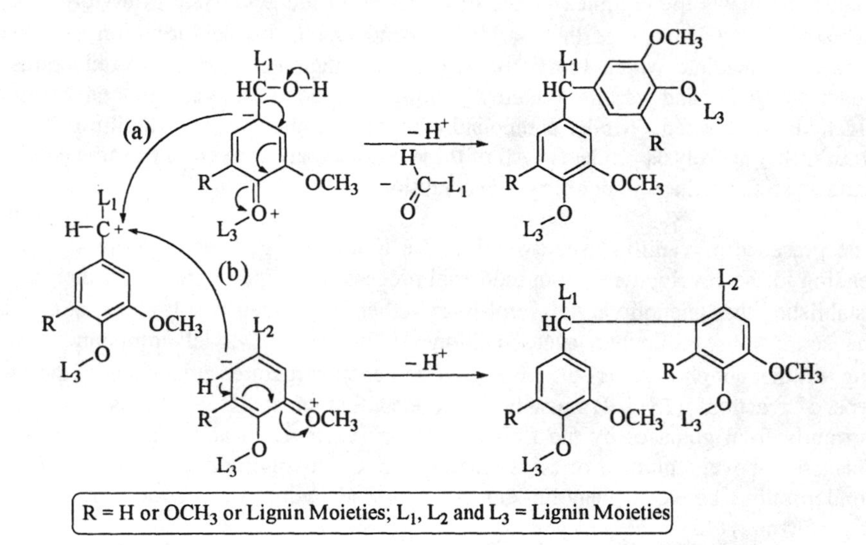

Figure 8. Condensation reactions; Formation of dipehenylmethane substructures.

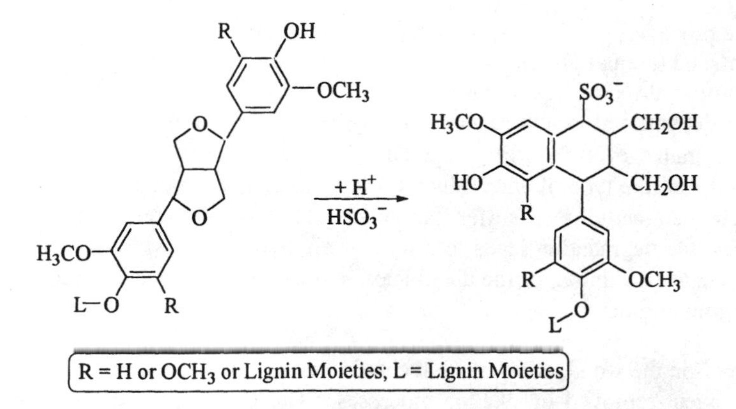

Figure 9. Condensation reaction; Intramolecular condensation of pinoresinol substructures. Adapted with permission from reference 2. Copyright 1971, Arbor Publishing.

(monomeric) lignosulfonates seem to originate from phenols formed by acid-induced hydrolysis of phenolic arylglycerol-β-arylether substructures.

Figure 10 shows the complex nature of reactions in the acidolysis of arylglycerol-β-guaiacylether. Cleavage of the β-arylether bond results in the formation of a ketol (enol) intermediate which further undergoes rearrangement and oxido-reductions to guaiacylacetone and vanilloyl methyl ketone. In addition, vanillin and coniferyl aldehyde are formed. Model compound experiments showed that vanillin originates from C_6-C_3 hydrolysis products. All of these products are converted to some extent to condensation products under the given conditions (*12*).

The presence of vanillin in softwood sulfite liquor has generated extensive studies leading to the development of an industrial process for its production. Further, it was established that phenolic arylglycerol-β-aryl ether substructures sulfonated at C-α are the source of vanillin and aceto-vanillone (Figure 11). Model compound studies suggest that nonphenolic α-sulfonated β-arylether substructures are subject to the same type of reactions (*13*). It must be mentioned that most of vanillin is synthesized currently from guaiacol by the Reimer-Tieman reaction In addition to the reaction discussed above, a number of "side-reactions" occur involving bisulfite ions (*14*), and condensations between phenolic substructures and hydrolysis products from xylans (e.g., furfural) (*15*).

Alkaline Processes

Alkaline processes are the most dominant for the production of pulps from both wood and nonwood (annual plant) resources. The chips are treated with an aqueous solution of sodium hydroxide, most commonly in the presence of additives, which both enhance delignification and reduce carbohydrate degradation. Pulping temperatures are in the range of 140-170°C, depending on the nature of the raw material to be processed, and the type of endproduct desired. As in acid processes, carbohydrates, in particular hemicelluloses, suffer partial degradation and removal. In alkaline processes, the degradation leads to a host of aliphatic acids which consume a major portion, up to two thirds, of the alkali leaving the rest for degradation and dissolution of the lignin fragments.

Extractives in the wood, which may in certain cases impede acid delignification, are to a large extent removed in alkaline processes. The good solubility of extractives in alkali allows the processing of tropical wood species which typically have a high content of complex mixtures of extractives. Alkaline pulping is also the technology of choice for processing nonwood resources, annual plants and agricultural residues (e.g., bamboo, bagasse, wheat and rice straws), which are known for having substantial silica contents.
Figure 12 shows an overview of alkaline processes and the corresponding reactive chemical species. Concentrations vary with pH and temperature. The additives are

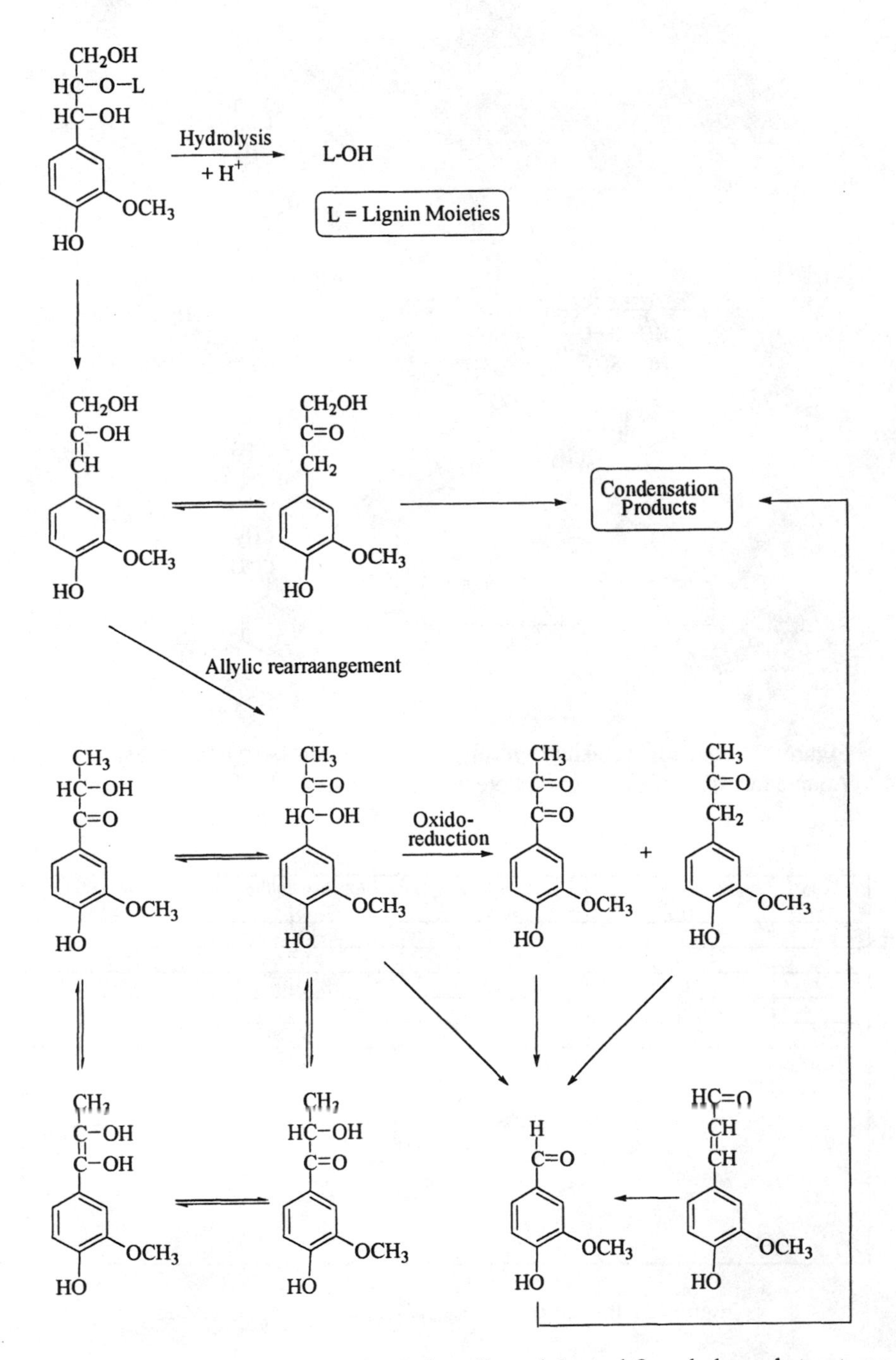

Figure 10. Acid-induced hydrolysis of phenolic arylglycerol-β-arylether substructures.

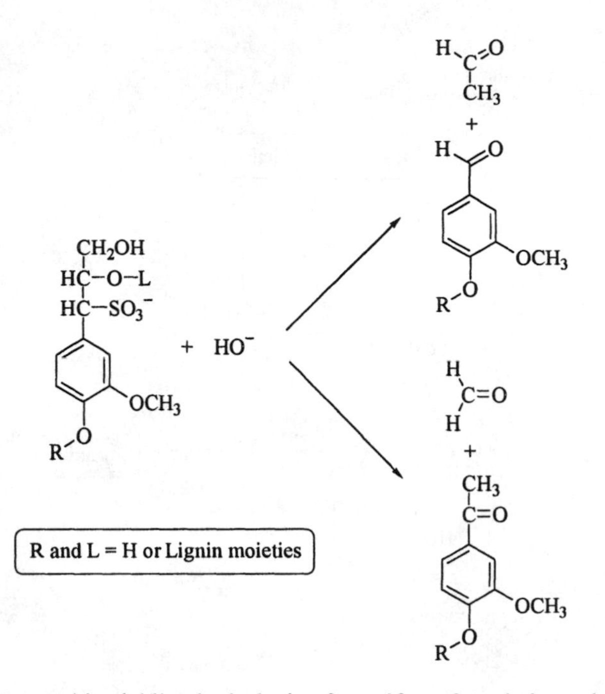

Figure 11. Anaerobic alakline hydrolysis of α-sulfono-β-arylether substractures. Adapted with permission from reference 12. Copyright 1959.

Soda [S]	Kraft [K]	Polysulfide [PS]	Alkaline Sulfite [AS]	Neutral Sulfite [NS]
HO^-	HO^-	HO^-	HO^-	HO^-
	HS^- / S_2^-	HS^- / S_2^-	SO_3^{2-}	HSO_3^-
	(S_n^{2-})	S_n^{2-}	(HSO_3^-)	(SO_3^{2-})
	$S_2O_3^{2-}$	($S_2O_3^{2-}$)		
	(CH_3S^-)	(CH_3S^-)		

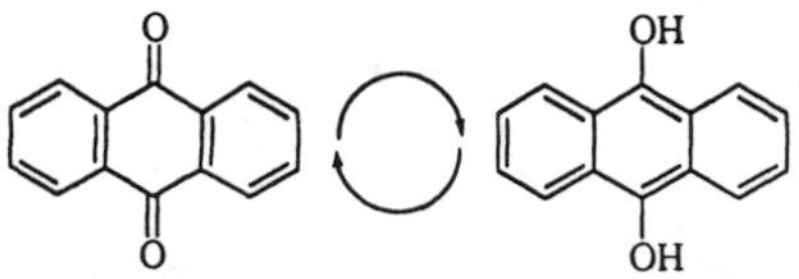

AQ ⇄ AHQ

Figure 12. Reactive species in alkaline pulping processes.

capable of promoting reactions leading to improvements in extension and selectivity of lignin removal by the following:

(a) promoting lignin degradation,
(b) suppression of lignin condensations,
(c) improvements in lignin solubility (e.g. introduction of sulfonic acid groups),
(d) suppression of carbohydrate degradation, and finally
(e) promotion of carbohydrate stablization.

In Figure 13, the reactivities of pulping chemicals in terms of process improvements are summarized.

As in the case of acid processes, lignin reactions may be divided into reactions leading to fragmentations and to condensations proceeding via common types of intermediates with electron deficient sites, i.e., quinone-methide intermediates.

Formation and Reactions of Quinone-Methide Intermediates

Phenolic substructures with an α-carbinol or a corresponding α-arylether group generate quinone-methides when heated with aqueous sodium hydroxide (Figure 14). Non-cyclic *p*-hydroxybenzyl ethers are the most reactive, followed by *p*-hydroxyphenyl coumarans (cyclic benzyl ethers) and the corresponding benzyl alcohols (*16*). The reactions can be classified as vinylogous β-elimination.

The high reactivity (instability) of quinone-methides is the result of their tendency to revert to the more stable aromatic structures (lower energy state) by reacting with functional groups capable of donating electrons (nucleophiles). Electrons may also become available from the very same side-chain by elimination of a proton from the adjacent C-β. Competing with the β-proton elimination is the removal of terminal hydroxymethyl groups, such as formaldehyde, and under more drastic conditions, elimination of the side-chain with concomitant oxidation of an α-carbinol group to the corresponding aldehyde (*17*). These reactions constitute reverse aldol reactions and involve cleavage of carbon-carbon bonds in the side-chain (Figure 15).

Cleavage of Phenolic β-Arylether Substructures

In the presence of additives, such as reduced sulfur compounds (kraft pulping), sulfite/bisulfite (alkaline/neutral sulfite pulping), and anthraquinone (AQ), the corresponding nucleophiles react preferentially with quinone-methide intermediates, competing with the aforementioned reactions initiated by internal electron shift. By applying molecular orbital calculations, it can be shown that reactions of quinone-methides are mediated by frontier molecular orbital considerations rather than by Coulomb attraction (*18*). Model compound studies have defined the following ranking of pulping additives in terms of reactivity towards quinone-methides:

Anthrahydroquinone (AHQ) > Sulfite ion (SO_3^{2-})
> Bisulfite ion (HSO_3^-) > Hydrosulfide (HS^-) > hydroxide ion.

HO^- (Base):	Carbohydrates and Lignin Degradation
HS^-, S_2^- (Reduced Sulfur Compounds in Kraft and Polysulfide Pulping):	Lignin Degradation
S_n^{2-} (Polysulfides):	Carbohydrate Stabilization, (Lignin Degradation)
HSO_3^-, SO_3^{2-} (Bisulfite and Sulfite ions):	Lignin Solubilization, Lignin Degradation
AQ/AHQ;	Carbohydrate Stabilization, Lignin Degradation

Figure 13. Reactivity of pulping chemical species in alkaline pulping.

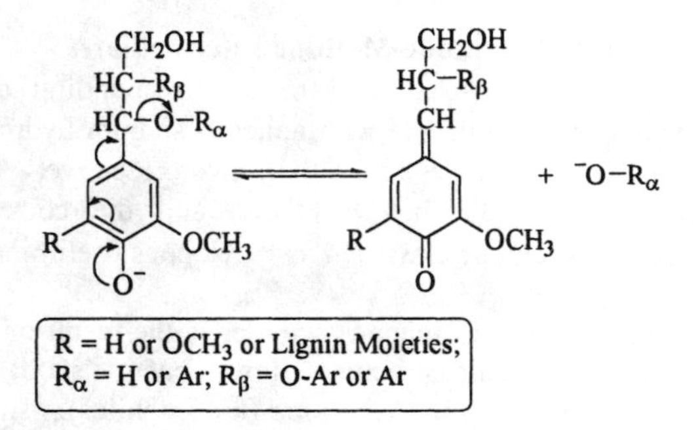

Figure 14. Formation of quinone-methide intermediate.

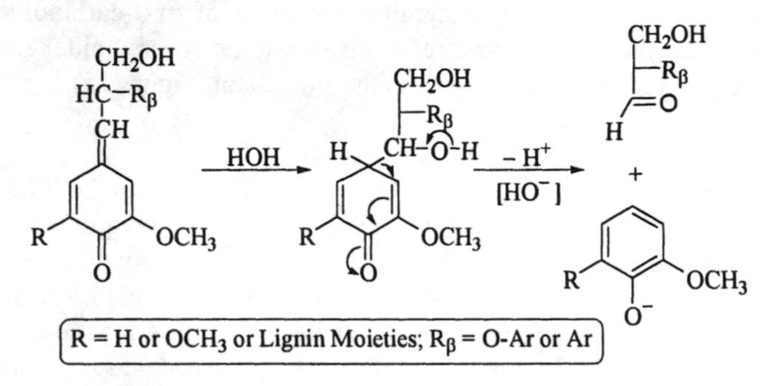

Figure 15. Reaction of quinone-methides with cleavage of side-chain via reversed aldol reaction.

Cleavage of the phenolic β-arylether bond is initiated by the addition of nucleophiles to the C-α of the corresponding quinone-methide restoring the aromatic structure. This is followed by interaction of this α-substituent with the C-β via neighboring group participation, resulting in the elimination of the aroxyl substituent at C-β.

Reactions with hydrosulfide/sulfide ions lead to an α-sulfide, followed by an intramolecular substitution (S_Ni) reaction which cleaves the β-arylether bond and forms a thiirane intermediate. This intermediate reacts further to form a styrene structure and elemental sulfur (Figure 16, Pathway a).

In the presence of the AQ-AHQ redox system, a quinone-methide undergoes nucleophilic attack on C-α by the tautomer of the reduced form (AHQ) to give an adduct, which leads to cleavage of the β-arylether bond via Grob-type fragmentation. This results in the formation of a *p*-hydroxy-styrene type substructure and AQ (*19,20*) [Figure 16, Pathway (b)]. The conjugated phenolic substructures thus formed may be subject to condensation reactions via extended quinone-methides (secondary condensation). Model compound studies have also shown enhanced base-induced cleavage of phenolic β-arylether bonds in the presence of reducing sugars, such as glucose. The reaction involves nucleophilic addition of the C-2 of the 1,2-enediol tautomer of glucose to the C-α of a quinone-methide. The resulting intermediate then undergoes a Grob-type fragmentation producing a *p*-hydroxystyrene type substructure and oxidized sugar (*21*). It was also found that this reaction peaks at temperatures around 150°C, suggesting instability of the reducing sugar. This mechanism is analogous to the one described above for the reductive cleavage of phenolic β-arylether subunits by AHQ. All of the above processes constitute reductive cleavages of β-arylether substructures, whereas the additives are oxidized.

In sulfite pulping, the quinone-methide with a β-O-4 bond will be first sulfonated at C-α. This is then followed by a nucleophilic substitution at the C-β by a sulfonate anion via neighboring participation of the C-α sulfonate group with concomitant cleavage of the β-ether bond. These reaction mechanisms were discussed previously for the reactions with sulfides and AHQ, as well as for the formation of an α,β-disulfonate, which is a minor reaction. Upon further reaction, the sulfonate group in C-α is eliminated to form a quinone-methide intermediate (Figure 17). A subsequent elimination of a proton in C-β produces a coniferyl alcohol β-sulfonic acid structure. Alternatively, an elimination of the terminal hydroxymethyl group as formaldehyde via reverse aldol condensation of the extended conjugate unsaturated ketone system gives a styrene-β-sulfonic acid structure (*4*).

As shown in Figure 18, cleavage of β-arylether bonds also may be accomplished by an electrophilic 1,4-oxygen-to-oxygen migration of the aryl group in β-O-4 type substructures with α-hydroxyl and α'-carbonyl groups in a temperature range of 20-40°C. This arrangement was shown to proceed via a spiro-Meisenheimer complex that is formed via enolization of the α'-carbonyl group to the cyclohexa-2,5-dinenylidenol structure, with concomitant nucleophilic attack of the corresponding α-

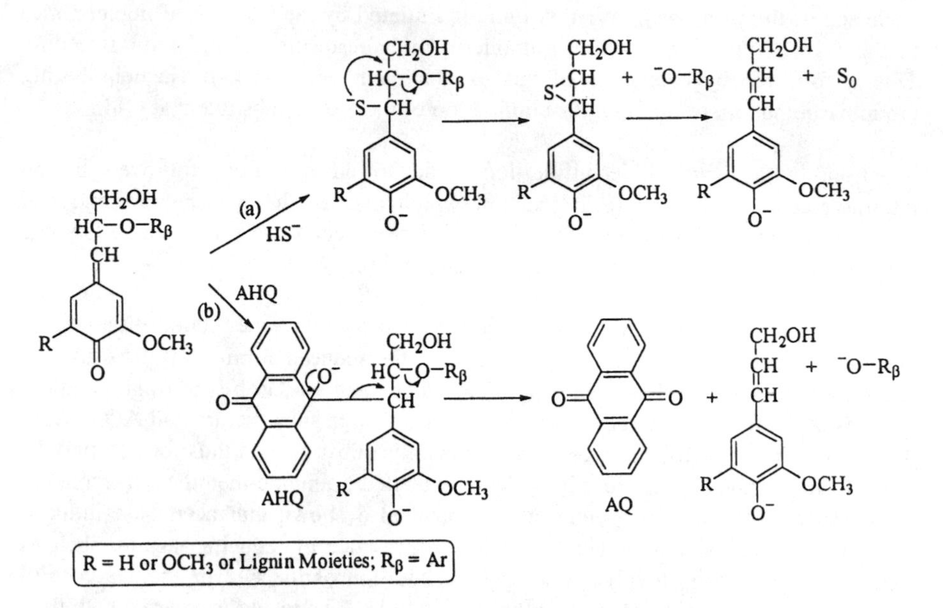

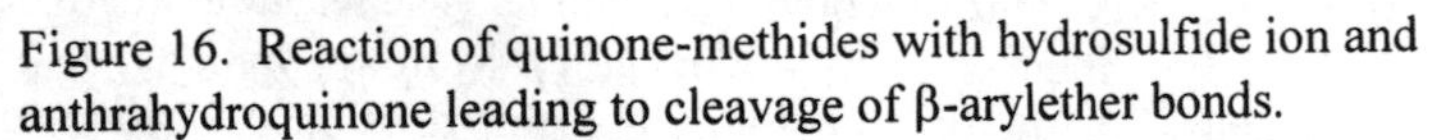

Figure 16. Reaction of quinone-methides with hydrosulfide ion and anthrahydroquinone leading to cleavage of β-arylether bonds.

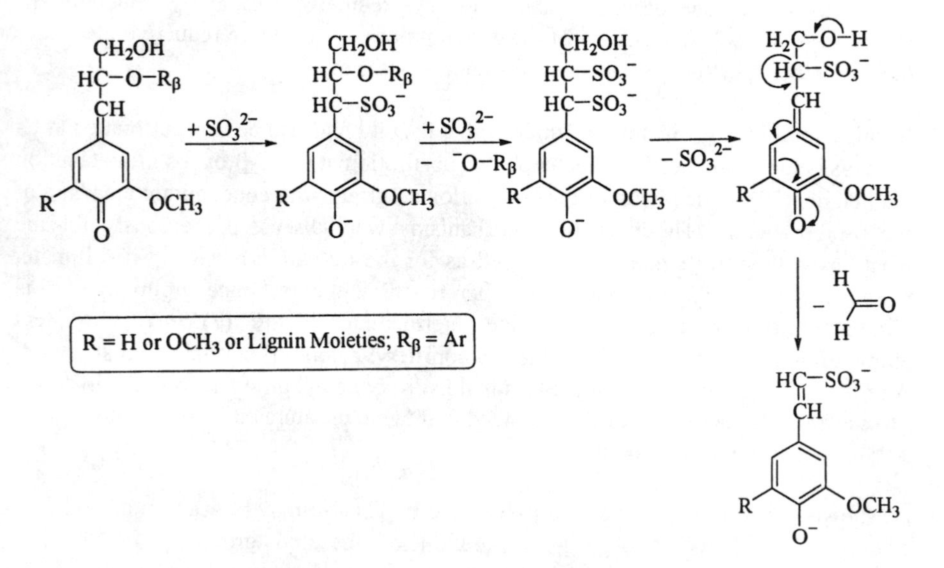

Figure 17. Reaction of quinonemethide with formation of lignosulfonates.

hydroxide anion on C-4 of the structure (*22*). Since the resulting α-O-4 arylether linkage is more susceptible to base-catalyzed hydrolysis than the β-O-4 type arylether linkage, any pretreatment of wood chips that introduces the α'-carbonyl groups would be beneficial to improving the pulping process.

Cleavage of Nonphenolic β-Arylether Substructures

Nonphenolic β-arylether substructures are much more stable than the corresponding phenolic subunits. Cleavage of the ether bonds usually occurs in the later stages of pulping, and depends primarily on the concentration of the base. The α-hydroxyl group in benzyl alcohol moieties is ionized, and reacts with the adjacent C-β via neighboring group participation resulting in elimination of the β-arylether bond with the accompanying formation of an oxirane intermediate (S_Ni reaction) which undergoes hydrolysis to the corresponding diol (*4*) (Figure 19). The mechanism is the same as the one operating in the base-induced hydrolysis of polysaccharides.

Condensation Reactions

Quinone-methides react with carbanions to generate C-C linkages between the reactants. One may distinguish between primary and secondary condensations. Examples of primary condensation are shown in Figure 20. A diarylmethane substructure will be formed via nucleophilic addition of an *ortho*-carbanion derived from a phenolate anion on C-α of a quinone-methide originating from a phenolic structure by elimination of the substituent at C-α (*4*). By contrast, when the quinone-methide intermediate undergoes a nucleophilic addition of a *para*-carbanion derived from a phenolate anion on C-α, a diarylmethane substructure will be formed with the concomitant elimination of a lignin fragment with an aldehyde group involving reverse aldol condensation of a side-chain with an α-hydroxyl group.

Figure 21 shows secondary condensation involving extended quinone-methides formed from conjugated phenolic moieties generated after initial cleavage reactions (e.g., coniferyl alcohol type structures). An extended quinone-methide of a coniferyl alcohol substructure undergoes a nucleophlilic attack by a carbanion at C-β of a second quinone-methide to form a C-C linkage between C-γ and C-β in the side-chains of two coniferyl alcohol units, with a concomitant elimination of formaldehyde via reverse aldol condensation.

As shown in Figure 22, nonphenolic β-arylether substructures may undergo condensation reaction via β-carbonyl intermediates (*23*). The rearrangement of the oxirane intermediate leads to homologues of etherified homovanillin (also see Figure 19). Aldol condensation between the keto and enol forms of the intermediates leads to the formation of 1,3-diaryl-1-propene structures.

Cleavage of β-Arylether Bonds in the β-O-4 type Condensed Substructures

As shown in Figure 23, the β-O-4 type *para* and *ortho* condensed diarylmethane substructures are susceptible to cleavage of the β-arylether bonds via neighbouring group participation (*24*). In both cases, the phenolate anions produced by condensation (also see Figure 20) initiate the cleavage, leading to stable aromatic

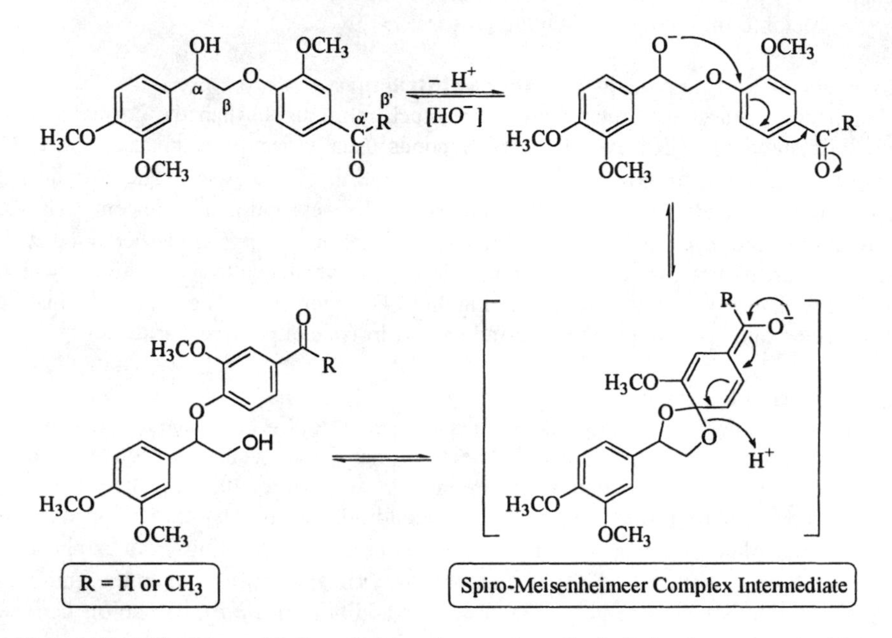

Figure 18. Migration of β-O-aryl group in nonphenolic β-O-4 substructures with α-hydroxyl and α'-carbonyl groups to α-O via spiro-Meisenheimer complex intermediate in alkaline solution at temperature range of 20-40°C. Adapted with permission from reference 22. Copyright 1997, American Chemical Society.

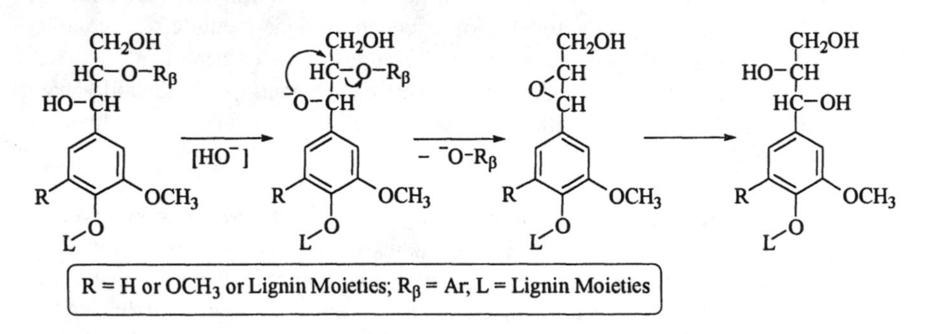

Figure 19. Cleavage of nonphenolic β-arylether substructures by base.

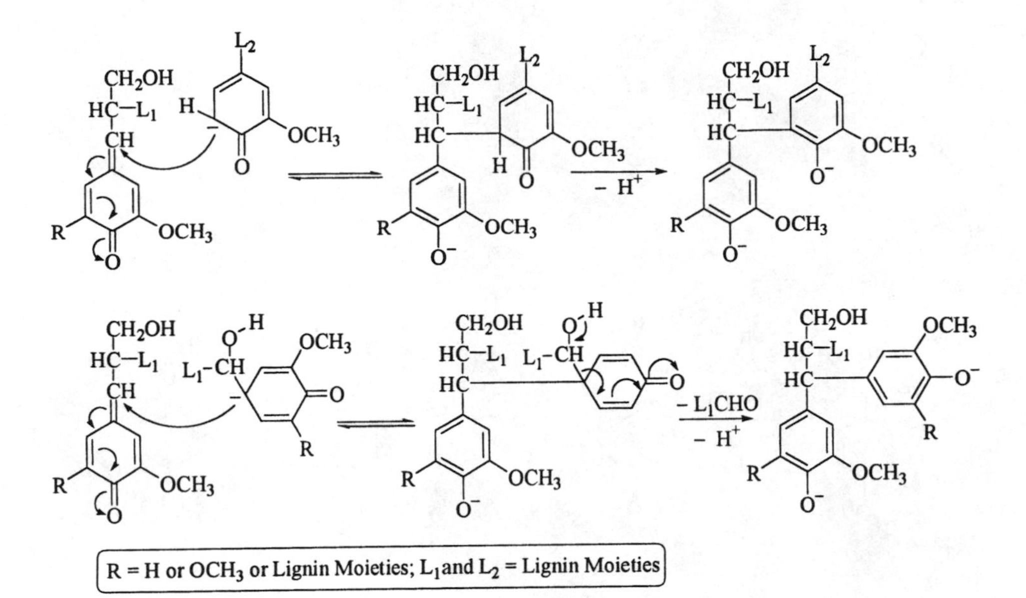

Figure 20. Reaction of quinone-methides involving lignin condensation to diarylmethane substructures.

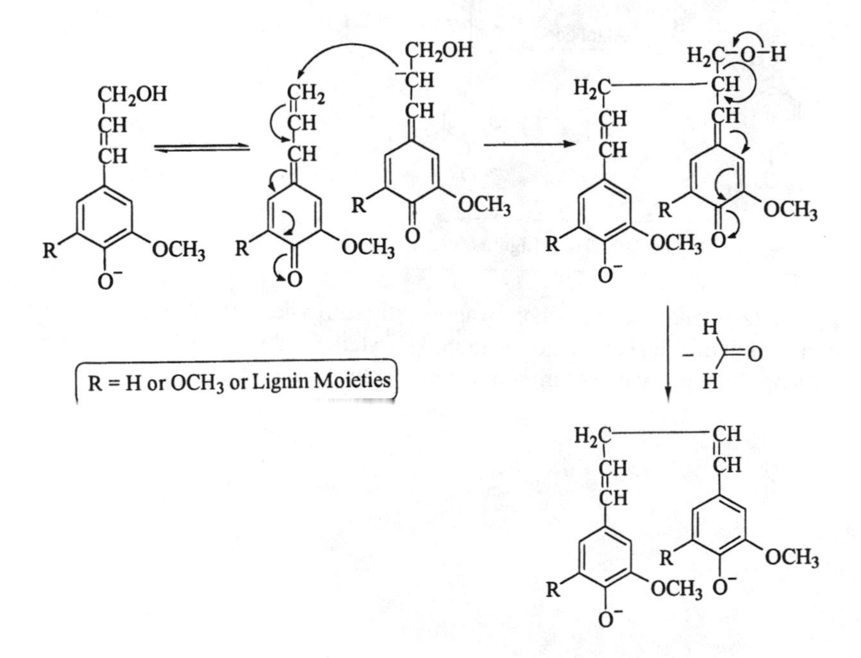

Figure 21. Reaction of quinone-methides, secondary condensation.

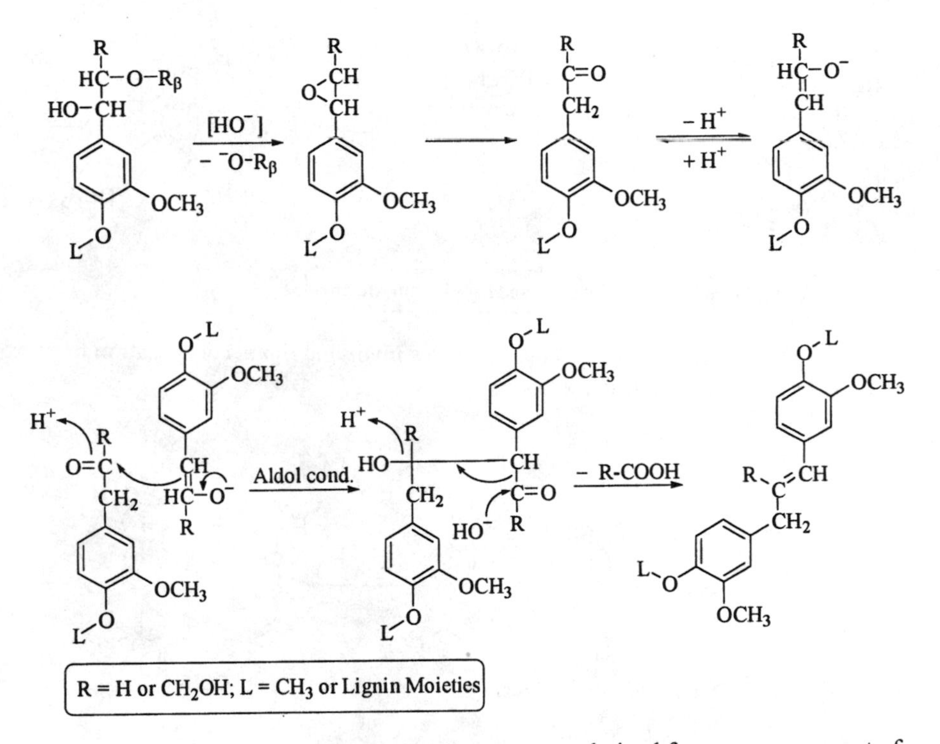

Figure 22. Condensation of homovanillin structures derived from rearrangement of oxirane intermediates from nonphenolic β-arylether substructures during alkaline pulping. Adapted with permission from reference 23. Copyright 1987.

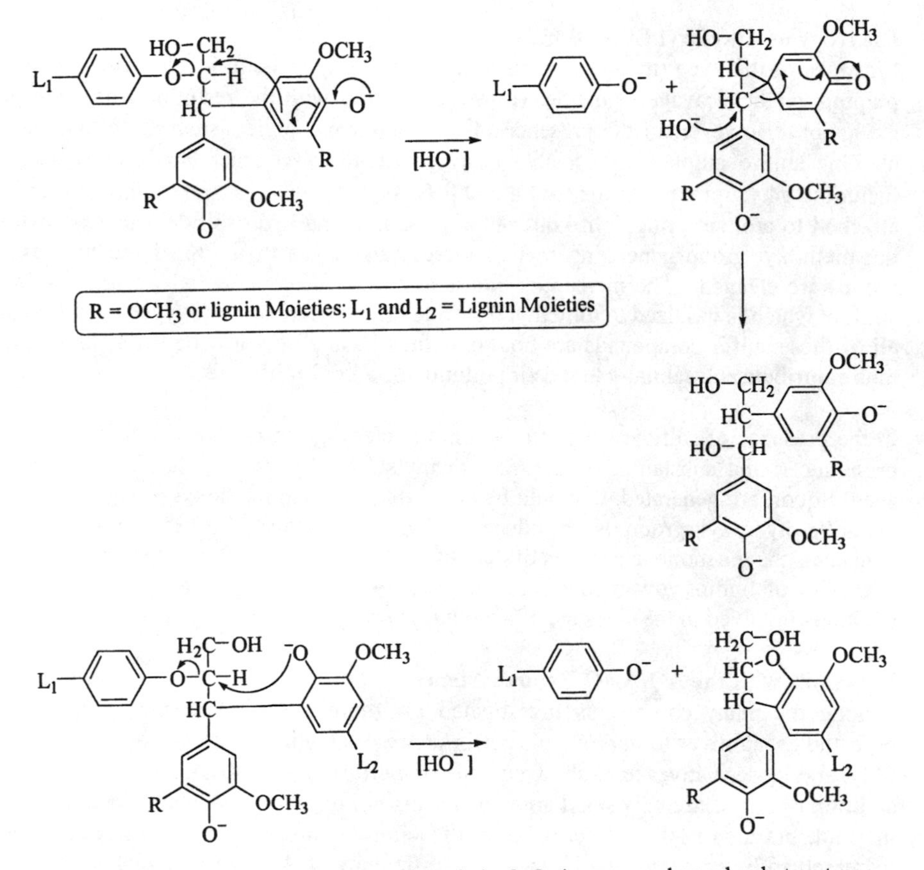

Figure 23. Cleavage of β-arylether bonds in β-O-4 type condensed substructures. Adapted with permission from reference 24. Copyright 1983, Arbor Publishing.

structures. In the case of *para* diarylmethane substructures, the resulting cyclopropane intermediate undergoes nuclophilic attack by hydroxide anion on C-α, leading to aromatization of the cyclopropane-4'-spiro-cyclodienone moiety. In the case of *ortho* diaryl-methane substructures, by contrast, the *ortho*-phenolate anion attacks C-β as a nucelophile resulting in the formation of an α-6 type coumaran structure with the concomitant cleavage of β-aryl ether bond. Thus, some of the condensation reactions of lignin in the alkaline pulping processes are beneficial to delignification.

Cleavage of Alkylaryl Ether Bonds

Linkages of this type (methoxyl groups) are reasonably stable towards base. In soda pulping, slight cleavage of methoxyl groups occured with the resulting formation of methanol. However, in the presence of stronger nucleophiles, such as in kraft and alkaline sulfite pulping, appreciable cleavage of alkylaryl ether bonds does occur, including alkylaryl ether of the α-O-4 and β-O-4 type structures, and methoxyl groups attached to aromatic rings. In kraft pulping, sulfide and hydrosulfide ions react with the methoxyl group generating methyl mercaptan. Less than 5% of the methoxyl groups are cleaved. The mercaptide ion is a strong nucleophile and forms dimethyl sulfide, which is oxidized to dimethyl disulfide on exposure to air (oxygen). Although all of these sulfur compounds are nontoxic, they have a very low odor threshold and thus contribute substantially to the air pollution of a kraft mill.

In the presence of sulfite or bisulfite anion, the cleavage of methoxyl groups is more pronounced and substantial amounts of methylsulfonic acid, soluble in the alkaline spent liquor, are generated. It should be noted that for each methoxyl group cleaved a phenolic hydroxyl group is introduced. Thus, the cleavage of methoxyl groups enhances the phenolic character of the lignin, which will affect significantly the reactivity of lignins toward the base. Figure 24 shows the general scheme of the reactions involved in the cleavage of a methoxyl group.

Reactions with the AQ/AHQ Redox system

Among the many compounds investigated for more than half a century for their potential as additives to improve the performance of alkaline processes, anthraquinone (AQ) and its derivatives have shown unique properties. Most intriguing is the fact that addition of only extremely small amounts results not only in substantial improvements in yield, but also in drastic acceleration of delignification (bulk delignification) and increased in lignin removal. The superior efficiency of AQ in improving both yield and lignin removal suggests that it is regenerated continuously and that its reduction product, the anthrahydroquinone (AHQ), is contributing in a major way to the process improvements. This means, we are dealing with a rather alkali-stable redox system with both the oxidant (AQ) and the reductant (AHQ) reacting throughout the cook. The amounts of additive required to achieve optimum results are extremely small, about 0.05-0.1 % based on wood. The recovery of about 30% of the initial charge suggests that the redox processes operate throughout the cook. With the discovery of the reductive cleavage of phenolic β-arylether substructures by AHQ and concomitant regeneration of AQ, a very important lignin fragmentation reaction that drives the cycle of redox processes was established (*19,20*) (Figure 16, Pathway b). Figure 25

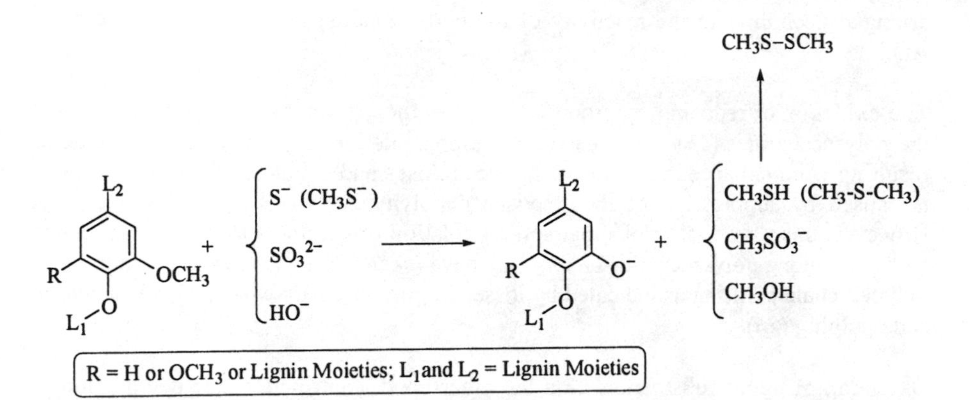

Figure 24. Cleavage of arylmethylether bonds; Formation of phenolic hydroxyl groups.

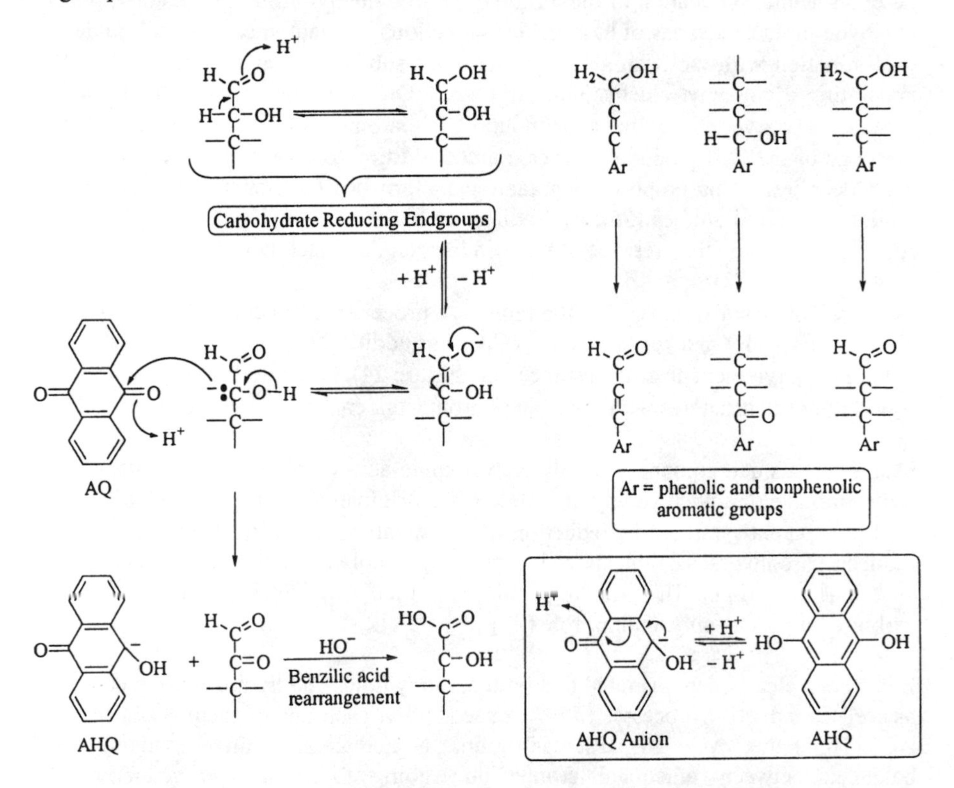

Figure 25. Oxidation of carbohydrate reducing endgroups and their homologues by AQ; Formation of AHQ by reduction of AQ.

shows the functional groups reacting with AQ. Their final oxidation products are arranged according to the reactivity of the corresponding functional groups towards AQ.

The oxidation of reducing endgroups constitutes the predominate oxidative process of the polysaccharides (*25*). This reaction is responsible for the pulp yield improvements resulting from enhanced retention of hemicelluloses and cellulose. The improvement is caused by suppression of the stepwise depolymerisation of the polysaccharides. However, there are indications that some oxidation of carbinols to the corresponding carbonyls along polysaccharide chains may have taken place. This would lead to base-induced chain splitting as indicated by losses in pulp viscosity when AQ was added to soda pulping (*26*).

Oxidation of lignin substructures, on the other hand, contribute to the degradation of lignin. Figure 26 shows the degradation of nonphenolic α-arylether substructures with an aroxyl or aryl sustituent at C-β by AQ. In the absence of AQ, such structures would be quite stable. Oxidation of the terminal hydroxymethyl group to the corresponding aldehyde initiates a series of base induced reactions. Its tautomer, the enol, undergoes β-elimination with the α-aroxyl group or other substituents at C-α as a nucleofugal generating a cinnamyl aldehyde substructure. On hydration, the coniferyl aldehyde subunit is converted to the corresponding α-carbinol structure, which undergoes cleavage of the C_α-C_β bond via reverse aldol addition to give vanillin and substituted acetaldehydes. This proposed degradation pattern has been supported by extensive model compound studies (*26*), by findings of higher carbonyl group content in soda AQ-lignins, and by the presence of vanillin in soda-AQ black liquors (*28*).

Figure 27 shows a summary for the reductive processes of phenolic β-arylether bond cleavage by AHQ and regeneration of AQ. In addition to the reductive cleavage of phenolic β-arylether substructures (see also Figure 24), both phenolic and nonphenolic β-arylether substructures with α-carbonyl groups undergo reductive cleavage.

Moreover, a small amount of 3,4-dideoxyhexonic acid was detected in addition to the large amounts of a wide variety of simple acids originating from polysaccharides. The acid is apparently formed by reduction of α,β-dicarbonyl structures transient in base-induced carbohydrate degradation, by stepwise depolymerisation, also known as the "peeling" reaction. This finding suggests that reductive processes involving carbohydrates play only a minor role (*29*).

It has been clearly demonstrated that both carbohydrates and lignins undergo oxidative as well as reductive processes (*30*). Consequently, each one of them could drive the AQ-AHQ redox cycle, in particular lignin. Nonetheless, lignin is evidently better balanced between functional groups undergoing oxidation and reduction than carbohydrates. One can visualize the operation of redox processes involving electron transfer from one to the other system with AQ-AHQ serving as "electron carriers." Under the conditions of pulping, the matrix polymers - lignin and hemicelluloses - are swollen gels thus limiting the mobility of the additives. Anthraquinones and their

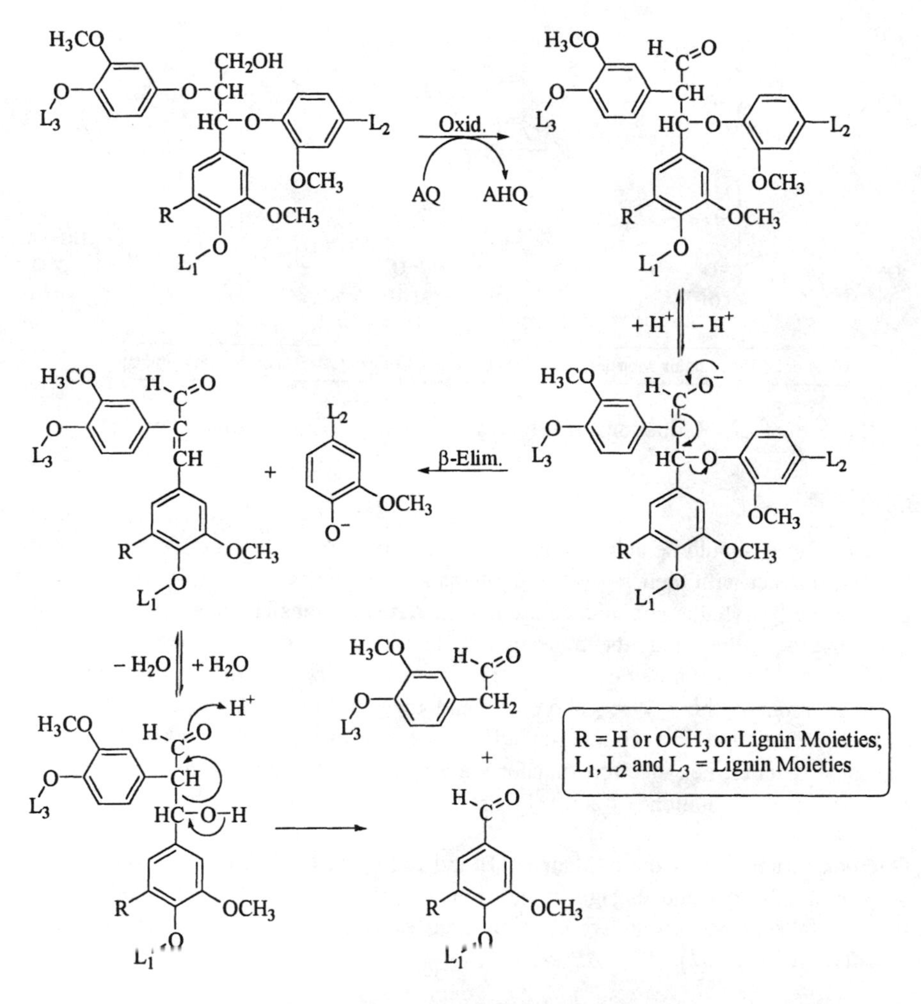

Figure 26. Reaction of AQ-AHQ redox system; Oxidation of a terminal hydroxymethyl group, β-elimination and subsequent reverse aldol condensation.

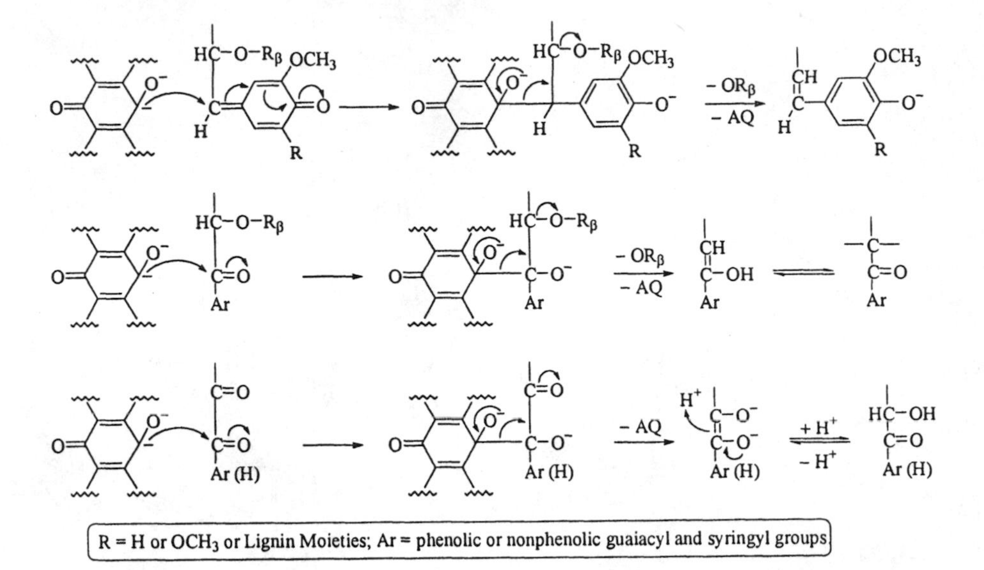

Figure 27. Reductions of ketone structures by AHQ; Formation of AQ.

reduced forms should be able to come in close proximity of the reactive groups in the gel and interact with their π-electron systems (π-complexes). The reductive cleavage of phenolic β-arylether subunits could be formulated as transferring electrons from the anthrahydroquinone and its transannular tautomer hydroxyanthrone (π-base) to appropriate lignin moieties (π-acid) via a π_{acid}-π_{base}-complex, followed by fragmentation of the π_{acid}-π_{base}-complex and regeneration of anthraquinone (Figure 28). The oxidation of a reducing endgroup, for example, by AQ can be described in the same manner, i.e., electron transfer via a π_{acid}-π_{base}-complex from the enolate (π-base) to the anthraquinone (π-acid) (Figure 29).

One can visualize that the oxidant (AQ) and reductant (AHQ) move along polymer surfaces in the gel and engage in the transfer of electrons. It is debatable, if the transfer takes place by a two-electron transfer (ionic) or by two single-electron transfers (radical) (*31*).

In the recent years, increasing attention has been given to the role of anthraquinone in sulfite pulping (alkaline and neutral sulfite processes), where the additive has shown promising potential in improving substantially the performance of the process and the quality of the products. The development of the ASAM, Alkaline Sulfite-Anthraquinone-Methanol, process is an excellent example of combining the process improving qualities of the AQ-AHQ redox system with those obtained by addition of methanol. The process results in improved penetration by pulping chemicals, reducing condensations and enhancing dissolution of lignin degradation products.

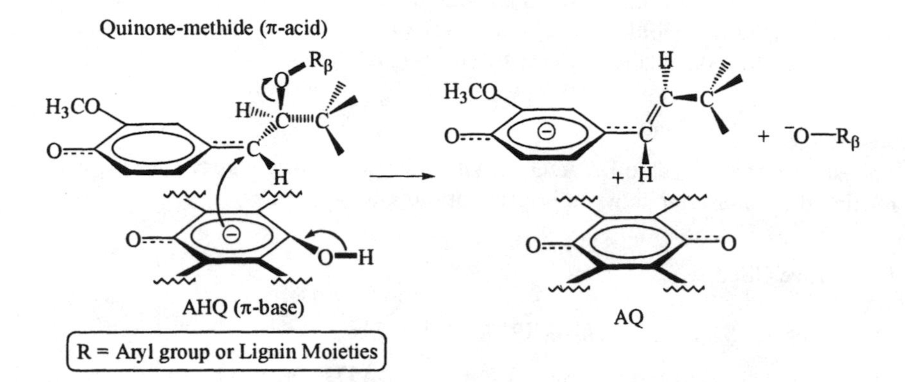

Figure 28. Reductive cleavage of β-arylether bond in phenolic β-O-4 substructures via fragmentation of a hypothetical π-complex formed by interaction of anthrahydroquinone (AHQ; π-base) and quinone-methide intermediate (π-acid).

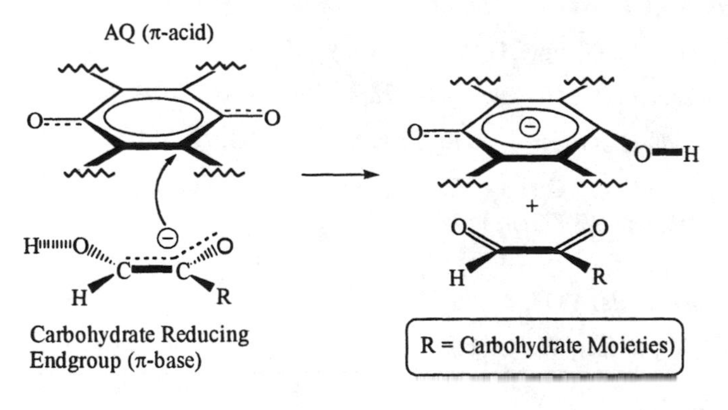

Figure 29. Reduction of anthraquinone (AQ; π-acid) to anthrahydroquinone (AHQ) by carbohydrate

Excellent results have also been obtained by adding AQ to polysulfide pulping, which have resulted in extended lignin removal with optimum preservation of yield and strength properties. Polysulfide-AQ pulping does represent substantial progress in alkaline pulping technology.

In summary, application of additives in alkaline pulping has resulted in substantial improvements in yield and in extending delignification without compromising strength properties and bleachability. In addition to the literature cited here, we would like to refer to the series on alkaline pulping by Gierer (*32,33*).

Acknowledgments
The authors are grateful to Dr. A. G. Kirkman for discussion, suggestions, advice and reading the manuscript in the preparation of this Chapter.

Literature Cited

1. Alder, E. *Wood Sci. Technol.* **1977**, *11*, 169-218.
2. Gellerstedt, G.; Gierer, J. *Svensk Papperstid.* **1971**, *74*(5), 117-127.
3. Gellerstedt, G. *Svensk Papperstid.* **1976**, *79*, 537-543.
4. Gierer, J. *Svensk Papperstid.* **1970**, *73*, 571-596.
5. Conners, W.J.; Sarkanen, S.; McCarthy, J.L. *Holzforschung* **1980**, *34*, 80-85.
6. Gierer, J. *J. Wood Chem. and Technol.* **1992**, *12*, 367-386.
7. Felicetta, V.F.; Glennie, D.W.; McCarthy, J.L. *Tappi* **1967**, *50*(4), 170-173.
8. Mothershead, J.S.; Glennie, D.W. *Tappi* **1964**, *47*(8), 519-524.
9. Gellerstedt, G.; Gierer, J. *Acta Chem.Scand.* **1968**, 22, 2510-2518.
10. Glasser, W.G.; Gratzl, J.S., Collins, J.J.; Forss, K.; McCarthy, J.L. *Macromolecules* **1973**, *6*, 114-128.
11. Glasser, W.G.; Gratzl, J.S.; Collins, J.J., Forss, K.; McCarthy, J.L. *Macromolecules* **1975**, *8*, 565-573.
12. Kratzl, K.; Kisser, W.; Gratzl, J.; Silbernagel, H. *Monatsh.Chem.* **1959**, *90*, 771-782.
13. Kratzl, K.; Risnyovszky, E. In *Chimie et Biochimie de la Lignine, de la Cellulose et des Hemicelluloses*, Int.Symp.Grenoble, France, Juli 1964, pp 151-163.
14. Goliath,M.; Lindgren, B.O. *Svensk Paperstid.* **1961**, *64*, 469-471.
15. Oblak-Rainer, M.; Budin, D.; Lipic, B. *Zellstoff und Papier* **1991**, *40*, 131-135.
16. Miksche, G.E. Lignin Reactions in Alkaline Pulping Processes (Rate Proceses in Soda Pulping); In *Chemistry of Delignification with Oxygen, Ozone and Peroxides*; Gratzl, J.S.; Nakano, J. Singh, R.P. Eds; Uni Publishers Co. Ltd, Tokyo, Japan, 1980, pp 107-120.

17. Gierer, J.; Noren, I. *Acta Chem. Scand.* **1962**, 16, 1713-1729.

18. Elder, T.J.; McKee, M.L.; Worley, S.D. *Holzforschung* **1988**, *42*, 233-240.

19. Landucci, L.L. *Tappi* **1980**, *63*(7), 95-99.

20. Gierer, J.; Lindeberg, O.; Noren, I. *Holzforschung* **1979**, *33*, 213-214.

21. Fullerton, T.J.; Wilkins, A.L.J. *J. Wood Chem. and Technol.* **1985**, *7*, 189-201.

22. Criss, D.L.; Fisher, T.H.; Schultz, T.P.; Ingram, Jr. L.L.; Saebo, D.B. *J. Org. Chem.* **1997**, *62*, 7885-7887.

23. Gierer, J.; Noren, I., Wannstrom, S. *Holzforschung* **1987**, *41*, 79-82.

24. Gierer, J; Ljungger, S. *Svensk Paperstid.* **1983**, *86* (9) , R100-R106.

25. Heikkila, H.; Sjostrom, E. *Cell.Chem.Tech.* **1975**, *9*, 3-11.

26. Wallis, F.A.; Wearne, R.H. *J. Wood Chem. and Technol.* **1985**, *7*, 513-526.

27 Hise, R.G.; Seyler, D.K.; Chen, C.-L.; Gratzl, J.S. *Proc. 5th. Int.Symposium on Wood and Pulping Chemistry*, Paris, France, April 27-30, **1987**, *Vol. 1*, 139-145.

28. Venica, A.; Chen, C.-L.; Gratzl, J.S. *Proc. the 6th Int. Symposium on Wood and Pulping Chemistry*, Raleigh, NC, USA, May 22-25, **1989**, *Oral Presentations*, 263-269.

29. Samuelson, O.; Sjoeberg; L.A. *Cell. Chem. and Tech.* **1978**, *12*,463-472.

30. Gratzl, J.S.; Chen, C-L. Redoxprocesses in Alkaline Pulping in Presence of Anthraquinone Compounds - An Overview. *Proc. 8th Int. Symposium on Wood and Pulping Chemistry*, Beijing, P.R.China, May 25-28, **1993**, *Vol.1*, 1-8.

31. Dimmel, D.R.; Schuller, L.F. *J. Wood Chem. and Technol.* **1986**, *6*, 565-590.

32. Gierer, J. *Wood Sci. Technol.* **1980**, *14*, 241-266.

33. Gierer, J. *Holzforschung* **1982**, *36*, 43-51.

Chapter 21

The Interplay between Oxygen-Derived Radical Species in the Delignification during Oxygen and Hydrogen Peroxide Bleaching

Josef Gierer

Royal Institute of Technology, Division of Wood Chemistry, SE–100 44 Stockholm, Sweden

Superoxide and hydroxyl radicals are important intermediates in oxygen and hydrogen peroxide bleaching. They play essential roles in the degradation of wood constituents during these processes both by reacting directly with the substrate and by participating in the generation of each other. Thus, superoxide adds to substrate radicals, bringing about cleavage of carbon-carbon bonds (fragmentation), and dismutates, affording hydrogen peroxide and subsequently - by reductive cleavage of the latter - hydroxyl radicals. Hydroxyl radicals, on the other hand, attack substrate molecules which results not only in direct degradation, but also in generation of substrate radicals, required for the formation and the reactions of superoxide.

About 20 years ago, it was found that during oxygen bleaching hydrogen peroxide is formed in varying amounts, depending on the bleaching conditions used (*1*). After this discovery, the chemistry of oxygen bleaching has been repeatedly interpreted in terms of the reactions of these two oxidants with model compounds, representing the most prominent lignin structures (*2-4*). Undoubtedly, this way of describing the oxygen and hydrogen peroxide bleaching processes can only give an incomplete picture, since it does not take into account that two further reactive species, superoxide and hydroxyl radicals, arise during the processes and participate in the oxidative degradation of lignin. These two species are intermediates in the four-step reduction of oxygen to water (Figure 1):

Parts of this work have also been presented at the ISWPC meeting in Helsinki, Finland, June 1995 and at the ACS conference in New Orleans, 1996.

Superoxide, $O_2^{\cdot-}$, represents the first reduction product from oxygen, and hydroxyl radical, $HO^{\cdot}$, arises by reductive cleavage of hydrogen peroxide.

The reactions of these two radical species with various model compounds have been intensively studied during the past fifteen years (*5-8*). The radicals were generated by γ-radiolysis under carefully controlled conditions and the reaction mixtures were analysed using gas chromatography-mass spectrometry. The analyses showed that the two radical species, due to their different chemical nature, exhibit different selectivities.

The results from these studies with "artificially generated" radicals were now compared with those obtained when the same model compounds were treated with oxygen under bleaching conditions. On the basis of this comparison, feasible reaction pathways for autoxidation processes in general and for oxygen bleaching in particular may be proposed which reveal the intimate cooperation between these two radical species, and between them and oxygen and hydrogen peroxide during oxygen and hydrogen peroxide bleaching.

Experimental

Model Compounds. The lignin models creosol (CR), veratrylglycol (VG), *t*-butylguaiacol (TBG) and *t*-butylsyringol (TBS), as well as the carbohydrate models ß-*D*-glucopyranose (GL) and methyl ß-*D*-glucopyranoside (MeGL) were used (Figure 2).

Analytical Procedures. The consumption of the model compounds was followed by gas-liquid chromatography and the formation and consumption of hydroxyl radicals by using the chemiluminescence method, worked out (*9*) and recently improved (*10*) in this laboratory. This method is based on the hydroxylation of phthalic hydrazide, followed by oxidation of the resultant hydroxyphthalic hydrazide to give hydroxyphthalic acid with concomitant emission of strong chemiluminescence which is a measure for the hydroxyl radicals formed.

Results and Discussion

Oxygen Bleaching

Formation and Reactions of Superoxide during Oxygen Bleaching. Superoxide is formed during oxygen bleaching by a single electron transfer from an activated substrate to oxygen (Figure 3).

This very first step of autoxidation is rate-determining and dependent on the nature of the substrate, in particular its reduction potential, and on the conditions used, such as temperature, pH and presence of metal ions. The electron transfer may either take place directly, e.g. from a phenolate anion, as shown in the upper part of Figure 3 or it may involve addition of oxygen to an intermediary substrate radical, followed by elimination of superoxide from the addition product, as shown in the lower part of Figure 3.

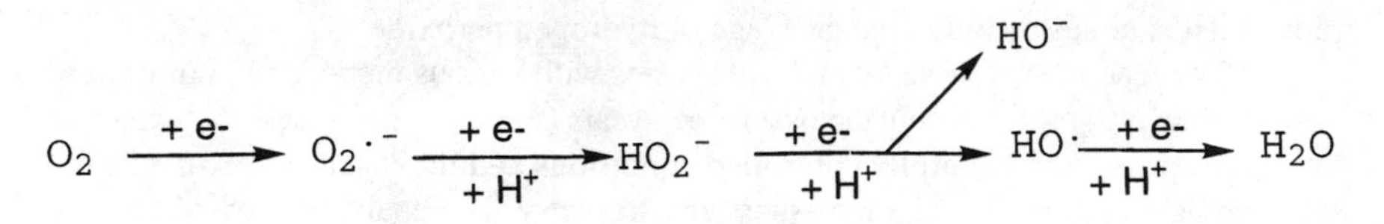

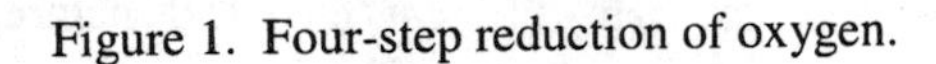
Figure 1. Four-step reduction of oxygen.

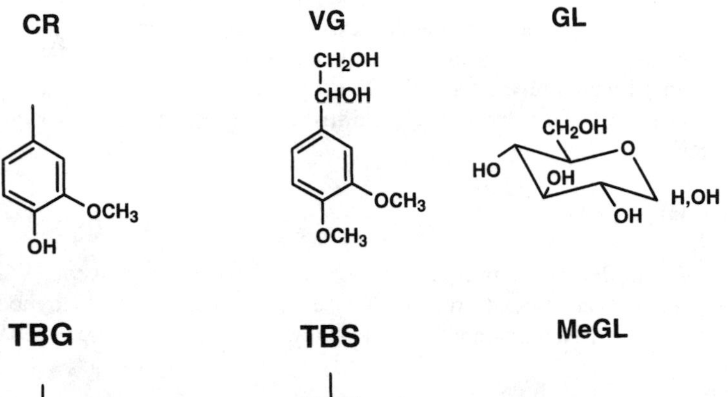

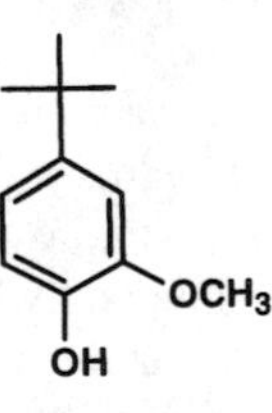

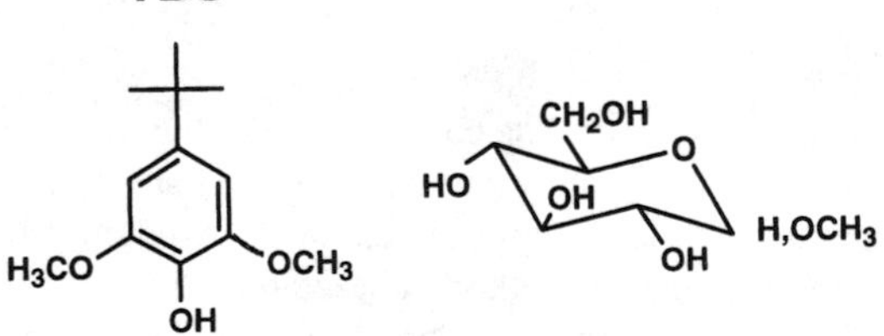

Figure 2. Model compounds used.

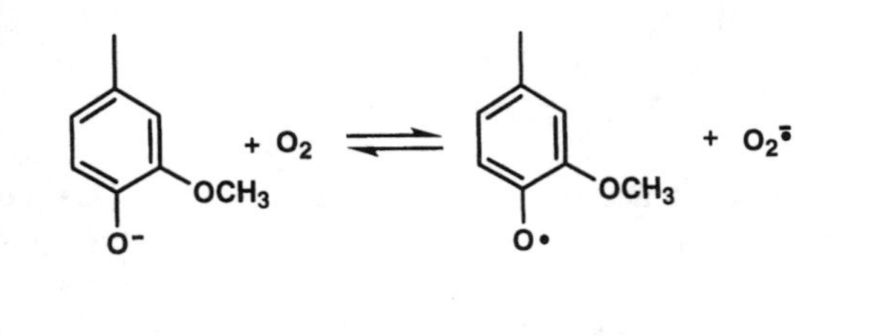

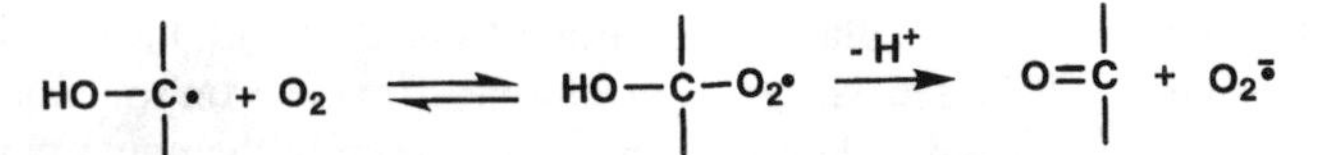

Figure 3. One-electron transfer to oxygen, formation of superoxide.

Under oxygen bleaching conditions, superoxide is relatively long-lived and is therefore able to penetrate into fibres. It is predominantly a reducing species but may also act as an oxidant towards compounds with a sufficiently low redox potential, such as catechols, hydroquinones and phenolic stilbenes. From the point of view of lignin degradation, the combination of superoxide radicals with substrate radicals, generated by hydroxyl radicals (Figure 4) is of extreme importance. This combination is followed by dioxetan formation and re-arrangement, leading to ring opening, oxirane formation or side chain cleavage. The primary reaction products may then be further degraded.

Besides the combination with substrate radicals, the dismutation of superoxide to give oxygen and hydrogen peroxide is the most important reaction mode (Figure 5). Thus, superoxide represents an important intermediate in the formation of hydrogen peroxide.

The direct dismutation of superoxide, shown in the upper equation in figure 5, is negligible. In practice, the reaction requires the presence of protons or metal ions to proceed at a reasonable rate. There is a competition between combination with substrate radicals (Figure 4) and dismutation of superoxide (Figure 5). In alkaline solution the combination alternative is strongly favoured.

Formation and Reactions of Hydroxyl Radicals during Oxygen Bleaching. Hydrogen peroxide, formed by acid-catalyzed disproportionation of superoxide anion (Figure 5), may undergo metal-catalyzed reductive cleavage, particularly by superoxide, yielding - in a Fenton-like reaction - hydroxyl radicals (Figure 6, upper part).

Two reaction modes of hydroxyl radicals can be distinguished: The first is electrophilic addition to aromatic systems to give *via* primary adducts isomeric hydroxycyclohexadienyl radicals (Figure 7). Oxidation of these radicals by oxygen affords - depending on the site of the radical center - hydroxylation-, demeth-oxylation- or C_{α}- C_{β} cleavage products, together with superoxide anions.

The second reaction mode of hydroxyl radicals is hydrogen abstraction, followed by oxygenation of the resultant aliphatic radical and elimination of superoxide, yielding carbonyl structures in lignin side chains (Figure 8) and in carbohydrates (see also Figure 21).

The different rates of these two reaction modes, are exemplified by the rate constants of two lignin models (K_L) and two carbohydrate models (K_C), given in Figure 9: They differ by a factor of 5 - 6, the so-called selectivity factor (S) which may be regarded as a measure for the limited selectivity of all TCF processes (*11*) under these conditions.

The question arises: How do these reaction modes of superoxide and hydroxyl radicals fit into the overall process of lignin degradation and how do these radical species co-operate during autoxidation ?

The interplay between superoxide and hydroxyl radicals is illustrated by a description of the course of events, when a simple phenolic model is subjected to

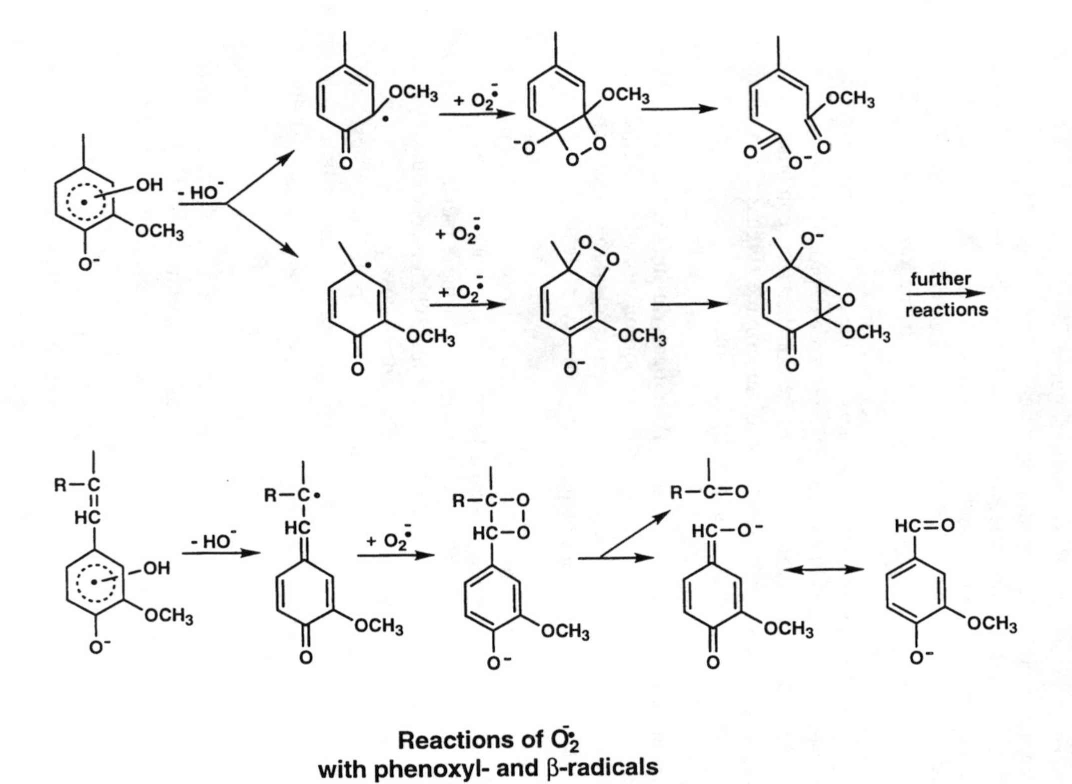

Figure 4. Addition of superoxide to phenoxyl radicals and subsequent reactions (Reproduced with permission from Reference number 8).

$$2\,O_2^{\cdot-} + H_2O \longrightarrow O_2 + HO_2^- + HO^-$$

$$O_2^{\cdot-} + HO_2^{\cdot} \longrightarrow O_2 + HO_2^-$$

Figure 5. Dismutation of superoxide.

$$H_2O_2 + Fe^{++} \longrightarrow HO^{\cdot} + HO^- + Fe^{+++}$$

$$Fe^{+++} + O_2^{\cdot-} \longrightarrow Fe^{++} + O_2$$

$$H_2O_2 + O_2^{\cdot-} \longrightarrow HO^{\cdot} + HO^- + O_2$$

$$RO{-}OH \longrightarrow RO^{\cdot} + HO^{\cdot}$$

Figure 6. Reductive cleavage of hydrogen peroxide and homolytic cleavage of hydroperoxide.

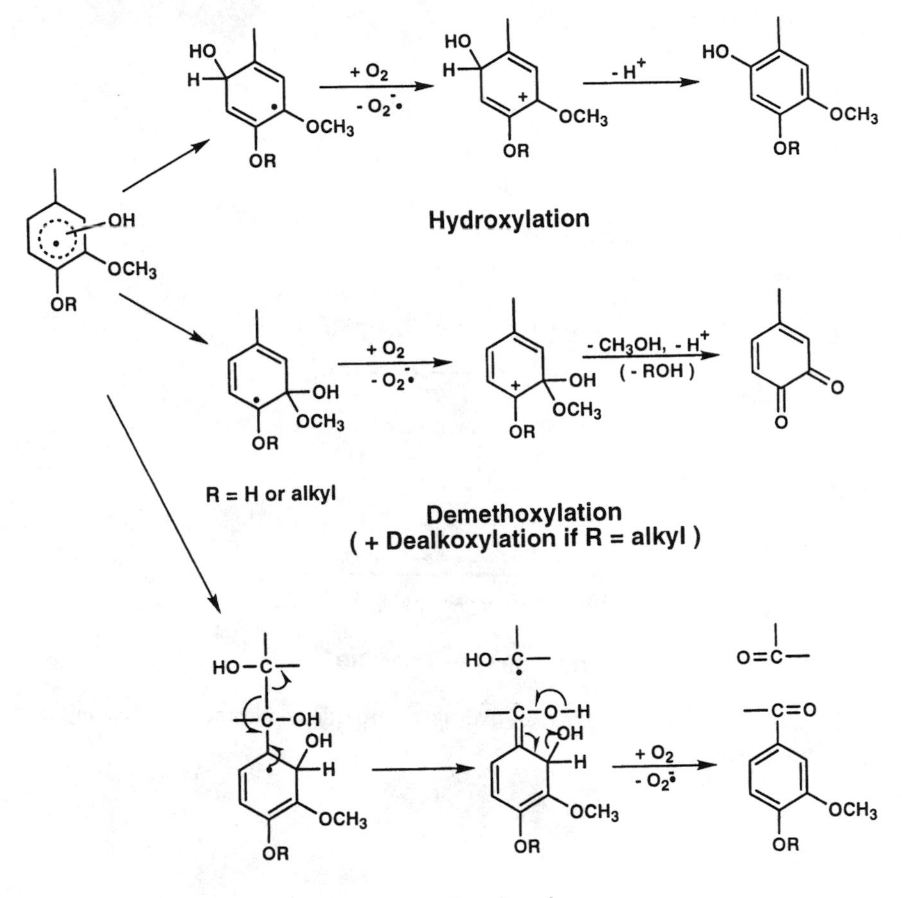

Figure 7. Reaction modes of hydroxyl radicals with aromatic structures (Reproduced with permission from Reference number 8).

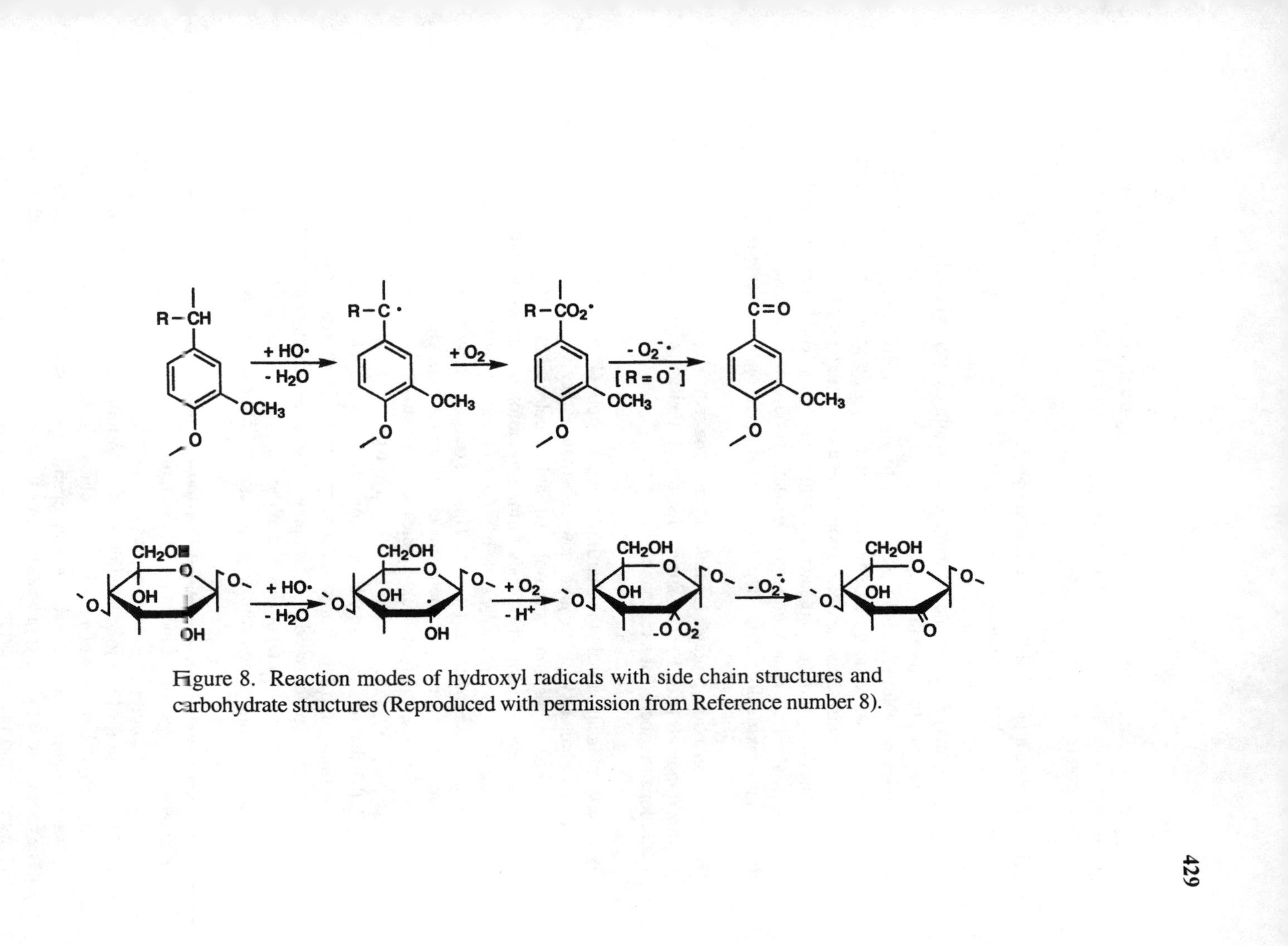

Figure 8. Reaction modes of hydroxyl radicals with side chain structures and carbohydrate structures (Reproduced with permission from Reference number 8).

oxygen bleaching conditions (Figure 10). This figure illustrates the generation, the main reactions and the cooperation of the two radical intermediates during the autoxidation of creosol:

The process starts with the single electron transfer from the substrate to oxygen (step **A**) to give a substrate radical and superoxide. These two radical species may react further in three different ways:

1) They may revert to the reactants, creosol and oxygen which limits the further course of the autoxidation,
2) They may combine with each other to give a hydroperoxide anion (step **B**) which may undergo intramolecular nucleophilic attack and rearrangement, resulting in cleavage of a carbon-carbon bond (e.g. step **C**) or
3) They may diffuse apart and react with themselves. The latter way implies that the substrate radical undergoes oxidative coupling (dimerisation, step **D**) and superoxide dismutates to give oxygen and hydrogen peroxide (step **E**).

Hydrogen peroxide, produced in this way, may be reductively cleaved, giving rise to hydroxyl radicals (step **F**). This superoxide-driven Fenton reaction requires transition metal ions to proceed at a reasonable rate.

Hydroxyl radicals attack the starting creosol (step **G**) much more rapidly and effectively than oxygen to give new substrate radicals and water, whereby the continued decomposition of creosol is strongly accelerated.

However, hydroxyl radicals formed by autoxidation of phenolic structures attack not only further phenolic structures (step **G**) and phenoxyl radicals (step **H**), but also non-phenolic structures (step **I**). This is exemplified by a description of the two reaction modes of the non-phenolic model compound veratrylglycol in Figure 11: The first mode of reaction of hydroxyl radicals, the electrophilic addition to the aromatic nucleus (step **A**), gives rise to isomeric hydroxy-cyclohexadienyl radicals and the second, the abstraction of a hydrogen atom from the side chain, predominantly from the α-position (step **B**), affords the corresponding C_{α}- centered radical. These two reaction modes are followed by oxidation with oxygen resulting in hydroxylation (step **C**), demethylation (step **D**) or cleavage of the C_{α} - C_{β} bond (step **E**), depending on the particular structure of the hydroxycyclohexadienyl radical, or in oxidation of the side chain (step **F**).

Thus, it may be concluded that the degradation of phenolic structures is initiated by oxygen in co-operation with superoxide radicals (Figure 10), whereas the degradation of non-phenolic structures is introduced by hydroxyl radicals, generated in the oxidation of phenolic structures, and continued by oxygen (Figure 11). This description of the course of autoxidation includes the four reduction steps from oxygen to water (Figure 1) and reveals the interdependence between the two radical species: The formation of hydroxyl radicals requires superoxide anions and hydrogen peroxide (Figure 6) and the formation of superoxide radicals by indirect electron transfer to

Model compound	Rate constant ($M^{-1}.sec^{-1}$)
Lignin model compound 1	K_{L1} 1.5×10^{10}
Lignin model compound 2	K_{L2} 1.7×10^{10}
Carbohydrate model compound 1	K_{C1} 3.2×10^{9}
Carbohydrate model compound 2	K_{C2} 2.6×10^{9}

Figure 9. Rate constants of the reactions between hydroxyl radicals and lignin and carbohydrate models.

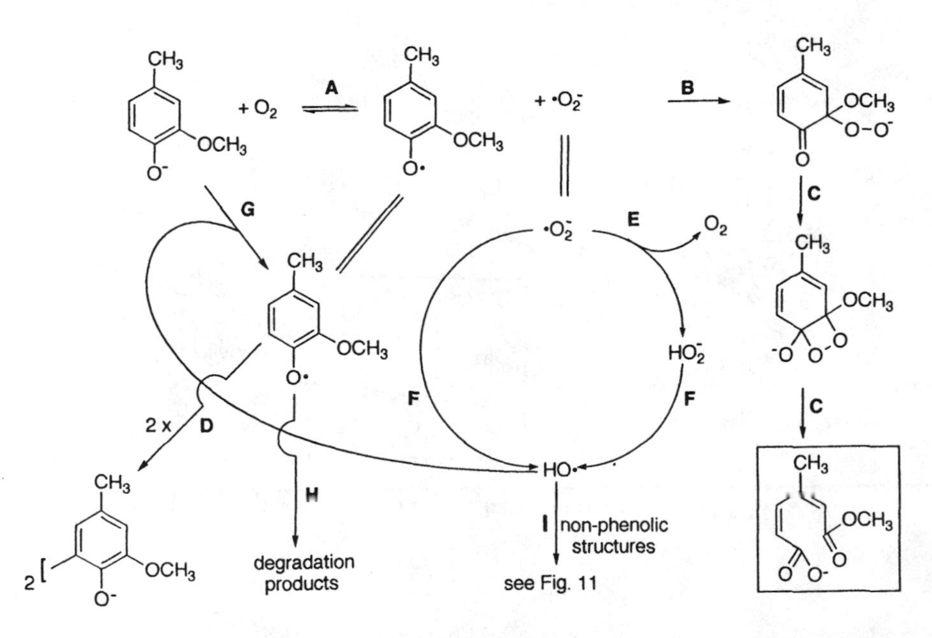

Figure 10. Scheme describing the autoxidation of creosol.

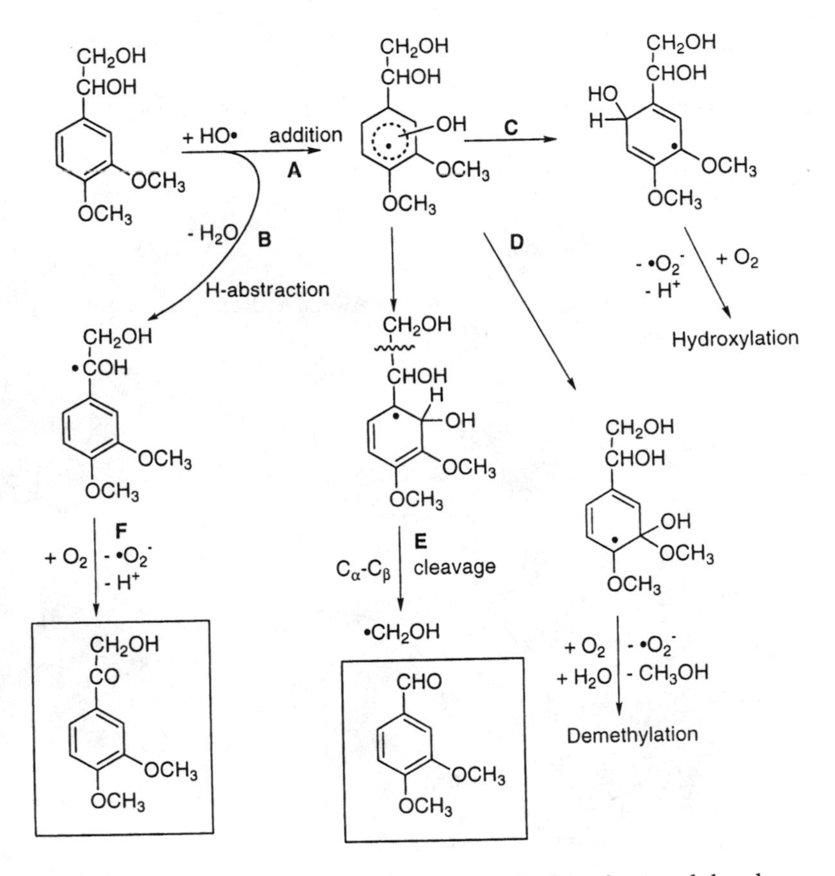

Figure 11. Reactions between hydroxyl radicals and veratrylglycol.

oxygen (Figure 3) requires hydroxyl radicals for the generation of substrate radicals.

The validity of these two schemes, describing the formation and involvement of the two radical species in oxygen bleaching is supported by a number of experimental findings, some of which are briefly described in the following:

1) The coupling between phenol consumption and formation of hydroxyl radicals is clearly demonstrated in Figure 12. As can be seen, the consumption of creosol and of (small amounts of) phthalic hydrazide is accompanied by the emission of chemiluminescence, given in arbitrary units, i.e., by a formation of hydroxyl radicals, in acccordance with Figure 10. Based on the consumption of phthalic hydrazide, the rate of production of hydroxyl radicals in the treatment of 1.5 mmoles creosol with oxygen under bleaching conditions was calculated to amount to about 1 μmole/min. This rate of hydroxyl radical production is of the same order of magnitude as that previously obtained in an oxygen bleaching experiment using kraft pulp (unpublished result).

2) The generation of hydroxyl radicals during treatment of creosol with oxygen was also substantiated in another way: Veratrylglycol, the non-phenolic model, is practically stable towards oxygen under bleaching conditions. However, when creosol, the phenolic model compound, is admixed to veratrylgycol before the treatment with oxygen, veratrylglycol is degraded to a small, but significant extent and veratraldehyde (VA) and "veratrylketol", [1-(3,4-dimethoxyphenyl)-2-hydroxyethanone-1, VK] are formed as main oxidation products from veratrylglycol (Figure 13). The figure shows the stability of veratrylglycol, the partial degradation of veratrylglycol in the presence of creosol and the formation of small amounts of veratraldehyde (VA) and veratrylketol (VK).

Considering the stability of veratrylglycol towards oxygen and hydrogen peroxide during the oxygen treatment, the oxidative attack on veratrylglycol in the presence of creosol must be attributed to hydroxyl radicals, formed during the autoxidation of creosol. This is also evidenced by the emission of chemiluminescence (Figure 14): Oxygen treatment of only veratrylglycol does not produce any noticeable chemiluminescence, whereas the same treatment of a mixture consisting of veratrylglycol and creosol (1:1) gives chemiluminescence of almost the same intensity as the treatment of only creosol.

3) The coupling between substrate consumption and hydroxyl radical formation is further strengthened by the pH-dependence of the rates of creosol (CR) degradation and chemiluminescence (CL) formation (Figure 15): The maximum at about 11.6 may be explained by the pKa values of creosol (= 11.5) and of the hydroxyl radical (=11.8): At pH values lower that 11.5, the initial step of creosol autoxidation, the single electron transfer from the creosolate anion to oxygen, proceeds more slowly, due to incomplete dissociation of creosol. On the other hand, when the pH is close to, or over, the pKa of the hydroxyl radical, i.e.,11.8, the hydroxyl radical increasingly dissociates,

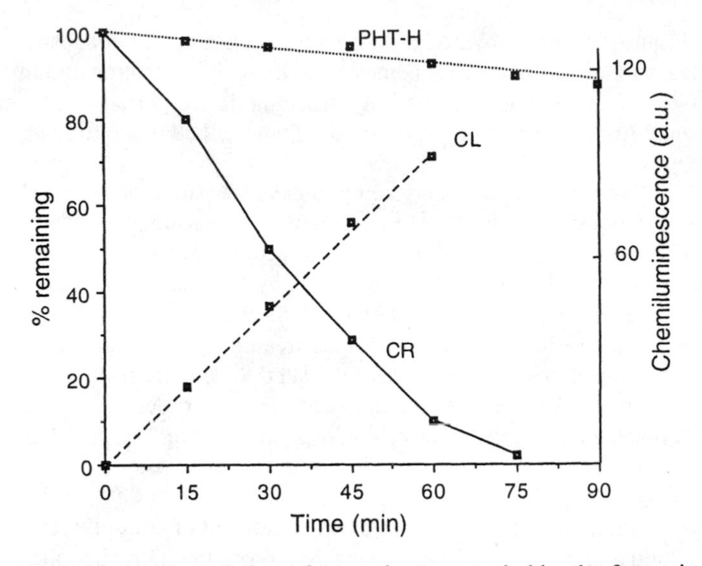

Figure 12. Oxidative degradation of creosol accompanied by the formation of hydroxyl radicals. PHT-H refers to phthalic hydrazide, CR to creosol and CL to chemiluminescense (in arbitrary units).

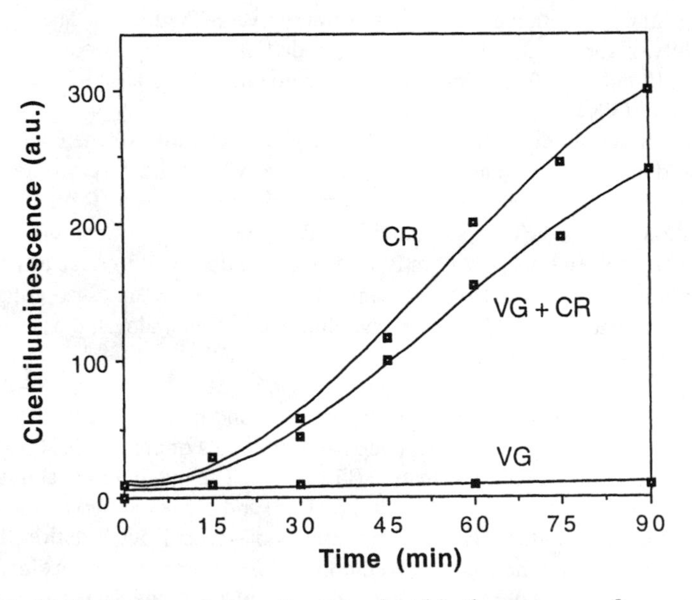

Figure 13. Oxidative degradation of veratryglycol in the presence of creosol. CR refers to Creosol, VG to veratrylglycol and VG + CR to a mixture of veratrylglycol and creosol (1:1). VA + VK indicates the formation of veratryaldehyde and veratrylketol.

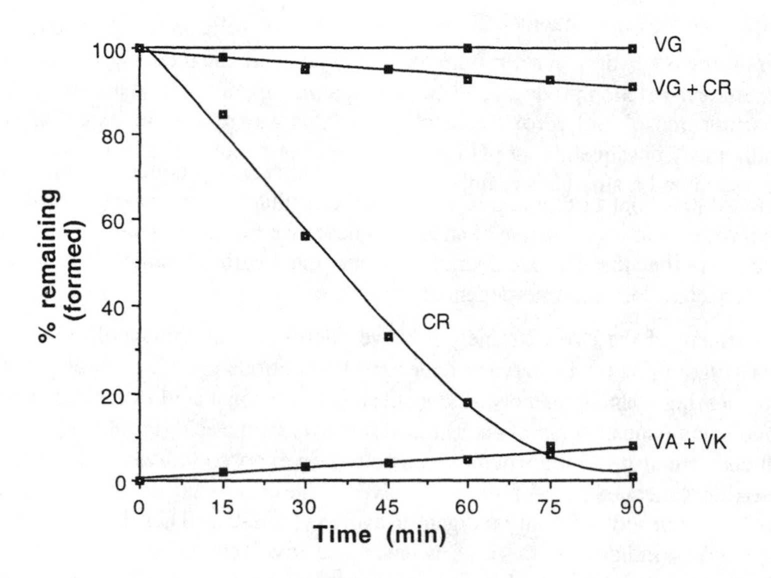

Figure 14. Formation of chemiluminescense (hydroxyl radicals) during the oxidative degradation of creosol + veratryglycol. CR refers to creosol, VG to veratrylglycol and VG + CR to a mixture of veratryglycol and creosol (1:1).

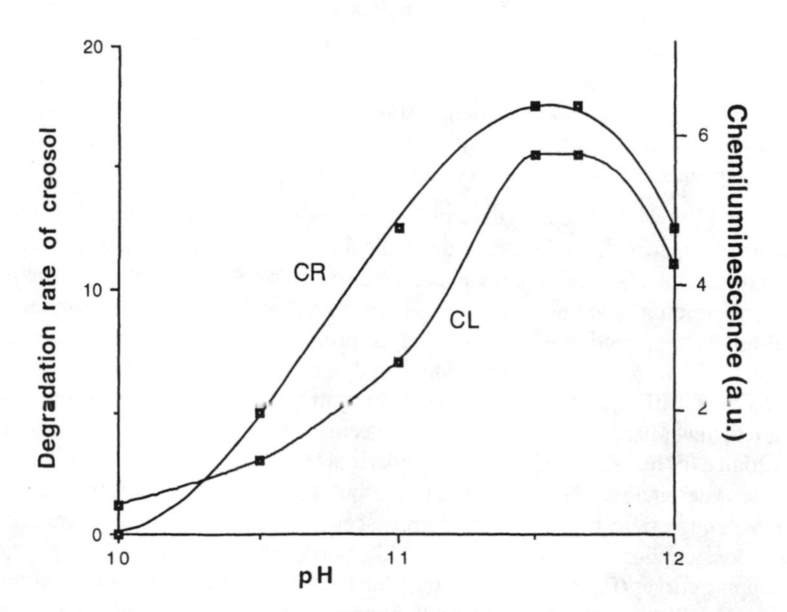

Figure 15. pH-dependence of the degradation of creosol and of the formation of chemiluminescence (hydroxyl radicals).

affording the oxyl anion radical, $O^{\cdot -}$ which has less ability to add to aromatic rings, but rather abstracts a hydrogen atom from the side chain, here the methyl group of creosol. This results in reaction products still containing an aromatic ring, which compete with the starting creosol for hydroxyl radicals and in this way reduce the extent of creosol degradation. Consequently, at pH lower than 11.5 and higher than 11.8, less creosol is degraded and a smaller amount of hydroxyl radicals is produced, explaining the maxima at pH about 11.6 in Figure 15. Thus, the similar pH-dependence of the rate of creosol degradation and the rate of chemiluminescence formation lends further support to the view that the creosol degradation and the hydroxyl radical formation are interconnected - or even interdependent processes.

4) Treatment of the two extremely reactive phenols, *t*-butylguaiacol (TBG) and *t*-butylsyringol (TBS), with oxygen under mild conditions gives analogous products, arising *via* ring opening, namely *t*-butyl-substituted muconic acid esters and, from the syringol compound also, *t*-butyl-substituted furane carbonic acid (Figure 16). The same products were also formed when the phenols were exposed to hydroxyl radicals and superoxide, generated by γ-irradiation. However, these products do not arise when the phenols are reacted with only oxygen or hydroxyl radicals. Therefore, they may be interpreted as indicating the intermediacy and involvement of superoxide in the oxidative degradation of the phenols, in agreement with reactions illustrated in Figure 10.

5) In Figure 6, the formation of hydroxyl radicals is proposed to be due to a reductive cleavage of hydrogen peroxide by superoxide according to a Fenton mechanism. Another mode of formation of hydroxyl radicals would imply homolytic cleavage of the peroxy linkage in hydroperoxy intermediates, as shown in the lower part of Figure 6.

In order to obtain experimental support for the latter alternative, the rate of hydroxyl radical formation during autoxidation of *t*-butylguaiacol was compared with that during autoxidation of *t*-butylsyringol (Figure 17).

In *t*-butyl-substituted guaiacol, the hydroperoxide intermediate formed by addition of superoxide to the free ortho-radical has two options to react further: The first is intramolecular nucleophilic attack resulting in ring opening with formation of the corresponding muconic acid ester (Figure 4) and is similar to the corresponding reaction of *t*-butylsyringol. The second is proton elimination with concomitant aromatization to give the corresponding hydroperoxyphenol which should readily undergo homolytic cleavage of the oxygen-oxygen bond affording a hydroxyl radical (Figure 6, lower part) and a highly stabilized semiquinone radical. On account of this extra source for hydroxyl radicals, *t*-butylguaiacol should be expected to produce these radicals faster and in a larger amount than *t*-butylsyringyl. However, this is not the case. When the ratio between chemiluminescence (hydroxyl radical) formation and phenol consumption is plotted versus time, the phenols TBS and TBG gave practically overlapping curves (Figure 18), indicating that the mode of hydroxyl radical formation during autoxidation of the two phenols is practically the same. This means that it

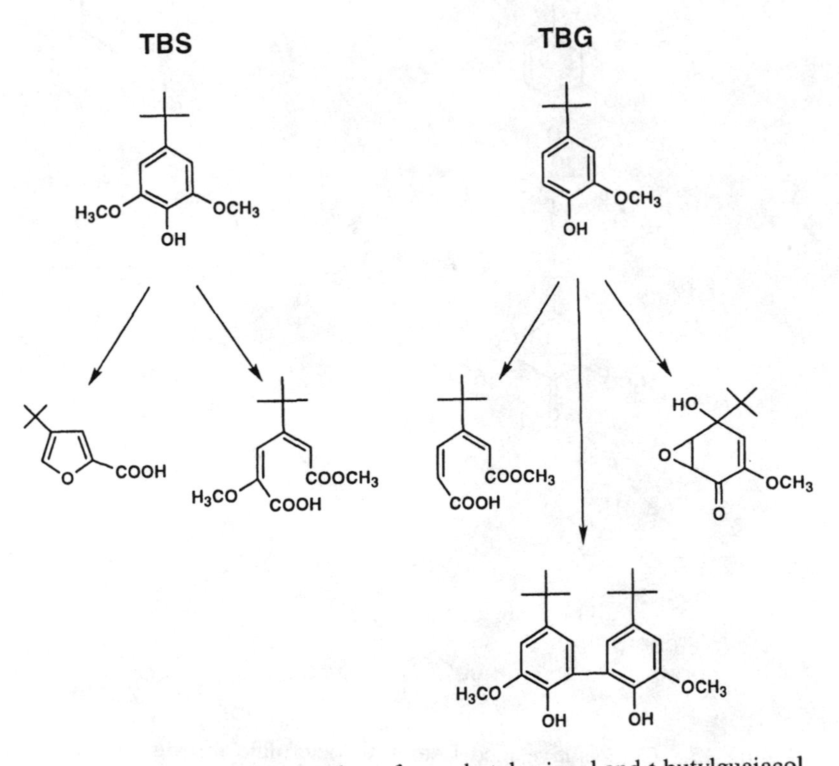

Figure 16. Autoxidation products from *t*-butylsyringol and *t*-butylguaiacol.

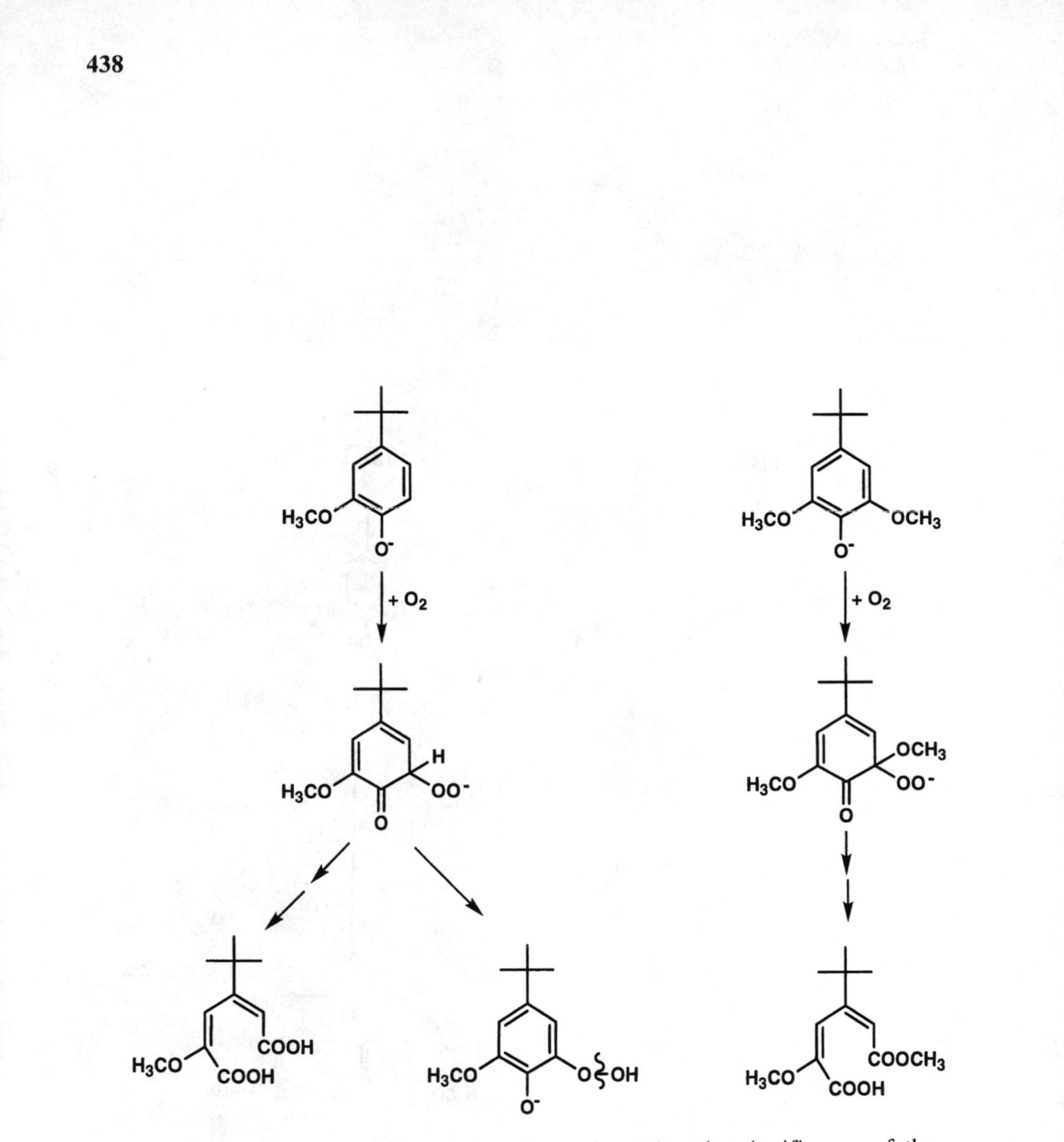

Figure 17. Comparative model study to determine the significance of the homolytic cleavage of hydroperoxides.

proceeds *via* reductive cleavage of hydrogen peroxide, according to the Fenton mechanism (Figure 6, upper part).

6) The initial step of autoxidation deserves a special comment. It has been suggested (*12*) that the electron transfer from an activated substrate to oxygen and the combination of the two resultant radicals may be facilitated by taking place in a solvent cage as indicated in Figure 19. In order to determine the significance of such a solvent cage in the initial step of the autoxidation, *t*-butylsyringol was reacted with oxygen under bleaching conditions in the absence and in the presence of a large excess of creosol and the rates of consumption of *t*-butylsyringol were compared. The kinetic data from this competition experiment revealed that only about 15% of the *t*-butylsyringyl radicals combine with superoxide anions in an initial solvent cage. The high yield of the 5,5-coupling product in the autoxidation of *t*-butylguaiacyl which contains one unsubstituted ortho-position (Figure 16) is also in agreement with the above result that only a small amount of *t*-butylguaiacyl radicals combine with superoxide in a cage. From these experiments, it may be concluded that the autoxidation of phenolic substrates in alkaline solution does not proceed in solvent cages to any greater extent.

7) The autoxidation mechanism proposed for the degradation of phenolic (Figure 10) and non-phenolic (Figure 11) lignin structures can also be applied to the degradation of carbohydrate structures. In this particular case, the initiation of the process is brought about by reducing end groups.

For example, under appropriate conditions, the reducing sugar ß-*D*-glucopyranose is degraded by oxygen under bleaching conditions and, as is true for creosol, the degradation is accompanied by the formation of hydroxyl radicals. In Figure 20, these two variables are plotted *versus* time.

A possible mechanism for the initial steps of the degradation of glucose, involving the formation and consumption of superoxide, is presented in the lower part of Figure 21: In alkaline solution, the reducing end group is ionized and the ionized (enolate) form is oxidized *via* a single electron transfer to molecular oxygen, giving rise to a substrate (= enoxyl) radical and superoxide. Thus, this initial step of autoxidation of the carbohydrate model parallels that of the lignin model creosol, i.e., the oxidation of an enolate corresponds to the oxidation of a phenolate. The subsequent reaction steps of the substrate radical, that is addition of superoxide anion radical, followed by cleavage of a carbon to carbon bond, are also analogous to the corresponding reaction steps of phenoxyl radicals outlined in the upper part of the Figure. However, carbohydrate-derived (enoxyl) radicals, unlike phenoxyl radicals, react also with oxygen to give carbonyl structures (glucosones) (*13*) and further oxidation products.

The dismutation of superoxide to give hydrogen peroxide and the reductive cleavage of the latter affording hydroxyl radicals proceed, of course, in the same way as described for the autoxidation of creosol. Hydroxyl radicals thus generated by

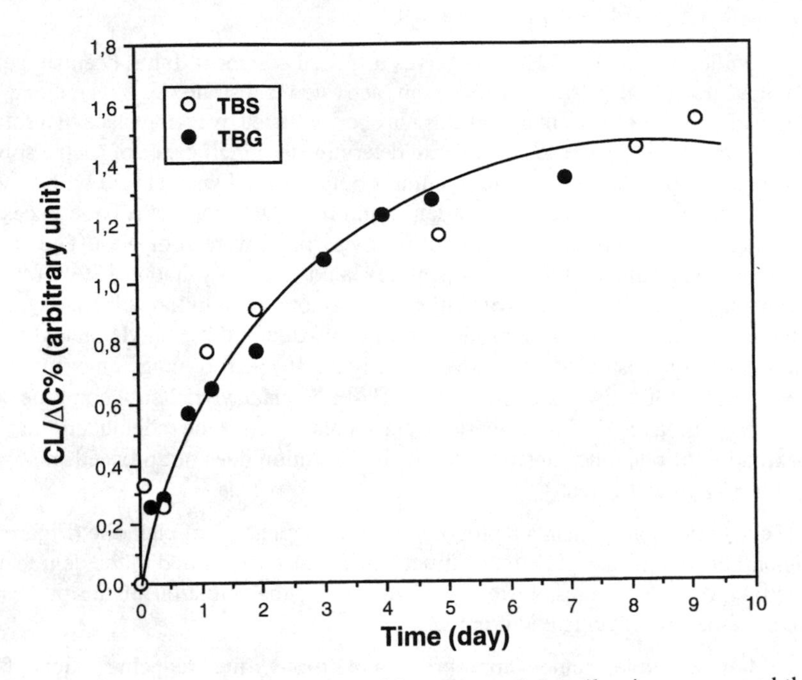

Figure 18. Ratio between the rate of formation of chemiluminescence and the rate of degradation of *t*-butylsyringol and *t*-butylguaiacol during the autoxidation.

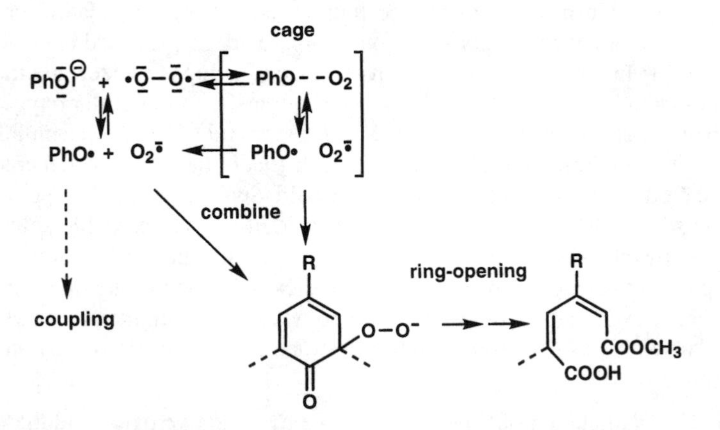

Figure 19. One-electron transfer from phenolate to oxygen and combination of phenoxyl radical with superoxide inside and and outside of a cage.

autoxidation of reducing end groups in carbohydrates, like those generated by autoxidation of phenolic groups in lignins, may add to aromatic structures in lignins or may abstract a hydrogen atom from lignin side chains or from any carbon atom in reducing and non-reducing carbohydrate structures, to give the corresponding carbon-centered radical(s). These radicals together with the primary substrate radicals, react then further to give the final degradation products.

In contrast to the degradation of creosol which goes to completion (Figures 12 and 13), the oxidative degradation of glucose under similar conditions seems to be limited to about 50% conversion (Figure 20). The formation of hydroxyl radicals, as measured by chemiluminescence, ceases at about the same extent of conversion. These findings indicate that the coupling between degradation of the substrate and the formation of hydroxyl radicals is much stronger in the case of the carbohydrate model (G) than in the case of the lignin model (C). An explanation for the fact that the degradation of glucose (G) and the formation of chemiluminescence (CL) cease at about 50 % conversion could be that metal ions, the autoxidation catalysts, are trapped by hydroxy acids, formed during the degradation of the carbohydrate model.

This would block the initial electron transfer and the Fenton reaction, i.e., the degradation of glucose and the formation of chemiluminescence (hydroxyl radicals).

The suggested mechanism of autoxidation of ß-*D*-glucopyranose (G) (Figure 21) is supported by the behaviour of its non-reducing derivative, methyl ß-*D*-glucopyranoside (MeG). It is stable and does not produce any detectable amount of hydroxyl radicals, when treated under similar conditions (Figure 20). However, when the treatment of the methyl glucoside is performed after addition of a phenolic compound, e.g. creosol, an appreciable degradation of the non-reducing sugar model is observed (see also Ref. *14*). Again, this result confirms the supposition that during the autoxidation of creosol hydroxyl radicals are generated which bring about the partial degradation of the (non-reducing) methyl glucoside, in analogy to the degradation of (the non-phenolic) veratrylglycol in the presence of creosol (Figure 13). It can be expected that the degradation of non-reducing sugars can also be initiated by the presence of reducing representatives.

Evidently, the autoxidation of carbohydrate structures takes a similar course as that of lignin structures. Both processes may therefore be outlined in a common Scheme (Figure 22) which describes the generation and reactions of superoxide and hydroxyl radicals and their interplay in general terms. As is true for the special schemes for the autoxidation of creosol (Figure 10) and veratrylglycol (Figure 11), this general scheme includes the four one-electron reduction steps I to IV from oxygen to water and illustrates the involvement and the cooperation of the two radical intermediates. The following course of events given in a chronological order is proposed:

Step I. One electron is transfered from an activated substrate, SB^-,particularly from phenolates, stilbene-derived phenolates or enolates, to oxygen with formation of

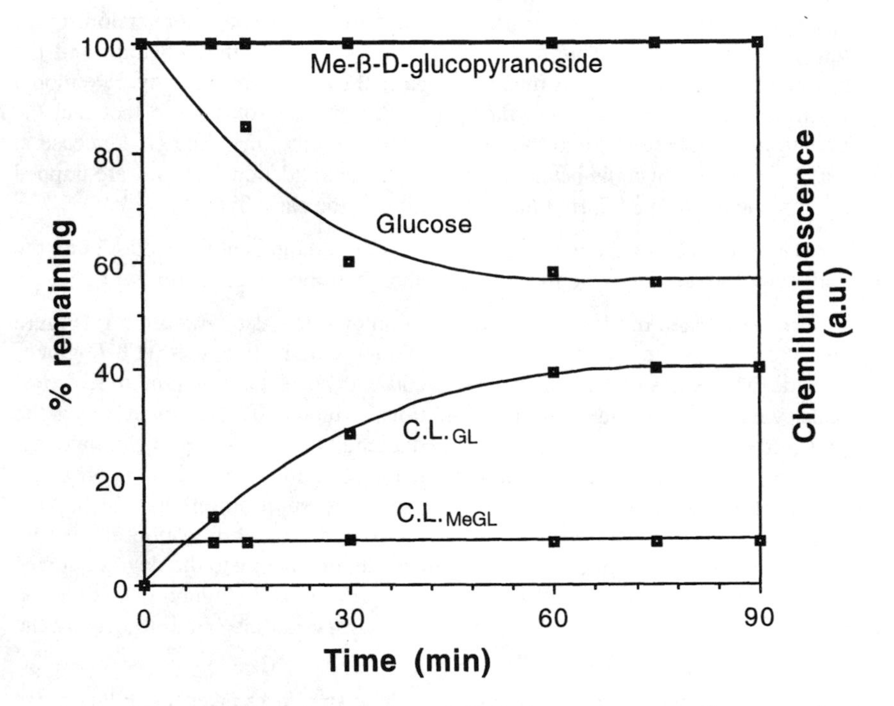

Figure 20. Oxidative degradation of β-*D*-glucopyranose with concomitant formation of chemiluminescence (hydroxyl radicals) and stability of methyl β-*D* -glucopyranoside.

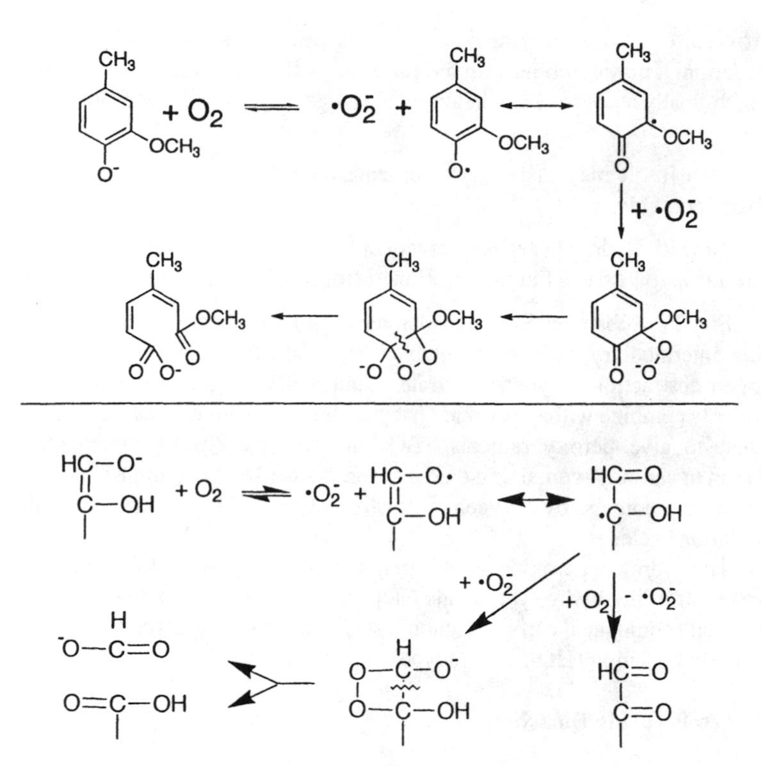

Figure 21. Mechanism of autoxidation of phenolate and enolate anions.

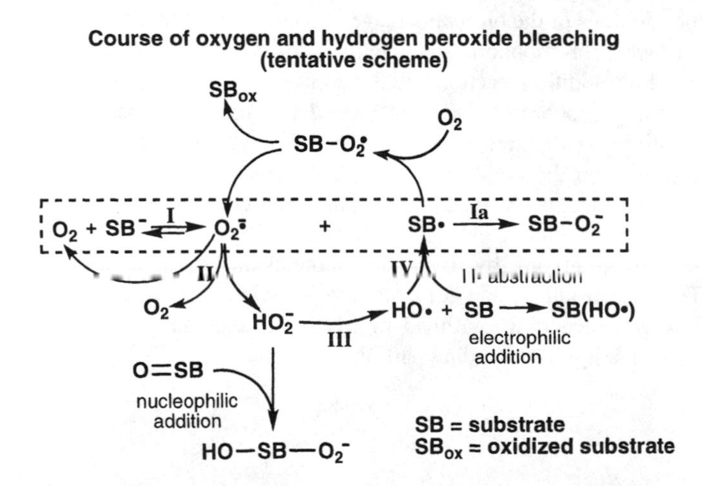

Figure 22: General scheme describing the course of oxygen and hydrogen peroxide bleaching (Reproduced with permission from Reference number 8).

superoxide, $O_2^{\cdot-}$. This intermediate occupies a key position in the course of the autoxidation. It may combine with the substrate radical to give a peroxide anion (step Ia) which finally leads to degradation by cleavage of carbon-carbon bonds (Figure 4) or -

Step II - it may undergo proton-catalyzed dismutation to give oxygen and hydrogen peroxide.

Step III. Hydrogen peroxide is reductively cleaved, in a heavy metal-catalyzed, superoxide anion-driven Fenton reaction, affording hydroxyl radicals and finally,

Step IV. Substrates SB are oxidatively attacked either *via* electrophilic addition to give intermediary hydroxycyclohexadienyl radicals $SB(HO^{\cdot})$(Figure 7), or *via* hydrogen abstraction to yield substrate radicals $SB^{\cdot}$ (Figure 8). Substrate radicals cannot only combine with superoxide (step Ia) but also with oxygen (except phenoxyl radicals) to give peroxy radicals $SBO_2^{\cdot}$ which may eliminate superoxide with formation of carbonyl-containing oxidation products (SB_{ox}) (Figure 8). This oxidation of substrate radicals by oxygen with formation of superoxide completes the autoxidation cycle.

The hydrogen peroxide anion can not only undergo metal-catalyzed reductive cleavage, affording hydroxyl radicals (step III), but may also nucleophilically attack carbonyl and conjugated carbonyl (enone) structures (O=SB), affording intermediates of the peroxide anion ($HO\text{-}SB\text{-}O_2^-$) type.

Hydrogen Peroxide Bleaching

The general scheme (Figure 22) may also be used to describe the course of hydrogen peroxide bleaching. Instead of starting from oxygen, the process starts from hydrogen peroxide. Hydrogen peroxide bleaching consists of a main "nucleophilic part", i.e., nucleophilic additions of the bleaching reagent to enone and other carbonyl structures.

In this way, chromophoric and potential chromophoric systems are destroyed. These nucleophilic addition reactions are therefore regarded as the lignin-retaining part of the bleaching process and are formulated in Figure 23 for all carbonyl and conjugated carbonyl structures, as well as for some representative examples.

Depending on the conditions used, hydrogen peroxide bleaching may also include, to varying extents, an "electrophilic part", caused by hydroxyl radicals, formed via metal-catalyzed reductive cleavage of the bleaching reagent (Figure 22, step III), and by superoxide, arising by oxidation of the bleaching reagent with hydroxyl radicals. This "electrophilic part" of hydrogen peroxide bleaching should follow the routes of oxygen bleaching, outlined in figure 22 and may be considered to be responsible for the lignin-degrading part of the process.

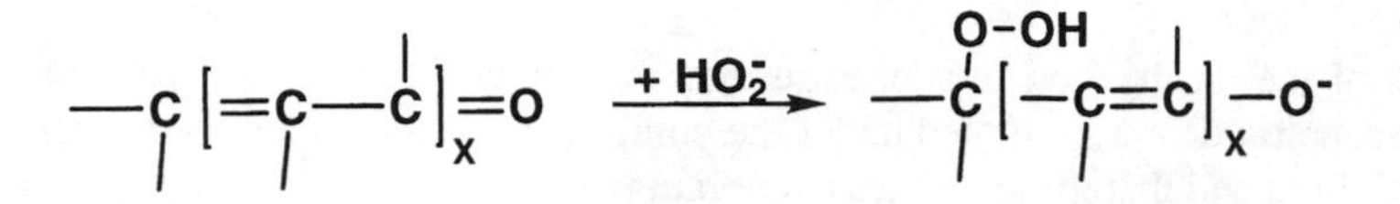

Nucleophilic addition of HO_2^- to enone structures

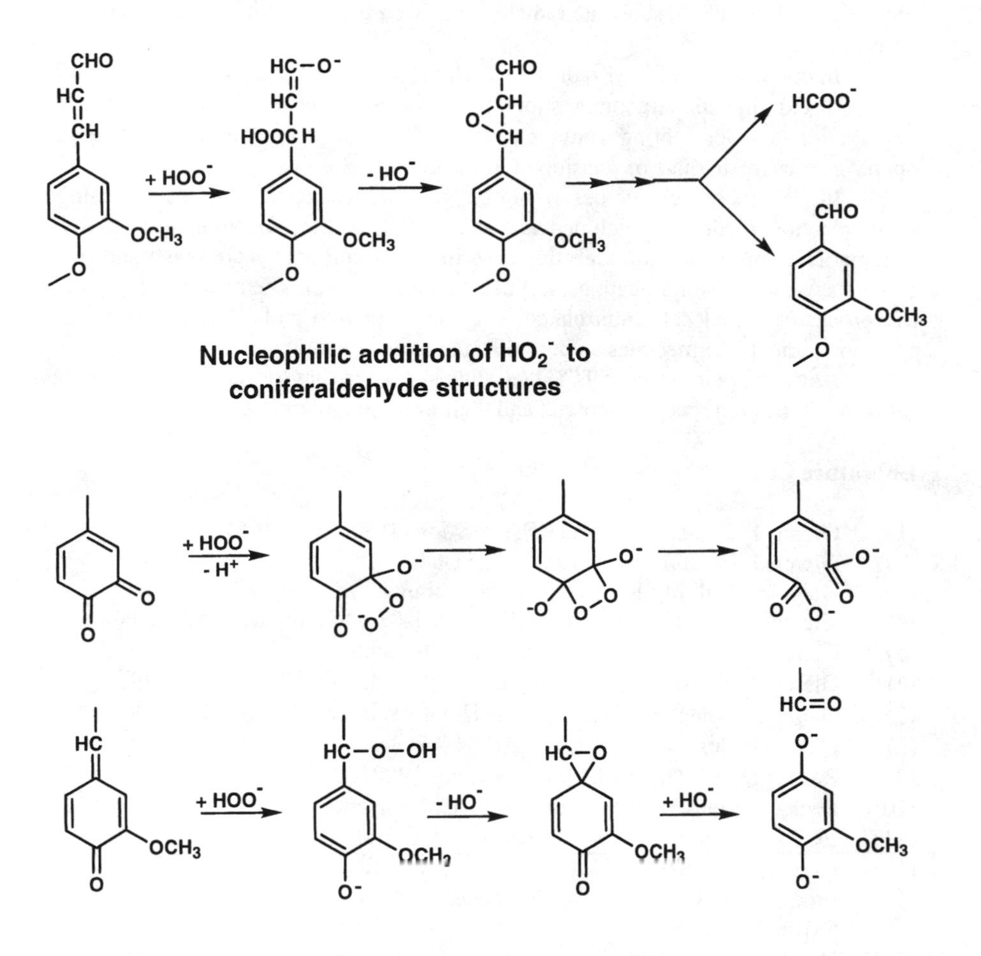

Nucleophilic addition of HO_2^- to quinone and quinone methide structures

Figure 23. Nucleophilic addition of hydrogen peroxide anions to carbonyl and enone structures.

Summary and Conclusion

On the basis of the results from model experiments, it is concluded that hydroxyl radicals and superoxide are generated under the conditions of oxygen and hydrogen peroxide bleaching and that these species cooperate in the degradation of lignin and carbohydrate structures during the processes. On the one hand, the formation of hydroxyl radicals presupposes the intermediacy of superoxide and hydrogen peroxide, and on the other, the genesis of superoxide requires hydroxyl radicals for the oxidation of substrate molecules to substrate radicals, followed by oxygenation and superoxide elimination.

In contrast to hydroxyl radicals which react fast and unselectively with both aromatic and aliphatic structures, superoxide radicals react slowly and, in specific degradation reactions, bring about cleavage of carbon-carbon bonds, leading to opening of aromatic rings or scission of aliphatic side chains.

In the general scheme describing oxygen and hydrogen peroxide bleaching, these reaction modes are included and the course of the bleaching processes is schematically outlined. Although this description is certainly incomplete and may contain several oversimplifications, it illustrates that, from the chemical point of view, the processes are closely interrelated, due the formation and involvement of the common radical intermediates, superoxide and hydroxyl radicals.

Any attempt to improve TCF-bleaching sequences has to take into account the roles av all reacting species involved and their complex interplay.

Literature Cited

(1) Gierer, J.; Imsgard,F. Svensk Papperstidn. 1977, 16, 510-518.
(2) Gierer, J. Holzforschung 1982, 36, 55-64.
(3) Holocher-Ertl, M.; Kratzl, K. Holzforschung 1982, 36, 11-16.
(4) Gierer, J. Wood Sci. Technol. 1986, 20, 1-33 and references cited therein.
(5) Gierer, J.; Reitberger, T.; Yang, E. Holzforschung 1992, 46, 495-504.
(6) Gierer, J.; Reitberger, T.; Yang, E. Holzforschung 1994, 48, 405-414.
(7) Gierer, J.; Yang, E.; Reitberger, T. Holzforschung 1996, 50, 342-359.
(8) Gierer, J. Holzforschung 1997, 51, 34-46.
(9) Reitberger, T.; Gierer, J. Holzforschung 1988, 42, 351-356.
(10) Backa, S.; Jansbo, K.; Reitberger, T. Holzforschung 1997, 51, 557- 564.
(11) Ek, M.; Gierer, J.; Jansbo, K.; Reitberger, T. Holzforschung 1989, 43, 391-396.
(12) Kratzl, K.; Gratzl, J.; Claus, P. Adv. Chem. Ser. 1966, 59, 157-176.
(13) Eriksson, B.; Lindgren, B.O.; Theander, O. Cellulose Chem. Technol. 1973, 7, 581-591.
(14) Yasumoto, M.; Matsumoto, Y.; Ishizu, A. J. Wood Chem Technol. 1996, 16, 95-107.

Chapter 22

A Comparison of the Structural Changes Occurring in Lignin during Alcell and Kraft Pulping of Hardwoods and Softwoods

Ying Liu[1], Sandra Carriero[1], Kendall Pye[2], and Dimitris S. Argyropoulos[1,3]

[1]Department of Chemistry and [3]PAPRICAN, Pulp and Paper Research Centre, McGill University, 3420 University Street, Montreal H3A 2A7, Quebec, Canada
[2]Repap Technologies Inc., 2650 Eisenhower Avenue, Valley Forge, PA 19482

During this investigation, softwood and hardwood chips were pulped to different extents using conventional kraft and Alcell protocols. The dissolved and residual lignins were then isolated and their functional group distributions examined using quantitative ^{31}P NMR. For both wood species, the solubilized kraft lignins contained the highest abundance of phenolic hydroxyl groups at all degrees of delignification, while the residual kraft lignins contained the lowest. This may be related to the considerably greater solvating abilities of alkaline aqueous media as opposed to those of ethanol toward inducing solubilization of the phenolic moieties. Consequently, at a given degree of delignification, a greater proportion of phenolic units are retained in the Alcell pulps, contributing to their documented higher reactivity with oxygen. As far as residual lignins are concerned, the data showed that condensed phenolic units are formed in greater abundance within softwood than in hardwood pulps for both processes. More specifically, condensed phenolic structures were formed most rapidly in the residual lignins of the softwood Alcell pulps and in particular during the early and latter phases of delignification. The stability of carbonium ions induced under the acidic conditions and the elevated temperatures of the Alcell process, may cause the formation of stable carbon-carbon condensed structures. These condensation reactions may be responsible for the deceleration of the delignification observed for softwoods during the Alcell process. This work also verified the presence of ester linkages that link p-hydroxyphenyl units within aspen milled wood lignin.

Introduction

The kraft pulping process has, for a long time, been the dominant method of production of strong chemical pulp for printing and writing papers. In recent years, however, kraft mills have been under public pressure to reduce or even eliminate their air and effluent discharges with mercaptans and other reduced sulfur compounds representing the majority of the odorous emissions of the process. In response to these issues a number of research efforts have been carried out aimed at the development of an economically viable and ecologically acceptable pulping process. Most alternative pulping processes that have been suggested use mixtures of organic solvents with water as the pulping media and as such they have been termed "organosolv pulping processes". Several literature accounts on this subject have reviewed the various solvent mixtures available, the resulting pulp characteristics and the overall advantages and limitations of these processes (1-4).

The Alcell process is an organosolv process that uses aqueous ethanol as the pulping liquor, qualifying it as a sulfur free and environmentally benign process. This process offers a fully bleachable chemical pulp at high yields with physical and optical properties comparable to those of kraft pulp. For this process solvent recovery is feasible via distillation, offering significant capital cost savings compared to the recovery equipment essential for a kraft mill i.e. recovery boiler, lime kiln and other pollution control equipment. The lignin fraction, the major by-product of the process, can potentially be utilized as either fuel or feedstock for chemical conversion (5-7). In a recent report by Lora (8), wheat straw and reed were successfully pulped by the Alcell process yielding pulps of competitive quality profiles to hardwood kraft pulps. The Alcell process, however, in its current state of development, is suitable only for the pulping of hardwoods and annual plants. Softwoods seem to be very resistant to the acidic delignification that takes place under Alcell conditions, with the precise reason remaining unclear (9).

The pH of the Alcell cooking liquor is relatively low, due to the organic acids that are generated by the hydrolysis of the hemicelluloses. Therefore, in contrast to the kraft pulping process, the Alcell process is an acidic delignification process (10). Fundamental enquiries aimed at examining the underlying chemistry occurring during the Alcell process, have shown that the cleavage of ether linkages in syringyl lignins is faster than their guaiacyl counterparts present in softwood lignin (6,7). This difference was offered as an explanation for the difficulties encountered in dissolving softwood lignin out of the wood with the Alcell process. The observations of Goyal and Lora (6,7) were later confirmed by Lai (11, 12) who studied the acidic delignification of aspen and Norway Spruce wood meals and demonstrated that the hydrolysis of the ß-aryl ether units had a direct impact on the initial delignification rates. The ß-aryl ethers present in the lignin of Norway Spruce were considerably more resistant to acid hydrolysis than those present in aspen.

Under acidic conditions, lignin can be subjected to condensation reactions mainly due to the instability of the resulting C_α carbonium ions. Obviously such reactions will be counterproductive to delignification and may eventually offer further difficulties during subsequent bleaching operations. Although model compound studies by Shimada (14, 5) have shown that the higher rate of lignin condensation occurring in softwood lignin models may be partially responsible for the slower softwood delignification rate, the role of these reactions during organosolv pulping remains unclear (13). In relation to these issues, a number of questions remain unanswered: Are condensation reactions of real significance during the Alcell process of softwood and hardwood? In case they do occur, are they more significant for softwoods than hardwoods and what are their formation profiles compared to the condensation reactions operating during kraft pulping? Questions of this nature as well as issues pertaining to the fundamental differences between the two processes have prompted the present investigation. In particular, softwood and hardwood chips were pulped to different extents using conventional kraft and Alcell protocols. The dissolved and residual lignins were then isolated and their functional group distributions examined using quantitative ^{31}P NMR (16-25).

Experimental

Alcell and Kraft Cooks

The Alcell cooks were carried out on aspen (*populus tremuloides*) and black spruce (*picea mariana*) chips by Repap Technologies Inc. The Alcell pulping conditions were: liquor to wood ratio 4.5, alcohol concentration 46%, maximum temperature 195°C, time at maximum temperature 0-60 min. Kraft cooks were carried out on the same chips in a laboratory batch digester, using a liquor to wood ratio of 4:1, a white liquor composition of 18.0% active alkali and 30% sulphidity. Other cooking conditions were: maximum temperature 120-170 °C, time to maximum temperature 58-90 min, time at maximum temperature 0-93 min. After pulping, the pulps were thoroughly washed with water and air-dried. The pulp yields and lignin contents were determined according to standard CPPA methods.

Isolation of Residual and Solubilized Kraft and Alcell Lignins

Residual lignins were obtained by extraction of kraft and Alcell pulps using a slightly modified acidolysis procedure (0.1 mol/L HCl in a 8.5:1.5 dioxane-water) for 2 hours described elsewhere (25). The extract was then concentrated, reprecipitated into a large volume of water whose pH was adjusted to 2-3. Residual lignins were dried under vacuum for 24 hours prior to quantitative ^{31}P NMR analyses.

The lignins from the kraft and Alcell black liquors collected at the various degrees of delignification were precipitated by acidification (pH ≅3-4). The liquor was centrifuged, washed with acidified water until the supernatant liquor was clear. The lignins were then dissolved in a mixture of dichloro- methane:ethanol (2:1), filtered and precipitated in ether, air-dried, and further dried under vacuum for 24 hours, prior to characterization.

Lignin Alkaline Hydrolysis

Aspen milled wood lignin (100 mg) were dissolved in 10 mL of 2M NaOH and stirred at 25°C for 48 hours. After the hydrolysis, the reaction mixture was acidified to pH 3 using 0.2 mol/L HCl and centrifuged. The solid residue was washed with water, centrifuged, freeze-dried and dried to constant weight at 40°C under reduced pressure.

Treatment of Aspen Milled Wood Lignin under Kraft Pulping Conditions

Initially a white liquor solution was prepared containing 7.0 g of sodium hydroxide and 6.2 g of sodium sulfide ($Na_2S.9H_2O$) per 100 mL deionized water. This solution was then mixed with an equal volume of absolute dioxane. Milled wood lignin (100 mg) was dissolved in 1.33 g of the above solution (approximately simulating a liquor/wood ratio of 4/1). The reaction was carried out in a 10 mL stainless steel bomb in an oil bath at 160°C for 20 min. After the reaction, the mixture was acidified to pH 3 using 0.2 mol/L HCl. The precipitate formed was collected by centrifugation, washed with acidified water, and dried to constant weight at 40°C under reduced pressure. The solution resulting from the aqueous acidic precipitation was then freeze-dried, extracted twice with tetrahydrofuran, dried over Na_2SO_4 and evaporated under reduced pressure. The residue was then dried in a vacuum oven at 25°C.

Quantitative ^{31}P NMR Analyses

Lignin samples were prepared by dissolving approximately 40 mg in 800 μL of a mixture of 1.6:1 pyridine/$CDCl_3$ (1.6/1 v/v) at room temperature. The lignin solution was then phosphitylated with 100 μL of 2-chloro-4,4,5,5-tetramethyl-1,3,2-dioxaphospholane. Solutions (100 μL) of cyclohexanol (1.1 mg) and chromium acetylacetonate (0.55 mg) in 10 mL of pyridine/$CDCl_3$ (1.6/1 v/v) were also added to serve as the internal standard and relaxation reagent, respectively. The quantitative ^{31}P NMR spectra were acquired by inverse gated decoupling on a Varian XL-300 MHz spectrometer operating at 121.5 MHz, by methods identical with those described previously (20, 23, 24-25). An observation sweep width of 6600 Hz was used. All chemical shifts reported were relative to the reaction product of water with the phosphitylation reagent, which has been reported to give a sharp signal at 132.2 ppm (22-24). For each spectrum 128 transients were acquired with a delay time of 25 sec between successive pulses.

Results and Discussion

Typical quantitative ^{31}P NMR spectra and signal assignments for residual and solubilized kraft and Alcell lignin samples isolated at approximately 40% extend of delignification are shown in Figure 1 for both the softwood and the hardwood species. The integration ranges applied and signal assignments are based on model compound and lignin studies carried out previously (16-20, 22, 24). Significant differences are apparent for all functional groups between the two species and the two processes. In the following account these data will be critically examined and discussed in the light of the underlying principles and chemistry for each process and species.

Delignification Profiles

The total phenolic hydroxyl group data (sum of syringyl, guaiacyl and condensed phenols) obtained for residual and solubilized lignins during the Alcell and kraft cooks for both species, as a function of the delignification extent, are plotted in Figure 2 . The delignification degree has been calculated by taking into account the pulp yields and lignin contents determined as Klason and UV lignins.

The enumerated limitations of pulping softwoods under Alcell conditions become obvious in the extend of delignification data being sampled during this effort. The softwood could only be delignified up to about 60% using the Alcell process. As such the collected data is limited in the 10-60% delignification range for the same of cooking time range.

With the exemption of the plot describing the phenolic group contents of the residual Alcell lignins, the total amount of phenolic groups present in all other lignins was found to increase as the degree of delignification was increased. This demonstrates that the gradual scission of aryl ether bonds present in lignin releases free phenolic hydroxyl groups, in agreement with previous accounts that have examined kraft and Alcell lignins using quantitative ^{31}P-NMR (21, 22, 24, 25) .

For both wood species, the solubilized kraft lignins seem to contain the highest abundance of phenolic hydroxyl groups at all degrees of delignification, while the residual kraft lignins the lowest (Figure 2). This may be related to the considerably greater solvating abilities of alkaline aqueous media as opposed to those of ethanol toward inducing solubilization of the phenolic moieties. This is also confirmed by the fact that for both wood species and at all degrees of delignification the Alcell residual lignins contained greater amounts of phenolic hydroxyl groups than the kraft residual lignins. Consequently the Alcell solubilized lignins contained lower amounts of phenolic hydroxyl groups than the kraft solubilized lignins. The fact that for both species the residual lignin present in Alcell pulps is enriched in phenolic hydroxyl groups as compared to residual kraft, may explain the observation that Alcell pulps can be oxygen delignified more extensively than their kraft counterparts (5, 26). The retained phenolic hydroxyl groups in Alcell pulps are easily attacked by molecular oxygen under

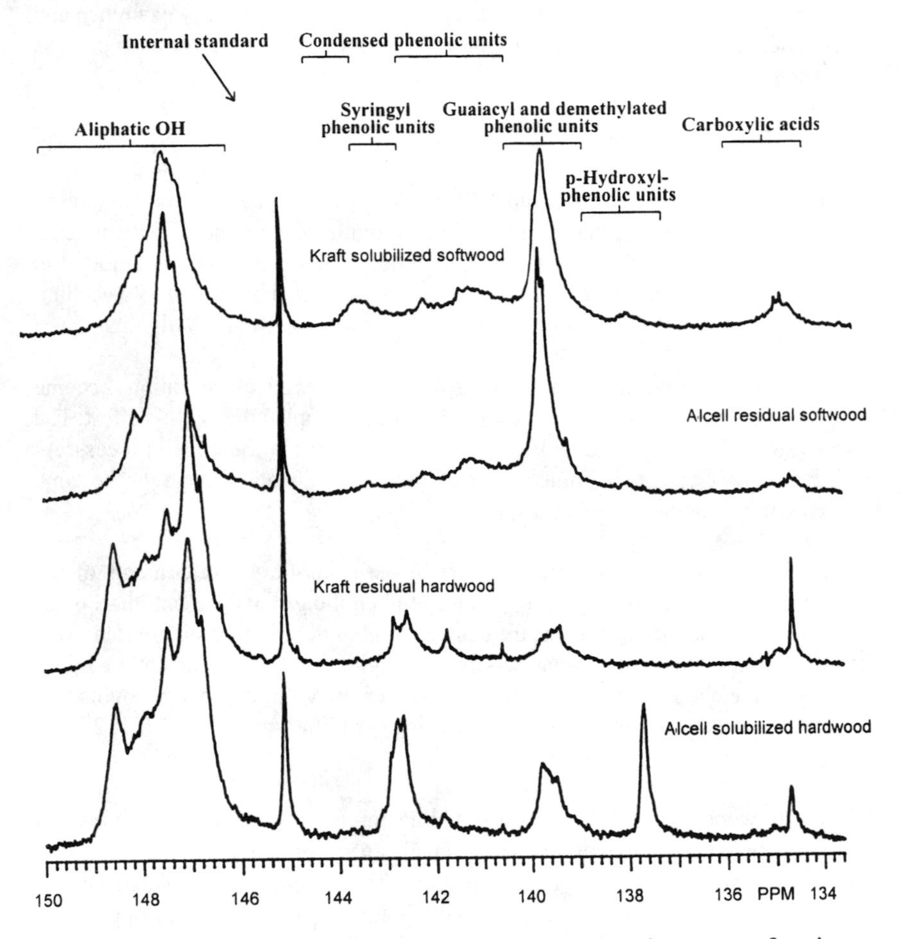

Figure 1. Quantitative ^{31}P NMR spectra and signal assignments of various lignins examined during this effort. The extend of delignification for all spectra displayed was approximately 40%.

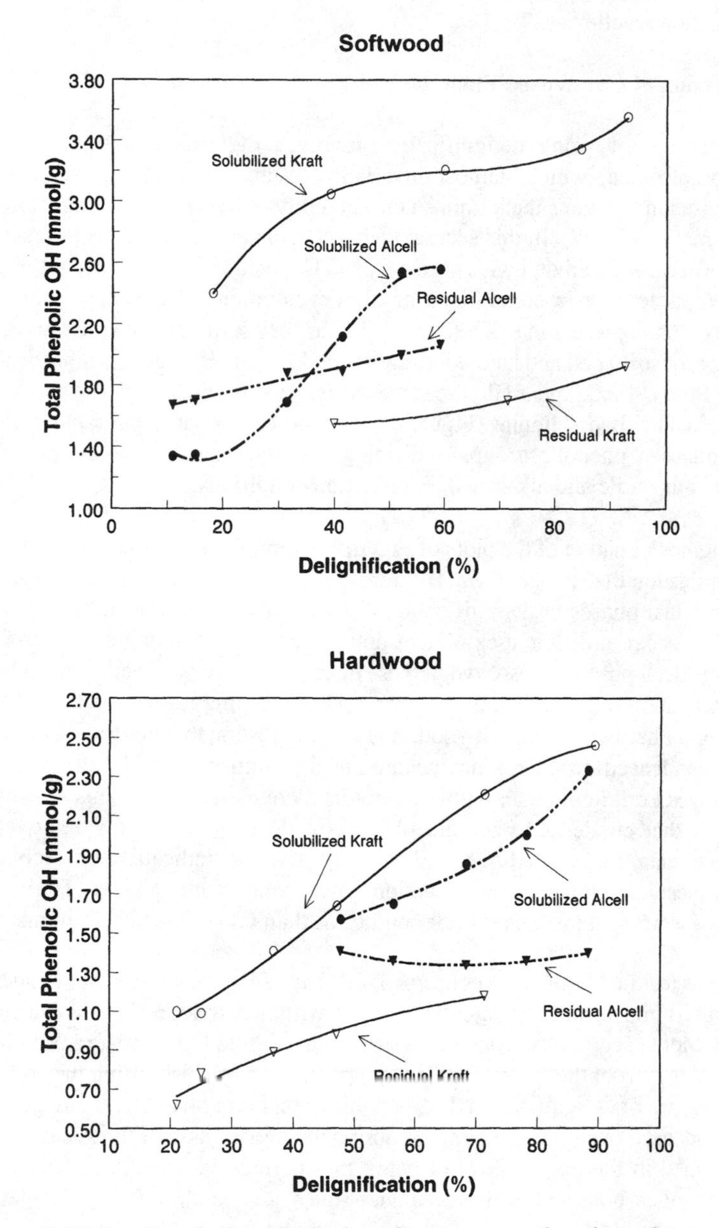

Figure 2. Plots of total phenolic hydroxyl groups as a function of extent of delignification for residual and solubilized lignins obtained after Alcell and Kraft cooks of softwood (upper) and hardwood (lower).

alkaline conditions, causing demethoxylation, ring opening and side chain elimination reactions (27).

Formation of Condensed Phenolic Units

The process of wood delignification involves the fragmentation of lignin macromolecules, which, almost invariably, is accompanied by competing condensation reactions that diminish the solubility of the residual lignin from the wood matrix (13, 28). In this section the term "condensation" refers to reactions that form carbon-carbon bonds at free C_5 and C_6 positions of a lignin unit. Figure 3 attempts to offer a comparison of the development of condensed phenolic hydroxyl units present in solubilized lignins, as they form during kraft and Alcell pulping for softwood and hardwood species at different degrees of delignification, while Figure 4 attempts to do the same for residual lignins. With the exemption of residual hardwood lignins (Figure 4), both processes cause the rapid build up of condensed phenolic groups, indicating that such species consistently form under both acidic and alkaline delignification conditions.

The sigmoidal nature of the plots of Figure 3 is reminiscent of the three phases of delignification that coincide with the transition points between the initial, the bulk and the final phases of delignification (29). However, the transition points are found to occur at different extents of delignification depending on the process. During Alcell pulping for softwood these three phases are observed at much lower delignification degrees than those during Alcell pulping for hardwood and kraft pulping for hardwood and softwood. It is also interesting to note that significantly more condensed structures are accumulated within softwood and hardwood solubilized kraft lignin than within their Alcell counterparts. This may be due to the fact that condensed phenolic units formed are more readily removed in alkaline than in acidic liquors. This may also be indicative of secondary intermolecular condensation reactions occurring with greater facility in homogeneous systems under kraft conditions than under Alcell conditions.

As far as residual lignins are concerned, the data of Figure 4 show that condensed phenolic units form in greater abundance within softwood than in hardwood pulps. More specifically, condensed structures seem to build most rapidly in the residual lignins of the softwood Alcell pulps and in particular during the early and latter phases of delignification. It is very likely that for softwoods the majority of the condensed phenolic units formed during the initial phase of the Alcell process is retained in the pulp instead of being transported into the Alcell liquor. The stability of carbonium ions induced under the acidic conditions and the elevated temperatures of the Alcell process, may cause the formation of stable C_α-C_6 and C_α-C_1 structures bearing phenolic hydroxyl groups. This is evidenced by the pronounced signals between 140-141.7 ppm in the spectra of Figure 1 (22,24).

7✓1

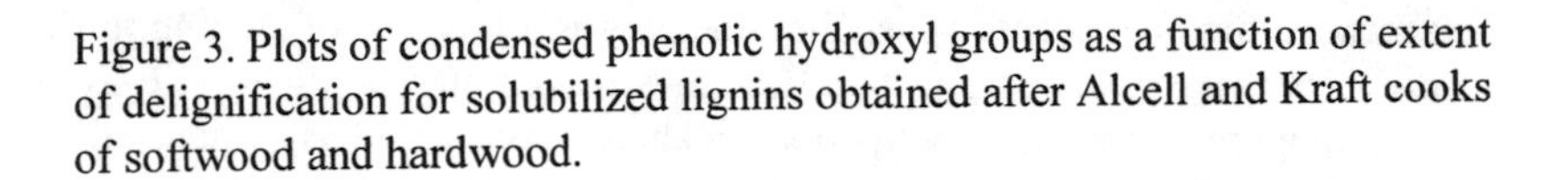

Figure 3. Plots of condensed phenolic hydroxyl groups as a function of extent of delignification for solubilized lignins obtained after Alcell and Kraft cooks of softwood and hardwood.

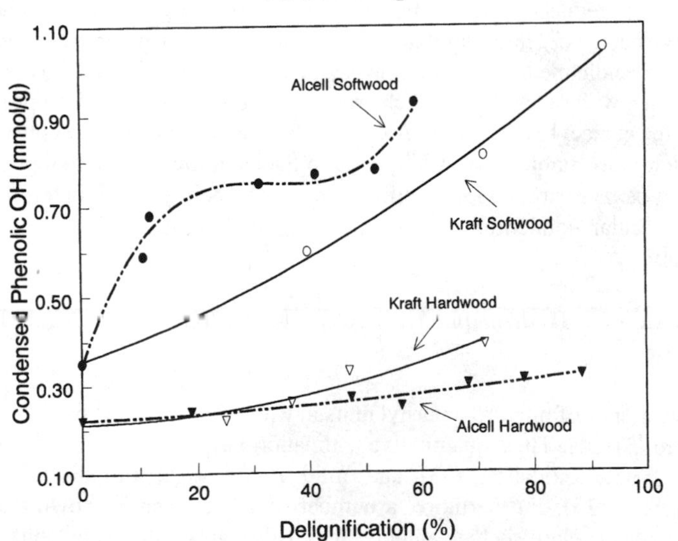

Figure 4. Plots of condensed phenolic hydroxyl groups as a function of extent of delignification for residual lignins obtained after Alcell and Kraft cooks of softwood and hardwood.

In an effort to gain further insight into the profiles of formation of condensation reactions under Alcell pulping conditions, the amounts of condensed phenolic structures present within the residual lignins were correlated with the extent of delignification and the respective pulping time profiles (Figure 5). The negative values on the x-axis indicate actual reaction times below the desired temperature of the Alcell process (195°C).

While aspen wood was significantly delignified (40% delignification) even before optimum temperature was reached (time 0 and beyond), almost no delignification was observed for the case of black spruce wood. At the same time, during the heat-up period and well beyond it, condensed structures were found to rapidly accumulate within residual black spruce lignins. In contrast, the condensed structures were found to accumulate at considerably lower rates within aspen residual lignins. These condensation reactions may also be responsible for the deceleration of the delignification rate observed for the softwood after 40 minutes at temperature (195°C). In contrast, the rate of delignification of the hardwood was considerably higher throughout the process.

It is also interesting to observe that for softwood the sigmoidal plot observed for the accumulation of condensed phenolic structures as a function of pulping time, is in concurrency with the plot that describes its delignification profile as a function of pulping time. This behaviour can also be considered as evidence that the condensation reactions that occur in softwood inhibit its efficient delignification under Alcell pulping conditions. Meshgini and Sarkanen (30) have showed that the rate of acid hydrolysis of α-aryl ether softwood model dimers was dramatically reduced when the benzyl moiety was changed from a guaiacyl to a syringyl nucleus. This is probably due to the fact that guaiacyl carbocations formed in acidic media are more reactive than their syringyl counterparts, rapidly condensing with other electron-rich aromatic carbons. Condensation products in which the guaiacyl nuclei act as electron-rich aromatic carbons are considered to be much more stable (15). Additives of a nucleophilic nature may prevent the reactive benzyl carbonation intermediates from undergoing counterproductive intermolecular condensations with other lignin fragments during the Alcell process.

The Fate of p-Hydroxyphenyl Groups Under Kraft and Alcell Pulping Conditions

The occurrence of p-hydroxyphenyl units in wheat straw lignins has been widely reported (31-34). Their quantitative estimation has been made on the basis of permanganate oxidation (33) and mild hydrolysis/quantitative ^{31}P NMR experiments (35). Furthermore, a number of workers have shown that aspen wood contains relatively high amounts of p-hydroxyphenylpropane units (31, 36, 37). For straw lignin such units have been considered to be mainly due to the presence of esterified p-coumaric acid (34, 35), but their bonding patterns in

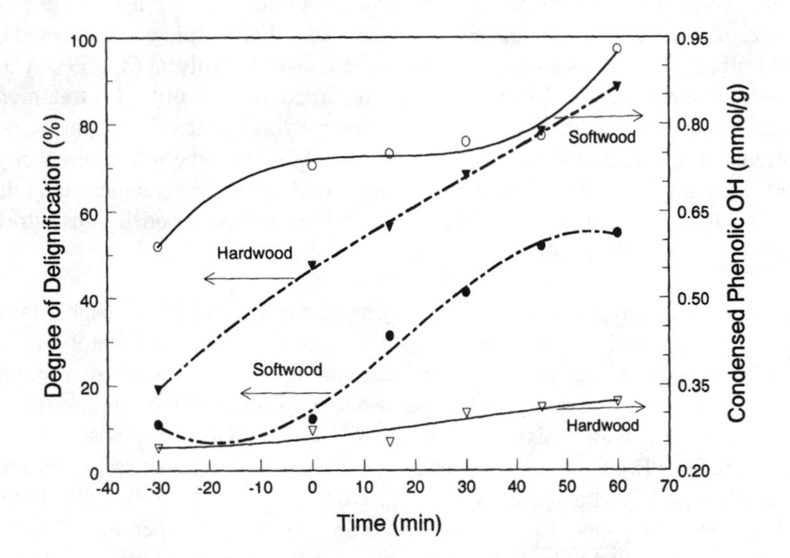

Figure 5. Profiles for the formation of condensed phenolic structures present within residual lignins under Alcell pulping conditions and correlation with the rates of delignification for softwood and hardwood.

aspen are not yet clear. One possibility is that enzymatically pre-esterified p-coumaric acid with p-hydroxycinnamyl alcohol monomers, may cause the formation of p-hydroxycinnamyl p-coumarates which could participate in the formation of the lignin macromolecule by conventional oxidative coupling reactions to yield γ-p-coumaroylated lignin (38).

In relation to the objectives of this effort, however, it was thought appropriate to examine the role and fate of such units under kraft and Alcell pulping conditions. As expected, the ^{31}P NMR spectrum of aspen milled wood lignin (Figure 6a) shows signals characteristic of the phosphitylated hydroxyls groups of p-hydroxyphenyls (137.4-138 ppm). On the basis of this spectrum the aspen milled wood lignin sample was determined to contain about 4 p-hydroxyphenyl groups per 100 phenyl propane units. Furthermore, this sample was subjected to mild alkaline hydrolysis resulting in complete ester hydrolysis (35, 39). After quantitative recovery its ^{31}P spectrum was recorded (Figure 6b). This treatment caused a distinct decrease in the amount of p-hydroxyphenyl groups and an increase in the amount of carboxylic acids, implying that the p-hydroxyphenyl units are partly esterified within aspen milled wood lignin, in agreement with the results of previous accounts related to hardwood milled wood lignins, straw lignins and grasses (37, 40, 41).

In accordance with previous accounts (21), signals responsible for phosphitylated hydroxyl groups of p-hydroxyphenyl groups were found to be present in the ^{31}P NMR spectra of Alcell solubilized and residual lignins. The presence of free p-hydroxyphenyl groups in organosolv hardwood lignins has also been reported by Goyal and Lora on the basis of FTIR studies (7) and Gallagher on the basis of ^{13}C-NMR studies (42). However, only trace amounts of such species were apparent in the ^{31}P NMR spectra of kraft residual lignins (Figure 1). To further clarify the role of p-hydroxyphenyl groups under pulping conditions, aspen milled wood lignin was subjected to kraft delignification conditions and the resulting liquor after acidification gave a precipitate whose ^{31}P NMR spectrum (Figure 6c) was free of p-hydroxyphenyl moieties. However, the remaining aqueous fraction (Figure 6d) contained all the p-hydroxyphenyl units that were present in the original milled wood lignin (0.18 mmol/g lignin). These experiments verified the presence of ester linkages that link p-hydroxyphenyl units within aspen milled wood lignin. In addition they supplied convincing evidence that such units do not participate in alkali promoted condensation reactions during kraft pulping of aspen, since they were quantitatively recovered in the aqueous fraction. It can also be concluded that the linkages of the p-hydroxyphenyl units within aspen wood lignin are more resistant to the Alcell process than the kraft process.

Carboxylic Acid Group Formation

The formation of carboxylic acids within the polymeric matrix increases the hydrophilicity of lignin macromolecules and promotes their dissolution in the

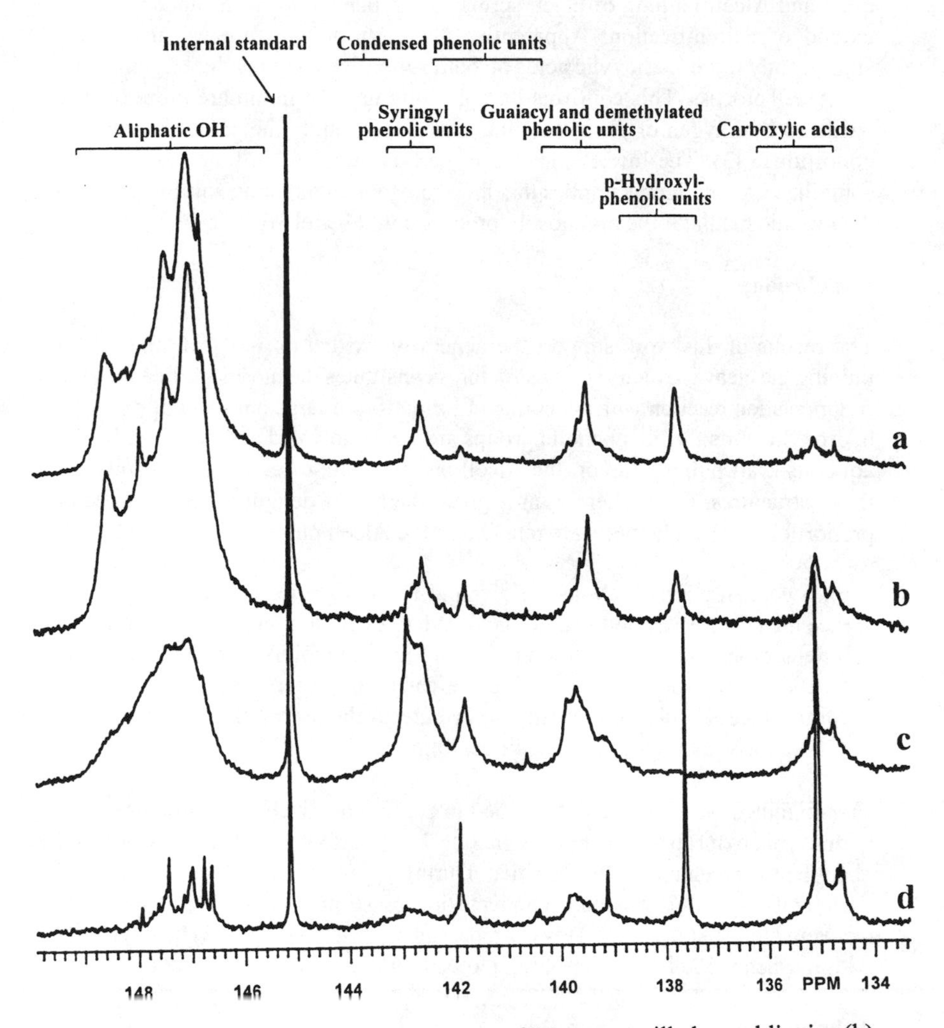

Figure 6. Quantitative ^{31}P NMR spectra of (a) aspen milled wood lignin; (b) aspen milled wood lignin after alkaline hydrolysis; (c) acid precipitated fraction after treating aspen milled wood lignin under Kraft pulping conditions; (d) dissolved fraction after treating aspen milled wood lignin under Kraft pulping conditions.

pulping liquor. For this reason it was thought appropriate to comparatively survey the profiles of formation of these groups during Alcell and kraft processes. Figure 7 shows the carboxylic group contents of residual and solubilized lignins for the kraft and Alcell pulping of black spruce and aspen woods as a function of the extend of delignification. Apparently, the kraft pulping process introduces significantly more carboxylic acids on both solubilized and residual lignins than the Alcell process. This confirms that phenolic units in lignin are more readily oxidized by oxygen or air in alkaline media than under the acidic organosolv conditions (43). The low abundance of carboxylic acids in both residual and solubilized Alcell lignins implies that their role toward inducing solubilization in the organic media of the organosolv process could be relatively minor.

Conclusions

The results of this work support the general view that during kraft and Alcell pulping the cleavage of aryl ether structures constitutes the most significant lignin fragmentation reaction with concomitant formation of large amounts of phenolic hydroxylic units. While phenolic groups are easily solvated and removed by the aqueous kraft pulping liquor, the Alcell pulping liquor has a lower affinity for these structures. Consequently, at a given degree of delignification, a greater proportion of phenolic units are retained in the Alcell pulps.

Lignin condensation reactions were found to be significantly more facile for spruce than their hardwood counterparts under Alcell delignification conditions. Condensed phenolic lignin structures were found to rapidly accumulate within the residual softwood lignins during the early and latter stages of the Alcell pulping process. This may greatly contribute to the slow delignification rates observed for softwoods under Alcell conditions.

Aspen milled wood lignin was verified to contain relatively high amounts of p-hydroxyphenylpropane structures (about 4 p-hydroxyphenyl groups per 100 phenyl propane units), partly esterified. During the kraft cook these units did not participate in alkali promoted condensation reactions but were quantitatively recovered in the kraft liquor. However, the linkages of the p-hydroxylphenyl units within aspen wood lignin were found to be more resistant to Alcell conditions.

Acknowledgements

The suggestions of Dr. J. Bouchard of Paprican are gratefully acknowledged. This work was supported by the Natural Sciences and Engineering Research council of Canada in the form of a Strategic Grant.

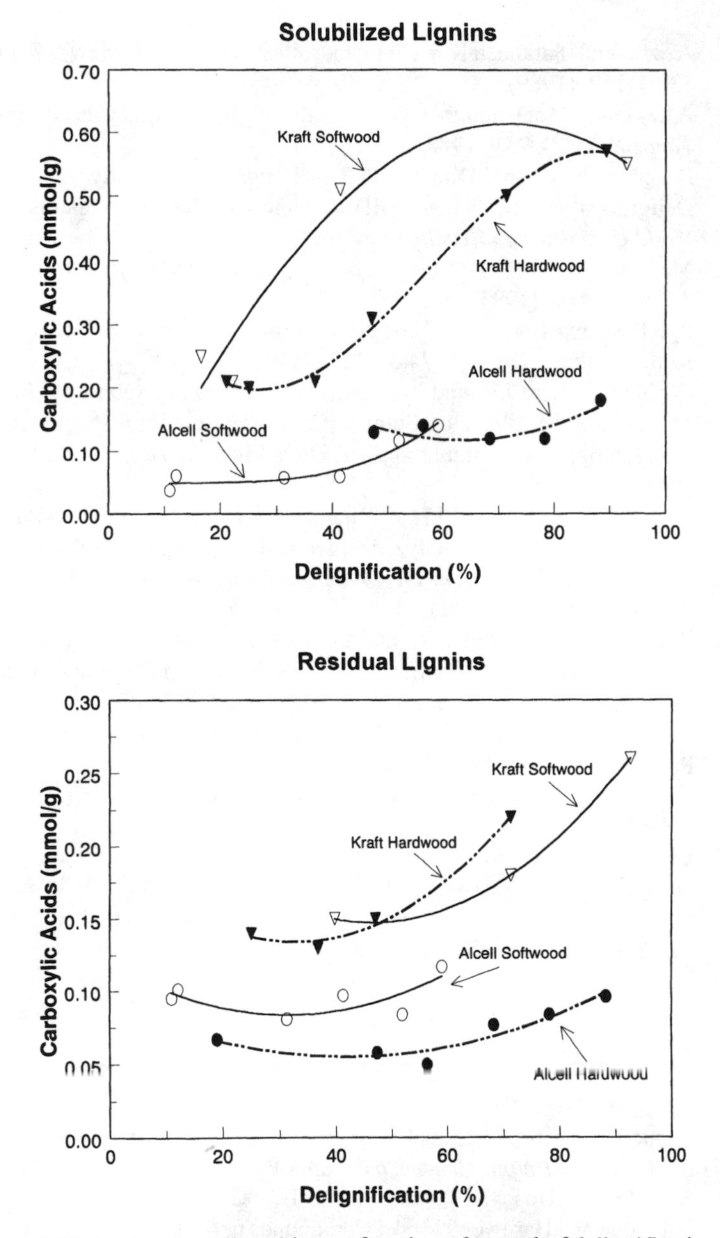

Figure 7. Plots of carboxylic acids as a function of extend of delignification for solubilized (upper) and residual (lower) lignins obtained after Alcell and Kraft cooks of softwood and hardwood.

Literature Cited

1. Aziz S. and Sarkanen K.V., "Organosolv Pulpings-A Review", *Tappi J.*, 73(3), 169 (1989)
2. Aziz S. and McDonough T.J., "Solvent Pulping-Promise and Progress", *Tappi J.*, 71(2), 251 (1988)
3. Brogdon B.N. and Dimmel D.R., "Fundamental Study of Relative Delignification Efficiencies (III): Organosolv Pulping Systems", *J. of Wood Chemistry and Technology* 16(3), 297 (1996)
4. McDonough T.J., "The Chemistry of Organosolv Delignification", *Tappi J.*, 76(8), 186 (1993)
5. Pye P.K. and Lora J.H., "The Alcell Process: A Proven Alternative to Kraft Pulping", *Tappi J.*, 74(3), 113 (1991)
6. Goyal G.C., Lora J.H. and Pye E.K., "Autocatalyzed Organosolv Pulping of Hardwoods: Effect of Pulping Conditions on Pulp Properties and Characteristics of Soluble and Residual Lignin", *Tappi J.*, 75(2), 110 (1992)
7. Goyal G.C. and Lora J.H., "Kinetics of Delignification and Lignin Characteristics in Autocatalyzed Organosolv Pulping of Hardwoods" *6th International Symposium on Wood and Pulping Chemistry*, Finland, Proceedings I, 205 (1991)
8. Winner S.R., Minogue L.A. and Lora J.H., "Alcell Pulpings of Annual Fibers", *9th International Symposium on Wood and Pulping Chemistry*, Montreal, Proceedings, Poster presentations, 120 (1997)
9. Schroeder M.C., "Possible Lignin Reactions in the Organocell Pulping Process", *Tappi J.*, 74(10), 197 (1991)
10. Sarkanen K.V., "Chemistry of Solvent Pulping",*Tappi J.*, 73(10), 215 (1990)
11. Lai Y.-Z. and Mun S.-P., "The Chemical Aspects of Acidic Delignification Processes. 1. Role of Aryl-Ether Hydrolysis in Aspen", *Holzforschung*, 48, 203 (1994)
12. Lai Y.-Z. and Guo X.-P., "Acid-Catalyzed Hydrolysis of Aryl Ether Linkages in Wood: Kinetics and Influence on Lignin Reactivity", *6th International Symposium on Wood and Pulping Chemistry*, Proceedings, 199 (1991)
13. McDonough T. J., "The Chemistry of Organosolv Delignification", *Tappi J.*, 76(8), 186 (1993)
14. Shimada K., Hosoya S., and Tomimura, Y., *6th Internatinal Symposium on Wood and Pulping Chemistry*, Tappi Press, Atlanta, 183 (1991)
15. Shimda K., Hosoya S., and Ikeda T., "Condensation Reactions of Softwood and Hardwood Model Compounds under Organic Acid Cooking Conditions", *J. of Wood Chemistry and Technology*, 17(1&2), 57 (1997)
16. Archipov Y., Argyropoulos D.S., Bolker H.I. and Heitner C., "^{31}P NMR Spectroscopy in Wood Chemistry. Part I: Model Compounds", *J. of Wood Chemistry and Technology*, 11(2), 137 (1991)

17. Archipov Y., Argyropoulos D.S., Bolker H.I. and Heitner C., "^{31}P NMR Spectroscopy in Wood Chemistry. Part II: Phosphite Derivatives of Carbohydrates", *Carbohydr. Res.*, 220, 49 (1991)
18. Argyropoulos D.S., Archipov Y., Bolker H.I. and Heitner C., "^{31}P NMR Spectroscopy in Wood Chemistry. Part IV: Lignin Models, Spin-Lattice Relaxation Times and Solvent Effects in ^{31}P NMR", *Holzforschung*, 47, 50 (1993)
19. Argyropoulos D.S., Bolker H.I., Heitner C. and Archipov Y., "^{31}P NMR Spectroscopy in Wood Chemistry. Part V: Quantitative Analysis of Lignin Functional Groups", *Journal of Wood Chemistry and Technology*, 13(2), 187 (1993)
20. Argyropoulos D.S., "Quantitative Phosphorus-31 NMR Analysis of Lignin; A New Tool for the Lignin Chemist", *Journal of Wood Chemistry and Technology*, 14(1), 45 (1994)
21. Argyropoulos D.S., "Quantitative Phosphorus-31 NMR Analysis of Six Soluble Lignins", *Journal of Wood Chemistry and Technology*, 14(1), 65 (1994)
22. Jiang Z.-H., Argyropoulos D.S. and Granata A., "Correlation Analysis of ^{31}P NMR Chemical Shifts with Substituent Effect of Phenols", *Mag. Res. Chem.*, 33, 375 (1995)
23. Sun Y. and Argyropoulos D.S., "Fundamental of High-Pressure Oxygen and Low-Pressure Oxygen-Peroxide (Eop) Delignification of Softwood and Hardwood Kraft Pulps: A Comparison", *J. of Pulp and Paper Science*, 21(6) (1995)
24. Granata A. and Argyropoulos D.S., "2-Chloro-4,4,5,5-Tetramethyl-1,3,2 Dioxaphospholance, a Reagent for the Accurate Determination of the Uncondensed and Condensed Phenolic Moieties in Lignins", *J. of Agricultural and Food Chemistry*, 43(6), 1538 (1995)
25. Jiang Z.-H. and Argyropoulos D.S., "The Stereoselective Degradation of Arylglycerol-β-acryl Ethers during Kraft Pulping", *J. of Pulp and Paper Science*, 20(7), 183 (1994)
26. Cronlund M. and Powers J., "Bleaching of Organosolv Pulps Using Conventional and Non-chlorine Bleaching Sequences", *Tappi J.*, 75(6), 189 (1992)
27. Ljunggren S., Gellerstedt G. and Petterson M., "Chemical Aspects on the Degradation of Lignin During Oxygen Bleaching" *6th International Symposium on Wood and Pulping Chemistry*, Proceedings I, 299 (1991)
28. Gierer J., "The Reactions of Lignin during Pulping", *Svensk Paperstid.*, 73, 571(1970)
29. Gellerstedt G. and Gustafsson K., "Structural Changes in Lignin During Degradation", *J. of Wood Chemistry and Technology,* 7(1), 65 (1987)
30. Meshgini M. and Sarkanen K.V., "Synthesis and Kinetics of Acid-Catalyzed Hydrolysis of some αAryl Ether Lignin Model Compounds", Holzforschung, 43, 239 (1989)

31. Nimz H.H., Robert D., Faix O. and Nemr M., "Carbon-13 NMR Spectra of Lignins, 8. Structural Differences between Lignins of Hardwoods, Softwoods, Grasses and Compression Wood", *Holzforschung*, 35, 16 (1981)
32. Higuchi R. and Kawamura I.,"Occurrence of p-Hydroxyl-Phenylglycerol-ß-Aryl Ether Structure in Lignins", *Holzforschung*, 20, 16 (1966)
33. Erickson M., Miksche G.E. and Somfai I., "Characterisierung der Lignine von Angiospermen durch Oxidativen Abbau. II. Monokotilen", *Holzforschung,* 27, 147 (1973)
34. Higuchi R., Ito Y., Shimada M. and Kawamura I., "Chemical Propertities of Milled Wood Lignin of Grasses", *Phytochemistry*, 6, 1551 (1997)
35. Crestini C. and Argyropoulos D.S., "Structural Analysis of Wheat Straw Lignin by Quantitative ^{31}P-NMR and 2D NMR Spectroscopy; The Occurrence of Ester Bonds and α-O-4 Substructures", *J. of Agricultural and Food Chemistry*, 45(4), 1212 (1997)
36. Venverloo C.J., "The Lignin of Populus nigra L.cv. 'Italica' and some other Salicaceae", *Holzforschung*, 25, 18 (1971)
37. Whiting P. and Goring D.A.I., "Chemical Characterization of Tissue Fractions from the Middle Lamella and Secondary Wall of Black Spruce Tracheids", *J. Wood Science and Technolnology*, 16, 261 (1982)
38. Ralph J., Hatfield R.D., Quideau S., Helm R.F., Grabber J.H. and Jung H.J.G., "Pathwaysof p-Coumaric Acid Incorporation into Maize Lignin as Revealed by NMR", *J. Am. Chem. Soc.*, 116, 9448 (1994)
39. Scalbert A., Monties B., Lallemand J.Y., Guittet E. and Rolando C., "Ether Linkage between Phenolic Acids and Lignin Fractions from Wheat Straw", *Phytochemistry*, 24, 1359 (1985)
40. Helm R.F. and Ralph J., "Lignin-Hydroxycinnamyl Model Compounds Related to Forage Cell Wall Structure. 2. Ester Linked Structures", *J. of Agricultural and Food Chemistry*, 41, 570 (1993)
41. Ralph J., Helm R.F., Qideau S. and Hatfield R.P., "Lignin-Feruloyl Esters Cross-Links inGrasses. Part I. Incorporation of Feruloyl Esters into Dehydrogenation Polymers", *J. Chem.Soc.*, Perkin Trans. I, 2961 (1992)
42. Gallagher D.K., Hergert H.L., Cronlund M. and Landucci L., "Mechanism of Delignification in an Autocatalyzed Solvolysis of Aspen Wood", *6th International Symposium on Wood and Pulping Chemistry*, Raleigh, NC, USA, 1989
43. Brink D.L., Bicho J.G. and Merriman M.M., "Oxidation Degradation of Wood III" In: "Lignin, Structure and Reactions", *Adv. Chem. Series*, 59, 177 (1966)

Chapter 23

A Room-Temperature β-O-4 to α-O-4 Rearrangement

D. L. Criss[1], T. H. Fisher[2], L. L. Ingram, Jr.[1], D. J. Beard[2], and Tor P. Schultz[1]

[1]Mississippi Forest Products Laboratory and [2]Department of Chemistry, Mississippi State University, Mississippi State, MS 39762

Nonphenolic β-aryl ether lignin models containing 4'-CHO, 4'-$COCH_3$, and 4'-NO_2 substituents rearranged in base at room temperature to give a mixture of the β- and α- aryl ethers. This unique base-catalyzed reversible rearrangement, which we investigated earlier by ^{17}O NMR and ^{18}O labeled MS, appears to react by an electrophilic 1,4 oxygen-to-oxygen aryl rearrangement (intramolecular S_NAr reaction). The rearrangement apparently occurs with resonance-stabilizing, electron-withdrawing groups, with the rate order being 4'-NO_2 > 4'-CHO > 4'-$COCH_3$.

The chemical pulping of wood to make paper requires the removal of lignin, a natural high molecular weight aromatic glue. Lignin roughly makes up 25% of the dry mass of wood and most of this is removed during pulping[1]. The chemical pulping process involves the depolymerization of lignin to form soluble low molecular weight products. The most frequent interunit linkage in native lignin is the arylglycerol-β-aryl ether (β-O-4) bond[2]. Cleavage of the β-O-4 bonds of non-phenolic lignins during the bulk delignification pulping stage has been implicated as the rate-limiting step[3] in chemical pulping and therefore has attracted a great deal of interest[4]. We have been looking at this reaction using substituted model compounds for structure-reactivity studies.[5a-d]

Prior work by this group uncovered an unusual base-catalyzed, room-temperature rearrangement[5e] of the 4'-CHO lignin model **1** (β-O-4) (Figure 1) to give an α-O-4 phenoxy compound (**2**). ^{17}O NMR studies and ^{18}O mass spectra data suggested a Meisenheimer complex [**8***] as an intermediate (Figure 2).

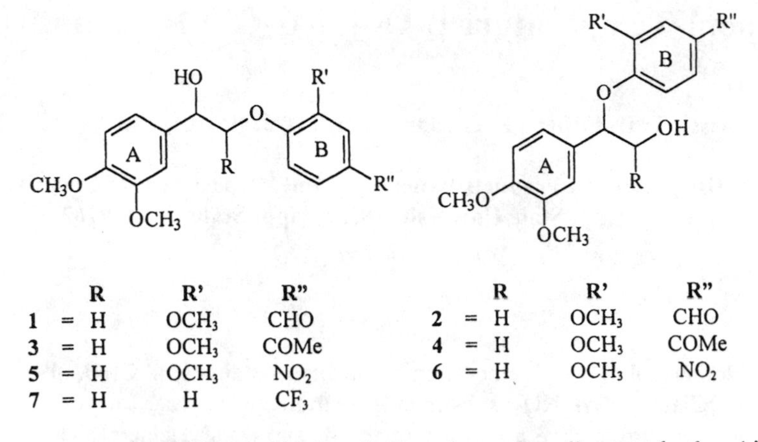

		R	R'	R"
1	=	H	OCH_3	CHO
3	=	H	OCH_3	COMe
5	=	H	OCH_3	NO_2
7	=	H	H	CF_3

		R	R'	R"
2	=	H	OCH_3	CHO
4	=	H	OCH_3	COMe
6	=	H	OCH_3	NO_2

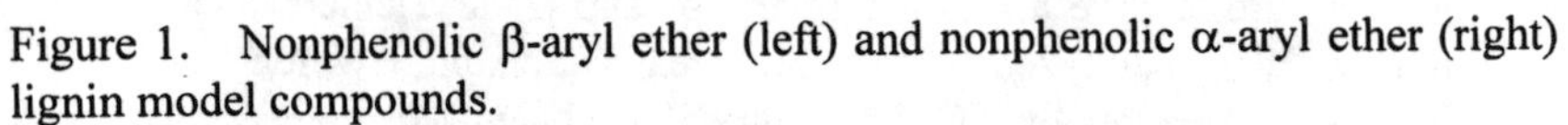
Figure 1. Nonphenolic β-aryl ether (left) and nonphenolic α-aryl ether (right) lignin model compounds.

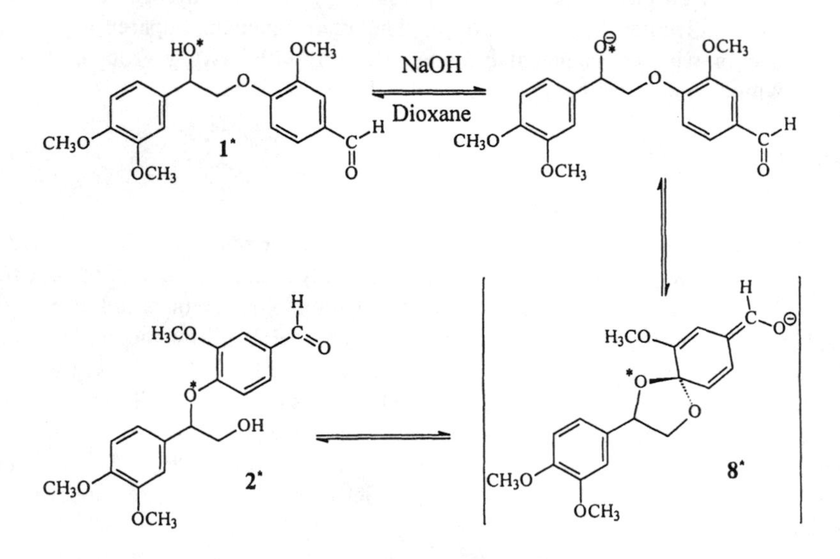

Figure 2. Possible Mechanism for Rearrangement: Spiro-Meisenheimer Complex Intermediate as determined using labeled $^{17,18}O$.[5e]

Using a model compound similar to those we studied but with a 4'-NO_2 substituent[6], Lindeberg and Walding reported that the β-aryl ether **5** rearranged in base to the α-aryl ether **6**. The authors also proposed that the rearrangement occurred via a Meisenheimer intermediate[7] but did not present any mechanistic evidence. Also, a stable spiro-Meisenheimer complex with a N_2^+ substituent has been recently reported.[8] We could locate no other literature reference and thus this rearrangement may be relatively unusual.

The purpose of this research was to further study this unusual reaction using lignin models **3** (4'-$COCH_3$) and **5** (4'-NO_2), which would have electron withdrawing capabilities less and greater than our prior 4'-CHO compound, respectively. We decided to use the relatively simple two carbon side chain models in our study since the three carbon models will introduce the diastereomeric effect. Also, a γ-OH should influence the rearrangement to give α-, β-, and γ-phenoxyls. However, future work with three-carbon side chain models would better represent native lignin.

Results and Discussion

Compounds **3** and **5** were treated in base at room temperature and the products isolated and purified as detailed in the experimental section. NMR analysis of the isolated compounds, and our earlier results for **1**, showed similarities in the ^1H and ^{13}C spectra for compounds **1**, **3**, and **5** (β-O-4). The same is true for compounds **2**, **4**, and **6** (α-O-4 compounds) (Table 1). NMR and MS of **4** and **6** showed the rearrangement product was the expected α-phenoxyl compound, as discussed below.

Table 1. ^1H and ^{13}C NMR data of side chain for β-O-4 compounds **1**, **3**, **5** and α-O-4 compounds **2**, **4**, **6**.

Cpd.	Substituent	H_α	H_β	C_α	C_β
			β-O-4		
1	4'-CHO	5.09, m	4.17, d	72.4	75.5
3	4'-COMe	5.07, m	4.17, d	72.5	75.5
5	4'-NO_2	5.10, t	4.21, d	72.3	75.6
			α-O-4		
2	4'-CHO	5.46, dd	3.8[a]	83.1	67.5
4	4'-COMe	5.47, dd	3.8[a]	83.2	67.7
6	4'-NO_2	5.49, dd	3.8[a]	83.6	67.4

[a]H_β–Peaks resonate beneath the methoxyl peaks.

In the ^{1}H NMR, the α-H in **3** [and **5**] absorbed at 5.07 [5.10] (m, 1H) ppm and the β-H at 4.17,[4.17] (d, 2H) ppm, while in **4** [**6**] the α-H was observed at 5.47 [5.49] (dd, 1H) and the β-H at 3.8 [3.8] ppm. The β-H of **4** and **6** overlaps the three OCH_3 H's but is clearly seen in the CHCOR spectrum. The chemical shifts of the α- and β-carbons in **3** [**5**] and **4** [**6**] have even more pronounced differences. The α-C absorbed at 72.5 (72.5) ppm in **3** [**5**] and at 83.2 (83.6) ppm in **4** [**6**], a nearly 11 ppm downfield shift. The β-C shifted in the other direction [from 75.5 [75.6] ppm on **3** [**5**] to 67.7 [67.4] ppm on **4** [**6**]] because the "B" aryl ring moved from the β- to the α-carbon. MS analysis of both **5** and **6** gave no parent peak at *m/e* of 349, but major peaks were observed at *m/e* 169 and 180. (We had previously observed that the α-phenoxy **2** fragmented during MS analysis). These masses were assigned to the two ions formed from a scission of the C_β-O of **5** and the C_α-O bond of **6**; one product being the "B" ring with a nitro group (*m/e* of 169 mass) and the other the "A" ring with a two carbon side chain and a *m/e* of 180.

To obtain a general idea of the relative rearrangement rate at room temperature, compounds **1**, **2**, **3**, and **5** were treated in base and samples withdrawn at various times for HPLC analysis. The 4'-CHO derivatives **1** and **2** gave a few percent (based on HPLC peak area) of the rearranged product within 2 minutes, about 30% rearrangement product in 60 min and an equilibrium of about 45:55 (**1**:**2**) after 1020 min. By contrast, the 4'-$COCH_3$ compound **3** rearranged much slower, forming a ratio (**3**:**4**) of only about 6:1 after 24 hours at room temperature. The 4'-NO_2 derivative **5**, conversely, reacted very fast with near equal amounts of **5** and **6** formed after only 1 min, in agreement with a prior paper[6]. Thus, it appears that the relative rearrangement rate [NO_2>CHO>$COCH_3$] follows σ^- (1.27, 1.03, and 0.84, respectively[9]) rather then σ values (0.78, 0.42, and 0.50 respectively[9]). Substituents capable of direct resonance appear to be necessary, as mentioned earlier, since no rearrangement product was observed with the 4'-CF_3 compound **7** even at relatively high temperatures. Interestingly, the rearrangement of compounds **1**, **3**, and **5** proceed with minimal by-product formation. For example, after **1** and **2** were individually left for 1020 min at room temperature the total peak area of the starting and rearrangement products accounted for about 99% of the total peak area.

This is an uncommon rearrangement. First it is an electrophilic rearrangement which involves a 1,4 oxygen-to-oxygen aryl migration. Second, the reaction occurs with only moderately strong electron-withdrawing groups ($COCH_3$ and CHO)[9]. While a similar reaction was previously reported[8] it involved an extremely strong electron-withdrawing N_2^+ substituent (σ^- of 3.43[9]). In addition, while Meisenheimer complexes are well known[7] they typically involve *di-* or *tri-*nitrosubstituted compounds; this reaction occurs with only a 4'-$COCH_3$ with an electron-donating *o*-methoxy group also present.

We hope to study this unusual reaction further by synthesizing 3-carbon side chain models with α- and γ-dihydroxyls to see if all three possible phenoxyls, with retention of configuration, are obtained. In addition, we are interested to see if some rearrangement might occur during isolation of various lignins which contain some α-carbonyl groups to give unrealistically high α-phenoxyl levels.

Experimental Section

The methods for the 4'-CHO compounds **1** and **2** were previously reported.[5e] The rearrangement of compound **3** to **4** was performed in a similar manner but with a longer reaction time, and isolated as before.[5e] Compounds **5** and **6** (4'-NO_2) were formed by reducing the appropriate carbonyl with $NaBH_4$ in dioxane, with the rearrangement also occurring under this slightly basic condition.

2-(4-acetyl-2-methoxyphenoxy)-1-(3,4-dimethoxyphenyl)-ethanol (3) [β-O-4]. Synthesized earlier.[5a] Mp 139-140 °C.

2-(4-acetyl-2-methoxyphenoxy)-2-(3,4-dimethoxyphenyl)-ethanol (4) [α-O-4]. Isolated in a similar manner as **2**.[5e] ^1H NMR (acetone-d_6): δ, ppm: 2.45 (s, 3H CH_3CO), 3.77 (s, 3H, OMe), 3.80 (s, 3H, OMe) 3.8 (m, 2H, $H_β$), 3.90 (s, 3H, OMe), 4.2 (m, 1H, OH), 5.4 (dd, 1H, J = 3.8, 7.8 Hz, $H_α$), 6.8-7.6 (6H, Ar-H). ^{13}C NMR (acetone-d_6): δ, 26.4 (CH_3CO), 29.92 (CD_3 of solvent is reference), 56.1 (2 peaks), 56.4, 67.7 ($C_β$), 83.2 ($C_α$), 111.3, 111.8, 112.6, 115.3, 119.8, 123.6, 131.6,131.9, 150.1, 150.4, 150.8, 153.0, 196.5 (C=O).

2-(4-nitro-2-methoxyphenoxy)-1-(3,4-dimethoxyphenyl)-ethanol (5) [β-0-4]. The appropriate carbonyl precursor was synthesized using standard procedures. The ketone, 88 mg (2.50 mmoles), and 50 mg (1.34 mmoles) of $NaBH_4$ were dissolved in 10 mL of EtOH. The mixture was stirred at room temperature for 12 hrs, acetone (3 mL) added to quench the reaction then 40 mL of water added and the solution extracted 3x with methylene chloride. The combined extracts were washed with water, dried over sodium sulfate and evaporated to afford 65 mg of an α-O-4 / β-O-4 mixture. Analysis of the HPLC data indicated two products. Chromatography of this mixture using the prior procedure[5e] yielded 35 mg of the β-O-4 isomer as bright-yellow needles. Mp 121-122 °C. ^1H NMR (acetone-d_6): δ, ppm: 3.80 (s, 3H, OMe), 3.83 (s, 3H, OMe), 3.97 (s, 3H, OMe), 4.21 (d, 2H, J = 5.7 Hz, CH_2), 4.85 (b, 1H, $OH_α$), 5.10 (t, 1H, J = 5.7 Hz, $H_β$) 6.93 (d, 1H, J = 8.2), 7.04 (d, 1H J = 8.2), 7.18 (s, 1H), 7.18 (d, 1H, J = 9.0), 7.77 (s, 1H), 7.86 (d, 1H, J = 9.0). ^{13}C NMR (acetone-d_6): δ, 27.47 (CH_3CO), 29.92 (CD_3 of solvent is reference), 56.14 (2 peaks, OCH_3), 56.64 (OCH_3), 72.28 ($C_α$), 75.61 ($C_β$), 107.61, 111.41, 112.52, 112.77, 118.22, 119.38, 135.08, 142.23, 149.86, 150.21, 155.24.

2-(4-nitro-2-methoxyphenoxy)-2-(3,4-dimethoxyphenyl)-ethanol (6) [α-0-4]. Preparation and separation of the α-O-4 isomer was in a similar manner as the β-O-4 isomer **5**. The α-O-4 isomer elutes after the β-O-4 isomer with near baseline separation to a give yellow oil. ^1H NMR (acetone-d_6): δ, ppm: 3.77 (s, 3H, OMe), 3.79 (s, 3H, OMe), 3.80 (d, 1H, J = 7.8 Hz, CH_2), 3.97 (d, 1H, J = 7.8 Hz, CH_2), 4.00 (s, 3H, OMe), 5.49 (dd, 1H, J = 7.7, 3.8 Hz, CH), 6.90 (d, 1H, J = 8.2), 6.97 (d, 1H J

= 8.2), 7.08 (d, 1H, J = 8.8), 7.09 (s, 1H), 7.71 (d, 1H, 8.8), 7.75 (s, 1H). ^{13}C NMR (acetone-d_6): δ, 29.80 (CD_3 of solvent is reference), 56.03 (OCH_3), 56.08 (OCH_3) 56.73 (OCH_3), 67.43 (C_β), 83.55 (C_α), 107.76, 111.31, 112.67, 114.88, 117.85, 119.81, 131.14, 142.26, 150.270, 150.46, 150.77, 154.52.

Acknowledgements

Special thanks goes out to Curry Tempelton for running GC/MS samples and Dr. Lynn Prewitt for HPLC assistance. This research was funded by USDA/WUR grant 96-34158-2557 and NSF grant EPS-9452857. Approved for publication as Journal Article No. FP 108 of the Forest and Wildlife Research Center, Mississippi State University.

Literature Cited

1. Pettersen, R.C. In *The Chemical Composition of Wood: The Chemistry of Solid Wood*; Rowell, R.M., Ed.; American Chem. Society, Washington, DC, 1984, pp. 57-126.
2. Sarkanen, K.V. In *Lignins-Occurrence, Formation, Structure and Reactions;* Ludwig, C. H., Ed.; Wiley Interscience: New York, NY, 1971, pp 205.
3. Gierer, J.; Ljunggren, S., *Svensk Papperstidn*, **1979**, *82*, 71. Ljunggren, S., *Svensk Papperstidn*, **1980**, *83*, 363.
4. (a) Sipilä, J.; Syrjänen K., *Holzforschung*, **1995**, *49*, 325. (b) Helms, R.; Ralph, J, *J. Wood Chem. Technology*, **1993**, *13*, 593. (c) Adler, E.; Lindgren, B.O.; Saeden, U.: *Svensk Papperstidn*, **1952**, *55*, 245.
5. (a) Hubbard, T.F.; Schultz, T.P.; Fisher, T.H., *Holzforschung*, **1992**, *46*, 315. (b) Collier, W.E.; Fisher, T.H.; Ingram, Jr. L.L.; Harris, A.L.; Schultz, T.P., *Holzforschung*, **1996**, *50*, 420. (c) Criss, D.L.; Collier, W.E.; Fisher, T.H.; Schultz, T.P., *Holzforschung*, **1998**, *52*, 171. (d) Criss, D.L.; Fisher, T.H.; Schultz, T.P., *Holzforschung*, **1998**, *52*, 57. (e) Criss, D.L.; Fisher, T.H.; Schultz, T.P.; Ingram, Jr. L.L.; Saebo, D.B., *J. Org. Chem.* **1997**, *62*, 7885.
6. Lindeberg, O.; Walding, J., *Tappi J.* **1987**, *70*, 119.
7. (a) Parker, K.A.; Coburn, C.A., *J. Org. Chem.* **1992**, *57*, 97. (b) Bernasxoni, C.F.; Wang, H., *J. Am. Chem. Soc.* **1976**, *98*, 6265. (c) Strass, M.J., *Acc. Of Chem. Res.* **1974**, *7*, 181.
8. Panetta, C.A.; Fang, Z.; Heimer, N.E., *J. Org. Chem.* **1993**, *58*, 6146.
9. Hansch, C.; Leo, A.; Taft, R.W., *Chemical Rev.* **1991**, *91*, 165.

Chapter 24

Peracids in Kraft Pulp Bleaching: Past, Present, and Future

K. Poppius-Levlin, A.-S. Jääskeläinen, A. Seisto[1], and A. Fuhrmann

KCL, P.O. Box 70, FIN–02151, Espoo, Finland

The main role of peracids in wood chemistry has long been in analytical applications. The TCF bleaching of chemical pulp has progressed rapidly during the 1990's. A lot of research has been carried out with different peracids. Industrial plants for peracetic acid production have been constructed and peracetic acid bleaching is currently in use, either permanently or occasionally, at several mills. The advantage of peracids over other oxygen-based bleaching chemicals is that they react selectively with lignin and leave pulp carbohydrates practically intact. The main parameters affecting the delignification kinetics are reaction temperature, peracid charge and pH. Peracids cause significant demethylation of pulp lignin leading to quinone and muconic acid type structures and to lower a biphenyl and biphenyl ether structure content than oxygen delignification. These chemical modifications activate the pulp lignin and improve pulp bleachability. A peracid bleaching sequence yields better quality pulp than an ozone sequence because of the lower content of alkali-labile carbonyl groups. The future of peracetic acid, which is already a serious competitor to ozone as a delignification agent, depends greatly on its cost.

The effect of peracetic acid as a delignifying agent has been known for 50 years (*1*). For a long time, however, the main role of peracids in wood chemistry was restricted to analytical applications, such as the preparation of holocellulose (*1,2,3*). It was not until the early 1980's that the search for less polluting pulping methods created a new interest in peracids, mainly peracetic and performic acid for pulping and further delignification (*4,5*). The formic acid/performic acid-based MILOX process (*5*) led from laboratory experiments to pilot plant applications. Recently, peroxonitric acid

[1]Current address: STFI, Box 5604, S–11486 Stockholm, Sweden.

(*6*) and peroxomonophosphoric acid (*7*) have been used to delignify aspen wood.

Increasingly stringent discharge regulations led not just to the development of totally new pulping processes but also to a search for new bleaching chemicals as alternatives to chlorine and chlorine dioxide. During the 1990's, progress in chemical pulp bleaching has been especially fast in Scandinavia, where all mills have already converted to ECF (Elemental Chlorine Free) or TCF (Totally Chlorine Free) processes. The early 1990's brought a breakthrough for peracids in pulp delignification and bleaching. Most studies have been carried out with peracetic acid (*8*), which is currently in use, either permanently or occasionally, at several mills (*9,10,11*). Performic acid has also given promising results on both laboratory and pilot scales (*12,13,14*). Very few reports are available for other organic peracids in pulp bleaching (*15,16*).

Peroxomonosulphuric acid (Caro's acid) is the most extensively studied inorganic peracid in pulp bleaching. Laboratory experiments with peroxomonosulphate under acidic conditions (*17*) led to mill trials a couple of years later (*18*). Recently, peroxomonophosphoric acid has been claimed to delignify pine kraft pulp even more rapidly and selectively than peroxomonosulphuric acid (*7*).

Preparation of Peracids

Numerous methods are known for the preparation of peracids (*19*). The most important and widely used method for both organic and inorganic peracids is the direct reaction of hydrogen peroxide with the corresponding acid. Carboxylic acids with two or more carbon atoms reach equilibrium quite slowly, and therefore a mineral acid catalyst is often used to speed up the reaction.

Peracids in equilibrium mixtures are often used as such for oxidation without isolation, which means that the oxidation mixture contains not just the peracid but also the corresponding acid, hydrogen peroxide and water. However, it is possible to distil some peracids out of their equilibrium mixtures. A distilled peracetic acid solution with a concentration of 30-40 wt.% and containing only a minimal amount of acetic acid and hydrogen peroxide is now being produced commercially. Distilled peracetic acid has been used in Sweden for some years. In Finland, Kemira Oy has started a peracetic acid distillation plant in Oulu producing 10,000 t/a (*20*).

Reactions of Lignin with Peracids

The most characteristic reaction of peracids is electrophilic oxygen transfer in acidic media. One oxygen atom of the peracid is tranferred to a nucleophilic substrate with a lone pair of electrons, such as lignin. Peracids can also act as nucleophilic oxidants in alkaline and neutral media (*21*).

The behaviour of lignin with peracetic acid under various conditions has been the subject of a number of studies carried out on appropriate lignin-related model compounds (*22*).

Peracetic acid oxidizes lignin under acidic, neutral or alkaline conditions. The reaction mechanisms and reaction products depend on pH. The most important reactions between lignin structures and peracetic acid are aromatic ring hydroxylation, oxidative demethylation, oxidative opening of the aromatic ring and the subsequent formation of muconic acid derivatives, cleavage of β–aryl ether bonds, cleavage of side chains, and epoxidation of aliphatic unsaturated bonds (*22*). In acid solution, ring hydroxylation of sufficiently activated phenolic and non-phenolic aromatic units is brought about by undissociated peracid, enhancing both the reactivity and hydrophilicity of the lignin compound. Under neutral and moderately alkaline reaction conditions mainly phenolic lignin structures are reactive leading to quinone structures. Olefinic double bonds are epoxidized in both acid and neutral solution, but the reaction is faster under neutral conditions. Epoxidation reactions probably also occur under alkaline conditions (*21*).

Delignification

Delignifying Efficiency of Different Peracids. The highest degree of delignification of a kraft pulp by a peracid is obtained with performic acid in concentrated formic acid solution (*15*) when the oxidant charge, calculated as oxidation equivalent, is equal (Figure 1). In aqueous solution, however, the delignification effect of performic acid is significantly weaker, because performic acid is easily hydrolysed to formic acid and hydrogen peroxide. On the other hand, peracids with longer carbon chains and Caro's acid are more stable and can be used effectively in aqueous solution. Peracetic

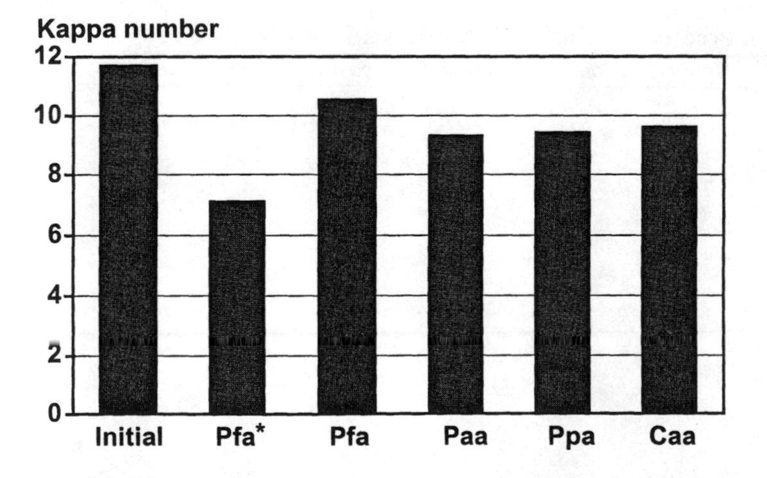

Figure 1. Comparison of different peracids in the delignification of oxygen pre-bleached softwood kraft pulp. The charge of oxidant (peracid + hydrogen peroxide), calculated as oxidation equivalent, was 1.18 OXE/kg pulp. (Caa= Caro's acid, Paa=peracetic acid, Pfa=performic acid, Pfa* = performic acid in 85% formic acid solution, Ppa=peroxypropionic acid.)

acid, peroxypropionic acid and Caro's acid all show equal delignification effects. Peracetic acid and Caro's acid resulted in the same degree of delignification when the peracid charge, calculated as active oxygen, was equal (*23,24*). Peroxomonophosphoric acid has been claimed to delignify pine kraft pulp more rapidly and selectively than peroxomonosulphuric acid (*7*).

Because the different peracids give essentially the same degree of delignification, other factors, such as pulp properties, price and recyclability, should be taken into account when choosing between them.

Factors Affecting Delignification. The process variables affecting delignification and the resulting pulp properties are listed in Table I. The rate of pulp delignification may be significantly increased by increasing either temperature or peracid charge (*25*). On the other hand, increasing peracid charge does not greatly contribute to pulp viscosity. With a peracetic acid charge of 5% on pulp, a delignification degree of 55% may be obtained with a viscosity drop of only 15% (*26*). However, increasing temperature, especially at acidic pH, also increases the rate of harmful reactions leading to a reduction in pulp viscosity (*27*).

The rate of pulp delignification may also be increased by increasing the delignification pH from acidic to only slightly acidic or neutral (*28*). Li and co-workers (*26*) suggested that an initial pH of 5 is optimal for peracetic acid delignification and an initial pH of 4 for Caro's acid delignification. However, conflicting results have also also been published. Hill and co-workers (*25*) showed that increasing the initial pH of peracetic acid delignification from 2 to 10 lowers the delignification effect of both softwood and hardwood.

Table I. Effect of Process Variables in Peracetic Acid Stage on Pulp Properties

	Delignification	*Brightness*	*Viscosity*	*Paa Consumption*
Paa charge 0.5→3.0%	++	++	*	±
temperature 50→90°C	++	++	*	-
pH 3→7	++	++	+	--
pretreatment none→Q	±	±	++	++
H_2O_2 0→1.5%	±	+	±	±
DTPA 0→0.25%	±	±	±	--
$MgSO_4$ 0→0.15%	± (+)	±	±	±

+/++ = positive effect; -/-- = negative effect; ± = no effect; * depends on pH; Paa = peracetic acid.

An important prerequisite for successful peracid delignification of kraft pulp is the removal of transition metals either by acid washing or by chelation (*29*). Transition metals cause the decomposition of peracids, thereby raising peracid consumption. Furthermore, the decomposition reactions also produce harmful compounds, possibly radicals, which may degrade carbohydrates and adversely affect pulp properties.

The rate of metal-catalysed peracetic acid decomposition in aqueous solution without the presence of chelating agents decreases in order Fe(III), Cu(II) and Mn(II) (*30*). The viscosity reduction during a peracetic acid stage follows the same order, i.e. the presence of manganese is less harmful than the presence of the same amount of iron (*31*).

The effect of free hydrogen peroxide in a peracid stage has been studied with peracetic acid, from which hydrogen peroxide can be eliminated relatively easily by distillation (*19*). When peracetic acid delignification is performed without a preceding chelation stage (*32)*, distilled peracetic acid gives higher pulp viscosities than equilibrium peracetic acid. This is because, in the presence of metal ions, hydrogen peroxide is degraded via the Haber-Weiss cycle (*33*) and the hydroxyl radicals formed can cause cellulose degradation. However, when harmful metal ions are removed prior to the peracid stage pulp viscosity is as high with equilibrium peracetic acid as it is with distilled peracetic acid (*8,28*).

Chelating agents are often added during alkaline peroxide treatment to hinder reactions between peroxide and metal ions and to improve pulp viscosity (*34*). Peracids, however, may react with DTPA (diethylenetriaminepenta acetic acid) or with DTMPA (diethylenetriaminepentamethylene phosphonic acid) causing an increase in peracid consumption (*30,35*) and probably also making the chelating agent less effective. Therefore, unlike alkaline peroxide bleaching, the addition of DTPA to the peracid stage does not improve the physical properties of pulps (*28*).

The addition of magnesium sulphate in peracetic acid bleaching stage of unchelated pulp has a negligible effect on the delignification degree and brightness gain (*30*) whereas a small increase in the delignification degree has been found when pulp was chelated prior to peracetic acid delignification (*28*). The cellulose degradation is not affected by the addition of magnesium sulphate (*28,30*). The presence of magnesium sulphate does not affect the stability of a neutral or acidic peracetic acid solution as much as it stabilises an alkaline hydrogen peroxide solution.

Selectivity. One of the main requirements for delignification and bleaching agents, if they are to compete with chlorine and chlorine dioxide, is selectivity, i.e. lignin degradation *versus* cellulose depolymerization. In other words, the agent should give a high degree of delignification yet leave pulp carbohydrates intact.

The degradation and dissolution of pulp carbohydrates during peracetic acid treatment under bleaching conditions depend on pH (*27*). Under neutral or slightly acidic conditions, the fall in pulp viscosity is very small. At acidic pH less carbohydrates dissolve than at neutral pH, although pulp viscosity is slightly more affected. Even at pH 5, significantly less carbohydrates are dissolved by peracetic acid than by ozone and the subsequent alkaline extraction performed after the OOP stages (*36*).

The total amount of carbohydrates dissolved during different peracetic acid treatments is negligible and the yield loss minor, even under conditions where a high degree of delignification was achieved (*27*).

Carbonyl groups introduced into cellulose by oxidation are responsible for depolymerization of the cellulose chain in the following alkaline stage and also during viscosity determination in strongly alkaline solution (*37*). If the pulp carbonyl groups are reduced to hydroxyl groups before viscosity determination, higher viscosities are obtained. The formation of alkali-labile carbonyl groups during ozonation is at least partly responsible for the decrease in pulp quality.

The carbonyl group content of an oxygen-delignified pine kraft pulp (17.0 mmol/kg) increased with performic acid and peracetic acid treatments to 21.0-23.4 and 22.2-28.7 mmol/kg, respectively (*15*), figures that are close to those of ozonated pulps (*36*) (Figure 2). On the other hand, as Figure 2 shows, no increase in pulp carbonyl group content can be seen when the peracetic acid stage was performed on pulp with low lignin content and high brightness, i.e. after the O/O(PO) stage. The carbonyl group content of pulp lignin may even be lowered by peracetic due to oxidation via the Baeyer-Villiger reaction. In the case of ozone treatment the number of pulp carbonyls increased in all cases.

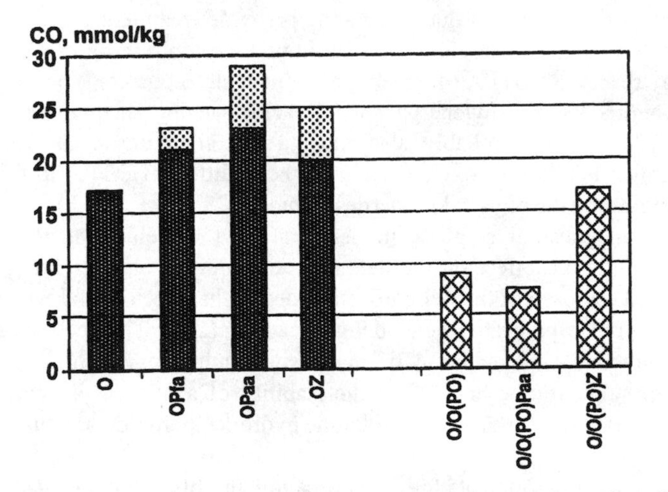

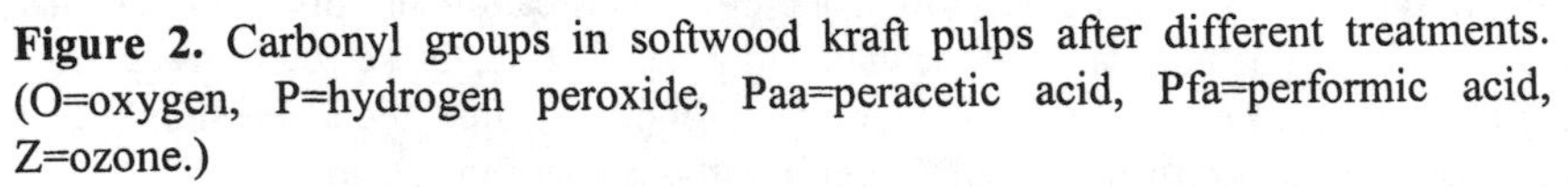

Figure 2. Carbonyl groups in softwood kraft pulps after different treatments. (O=oxygen, P=hydrogen peroxide, Paa=peracetic acid, Pfa=performic acid, Z=ozone.)

Figure 3, which shows the differences between borohydride viscosities and standard viscosities for softwood kraft pulp after different treatments, confirms that peracid and ozone differ from each other in relation to their reactions with pulp components. There was no difference between borohydride and standard viscosity for peracetic acid pulps with either low (OOPPaa) or high (OPaa) carbonyl group contents, but a significant difference for the ozone pulps. Oxidation by peracetic acid led to the formation of carbonyl groups largely on lignin structures, while oxidation by ozone produced carbonyl groups mainly on pulp carbohydrates. The results show that the aliphatic hydroxyl groups of polysaccharides do not provide suitable reaction sites for electrophilic reactions by peracetic acid, and are thus in agreement with earlier results (*38*). On the other hand, in the presence of metal ion catalysts, aliphatic hydroxyl groups can be oxidized to carbonyl groups, and can thus contribute to lowering the viscosity of peracid pulps (*37*).

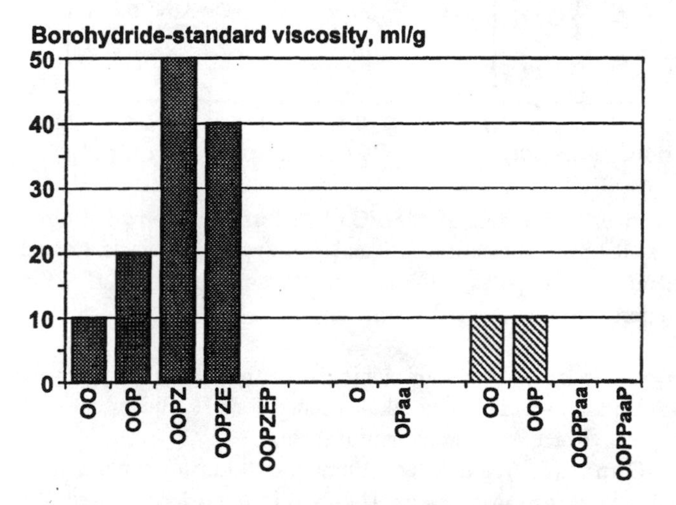

Figure 3. Differences between borohydride and standard viscosities of softwood kraft pulps after different treatments. (E=alkaline extraction, O=oxygen, P=hydrogen peroxide, Paa=peracetic acid, Z=ozone.)

Final Bleaching

Final Bleaching with Alkaline Hydrogen Peroxide. The reactivities of the residual lignins of both pine and birch kraft pulps are significantly increased by treating the pulps with peracids prior to alkaline peroxide bleaching. Performic acid-delignified birch kraft pulps with kappa numbers varying from 5 to 12 reached significantly higher final brightnesses with a lower consumption of peroxide than oxygen-delignified birch kraft pulp under the same conditions (Figure 4a) (*13*). A similar activation of peracetic acid-treated pulps with kappa numbers 7-10 is also shown (Figure 4b). However, the increase in responses towards alkaline peroxide was lower for peracetic acid-treated than for performic acid-delignified pulps.

TCF and ECF Bleaching Sequences with Peracetic Acid. Peracids and ozone have been used in the laboratory to activate pulp towards oxygen delignification before final bleaching (*26,39*), e.g. by using two-stage oxygen delignification with an activation stage between. Although the lignin contents of the pulps after O, OPaa, OCaa and OZ treatments were similar, a higher delignification after the second oxygen stage (carried out under identical conditions) was achieved for the peracid- and ozone-activated

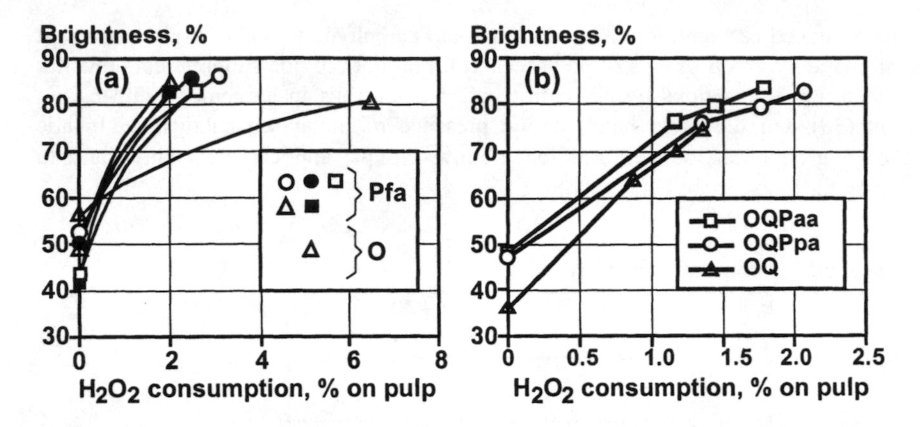

Figure 4. Alkaline peroxide bleaching a) at 80°C of performic acid-treated birch kraft pulps, and b) at 85°C of peracetic acid- and peroxypropionic acid-treated oxygen-delignified pine kraft pulps. (O=oxygen, Paa=peracetic acid, Pfa= performic acid, Ppa=peroxypropionic acid, Q=chelation.)

pulps. In spite of the greater kappa reduction achieved in the second oxygen stage using peracids, the large amounts of acid and alkali needed for pH adjustment do not make this combination very attractive on an industrial scale.

For some years Caro's acid was used at a Finnish mill during manufacture of TCF pulp. However, due to corrosion problems Caro's acid is no longer used at the mill. The difficulty of adjusting pulp pH and the high amounts of alkali needed for neutralization have further reduced the industry's interest in Caro's acid. Interest in performic acid is also low.

The benefits of Caro's acid compared with peracetic acid are that it can be prepared on site at the mill at low chemical costs and without any investments. Peracetic acid can also be prepared on site as an equilibrium product, although to reduce chemical costs unreacted acetic acid should be recovered and recirculated.

The industrial use of peracetic acid started in Sweden at several TCF pulp mills. Various bleaching sequences such as PaaQP, QPaaP, PaaQ(PO), QPaaQ(PO), QPaa(EOP)Q(PO) and Q(PaaQ)P have been used (*40*).

Peracetic acid has been used continuously at the Vallvik mill since 1995. Currently it is being used at the beginning of the sequence (QPaa/Q(PO)PPP) (*9*).

Taking costs into account, peracetic acid should preferably be used when the amount of residual lignin is quite low in order to keep the peracetic acid charge down. At mills not prepared to invest in an ozone plant, peracetic acid is often used in the middle of the sequence.

At the new bleaching plant under construction at Sunila, Finland, the peracetic stage will be situated in the middle of the sequence. The plant will be able to produce both ECF and TCF pulps as follows (*10*):

ECF pulp: D(EOP)/(EOP)D(PO)
TCF pulp: Q(EOP)/(EOP)Paa(PO)

Several mill trials have been performed in Finland using peracetic acid at various stages of both ECF and TCF sequences, but no details are yet available (*11*).

Recently laboratory studies have shown that the use of peracetic acid in the production of bleached pulp also has other benefits. By situating the peracetic acid stage at the end of the bleaching sequence, the final brightness of ECF bleached pulp can be improved, one such sequence being: OD(EO)DEDPaa.

In one laboratory study, treating an ECF bleached pulp with 0.15% peracetic acid raised the final brightness from 88.4% to 89.6% ISO. Mill scale results showed that this post-bleaching also gave a more even brightness and also improved runnability on the paper machine (*41*).

Effect of Peracids on Pulp Fibre Surface Lignin. ESCA (Electron Spectroscopy for Chemical Analysis) can be used to study the effect of different bleaching chemicals on the content of surface lignin in kraft pulp fibres (*14,42*).

In a comparison of the effect of different peracids on the surface lignin content of kraft pulp, oxygen-delignified pine and birch kraft pulps were bleached with a PXP sequence, in which X was peracetic acid (TCF_{Paa}), Caro's acid (TCF_{Caa}) or performic acid (TCF_{Pfa}), to a final brightness of 89% (*14,43*). The peracid charges as well as the hydrogen peroxide charges in the final peroxide stages were varied in order to reach the final brightness under optimum conditions.

After bleaching to full brightness, the total lignin contents were very similar for both birch and pine pulps with all the sequences used (Figures 5 and 6). However, the effect of different peracids on the surface lignin content of the pulps varied. As shown in Figure 7, for birch kraft pulp, performic acid was considerably more effective in reducing the lignin content on the fibre surfaces than peracetic acid and Caro's acid. In the performic acid stage, it was possible to remove a large proportion of the surface lignin and to activate the remaining lignin towards the final alkaline

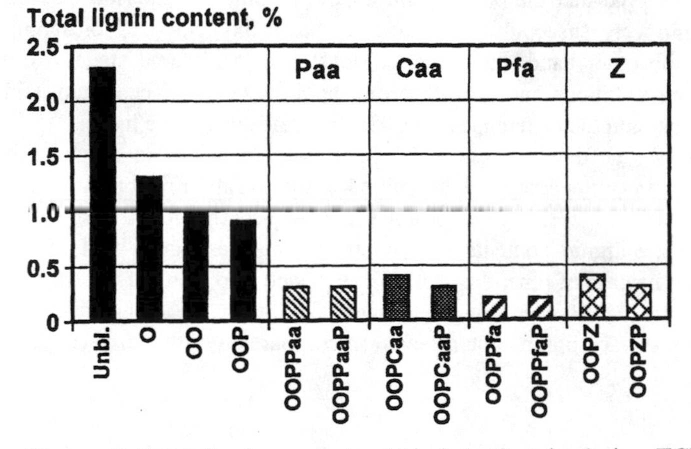

Figure 5. Total lignin content of birch kraft pulp during TCF bleaching with peracid and ozone sequences. (Caa=Caro´s acid, O=oxygen, P=hydrogen peroxide, Paa=peracetic acid, Pfa=performic acid, Z=ozone.)

peroxide stage. The reduction in the surface lignin content was from 8% to 2% with performic acid and further to about 1.5% with hydrogen peroxide. With TCF_{Paa} and TCF_{Caa} sequences the surface lignin contents were as high as 6% and 4.5%, respectively, after the final bleaching stage. In the case of pine (Figure 8), the surface lignin contents of the pulps were very similar after the different peracid stages, i.e. about 5%. However, performic acid was again able to activate the residual lignin on the fibre surfaces towards the final peroxide stage more than peracetic acid and Caro's acid.

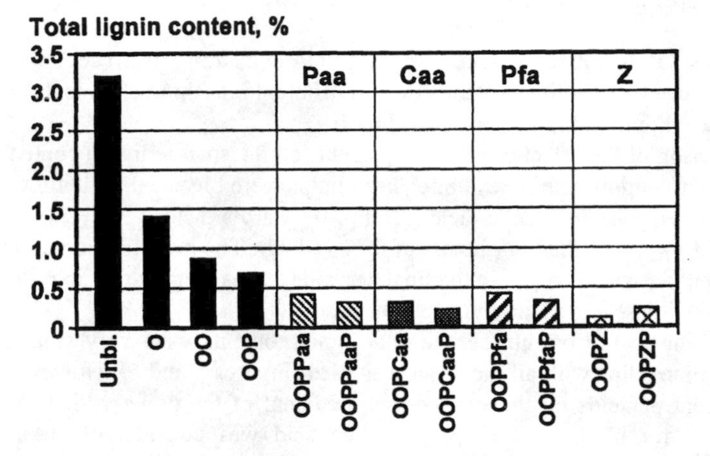

Figure 6. Total lignin content of pine kraft pulp during TCF bleaching with peracid and ozone sequences. (Caa=Caro's acid, O=oxygen, P=hydrogen peroxide, Paa=peracetic acid, Pfa=performic acid, Z=ozone.)

It has been observed that the pre-bleaching of kraft pulp with performic acid renders residual lignin very susceptible to attack by other oxygen-based bleaching chemicals (*12*). On the other hand, the peracetic acid and Caro's acid stages were carried out in aqueous solutions and the performic acid stage in concentrated acid solution, which may explain the difference in the reactivity of the surface lignin.

When ozone was used in the same bleaching sequence instead of peracids (OOPZP), the total lignin contents of birch and pine kraft pulps at a final brightness of 89% were the same (about 0.2%, Figures 5 and 6). However, the effect of a TCF_{Z} sequence on the surface lignin contents of the pulps more resembled that of the TCF_{Paa} sequence. In the case of pine, the TCF_{Caa} sequence also resulted in a similar surface lignin content (*43*). Hence, using performic acid in a bleaching sequence resulted in a very low surface lignin content, even in comparison with ozone (Figures 7 and 8).

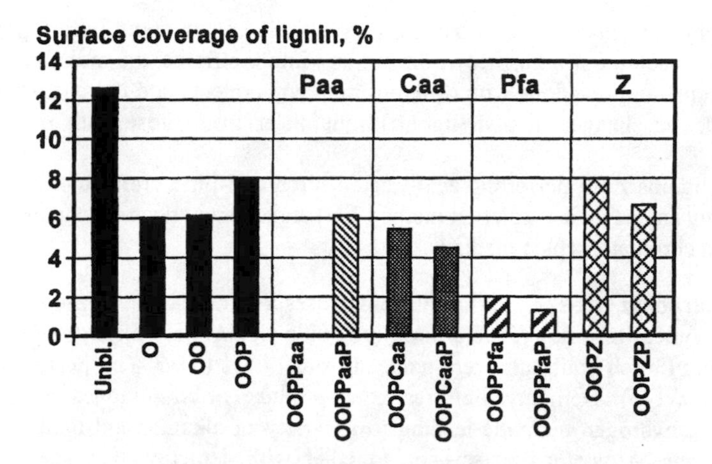

Figure 7. Surface coverage of lignin of birch kraft pulp during TCF bleaching with peracid and ozone sequences (OOPPaa sample contaminated). (Caa=Caro's acid, O=oxygen, P=hydrogen peroxide, Paa=peracetic acid, Pfa=performic acid, Z=ozone.)

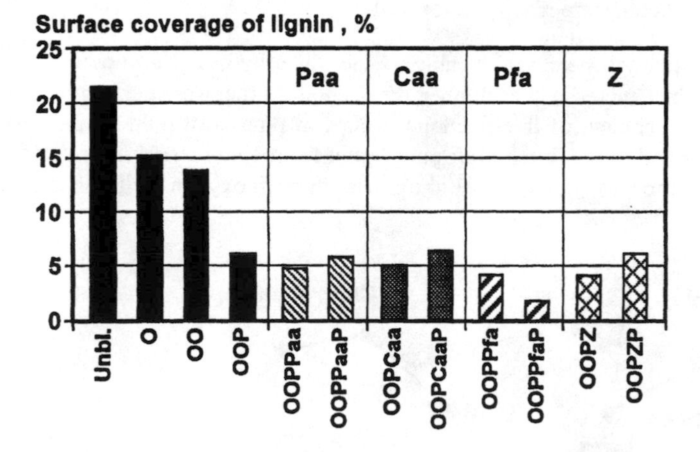

Figure 8. Surface coverage of lignin of pine kraft pulp during TCF bleaching with peracid and ozone sequences. (Caa=Caro's acid, O=oxygen, P=hydrogen peroxide, Paa=peracetic acid, Pfa=performic acid, Z=ozone.)

Structural Features of Pulp Lignins

The greater reactivities of the peracid-treated pulps towards oxygen delignification and alkaline peroxide bleaching can be explained in the light of structural differences in the residual lignins of peracid-treated pulps compared with the less reactive pulps. Although a lot of research has been carried out on low molar mass lignin model compounds in homogeneous reaction mixtures, very little is known about pulp lignin itself. Taking into account mass transfer of the oxidant into the fibres, the molar mass of pulp lignin, competing reactions with different pulp components and the charge of the oxidant used, the changes in pulp lignin might differ from those with model compounds.

Residual lignins from performic acid-treated birch and pine kraft pulps have been isolated using an enzymatic isolation method (*44*) and characterized by chemical, spectroscopic and chromatographic methods.

Methoxyl and Carboxyl Groups. Performic acid causes a significant decrease in the methoxyl group content of pine (*45*) and birch (*46*) pulp lignin. Methoxyl groups in the residual lignin of birch kraft pulp were reduced from 14.5% to 9.3% by performic acid delignification (*46*). Demethylation reactions produce ortho-quinones, which react with alkaline hydrogen peroxide leading to a variety of alkaline and oxidative degradation products. Aromatic ring scission together with demethylation leads to muconic acid structures, which increase lignin hydrophilicity and introduce new reactive aliphatic carbon-carbon double bonds into the lignin. Although no increase in carboxyl group content was observed in the isolated and precipitated part of the residual lignin from performic acid-treated birch kraft pulp, it can be assumed that the soluble part of residual lignin was enriched in carboxyl groups (*45*). Moreover, muconic acid-type structures are readily lactonized.

Phenol Groups. Although peracids are able to introduce hydroxyl groups into the aromatic ring and thus increase the number of phenolic structures, performic acid treatment lowered the content of free phenolic groups in pine kraft pulp lignin from 1.77 to 1.09 mmol/g and that in birch kraft pulp lignin from 1.40 to 1.14 mmol/g (*47*) (Figure 9). The decrease was, however, slightly less than in oxygen delignification.

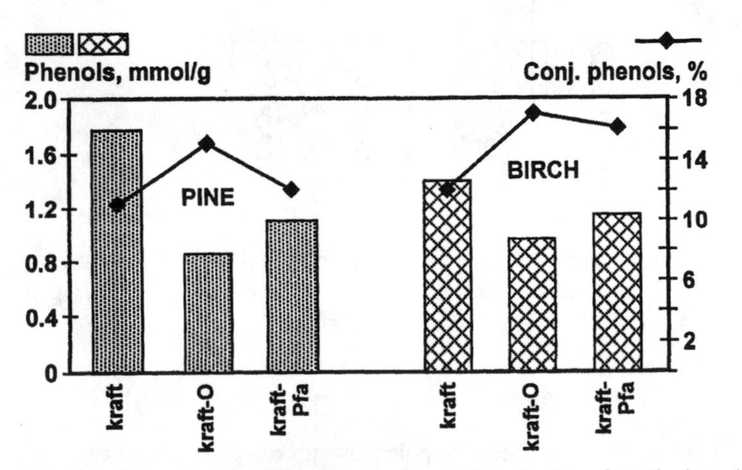

Figure 9. Total phenols and conjugated phenols (% of total phenols) in residual lignins from unbleached pine and birch kraft pulps, oxygen-delignified pine and birch kraft pulps, and performic acid-delignified pine and birch kraft pulps (O=oxygen, Pfa=performic acid.)

One reason for the decrease in phenols caused by performic acid, despite evidence of considerable demethylation, could be formylation. The esters would then hydrolyse to free phenols, for example during final bleaching under alkaline conditions. It is also possible that reactions leading to aromatic ring scission or oxidation consume the new phenolic groups formed through demethylation and ring hydroxylation. The results also indicate that the phenolic structures are reactive towards peracids.

During the peracid treatment of birch kraft pulp the proportion of conjugated phenols increased to the same extent as during oxygen treatment (Figure 9).

Condensed Structures. Degradation reactions, such as oxidative degradation by permanganate followed by hydrogen peroxide, provide information on the changes in lignin structures occurring during pulping and bleaching. Particularly valuable information is obtained about possible condensation reactions.

The compositions of the aromatic carboxylic acids obtained after oxidative degradation of the residual lignins isolated from pine kraft pulp, oxygen-delignified pine kraft pulp and performic acid-delignified pine kraft pulp are shown in Table II, with pine MILOX pulp as reference. The structures of the corresponding acids are shown in Figure 10. The content of acid **3**, which suggests C5 condensation reactions

Table II. Composition (mol %) of the Aromatic Carboxylic Acids Obtained after Oxidative Degradation of Residual Lignins Isolated from Pine Kraft Pulp, Oxygen-Delignified Pine Kraft Pulp, Performic Acid-Delignified Pine Kraft Pulp and Pine MILOX Pulp. (Structures of the acids are shown in Figure 10.)

Residual Lignin	**1.**	**2.**	**3.**	**4.**	**5.**	**6.**
kraft[1]	1.5	54.1	19.4	4.3	13.0	8.0
kraft-O[1]	0.7	36.0	28.5	4.4	21.0	9.0
kraft-Pfa	-	40.1	33.8	9.5	9.8	6.8
MILOX	-	23.6	34.4	28.5	9.7	3.7

[1]From ref. (48).

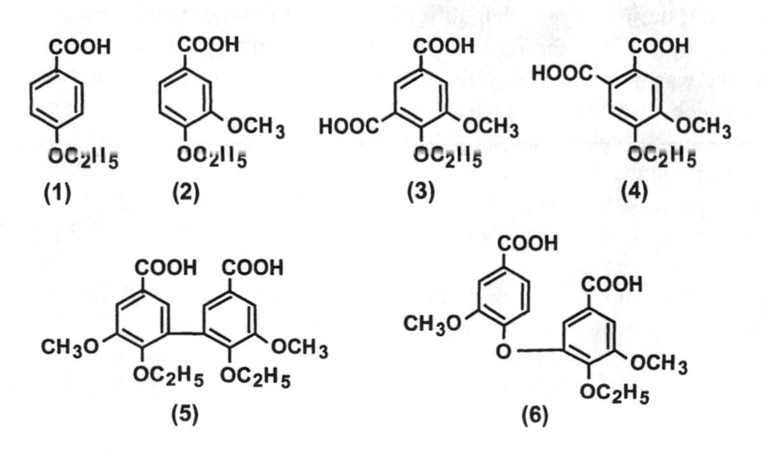

Figure 10. Carboxylic acids formed after oxidative degradation of lignins.

or extraction of less condensed part of the residual lignin, increased in the residual lignin from pine kraft pulp to oxygen-delignified and performic acid-delignified pine kraft pulps and was even slightly higher in the residual lignin of the MILOX pulp. The content of acid **4**, which represents C6 condensation reactions, was higher in the residual lignin from the performic acid-treated pine kraft pulp than in those from kraft and oxygen-delignified pulps and was highest in the residual lignin of the MILOX pulp. An increase in C6 condensation has also been observed in dissolved lignins during MILOX pulping (5). On the other hand, the contents of biphenyl (acid **5**) and diaryl ether structures (acid **6**), which are detrimental to pulp bleaching, were lower in the performic acid treated pulp residual lignins than in oxygen-delignified pulp residual lignins.

The greater activation of the residual lignins in performic acid-treated kraft pulps than in non-treated pulps can thus be explained by the significant chemical modifications caused by peracids - oxidation and increase in hydrophilicity - without simultaneous enrichment of detrimental biphenyl and diaryl ether structures. On the other hand, the differences in molar masses are rather small and apparently do not have any very significant effect on the bleachability of these pulps (*46*).

Physical and Optical Properties

Depending on the wood species used for pulping, chemical pulp fibres are generally treated mechanically to a certain degree to give the paper the desired strength and optical properties. The use of different bleaching chemicals can lead to pulps with very different properties, such as tensile and tear strength. On the other hand, the behaviour of the bleached pulps during beating can be difficult to predict. ECF and TCF bleached softwood pulps have shown both lower and higher energy consumptions during mechanical treatment as well as both lower and higher strengths than conventional chlorine-bleached pulps (*49*).

Figure 11 presents the tear indices for TCF bleached birch kraft pulps at tensile index 50 Nm/g, which is a typical value required for short fibred hardwood pulps. According to Figure 11, during oxygen delignification and the first peroxide stage, the tear index increased slightly. On the other hand, at the beginning of the bleaching sequence less beating was needed to reach the target tensile index than after the final bleaching stages, which also affects the tear index obtained. In the peracid stage, peracetic acid gave a much higher tear strength than Caro's acid and performic acid, and the difference between the pulps was still in evidence after the final peroxide stage (Figure 11). A tear index of 9.2 mNm^2/g was obtained with the TCF_{Paa} sequence, compared with 8.7 mNm^2/g with the TCF_{Caa} and 8.4 mNm^2/g with the TCF_{Pfa} sequence.

In the case of pine pulp, the typical tensile index required is 70 Nm/g. The tear indices of TCF bleached pine pulps beaten to this tensile index are shown in Figure 12 (*43*). Some fibre damage probably occurred in the first peroxide stage, as there was a reduction in tear strength. After the peracid stages, the tear strengths of pine pulps were very similar, but after the final peroxide stage the trend was similar to that for birch pulps, i.e. the highest strength was obtained with the TCF_{Paa} sequence. The difference in strength was somewhat larger in the case of pine than birch. The tear index of TCF_{Paa} bleached pulp was close to 14 mNm²/g and that of TCF_{Caa} and TCF_{Pfa} pulps about 12 mNm²/g. Replacing the peracid stage with ozone resulted in a tear strength similar to that obtained with a TCF_{Caa} sequence with both birch and pine pulps (Figures 11 and 12).

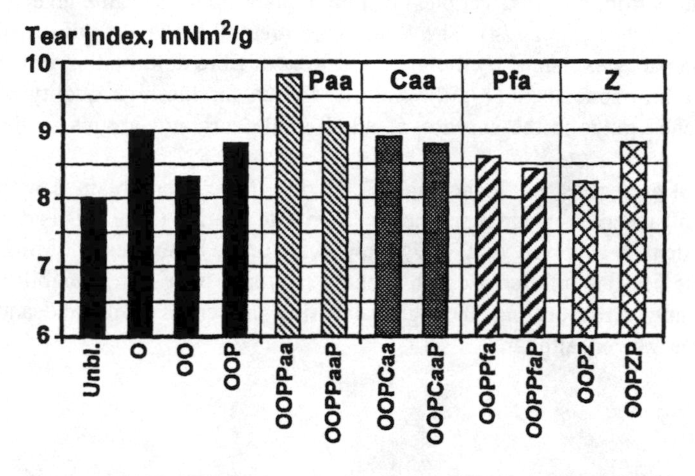

Figure 11. Tear index at tensile index 50 Nm/g for TCF bleached birch kraft pulps (Caa=Caro's acid, O=oxygen, P=hydrogen peroxide, Paa=peracetic acid, Pfa=performic acid, Z=ozone.)

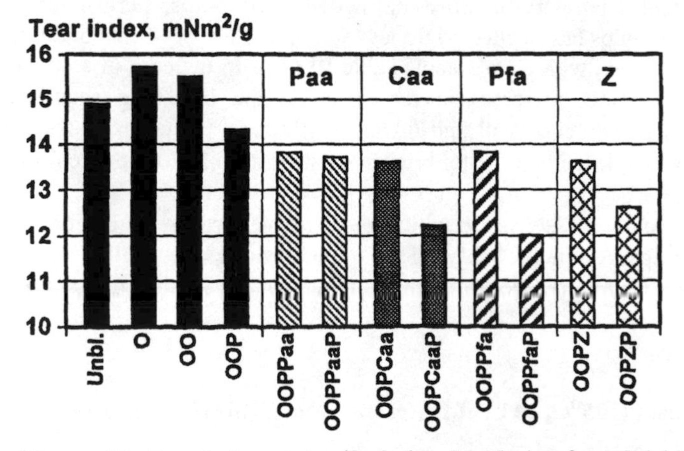

Figure 12. Tear index at tensile index 70 Nm/g for TCF bleached pine kraft pulps (Caa=Caro's acid, O=oxygen, P=hydrogen peroxide, Paa=peracetic acid, Pfa=performic acid, Z=ozone.)

One reason for the higher strength of peracetic acid over Caro's acid and performic acid pulps could be the difference in pH during bleaching. A lower pH in the peracid stage could increase the sensitivity of the fibres in the final peroxide stage, resulting in a reduction in fibre strength and thus paper strength. As the pH in the performic acid stage was the lowest (85% formic acid solution), the lowest tear strength was also obtained for that pulp. In the case of birch pulps, some fibre damage was seen as early as the Caro's acid and performic acid stages (*14*). It has been found that by reducing the mixing in the performic acid stage, the fibre strength (measured as zero-span tensile index from rewetted samples) can be brought up to the same level as with other peracid bleached pulps (*14*). The lower tear strength of TCF_Z pulp than TCF_{Paa} pulp can be partly explained by the greater degree of fibre deformation in the pulp bleached using an ozone stage (*43,50*). Another reason for the high quality of peracetic acid-bleached pulps is the absence of alkali-labile carbonyl groups in the cellulose.

Better optical properties, i.e. higher light scattering, have been obtained with peracid than ozone bleached birch and pine kraft pulp prior to beating (*14*). This is due to the higher paper density achieved with TCF_Z pulp, which may be the result of both the higher degree of fibrillation after ozone bleaching (*43*) and poor fibre flexibility after peracid bleaching. After beating, the light scattering properties of peracid and ozone bleached pulps are very similar.

Brightness Stability

The brightness stability of mill TCF kraft pulp bleached using a peracid is about the same as that of ECF bleached kraft pulp. Pulp bleached with Caro's acid shows better brightness stability against heat and moisture than an ozone-bleached softwood kraft pulp with the same kappa number, as can be seen (*51*) from the lower PC numbers.

Of the different laboratory-bleached hardwood kraft pulps, performic and peracetic acid bleached pulps have better brightness stability under dry-heat conditions than the same pulp bleached with Caro's acid, Table III (*51*). In the case of softwood kraft pulps, the PC number increases in the order of the bleaching sequences OPaaOPPaaP, OZOPZP, OCaaOPCaaP and OD(EP)DED (*26*), which indicates that pulp bleached with peracetic acid has better brightness stability than ozone or chlorine dioxide-bleached pulp.

Peracetic acid may also be used to inhibit the brightness reversion reactions of ECF bleached kraft pulps. Addition of only 0.15% of peracetic acid to a fully bleached ECF kraft pulp significantly reduces brightness reversion for both softwood and hardwood (*41*).

Table III. PC Numbers (105°C, 24 h) of Different Peracid-Bleached Hardwood Kraft Pulps (*51*).

Bleaching Sequence	*PC Number*
OAPfaPPP	0.41
OQPaa(pH 3)PPP	0.50
OQPaa(pH 5)PPP	0.45
OQCaaPPP	0.77

Future of Peracids in Kraft Pulp Bleaching

Today more and more oxygen-based bleaching chemicals, i.e. oxygen, hydrogen peroxide and ozone, are being used not only in TCF but also in ECF bleaching. Peracids, especially peracetic acid, will probably be the next generation bleaching chemicals, as they seem to be less susceptible to transition metal-induced decomposition than hydrogen peroxide and are more selective than ozone. Taking costs into account, peracetic acid should preferably be used when the amount of residual lignin is quite low in order to keep the peracetic acid charge down.

More basic research should be performed in order to improve the delignification effect of peracids. The role of reaction conditions in determining the bleaching result also needs further study. More knowledge is also needed about the components formed in the bleaching of kraft pulp with peracids and how they affect the treatment of bleaching effluents. Little is known about the recycling properties of pulps bleached with peracids. Combinations of different peracids and/or other bleaching chemicals may result in new and beneficial synergistic effects on the pulp and should be investigated.

Interest in the use of peracetic acid is expected to grow in the future both in the scientific community and in industry, especially as peracetic acid seems to have some benefits over ozone in the context of loop closure at bleaching plants (smaller amounts of carbohydrates and oxalic acid are dissolved in the effluents). Furthermore, if light stability and runnability on the paper machine are also improved by peracetic acid end-treatment, peracetic acid will benefit both TCF and ECF bleaching processes. The main question, however, still remains to be answered - will the price of peracetic acid be more competitive in the future than it is today?

Literature Cited

1. Poljak, A. *Angew. Chem.* **1948,** *60*, p 45.
2. Haas, H.; Schoch, W.; Ströle, U. *Das Papier* **1955,** *9*:19/20, pp 469-475.
3. Leopold, B. *Tappi* **1961**, *44*:3, pp 230-232.
4. Hergert, H. L. In *Environmentally Friendly Technologies for the Pulp and Paper Industry*; Editors, Young, R. A.; Akhtar, M.; John Wiley & Sons Inc., New York, 1997, pp 5-67.
5. Sundquist, J.; Poppius-Levlin, K. In *Environmentally Friendly Technologies for the Pulp and Paper Industry*; Editors Young, R. A.; Akhtar, M.; John Wiley & Sons Inc., New York, 1997, pp 157-190.
6. Springer, E. L. *Tappi J.* **1994**, *77*:6, pp 103-108.
7. Springer, E. L. *J. Pulp Pap. Sci.* **1997,** *23*:12, pp J582-J584.
8. Liebergott, N. *Pulp Pap. Canada* **1996**, *97*:2, pp T45-TT50.
9. Anon. *Svensk Papperstidn./Nordisk Cellulosa* **1996,** *99*:3, p 36 (in Swedish).
10. Anon. *Sunilan viesti* **1997,** *53*:1, pp 1,3 (in Finnish).
11. Anon. *PPC, Kemira Chemicals Oy:n asiakaslehti* **1996**, no 2, p 3 (in Finnish).
12. Poppius, K.; Hortling, B.; Sundquist, J. *International Symposium on Wood and Pulping Chemistry*, Raleigh, May 22-25, 1989, pp 145-150.

13. Poppius-Levlin, K.; Toikkanen, L.; Tuominen, I.; Sundquist, J. *International Symposium on Wood and Pulping Chemistry,* Melbourne, April 30-May 4, 1991, Vol. 1, pp 99-106.
14. Seisto, A.; Poppius-Levlin, K.; Fuhrmann, A. *Pap. Puu,* in press.
15. Poppius-Levlin, K.; Jääskeläinen, A.-S.; Sundquist, J. *Fourth European Workshop on Lignocellulosics and Pulp,* Stresa, September 8-11, 1996, pp 216-222.
16. Gruber, E.; Dintelmann, T. *Das Papier* **1998**, *52*:1, pp 9-15.
17. Springer, E. L.; McSweeny, J. D. *Tappi J.* **1993**, *76*:8, pp 194-198.
18. Seccombe, R.; Mårtens, H.; Haakana, A. *48th Appita Annual General Conference,* Melbourne, May 2-6, 1994, pp 139-146.
19. Swern, D. In *Organic Peroxides*; Editor, Swern, D.; John Wiley & Sons Inc., New York, 1970, pp 313-474.
20. Anon. *Kemia-Kemi* **1996**, *23*:1, p. 49 (in Finnish).
21. Sawaki, Y. In *Organic peroxides*; Editor, Ando, W.; John Wiley & Sons Ltd., Guildford, 1992, pp 430-446.
22. Gierer, J. *Wood Sci. Technol.* **1986,** *20*:1, pp 1-33.
23. Amini, B; Webster, J. *Tappi J.* **1995**, *78*:10, pp 121-133.
24. Nimmerfroh, N.; Süss, H. U. *International Non-Chlorine Bleaching Conference,* Orlando, March 24-28, 1996, paper 5-4, 38 p.
25. Hill, R. T.; Walsh, P. B.; Hollie, J. A. *Tappi Pulping Conference,* Boston, November 1-5, 1992, Book 3, pp 1219-1230.
26. Li, X.-L.; Fuhrmann, A.; Rautonen, R.; Toikkanen, L. *1996 International Pulp Bleaching Conference,* Washington, April 14-18, 1996, Book 1, pp 93-103.
27. Jääskeläinen, A.-S.; Poppius-Levlin, K. *1998 International Pulp Bleaching Conference,* Helsinki, June 1-5, 1998, Book 2, pp 423-428.
28. Jääskeläinen, A.-S.; Poppius-Levlin, K. *International Symposium on Wood and Pulping Chemistry,* Montreal, June 9-12, 1997, pp D5-1-D5-6.
29. McGrouther, K.; Allison. R. W. *47th Appita Annual General Conference,* Rotorua, April 19-23, 1993, Vol. 1, pp 191-199.
30. Yuan, Z.; d'Entremont, M.; Ni, Y.; van Heiningen A. R. P. *83rd Annual Meeting, Technical Section CPPA,* Montreal, January 28-31, 1997, pp B213-B220.
31. Dahl, O.; Niinimäki, J.; Tirri, T.; Jääskeläinen, A.-S.; Kuopanportti, H. *Tappi Pulping Conference,* San Francisco, October 19-23, 1997, Book 2, pp 1061-1067.
32. Basta, J.; Holtinger, L.; Hällström, A.-S.; Lundgren, P. *Tappi Pulping Conference,* San Diego, November 6-10, 1994, Book 3, pp 953-956.
33. Desprez, F.; Hoyos, M.; Devenyns, J.; Troughton, N. A. *Revue A.T.I.P.* **1995**, *48*:6, pp 172-178.
34. Bambrick, D. R. *Tappi* **1985**, *68*:6, pp 96-100.
35. Yuan, Z.; d'Entremont, M.; Ni, Y.; van Heiningen, A. R. P. *Tappi Pulping Conference,* San Francisco, October 19-23, 1997, Book 2, pp 931-949.
36. Fuhrmann, A.; Li, X.-L; Rautonen, R. *J. Pulp Pap. Sci.* **1997**, *23*:10, pp J487-493.
37. Bailey, C. W.; Dence, C. W. *Tappi* **1966**, *49*:1, pp 9-15.
38. Johnsson, D. C. *1st International Symposium on Delignification with Oxygen, Ozone and Peroxides,* Raleigh, May 27-29, 1975, pp 217-228.
39. Allison, R. W. McGrouther, K. G. *Tappi J.* **1995**, *78*:10, pp 134-142.

40. Basta, J.; Holtinger, L.; Lundgren, P.; Persson, C. *Tappi Pulping Conference*, Chicago, October 1-5, 1995, Book 1, pp. 53-57.
41. Jäkärä, J.; Parén, A. *51st Appita Annual General Conference,* Melbourne, April 28-May 2, 1997, Vol. 2, pp 363-367.
42. Laine, J.; Stenius, P.; Carlsson, G.; Ström, G. *Nordic Pulp Pap. Res. J.* **1996,** *11*:3, pp 201-210.
43. Seisto, A.; Poppius-Levlin, K.; Fuhrmann, A. *1998 International Pulp Bleaching Conference*, Helsinki, June 1-4, 1998, Book 1, pp 175-183.
44. Hortling, B.; Ranua, M; Sundquist, J. *Nordic Pulp Pap. Res. J.* **1990,** *5*:1, pp 33-37.
45. Hortling, B.; Poppius-Levlin, K.; Sundquist, J. *Second European Workshop on Lignocellulosics and Pulp*, Grenoble, September 2-4, 1992, pp 93-94.
46. Poppius-Levlin, K.; Hortling, B.; Sundquist, J. *International Symposium on Wood and Pulping Chemistry*, Beijing, May 25-28, 1993, Vol. 1, pp 214-222.
47. Tamminen, T.; Hortling, B.; Poppius-Levlin, K.; Sundquist, J. *International Symposium on Wood and Pulping Chemistry*, Beijing, May 25-28, 1993, Vol. 3, pp 228-232.
48. Hortling, B.; Turunen, E.; Sundquist, J. *Nordic Pulp. Pap. Sci.* **1992,** *7*:3, pp 144-151.
49. Lumiainen, J. *Pap. Technol.* **1996,** *37*:5, pp 22-26.
50. Seisto, A.; Hornatowska, J.; Lindström, M. *STFI BF-report* 11, 1998.
51. Suppanen, U. Master's Thesis, Helsinki University of Technology, 1997, pp. 46, 97, 99-100 (in Finnish).

Chapter 25

Chemical Modification of Lignin-Rich Paper

Light-Induced Changes of Softwood and Hardwood Chemithermomechanical Pulps: The Effect of Irradiation Source

Magnus Paulsson[1] and Arthur J. Ragauskas[2]

[1]Chalmers University of Technology, Department of Forest Products and Chemical Engineering, S–412 96, Göteborg, Sweden
[2]Institute of Paper Science and Technology, Chemical and Biological Sciences Division, 500 10th Street, N.W., Atlanta, GA 30318

Light-induced brightness reversion is currently a topic of considerable interest for the pulp and paper industry. The objective of this study was to describe the effect of using different irradiation sources on the accelerated photoyellowing properties of untreated and chemically modified bleached high-yield pulps. The wavelength distribution of the irradiation source strongly influenced the photochemistry of both untreated and acetylated pulps. UV/VIS absorption difference spectra of aspen chemithermomechanical pulp (CTMP), irradiated with broad-band UV-radiation, exhibited an apparent absorption maximum at 370 nm with a shoulder at 415 nm, whereas the aspen CTMP aged with the UV/VIS-fluorescent lamps exhibited an absorption peak at 360 nm and substantially less absorption in the entire visible region (λ>400 nm). Two major absorption peaks, at 330-345 nm and at 425 nm, were observed in the absorption difference spectra of irradiated spruce CTMP regardless of the irradiation source used. However, in addition to the photodiscoloration, photobleaching with λ_{max}~390 nm was observed when the spruce CTMP was irradiated with the UV/VIS-fluorescent lamps. Acetylation was found to slow down the UV-light induced reactions, but could also promote photobleaching reactions when the pulps were subjected to an irradiation source emitting light in the visible range. Generally, no further discoloration was observed when the acetylated CTMPs (acetyl content, 8-10%) were irradiated with the UV/VIS-fluorescent lamps under the aging conditions used in this work. This investigation has shown the importance of choosing a light source that resembles the actual reversion situation as closely as possible (e.g., contains both an ultraviolet and a visible component) if realistic accelerated light-induced reversion conditions are to be obtained.

The use of mechanical and chemimechanical pulps as constituents in higher grades of printing paper is severely restricted by the rapid color reversion (yellowing) that occurs upon exposure to daylight or indoor illumination. This phenomenon has been attributed to a light-induced oxidation of the lignin present in the pulp. Extensive and comprehensive research, performed during the last decade, has given not only new information about the photochemical reactions causing yellowing, but also information on the potential photostabilizing methods, although no single approach so far has become technically or economically feasible to meet all the needs of the paper industry. Heitner and Schmidt (*1,2*) and Leary (*3*) have summarized current knowledge in this field in comprehensive literature reviews.

The photochemical changes that occur during irradiation of lignin-rich pulps do not involve only discoloration reactions. Several researchers have reported a photobleaching effect when pulps were irradiated at longer wavelengths, i.e., visible light (*4-9*). If the color-forming reactions could be eliminated or slowed down, the bleaching reactions would predominate and result in a brightness increase for high-yield pulps when irradiated in daylight. However, the darkening reactions normally dominate over the brightening reactions, and the net result is a discoloration of the lignin-rich material.

Exposure of CTMP (chemithermomechanical pulp) and TMP (thermomechanical pulp) to ultraviolet radiation with a wavelength of 300-400 nm produced chromophores with absorption maxima at 350 and 420 nm (*10*). These chromophores were considered to be a methoxy-*p*-benzoquinone and an *ortho*-quinone with an aliphatic substituent in the 2-position. Structures of the *ortho*-quinonoid type have been detected in different types of high-yield pulps, both before and after irradiation. Lebo et al. (*11*) found that about 75% of the increase in color during irradiation of white spruce refiner mechanical pulp was caused by the formation of *ortho*-quinones. Zhu et al. (*12*) used fluorescence spectroscopy to confirm the presence of photochemically generated *ortho*-quinones in several hydrogen-peroxide-bleached high-yield pulps. Schmidt et al. (*13*) reported that both methoxylated and unmethoxylated *ortho*-quinones were likely to be introduced in peroxide-bleached mechanical pulps during irradiation. Monomeric *ortho*-quinones were suggested to be the major chromophores formed during light-induced yellowing of lignin-rich pulps based on UV/VIS reflectance spectroscopy of peroxide-bleached GWP (groundwood pulp), CTMP, and quinonoid lignin model compounds *(14*). Stilbene *ortho*-quinones were reported to be the only colored products formed when monohydroxystilbenes (deposited on the surface of a filter paper) were exposed to UV-radiation (*15*). Furthermore, the chromophores produced by UV-radiation were found to be almost completely removed by sodium borohydride reduction (*16-19*). This suggests that most of the colored substances formed contain carbonyl structures (quinones, conjugated ketones, aldehydes) that can be reduced by $NaBH_4$.

Argyropoulos et al. (*20*) used solid-state ^{31}P NMR spectroscopy on samples of irradiated unbleached and hydrogen-peroxide-bleached GWP (oxyphosphorylated) in order to determine the early photochemical changes that occur during light-induced yellowing. The data suggested that *ortho*-quinones initially produced during irradiation subsequently reacted to form more complex carbonyl chromophoric structures which do not have a quinonoid character. It was also found that *ortho*-quinones were converted faster in the unbleached pulp than in the peroxide-bleached pulp. The bleached pulp contains more stilbenes that could be converted to stilbene *ortho*-quinones during irradiation (*15*), and it is possible that these structures require longer irradiation times to be eliminated. The conversion of quinones to more complex chromophores could explain why no quinones or quinonoid structures were detected in irradiated milled wood lignin (MWL) (*21*).

Forsskåhl and Janson (*6*) reported that the two main chromophores (displaying bands at 370 and 430 nm), generated by irradiation of chemimechanical pulps, were interrelated. Irradiation at a short wavelength (373 nm) creates a colored chromophore, while irradiation at the longer wavelength (435 nm) leads to the formation of a colorless product. These authors suggest that there are several possible candidates for such a system; a hydroquinone-quinone system, charge transfer complexes, possibly quinone methides and also cis-trans isomerism of aromatic conjugated double bonds. The same effect was later found for irradiated unbleached and peroxide-bleached thermomechanical pulps (*22*). Schmidt et al. (*13*) reported that methoxylated *ortho*-quinones could be bleached by 420 nm irradiation, whereas unmethoxylated *ortho*-quinones sensitized destruction of aromatic groups. Robert and Daneault (*23*) observed two major absorption peaks at 360 and 425 nm when TMP paper (black spruce/balsam fir mixture) was exposed to UV light (λ, 300-400 nm). The peak at 360 nm was the result of the disappearance of one chromophore and the appearance of a different chromophore. The other peak (at 425 nm) was the result of the formation of three chromophores. It is important to bear in mind that structures of the quinonoid type are themselves photosensitizers and they can thus contribute to further chromophore formation (*24-28*).

The present communication describes the effect of irradiation sources with different wavelength distributions on the light-induced yellowing of untreated and acetylated hydrogen-peroxide-bleached aspen CTMP and spruce CTMP, the pulps most often suggested as replacement for chemical pulps in high-quality paper grades. The mechanism of stabilization (and yellowing) has been studied using solid-state UV/VIS diffuse reflectance spectroscopy performed on high basis weight (200 g/m^2) handsheets.

Experimental

Pulps and Paper Samples. Commercially produced hydrogen-peroxide-bleached spruce (*Picea abies*) CTMP and hydrogen-peroxide-bleached aspen (*Populus tremuloides*) CTMP were used as received for the studies described in this paper. The pulps were obtained as dried samples. The high-brightness hardwood CTMP was manufactured employing hydrogen peroxide both as a chemical pretreatment and as a bleaching stage. Handsheets [60 or 200 g/m^2 (for UV/VIS spectroscopy)] were prepared according to TAPPI Test Method T 205 om-88. The paper sheets were then conditioned at 23°C and 50% R.H. according to TAPPI Test Method T 402 om-88 before further treatment.

Acetylation Procedure. The handsheets [cut into strips (30 x 75 mm)] were acetylated (at 100 or 110°C) according to the procedure previously described (*29*). The acetyl content was calculated from the amount of acetate liberated after saponification with sodium hydroxide (*30*). The acetyl content is given as a percentage of the dry weight of the paper.

UV/VIS Diffuse Reflectance Spectroscopy. UV/VIS spectra were recorded on a Perkin Elmer Lambda 19 DM spectrophotometer equipped with a diffuse reflectance and transmittance accessory (Labsphere RSA-PE-90). The accessory is essentially an optical bench which includes double beam transfer optics and a six-inch diameter (154 mm) integrating sphere. Background correction was made with a SRS-99-010-7890 standard. The absorbance (*ABS*) was calculated from the diffuse reflectance (R_∞) using the following expression derived from the Beer-Lambert law (cf. *23*): $ABS = -\log R_\infty$. The resulting spectra were averaged from four to six measurements.

Accelerated Light-Induced Yellowing. The paper samples were subjected to accelerated photoyellowing in a Rayonet photochemical reactor (Model RPR 100, The Southern New England Ultraviolet Company, Branford, CT, USA) equipped with eight RPR 3500Å UV-fluorescent lamps ("blacklight," The Southern New England Ultraviolet Company) or eight RPR 5750Å UV/VIS-fluorescent lamps (The Southern New England Ultraviolet Company) and a merry-go-round apparatus for uniform irradiation. Accelerated reversion studies were also conducted with a SUNTEST CPS (Heraeus HANAU, Hanau, Ger.) light-aging tester equipped with a xenon lamp (nominal rating of xenon burner, 1.1 kW) and filters (ultraviolet and window-glass), which eliminate radiation of wavelengths below 310 nm. The irradiance was controlled by an optical sensor that compensates for possible main voltage fluctuations and burner aging. The temperatures during irradiation were 29, 27, and 30°C for the UV-fluorescent lamps, UV/VIS-fluorescent lamps, and xenon lamp, respectively. The use of controls in each irradiation experiment and the exchange of UV/VIS-fluorescent lamps between aging series were done to monitor and minimize the effects of aging of the light sources studied.

Optical Measurements. TAPPI brightness was measured using a Technidyne Brightimeter (Model S-5) according to TAPPI Test Method T 452 om-92. The reflectance of a single sheet of paper (60 g/m^2) over a completely black, nonreflecting surface (over a hollow black body, reflectance <0.5%) and the reflectance over a stack of paper (high enough to inhibit any transparence of light) were recorded. The specific light scattering coefficient (*s*, at 457 nm) was then calculated using the Kubelka-Munk theory. Optical properties of the SUNTEST CPS irradiated handsheets were measured using an Elrepho 2000 spectrophotometer. The post color (PC) number at 457 nm (*31*) was calculated for the acetylation treatment (PC_1) and for the light-induced reversion (PC_2) (cf. *32*). The sum of PC_1 and PC_2 represents the total effect of the treatment: PC = PC_1 + PC_2. The change in specific light-absorption coefficient (Δk, at 457 nm) during irradiation was calculated from the PC_2-value with the assumption of constant light scattering (*33*).

Results and Discussion

Light Sources. To assess the importance of the wavelength distribution of irradiation sources used for accelerated reversion tests, a series of handsheets made from untreated and acetylated high-yield pulps were irradiated with three different light sources and the optical properties monitored. The light sources chosen were UV-fluorescent lamps, UV/VIS-fluorescent lamps, and a xenon lamp, i.e., commonly used irradiation sources in many light-induced aging equipments. The UV-fluorescent lamps emit light in a band between 300 and 420 nm (approximate Gaussian spectral distribution, λ_{max}=350 nm), i.e., the visible component of diffuse sunlight or office light is practically missing (Figure 1a). The UV-fluorescent lamps give an assessment of the UV light-aging properties and could be useful when complications from other wavelengths are not desirable. The UV/VIS-fluorescent lamps emit light in the ultraviolet and visible range (from about 350 to 700 nm, λ_{max}=575 nm) and have a comparatively close match to the conventional standard "cool white" fluorescent color used in many commercial lighting installations (Figure 2a). The energy distribution is enriched at 575 nm, i.e., most of the light is emitted in the visible part of the spectrum. The spectral distribution of the transmitted light from the xenon lamp and filter combinations used was similar to that of average indoor daylight (Figure 3a). The spectral distribution of standard "cool white" fluorescent color (*34*) and natural daylight (*35*) is given, as a comparison, in Figures 2a (broken line) and 3a (dotted line), respectively.

Change in Optical Properties During Accelerated Photoyellowing. The pulps chosen for this study were a hydrogen-peroxide-bleached spruce CTMP and a hydrogen-peroxide-bleached aspen CTMP. The optical properties, before and after the acetylation treatment and after the accelerated aging procedure, are shown in Table I.

Unacetylated Pulps. Figures 1b, 2b, and 3b show the effects of different light sources on the photoyellowing properties of untreated pulps. The H_2O_2-bleached spruce CTMP was used as a control to determine the different periods of irradiation that were needed to obtain an approximately equivalent degree of light-induced

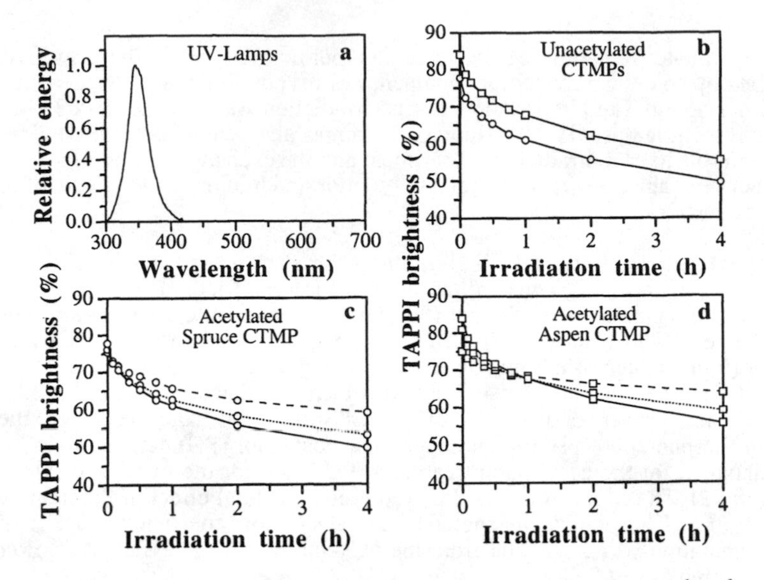

Figure 1. The change in TAPPI brightness of untreated and acetylated spruce CTMP (O) and aspen CTMP (□) after aging with the UV-fluorescent lamps. The spectral energy distribution of the UV-fluorescent lamps is given in **a**. Legends (acetyl content): (····O····), 4.6%; (---O---), 9.4%; (····□····), 4.1%; (---□---), 9.7%.

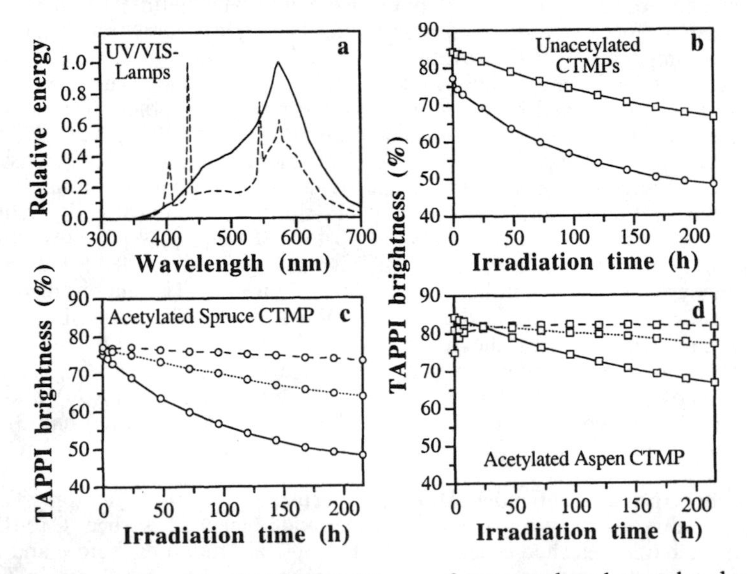

Figure 2. The change in TAPPI brightness of untreated and acetylated spruce CTMP (O) and aspen CTMP (□) after aging with the UV/VIS-fluorescent lamps. The spectral energy distribution of the UV/VIS-fluorescent lamps (solid line) and standard "cool white" fluorescent color (broken line) is given in **a**. Legends (acetyl content): (····O····), 4.6%; (---O---), 9.4%; (····□····), 4.1%; (---□---), 9.7%.

reversion under the test conditions used (cf. Table I). The comparison was made at 4, 40, and 216 hours (9 days) for the UV-fluorescent lamps, xenon lamp, and UV/VIS-fluorescent lamps aged sheets, respectively. As can be seen in Figures 1b and 3b, the yellowing of the UV-fluorescent lamps and xenon lamp irradiated sheets is characterized by a rapid initial phase followed by a slower phase. This is the general behavior of the discoloration process that has been reported by several researchers (*26,30,36,37*). When the UV/VIS-fluorescent lamps were used as an irradiation source the initial rapid phase was less pronounced, resulting in flatter brightness curves, especially for the high-brightness aspen CTMP (Figure 2b). The rate of light-induced chromophore formation was significantly lower for the aspen CTMP than for the spruce, and the variation obtained in aging results between the different exposure techniques used was larger. The change in light absorption coefficient (Δk) was between 2.7 and 6.2, i.e., ca. 40 to 75% lower than that for the spruce CTMP (cf. Table I). Hardwood pulps have previously been reported to be more resistant toward light-induced aging than softwood pulps (*32,33,38*). The lower rate of yellowing of hardwoods may be associated, in part, with their lower lignin contents compared to that of softwoods.

Acetylated Pulps. To establish if chemical modification (e.g., acetylation) of the pulps altered the aging response toward the tested irradiation sources, a series of test sheets were acetylated to various degrees. As can be seen in Table I, the acetylation treatment resulted in a small brightness loss for the hydrogen-peroxide-bleached spruce CTMP (about 2.5 brightness units). The brightness loss for the hydrogen-peroxide-pretreated high-brightness aspen CTMP was more severe; an acetylation time of 10 minutes (110°C) resulted in a brightness loss of about 10 brightness units. These findings agree well with results from earlier investigations (*29,* Paulsson, M.; Ragauskas, A.J. *Nord. Pulp Pap. Res. J.*, in press).

As can be seen in Figures 1 to 3 (and Table I), acetylation strongly inhibited the photoyellowing as previously reported (*30,39*). In Figures 1-3, **c** illustrates the effect of light source on the yellowing properties of acetylated, H_2O_2-bleached spruce CTMP sheets. When irradiated with the UV/VIS-fluorescent lamps, photobleaching of the acetylated sheets took place. As expected, the most severe yellowing (no photobleaching) was observed when the CTMP sheets were irradiated with the UV-lamps (emitted almost no light in the visible range). The difference in reverted brightness between the different light sources was about 11 brightness units for the low acetylated pulps (acetyl content of 4.6%) and more than 14 units for the high acetylated pulps (acetyl content, 9.4%, see Table I).

The degree of reversion was in the order UV>xenon>>UV/VIS lamps for the acetylated aspen CTMP (**d** in Figures 1-3), i.e., the same order as for the bleached spruce CTMP. The UV/VIS-fluorescent lamps induced rapid and strong photobleaching of the acetylated samples. The TAPPI brightness was ca. 82% after 24 hours of irradiation for both acetylation levels, which corresponds to a brightness gain of 1 to 7 brightness units, depending on the derivatization degree studied. Thus, most of the colored substances formed on acetylation of hydrogen-peroxide-pretreated aspen CTMP can easily be converted to colorless structures during irradiation with visible light.

The photobleaching (decrease in light absorption at λ_{max}~360 nm) observed on irradiation of MWL and high-yield pulps has been attributed to the photooxidation of coniferaldehyde end groups in lignin to produce vanillin-type end groups (*40,41*). Furthermore, the dissimilarities in the absorption difference spectra of peroxide-bleached softwood TMP and peroxide-bleached softwood CTMP have been explained in terms of coniferaldehyde photobleaching (*18*). It is also possible that quinonoid structures, absorbing in the blue-green region of the spectrum, photochemically convert to colorless structures (for general comments on light-induced reactions of quinones, cf. *42,43*). This has been demonstrated in the case of methoxy-*p*-benzoquinone incorporated in a solid carbohydrate matrix (i.e., 2-

Table I. Change in optical properties on acetylation and light-induced aging of hydrogen-peroxide-bleached pulps. Acetyl content (% by mass) are given within parentheses.

Pulp		TAPPI brightness (%)	PC_1[a]	UV-lamps Irradiation time: 4 h			UV/VIS-lamps Irradiation time: 216 h			Xenon lamp Irradiation time: 40 h		
				TAPPI brightness	PC_2[b]	Δk[c]	TAPPI brightness	PC_2[b]	Δk[c]	Brightness[d]	PC_2[b]	Δk[c]
CTMP, spruce												
Control	(0.5)	78.2	0.0	49.6	22.5	10.9	48.2	24.8	12.0	48.4	24.4	11.8
Acetylated	(4.6[e])	76.3	0.6	53.0	17.2		63.9	6.5		55.5	14.4	
	(9.4[e])	75.6	0.9	59.0	10.3		73.5	0.8		61.5	8.4	
CTMP, aspen												
Control	(0.6)	84.3	0.0	55.6	16.2	6.2	66.5	7.0	2.7	58.2	13.4	5.1
Acetylated	(4.1[e])	81.0	0.9	59.0	12.0		77.0	1.3		64.0	8.2	
	(9.7[f])	74.6	3.1	63.8	6.1		81.5	-2.1		70.7	2.3	

[a]The post-color (PC_1) number at 457 nm (due to acetylation) was calculated according to *(31)*.
[b]The post-color (PC_2) number at 457 nm (due to irradiation) was calculated according to *(31)*.
[c]The change in light absorption (Δk) was calculated from the PC_2-value with the assumption of constant light scattering *(33)*.
[d]Brightness measured with an Elrepho 2000 spectrophotometer. This instrument uses a diffuse light source and measures the reflected light at a 0°viewing angle, i.e., the "Elrepho" brightness cannot directly be translated to TAPPI brightness that is determined with an instrument employing a directional light source (incident angle of 45°) and measures the reflected light at a 0°viewing angle. The difference in measured brightness was, however, small (less than 1 brightness unit).
[e]Acetylation temperature, 100°C.
[f]Acetylation temperature, 110°C.

hydroxypropylcellulose); the photoreduction to methoxy-hydroquinone was a rapid process, whereas the photooxidation of the hydroquinone was comparatively slow (*27*). Ragauskas (*44*) reported that simple *para*- and *ortho*-quinones, impregnated onto cellulose sheets, did not generally cause any further darkening during irradiation (photolyzed with a xenon lamp). Several of the examined quinones instead exhibited a small brightness increase in the irradiated sheets, which could possibly be attributed to a photobleaching effect. Furthermore, Schmidt et al. (*45*) found that methoxylated *ortho*-quinones, introduced into thermomechanical pulp by treatment with Fremy's salt, were bleached by 420 nm irradiation, whereas unmethoxylated *ortho*-quinones, generated by treatment with sodium periodate, were not.

Several researchers have reported that methylated and, particularly, acetylated lignocellulosic materials, can be photobleached during irradiation (*39,46-51*). The photobleaching of acetylated materials has been ascribed to the formation of acetyl peroxide or peroxides formed from the carbohydrate acetates during irradiation (*47,50*). When comparing the degree of yellowing of acetylated high-yield pulps aged under different light sources, with the degree of yellowing for untreated high-yield pulps, it is evident that the photostabilization obtained by the acetylation treatment is not only an effect of retarding the color-forming reactions, but is also a result of promoting photobleaching reactions. The results from aging experiments with the UV-fluorescent lamps give an assessment of the UV-light aging properties, i.e., a measurement of the chromophore-retarding effect of acetylation. The photobleaching response of acetylated pulps is illustrated for the sheets aged with the UV/VIS-fluorescent lamps.

It is difficult to explain the photobleaching effect obtained for acetylated mechanical pulps in terms of coniferaldehyde or *ortho*-quinone photobleaching since the amount of these structural units in peroxide bleached lignin is very small. Several investigations have shown that hydrogen peroxide bleaching removes most (but not all) of the coniferaldehyde units in the lignin (*18,52-56*). Furthermore, model compound studies have shown that acetylation rapidly decomposes most of the *ortho*-quinone units, whereas the *para*-quinones were essentially unaffected (*57*). The significance of the remaining coniferaldehyde structures, *ortho*- and *para* -quinones, diguaiacylstilbenes (cf. *58-60*; the term guaiacyl refers to a 4-hydroxy-3-methoxyphenyl residue), or some other still unknown structure for the photobleaching (and brightness stability) needs to be investigated further.

Although the different light sources tested in this investigation gave a comparable degree of reversion for most of the untreated pulps, the effects of different irradiation techniques, i.e., spectral distribution of the light source used, considerably altered the photostabilizing effect obtained by acetylation. The same effect has also been shown for UV-screen treated pulps (Paulsson, M.; Ragauskas, A.J. *Nord. Pulp Pap. Res. J.*, in press). The discrepancy in effectiveness of an inhibiting treatment reported in the literature can, in many cases, be explained in terms of the different exposure techniques used by the researcher/laboratory.

Solid-State UV/VIS Diffuse Absorption Difference Spectroscopy. The interpretation of a UV/VIS absorption spectra at short wavelengths (e.g., below 340 nm) is difficult since the signal-to-noise ratio is low in this region. This is because the paper sample absorbs most of the light due to its thickness and high lignin concentration (*23*). A complement to this study, in the low wavelength range, could be performed using low basis weight sheets, i.e., 10 g/m^2 (cf. *61*). Care should also be taken when evaluating the results after high UV exposures since the assumption of exponential distribution of chromophores in the thickness of the sheet may no longer be valid (*23*). Nevertheless, solid-state UV/VIS diffuse reflectance spectroscopy, performed on thick handsheets and with the consideration of the above limitations, is a powerful technique for the study of light-induced yellowing of lignin-containing materials. The result that is obtained with this method is what the eye will perceive as "yellowing of a thick paper."

Irradiated Unacetylated Pulps. Figure 4 shows the UV/VIS absorption difference spectra (Δ*ABS* vs. wavelength) of hydrogen-peroxide-bleached aspen CTMP recorded after aging with various light sources. It is evident that the choice of irradiation source strongly influenced the photochemistry. Although both light sources generated a discoloration in the entire visible region (λ>400 nm), the shape of the absorption curves was different. Irradiation with the UV-fluorescent lamps generated an apparent absorption maximum at 370 nm with a shoulder at approximately 415 nm (Figure 4a), whereas the UV/VIS-fluorescent lamps generated an absorption maximum at 360 nm and showed substantially less absorption in the λ>400 nm range (Figure 4b). The spruce CTMP behaved somewhat differently as can be seen in Figure 5. Irradiation with UV-fluorescent lamps generated an apparent maximum at 330 nm and a distinct maximum at 425 nm. The UV/VIS-fluorescent lamps introduced two maxima; one at 330 nm that was shifted toward longer wavelengths (345 nm) after extensive reversion, and one at 425 nm. In addition to the photodiscoloration, photobleaching with λ_{max}~390 nm was observed for the last light source.

Several research groups have used UV/VIS diffuse reflectance spectroscopy to study the yellowing phenomena. Schmidt and Heitner (*18*) reported that hydrogen-peroxide-bleached aspen CTMP showed a single maximum at about 360 nm when exposed to broad-band UV-radiation, but no shoulder at ~415 nm was detected. The high-brightness aspen CTMP used in the present investigation was pretreated with hydrogen peroxide, and it is possible that this treatment changes the chromophoric precursors in such a way that new colored substances, absorbing in the 415 nm region, were formed when subjected to UV-exposure. Generally, UV/VIS diffuse reflectance spectroscopy of hydrogen-peroxide-bleached softwood pulps (GWP, TMP, and CTMP), performed on both thin and thick sheets, shows an increased absorption in the UV region at 320-360 mm and in the visible region at 410-435 nm when exposed to UV-radiation (*10,14,18,23,36,61-67*). The increase in absorption in the UV region has been attributed to the formation of aromatic carbonyl groups (λ_{max}~330 nm) and to the formation of quinones [possibly methoxylated *para*- or *ortho*-quinones (monohydrate adduct see *18*, cf. *68*), λ_{max}~350-370 nm]. It is also possible that coniferaldehyde with an absorption maximum at ~350 nm, generated through photooxidation of coniferyl alcohol, contributes to the UV-absorption peak. The increase in absorption in the visible region is attributed to the formation of chromophores of the *ortho*-quinonoid type, at least in an initial phase.

The present investigation showed that regardless of the wavelength distribution of the irradiation source, absorption peaks appeared in the above-mentioned UV and visible regions during aging of hydrogen-peroxide-bleached spruce CTMP. The relationship between the peak areas and the position of the maxima was, however, dependent on the light source used. The decrease in absorption at 360 nm observed during irradiation of unbleached lignin-containing materials has been attributed to the destruction of coniferaldehyde end groups or to the conversion of quinones to colorless structures as discussed above. The reduction in chromophore content at λ_{max}~390 nm cannot entirely be explained by the elimination of these structures since sulfonation and hydrogen peroxide bleaching eliminate most of the structures of the coniferaldehyde and quinonoid type (cf. *69*). Hydroxystilbenes, introduced in the lignin moiety during high-yield pulping and alkaline bleaching conditions (*59,70,71*), have been proposed as the leucochromophores that, to a large extent, are responsible for the initial discoloration of bleached high-yield pulps (*60*, cf. *58*). Simple stilbenes have an absorption maximum at about 330 nm in solution, but it is possible that this maximum can be shifted to higher wavelengths when incorporated in the lignin macromolecule due to steric and electronic effects of substituent groups. Zhang and Gellerstedt (*14*) reported a red shift of the UV/VIS absorption maxima of quinone and

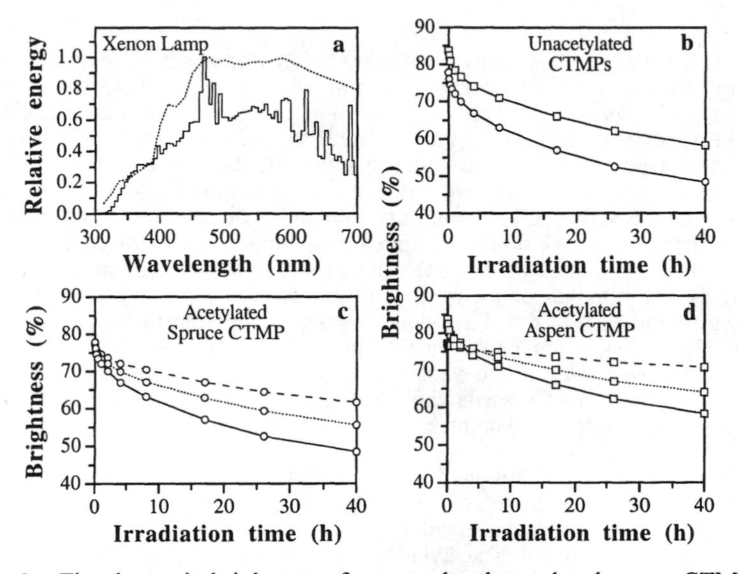

Figure 3. The change in brightness of untreated and acetylated spruce CTMP (O) and aspen CTMP (□) after aging with the xenon lamp. The spectral energy distribution of the xenon lamp (solid line) and natural daylight (dotted line) is given in **a**. Legends (acetyl content): (····O····), 4.6%; (---O---), 9.4%; (····□····), 4.1%; (---□---), 9.7%.

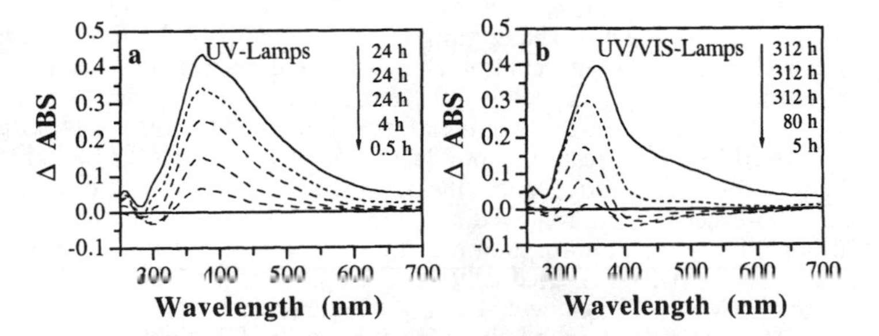

Figure 4. Absorption difference spectra ($\Delta ABS = ABS_{irradiated} - ABS_{unirradiated}$) of acetylated and untreated hydrogen-peroxide-bleached aspen CTMP after aging with UV-lamps (**a**) and UV/VIS-lamps (**b**). Legends: (—), control; (····), acetylated - 3.9%; (- -), acetylated - 8.1%.

stilbene model compounds in the solid state (on filter paper or on bleached GWP or CTMP) compared with the absorption in solution. The red shift was 26 nm for a hydroxystilbene model and between 32 and 148 nm for compounds of the quinonoid type, an effect charge transfer complexes (quinone-phenol) might account for (*14,* cf. *72,73*).

Acetylated Pulps. It is known that acetylation influences brightness differently depending on the type of mechanical pulp derivatized (*30,48,* Paulsson, M.; Ragauskas, A.J. *Nord. Pulp Pap. Res. J.*, in press). Figure 6 shows the UV/VIS absorption spectra (acetylated - unacetylated) for the two tested pulps. During acetylation of aspen CTMP, a substantial increase in absorption is seen in the entire wavelength range 350-700 nm with maxima at approximately 370 and 430 nm. Acetylation of the spruce CTMP not only induced a discoloration, indicated by the absorption increase at λ>410 nm, but also a bleaching at wavelengths below 410 nm (maximum at approximately 385 nm). The decrease at 385 nm could possibly be explained by the removal of a hydroxystilbene structure (cf. the discussion above dealing with a red shift of the absorption maxima of stilbenes in the solid state). The peaks at 370 and 430 nm could result from the characteristic UV absorption of methoxylated *para*-quinones and visible absorption of *ortho*-quinones, respectively. The absorption peak at 455 nm is close to the absorption peak of a stilbene *ortho*-quinone structure incorporated on bleached CTMP (λ_{max}=460 nm) (*14*).

Irradiated Acetylated Pulps. Figures 4 and 5 show the effects of different irradiation times and irradiation sources on the UV/VIS absorption difference spectra of acetylated aspen CTMP and acetylated spruce CTMP.

It was interesting to see if acetylation, an efficient way to retard yellowing of different types of high-yield pulps, fundamentally changed the photochemistry or only slowed down the rate of chromophore formation, as judged from the UV/VIS absorption spectra. The acetylated aspen CTMP, irradiated with the UV-fluorescent lamps, generated an apparent absorption maximum in the same region as the unacetylated CTMP (λ, 370 nm, Figure 4a). No shoulder at 415 nm was, however, observed, and the increase in absorption was less in the whole UV-visible region. The UV/VIS-fluorescent light source induced a photobleaching in the visible range. The absorption in the UV-part of the spectra was also less and shifted about 20 nm toward shorter wavelengths (Figure 4b). The brightness values were 79-80% for the acetylated sheets compared to 60% for the control after an intense aging for 312 hours. The reduced absorption at λ_{max} just below 400 nm (extending into the visible range) indicates that some of the colored structures formed on acetylation of high-brightness aspen CTMP most likely are readily photobleachable when subjected to a light source emitting light in the visible range. Also, the acetylated spruce CTMP, irradiated with the UV/VIS-fluorescent lamps, removed structures contributing to the absorption just below 400 nm. The apparent absorption maximum at 425 nm observed for irradiated unacetylated spruce CTMP was essentially nonexistent after acetylation, except for the low derivatized CTMP (acetyl content, 4.4%, see Figure 5b).

These results indicate that acetylation slows down the UV-light induced reactions as seen for the UV-lamps aged CTMPs. However, acetylation promotes photobleaching reactions when subjected to an irradiation source emitting light in both the ultraviolet and visible regions, an effect that contributes to the improved stability toward light-induced aging. It is evident that measurements of brightness do not give a complete and accurate description of a complicated photochemical process such as the discoloration of lignin-containing materials. These results also indicate that different photochemical reactions occur, depending on the light source used for reversion, for acetylated pulps as well as for untreated pulps. It is therefore extremely important to use a light source that mimics the actual aging conditions as closely as possible if an accurate picture of the yellowing phenomenon is going to be obtained. Further studies,

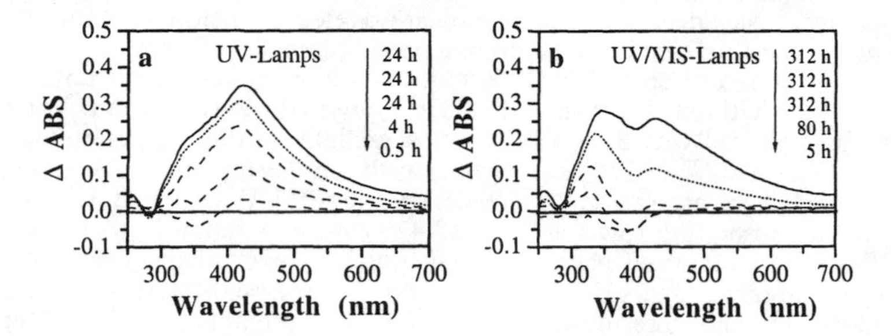

Figure 5. Absorption difference spectra ($\Delta ABS = ABS_{irradiated} - ABS_{unirradiated}$) of acetylated and untreated hydrogen-peroxide-bleached spruce CTMP after aging with UV-lamps (**a**) and UV/VIS-lamps (**b**). Legends: (—), control; (····), acetylated - 4.4%; (- -), acetylated - 9.6%.

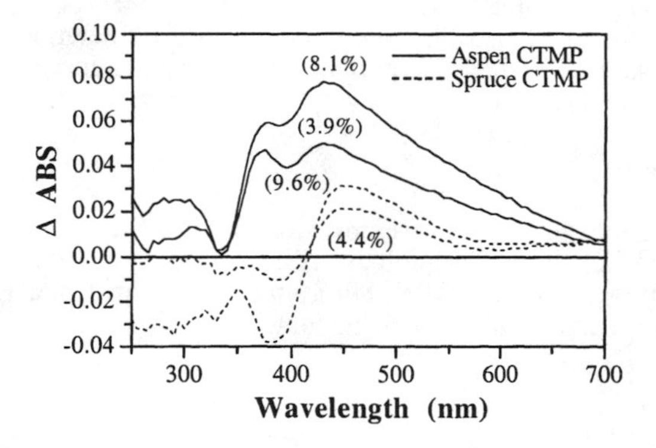

Figure 6. Absorption difference spectra ($\Delta ABS = ABS_{acetylated} - ABS_{unacetylated}$) of aspen and spruce CTMPs after acetylation. The numbers in brackets denote the acetyl content (% by mass).

using both unacetylated and acetylated model compounds incorporated on different types of high-yield pulps, are needed to fully explain the changes in the absorption difference spectra of acetylated and/or irradiated high-yield pulps.

Conclusions

Acetylation of aspen CTMP was found to generate a substantial increase in absorption in the entire wavelength region of 350-700 nm with maxima at approximately 370 and 430 nm, whereas the changes that took place during acetylation of the spruce CTMP were manifested as a decrease in absorption at wavelengths below 410 nm (maximum at 385 nm) and an increase in the absorption band centered at 455 nm.

UV-irradiation of aspen CTMP introduced chromophores with an absorption maximum near 370 nm and with a shoulder at approximately 415 nm. When aspen CTMP was irradiated with the UV/VIS lamps, a slight shift in the absorption peak to shorter wavelengths (λ_{max}~360 nm) together with less absorption in the visible range was observed. The absorption difference spectra of UV-irradiated spruce CTMP displayed two absorption bands at 330 and 425 nm. In addition to these two bands, UV/VIS-irradiation induced photobleaching in the UV-part of the spectra extending into the visible range (λ_{max}~390 nm). Acetylation of the pulps slowed down the UV-light induced chromophore formation, but could also promote photobleaching if the acetylated pulps were subjected to both UV-radiation and visible light.

It is difficult to assign the changes seen in the UV/VIS spectra to the formation/destruction of chromophoric and leucochromophoric structures in the lignin since (i) most absorption data are given for simple model compounds in solution, and it is likely that the position of the absorption peeks may be different in the solid phase, and (ii) the position of absorption maxima of the chromophoric and leucochromophoric structures may be shifted when incorporated in the lignin macromolecule. Further studies are therefore needed to determine the absorption characteristics of several classes of important structures when incorporated in different types of high-yield pulps.

Acknowledgments

Financial support for this work was provided by the Member Companies of the Institute of Paper Science and Technology, USDA Improved Utilization of Wood and Wood Fiber (Contract No. 96-35103-3800), and The Gunnar and Lillian Nicholson Graduate Fellowship and Faculty Exchange Fund.

Literature Cited

1. Heitner, C.; Schmidt, J.A. *Proc. 6th Int. Symp. Wood Pulp. Chem.*, Melbourne, Australia. April 29-May 3, **1991**; Vol. 1, 131.
2. Heitner, C. In *Photochemistry of Lignocellulosic Materials*; Heitner, C.; Scaiano, J.C., Eds.; ACS Symposium Series 531; ASC: Washington, D.C., USA, **1993**; Chapter 1, 2.
3. Leary, G.J. *J. Pulp Pap. Sci.* **1994**, *20*:6, J154.
4. Nolan, P.; Van den Akker, J.A.; Wink, W.A. *Paper Trade J.* **1945**, *121*:11, 33.
5. Andtbacka, A.; Holmbom, B.; Gratzl, J.S. *Proc. 5th Int. Symp. Wood Pulp. Chem.*, Raleigh, N.C., USA. May 22-25, **1989**; Vol. 1, 347.
6. Forsskåhl, I.; Janson, J. *Proc. 6th Int. Symp. Wood Pulp Chem.*, Melbourne, Australia. April 29-May 3, **1991**; Vol. 1, 255.

7. Andrady, A.L.; Song, Y.; Parthasarathy, V.R.; Fueki, K.; Torikai, A. *Tappi J.* **1991**, *74*:8, 162.
8. Forsskåhl, I.; Tylli, H. In *Photochemistry of Lignocellulosic Materials*; Heitner, C.; Scaiano, J.C., Eds.; ACS Symposium Series 531; ACS: Washington, D.C., USA, **1993**; Chapter 3, 45.
9. Andrady, A.L.; Searle, N.D. *Tappi J.* **1995**, *78*:5, 131.
10. Heitner, C.; Min, T. *Cellul. Chem. Technol.* **1987**, *21*:3, 289.
11. Lebo, S.E.; Lonsky, W.F.W.; McDonough, T.J.; Medvecz, P.J.; Dimmel, D.R. *J. Pulp Pap. Sci.* **1990**, *16*:5, J139.
12. Zhu, J.H.; Olmstead, J.A.; Gray, D.G. *J. Wood Chem. Technol.* **1995**, *15*:1, 43.
13. Schmidt, J.A.; Kimura, F.; Gray, D.G. *Res. Chem. Intermed.* **1995**, *21*:3-5, 287.
14. Zhang, L.; Gellerstedt, G. *Proc. 3rd European Workshop on Lignocellulosics and Pulp*, Stockholm, Sweden. August 28-31, **1994**; 293.
15. Zhang, L.; Gellerstedt, G. *Acta Chem. Scand.* **1994**, *48*:6, 490.
16. Hemmingson, J.A.; Morgan, K.R. *Holzforschung* **1990**, *44*:2, 127.
17. Capretti, G.; Castellan, A.; Choudhury, H.; Davidson, R.S.; Noutary, C.; Origgi, S.; Trichet, V. *Proc. 3rd European Workshop on Lignocellulosics and Pulp*, Stockholm, Sweden. August 28-31, **1994**; 159.
18. Schmidt, J.A.; Heitner, C. *J. Wood Chem. Technol.* **1995**, *15*:2, 223.
19. Pan, X.; Ragauskas, A.J. *Proc. 9th Int. Symp. Wood Pulp. Chem.*, Montréal, Canada. June 9-12, **1997**; 97-1.
20. Argyropoulos, D.S.; Heitner, C.; Schmidt, J.A. *Res. Chem. Intermed.* **1995**, *9*:3-5, 263.
21. Sjöholm, R.; Holmbom, B.; Åkerback, N. *J. Wood Chem. Technol.* **1992**, *12*:1, 35.
22. Forsskåhl, I.; Maunier, C. In *Photochemistry of Lignocellulosic Materials*; Heitner, C.; Scaiano, J.C., Eds.; ACS Symposium Series 531; ACS: Washington, D.C., USA, **1993**; Chapter 12, 156.
23. Robert, S.; Daneault, C. *J. Wood Chem. Technol.* **1995**, *15*:1, 113.
24. Gierer, J.; Lin, S.Y. *Svensk Papperstidn.* **1972**, *75*:7, 233.
25. Neumann, M.G.; Machado, A.E.H. *J. Photochem. Photobiol., B.* **1989**, *3*:4, 473.
26. Ek, M. *Some Aspects on the Mechanisms of Photoyellowing of High-yield Pulps*; Royal Institute of Technology, Stockholm, Sweden, **1992**; Ph.D. Thesis.
27. Castellan, A.; Nourmamode, A.; Jaeger, C.; Forsskåhl, I. *Nord. Pulp Pap. Res. J.* **1993**, *8*:2, 239.
28. Forsskåhl, I.; Tylli, H.; Olkkonen, C. *Proc. 7th Int. Symp. Wood Pulp. Chem.*, Beijing, P.R. China. May 25-28, **1993**, Vol. *2*, 750.
29. Paulsson, M.; Simonson, R.; Westermark, U. *Nord. Pulp Pap. Res. J.* **1994**, *9*:4, 232.
30. Paulsson, M.; Simonson, R.; Westermark, U. *Nord. Pulp Pap. Res. J.* **1996**, *11*:4, 227.
31. Giertz, H.W. *Svensk Papperstidn.* **1945**, *48*:13, 317.
32. Janson, J.; Forsskåhl, I. *Nord. Pulp Pap. Res. J.* **1989**, *4*:3, 197.
33. Johnson, R.W. *Tappi J.* **1989**, *72*:12, 181.
34. Waymouth, J.F. In *Encyclopedia of Physical Science and Technology*; Meyers, R.A., Ed.; Academic Press Inc.: San Diego, CA, USA, **1992**; Vol. 8; 697.
35. Merrigan, J.A. *Sunlight to Electricity*; The MIT Press: MA, USA, **1975**; Chapter 3, 41.
36. Lewis, H.F.; Reineck, E.A.; Fronmuller, D. *Paper Trade J.* **1945**, *121*:8, 44.
37. Francis, R.C.; Dence, C.W.; Alexander, T.C.; Agnemo, R.; Omori, S. *Tappi J.* **1991**, *74*:12, 127.
38. Forsskåhl, I.; Janson, J. *Pap. Puu* **1992**, *74*:7, 553.
39. Paulsson, M.; Simonson, R.; Westermark, U. *Nord. Pulp Pap. Res. J.* **1995**, *10*:1, 62.

40. Wang, J.; Heitner, C.; Manley, R.St.J. *Proc. 7th Int. Symp. Wood Pulp. Chem.*, Beijing, P.R. China. May 25-28, **1993**; Vol. 2, 740.
41. Wang, J.; Heitner, C.; Manley, R.St.J. *Proc. 8th Int. Symp. Wood Pulp. Chem.*, Helsinki, Finland. June 6-9, **1995**; Vol. 3, 27.
42. Bruce, J.M. *Quart. Rev.* **1967**, *21*:2, 405.
43. Bruce, J.M. In *The Chemistry of the Quinonoid Compounds*; Patai, S., Ed.; John Wiley & Sons Ltd.: Bristol, G.B., **1974**; Vol. 1, Chapter 9, 465.
44. Ragauskas, A.J. In *Photochemistry of Lignocellulosic Materials*; Heitner, C.; Scaiano, J.C., Eds.; ACS Symposium Series 531; ACS: Washington, D.C., USA, **1993**; Chapter 5, 77.
45. Schmidt, J.A.; Kimura, F.; Gray, D.G. *Proc. 8th Int. Symp. Wood Pulp. Chem.*, Helsinki, Finland. June 6-9, **1995**; Vol. 1, 443.
46. Callow, H.J. *Nature* **1947**, *159*:4035, 309.
47. Callow, H.J.; Speakman, J.B. *J. Soc. Dyers Col.* **1949**, *65*:12, 758.
48. Manchester, D.F.; McKinney, J.W.; Pataky, A.A. *Svensk Papperstidn.* **1960**, *63*:20, 699.
49. Andrews, D.H.; Des Rosiers, P. *Pulp Pap. Mag. Can.* **1966**, *67*:C, T119.
50. Lorås, V. *Pulp Pap. Mag. Can.* **1968**, *69*:2, 57.
51. Ek, M.; Lennholm, H.; Lindblad, G.; Iversen, T. *Nord. Pulp Pap. Res. J.* **1992**, *7*:3, 108.
52. Hirashima, H.; Sumimoto, M. *Mokuzai Gakkaishi* **1987**, *33*:1, 31.
53. Pan, X.; Lachenal, D.; Lapierre, C.; Monties, B. *J. Wood Chem. Technol.* **1992**, *12*:2, 135.
54. Pan, X.; Lachenal, D.; Lapierre, C.; Monties, B.; Neirinck, V.; Robert, D. *Holzforschung* **1994**, *48*:5, 429.
55. Agarwal, U.P.; McSweeny, J.D. *Proc. 8th Int. Symp. Wood Pulp. Chem.*, Helsinki, Finland. June 6-9, **1995**; Vol. 1, 435.
56. Agarwal, U.P.; Atalla, R.H.; Forsskåhl, I. *Holsforschung* **1995**, *49*:4, 300.
57. Paulsson, M.; Li, S.; Lundquist, K.; Simonson, R.; Westermark, U. *Nord. Pulp Pap. Res. J.* **1996**, *11*:4, 220.
58. Castellan, A.; Colombo, N.; Nourmamode, A.; Zhu, J.H.; Lachenal, D.; Davidson, R.S.; Dunn, L. *J. Wood Chem. Technol.* **1990**, *10*:4, 461.
59. Gellerstedt, G.; Zhang, L. *J. Wood Chem. Technol.* **1992**, *12*:4, 387.
60. Gellerstedt, G.; Zhang, L. In *Photochemistry of Lignocellulosic Materials*; Heitner, C.; Scaiano, J.C., Eds.; ACS Symposium Series 531; ACS: Washington, D.C., USA, **1993**; Chapter 10, 129.
61. Schmidt, J.A.; Heitner, C. *Tappi J.* **1993**, *76*:2, 117.
62. Hirashima, H.; Sumimoto, M. *Mokuzai Gakkaishi* **1986**, *32*:9, 705.
63. Fornier de Violet, P.; Nourmamode, A.; Colombo, N.; Castellan, A. *Cellul Chem. Technol.* **1989**, *23*:5, 535.
64. Forsskåhl, I.; Janson, J. *Nord. Pulp Pap. Res. J.* **1991**, *6*:3, 118.
65. Schmidt, J.A.; Heitner, C. *J. Wood Chem. Technol.* **1991**, *11*:4, 397.
66. Schmidt, J.A.; Heitner, C. *J. Wood Chem. Technol.* **1993**, *13*:3, 309.
67. Schmidt, J.A.; Heitner, C. *Proc. 9th Int. Symp. Wood Pulp. Chem.*, Montréal, Canada. June 9-12, **1997**; 98-1.
68. Adler, E.; Lundquist, K. *Acta Chem. Scand.* **1961**, *15*:1, 223.
69. Dence, C.W. In *Pulp Bleaching - Principles and Practice*; Dence, C.W.; Reeve, D.W., Eds.; TAPPI Press: Atlanta, GA, USA, **1996**; Section III, Chapter 4, 161.
70. Gellerstedt, G.; Agnemo, R. *Acta Chem. Scand.* **1980**, *B34*:6, 461.
71. Wu, Z.H.; Matsuoka, M.; Lee, D.Y.; Sumimoto, M. *Mokuzai Gakkaishi* **1991**, *37*:2, 164.
72. Furman, G.S.; Lonsky, W.F.W. *IPC Technical Paper Series 198*, **1986**.
73. Furman, G.S.; Lonsky, W.F.W. *J. Wood Chem. Technol.* **1988**, *8*:2, 165.

Chapter 26

Investigation of *ortho*- and *para*-Quinone Chromophores in Alkaline Extraction Stage Residual Lignins

Michael Zawadzki, Troy Runge, and and Arthur J. Ragauskas

Institute of Paper Science and Technology, 500 10th Street, N.W., Atlanta, GA 30318

The chromophoric properties of a series of residual lignins were studied in order to understand brightness development during pulp bleaching. This study focused upon lignins isolated from kraft softwood brownstock, chlorine dioxide delignified brownstock, and a series of oxidative alkaline extracted pulps. The chromophoric properties of the isolated lignins were assessed by both visible absorbance and ^{31}P-NMR spectroscopy. A ^{31}P-NMR spectroscopic method was employed for the quantification of the combined *ortho*- and *para*-quinone content in the isolated lignins. The ^{31}P-NMR method, modified from the literature, utilized the derivatization of lignin quinone structures by trimethylphosphite. The results suggest that chromophores, such as *ortho*- and *para*-quinones, may be important contributors to brightness ceiling development during chemical pulp bleaching.

A number of structures have been implicated as chromophores in mechanical and chemical pulps, including: catechol metal complexes [1-4], coniferaldehyde [1,2,5], quinone methides [1], stable radicals [1] and quinones [1,2,4,6-9]. Of the various possible chromophores, quinones have been suggested to be major contributors to the color of kraft lignin [6-8,10]. This study employed visible absorbance spectroscopy and a ^{31}P-NMR-based procedure to investigate the presence of quinone chromophores in residual lignin isolated from bleached kraft pulps. The generation of quinone structures during chlorine dioxide bleaching and their fate during oxidative alkaline extraction were explored.

Chemical Pulp Bleaching. Multistage bleaching consists of delignification and brightening stages. In the delignification bleaching stage, bulk residual lignin is degraded and removed. Chlorine dioxide (D), as a delignification agent, is replacing chlorine (C) or chlorine/chlorine dioxide (C/D) because of environmental pressures to reduce adsorbable organic halide (AOX) formation.

Contemporary bleaching sequences use alkaline extraction (E) after a chlorine dioxide (D) stage to remove oxidized lignin and increase the efficiency of a subsequent chlorine dioxide stage. The primary function of the alkaline extraction stage is thought to involve solubilization of oxidized lignin fragments by conversion of various functional groups to their ionized forms: carboxylate, phenolate, and enolate anions [11]. Oxidants, such as hydrogen peroxide (P) and oxygen (O), are often applied in the alkaline extraction stage to further assist with delignification and increase pulp brightness.

The final brightening stages are responsible for the elimination of residual chromophoric structures. The chromophoric structures may be present initially both in the pulp and/or formed during the preceding bleaching sequences. During the final brightening stages of bleaching, the residual lignin concentration is low. Therefore, during brightening, the elimination of the chromophoric structures must be highly selective or else cellulose damage will take place.

Quinone Chemistry. Lignin quinone structures are important because of their chromophoric properties and because they may be formed and destroyed concurrently during bleaching. For example, chlorine dioxide has been shown by several investigators [12-18] to react with phenolic lignin structures giving *ortho*- and *para*-quinone structures among its products. Conversely, hydroperoxide anion, generated during hydrogen peroxide bleaching, specifically removes conjugated carbonyl structures such as quinones [19-21]. Quinones may also be formed by the Dakin reaction of hydrogen peroxide with *para*-hydroxy carbonyl structures [19,22].

During alkaline oxygen bleaching, hydroxyl radicals may generate lignin-hydroxycyclohexadienyl radicals which lead to quinone formation *via* disproportionation or demethoxylation [23]. Alternatively, given the presence of superoxide anion, the hydroxycyclohexadienyl radical may be degraded to muconic acid structures.

Lignin model compound studies have also shown *ortho*-quinones to be susceptible to nucleophilic attack by hydroxide anions. *Ortho*-quinone structures may rearrange to an α-hydroxy-carboxylic acid cyclopentadiene structure by a benzylic acid type of rearrangement [19]. Also, hydroxide may add to quinone structures by nucleophilic addition to give precursors of chromophoric hydroxy-substituted quinones [19,24].

Materials and Methods

Chemicals. All chemicals, except 1,4-dioxane, were purchased and used as received. Before use, 1,4-dioxane was purified by distillation over sodium borohydride.

Pulps. Conventional kraft pulp was obtained from a single, 30-year-old, disease-free Loblolly pine (*Pinus taeda*) tree. Brownstock pulp (kappa number 30.5) was bleached in a D_0 stage under the following conditions: 2.3% chlorine dioxide charge, 10% consistency, 45°C, final pH=2.0, and 45 minute reaction. The bleached pulp was then washed with water and characterized for kappa number, Klason lignin content, and viscosity.

Alkaline Extraction. Chlorine dioxide (D_0) delignified pulp was alkaline extracted in a stirred pressure reactor under the following general conditions: 10% consistency, 70°C, and 75 minute reaction. **Table I** summarizes the specific conditions used for the oxidative alkaline extraction study. Washed alkaline extracted pulps were characterized in terms of kappa number, Klason lignin content, and pulp viscosity.

Table I. Alkaline Extraction Stage Conditions.

Stage	Bleaching Conditions[a]
E	2.0% NaOH; atmospheric pressure air.
E+O	2.5% NaOH; 60 psig O_2 decreased by 12 psig/5 minutes.
E+P	2.5% NaOH; 0.5% H_2O_2.
E+O+P	2.5% NaOH; 0.5% H_2O_2; 60 psig O_2 decreased by 12 psig/5 minutes.
E+Ar	Air removed by a freeze-thaw cycle; 2.0% NaOH; 10 psig Argon.

[a] final pH > 10.5

Brightness Ceiling Determination. The alkaline extracted pulps were further bleached with a D_1ED_2 sequence. The D_1 stage conditions were as follows: 0.75% chlorine dioxide charge, 10% consistency, 70°C, and 3-hour reaction. E_2 stage conditions were as follows: 1.0% sodium hydroxide charge, 10% consistency, 70°C, and 60-minute reaction. Washed E_2 stage pulps were bleached in a D_2 stage. Chlorine dioxide charge in the D_2 stage was varied from 0.2% to 0.8% charge in a series of separate experiments. A small amount of sodium hydroxide (25% of chlorine dioxide charge added) was added at the D_2 stage for pH adjustment.

Isolation of Residual Lignin. Residual lignin was isolated from the pulps by a mild acidic dioxane hydrolysis procedure modified from the literature [25-28]. Pulp was extracted using 90% 1,4-dioxane/0.1 N HCl (v/v) solution (8% consistency) by refluxing for 2 hours under an argon atmosphere. The extract was filtered, neutralized, and 1,4-dioxane was removed under reduced pressure at 40°C. The resulting aqueous lignin solution was acidified (pH 2.5) to precipitate the lignin. The precipitated lignin was purified by three cycles of a freeze-thaw-centrifuge-decant sequence. The purification sequence involved freezing the aqueous lignin sample (-20°C), slow thawing, centrifugation, decanting, and washing the lignin with water. Between each cycle the pH of the solution was adjusted to 2.5. After freeze drying, the yield of residual lignin, relative to Klason lignin, was 45-65%.

Visible Spectrum of Lignin. The visible absorbance spectra of isolated lignins were measured in 90% 1,4-dioxane/water (v/v) solvent. The visible absorbance spectra were acquired with a Shimadzu UV160U ultraviolet/visible spectrophotometer.

Pulp Characterization. The lignin contents were measured by both a ¼ kappa number test (TAPPI T 236 om-85) and a modified standard Klason lignin content test (TAPPI T 222 om-88). The modified Klason lignin content test used an autoclave to speed up the time required for pulp digestion. Viscosity of cupriethylenediamine (CED) dissolved pulp was measured by TAPPI method T 230 om-94. Standard TAPPI handsheets were prepared from D_2 stage pulp (basis weight of 150 g/m^2) and used to measure ISO brightness (TAPPI T 205 sp-95).

Quinone Determination. Dry residual lignin (30 mg) was derivatized with 500 μL 50% trimethyl phosphite/DMSO (v/v) under an argon atmosphere at room temperature for seven days. Lignin samples were previously dried under vacuum (3 millitorr) at 40°C for 24 hours.

Derivatized lignin samples were prepared for analysis by removing excess trimethyl phosphite under vacuum at 40°C for 3 hours. The treated lignins were dissolved in 400 μL of solvent consisting of DMSO-d_6, tri-*m*-tolyl-phosphate (2.5 mg/mL) and chromium acetylacetonate (1.0 mg/mL). Derivatized lignin quinone structures were hydrolyzed to the open-chain phosphate ester by the addition of 5 μL water (0.3 mmol per 30 mg lignin). After 12 hours, the ^{31}P-NMR spectrum of the resulting solution was acquired with a Bruker 400 MHz NMR spectrometer.

Phosphorus-NMR spectra were acquired under quantitative conditions at 305°K. A 90° pulse was utilized with a 5-second pulse delay along with inverse-gated broad-band proton decoupling. A line-broadening factor of 15 Hz was used and the time domain (TD) size was 64K. For each spectrum 1000-3000 scans were collected. The internal standard tri-*m*-tolyl-phosphate (-16.3 ppm) was used both for quantification and as a shift reference. The ^{31}P-NMR chemical shift of tri-*m*-tolyl-phosphate in DMSO-d_6 was determined with the aid of 85% H_3PO_4 as an external shift reference. Quantification of lignin quinone content was achieved by integrating the areas of the internal standard and the phosphate-ester (quinone adduct) resonance centered at -2.5 ppm.

Chromium acetylacetonate was used to reduce the T_1 (spin-lattice) relaxation of the components of interest including the internal standard. The T_1 value for the open-chain phosphate ester (quinone adduct) in trimethyl phosphite treated lignin was found to be 0.7 seconds. The T_1 relaxation time constant for the internal standard, tri-*m*-tolyl-phosphate, was found to be 0.9 seconds. A standard inversion-recovery experiment [29] was used to determine the T_1 parameters.

Results and Discussion

Pulp Characterization. Lignin contents of the bleached pulps were determined by both Klason lignin and kappa number tests and are shown in **Table II**. In general, a higher degree of delignification occurs with increased application of oxidant in the alkaline extraction stage. The CED viscosity data, which is an indirect measure of cellulose degradation, is also given in **Table II**. The viscosity data reveals that only minor carbohydrate damage occurs during the bleaching stages. In the alkaline extraction stage, hydrogen peroxide was more selective towards lignin removal than oxygen on the basis of Δkappa per Δviscosity (using brownstock for the initial values).

Table II. Pulp Characterization Data.

Pulp Description	Kappa Number	Klason Lignin	CED viscosity
Brownstock	30.4	1.895	30.27
D	14.6	1.583	26.40
D(E+Ar)	7.10	0.768	25.72
D(E)	6.37	0.709	22.57
D(E+O)	4.17	0.661	20.10
D(E+P)	4.80	0.671	22.64
D(E+O+P)	3.33	0.589	20.38

Brightness Ceiling. A brightness ceiling is the maximum brightness that can be achieved in a given bleaching stage after which further application of bleaching agent does not lead to an increase in brightness. The alkaline extracted pulps were further bleached with a D_1ED_2 sequence to generate a D_2 brightness ceiling (**Figure 1**).

The D_2 brightness ceiling data reveals that the use of hydrogen peroxide in the alkaline extraction stage, D(E+P) and D(E+O+P), results in the highest achievable brightness ceilings. When excluding all oxygen with argon, D(E+Ar), or incorporating air, D(E), or reinforcing with pressurized oxygen, D(E+O), in the alkaline extraction stage gives similar D_2 brightness ceiling values. **Figure 1** demonstrates that pulp properties altered in the first alkaline extraction stage directly impact the bleachability of the pulp. Specifically, hydrogen peroxide decreases the content of structures which have a detrimental influence on the final brightness ceiling value.

Previous chemical pulp bleaching studies have investigated parameters influencing brightness ceiling values. For example, McDonough found that brightness ceiling development during D_2 stage bleaching (of a D_0(E+O)D_1ED_2 sequence) is dependent upon D_1 stage brightness [30]. In further studies of the D_0(E+O)D_1ED_2 bleaching sequence, McDonough *et al.* found that at a constant kappa factor the D_2 brightness ceiling is affected by the unbleached kappa number and the effective alkali charge during pulping [31]. These results suggest that lignin structural features may influence final brightness ceiling values.

Senior *et al.* showed that the brightness ceiling of a DEDP sequence is greater than that of a DEPD sequence [32]. Senior *et al.* hypothesized that the higher brightness ceiling of the DEDP sequence is due to the presence of quinone or conjugated-carbonyl chromophores which survive the DED sequence only to be removed when hydrogen peroxide stage is subsequently applied [32]. Similarly, the results of this investigation suggest that hydrogen peroxide, applied in the alkaline extraction stage, removes quinone (or conjugated carbonyl) structures that would otherwise cause a lower D_2 brightness ceiling value.

Visible Spectrum. The visible absorbance difference spectra for the series of residual lignins isolated from chlorine dioxide delignified brownstock and oxidative alkaline extracted pulps were acquired (**Figure 2**). Difference spectra were calculated by subtracting the brownstock residual lignin absorption spectrum from the absorption spectra of the isolated lignins. Analysis of difference spectra allows for the identification of chromophore changes occurring in the alkaline extraction stage relative to the unbleached brownstock.

Clearly, the absorbance difference spectra are observed to cluster into groups based upon the oxidant applied to the alkaline extraction stage. Residual lignins arising from peroxide-treated pulps displayed considerably less visible absorbance than the initial unbleached brownstock residual lignin. Note that with the exception of hydrogen peroxide bleaching, D(E+P) and D(E+O+P), all residual lignins are darker than the initial brownstock residual lignin (**Figure 2**).

Quinone structures are potential chromophoric contributors to the brightness ceiling phenomena. The n–π^* transition for quinones occurs in the visible region and may contribute to the colored nature of pulps. In general, the n–π^* transition for *para*-quinones occurs in the 420-460 nm region and 500-580 nm for *ortho*-quinones [33]. According to Furman and Lonsky, the absorption maximum for kraft lignin quinone structures occurs at ~430 nm [8]. If the residual lignins are ordered in terms of absorbance at 430 nm the following series is derived: D(E+Ar) > D = D(E) > D(E+O) >> D(E+P) > D(E+O+P). It can be noted

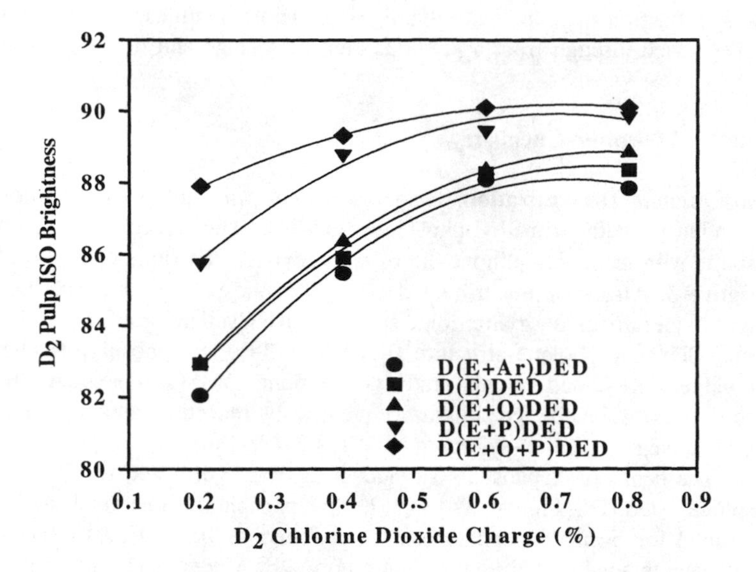

Figure 1. D_2 Stage Brightness Ceiling Data.

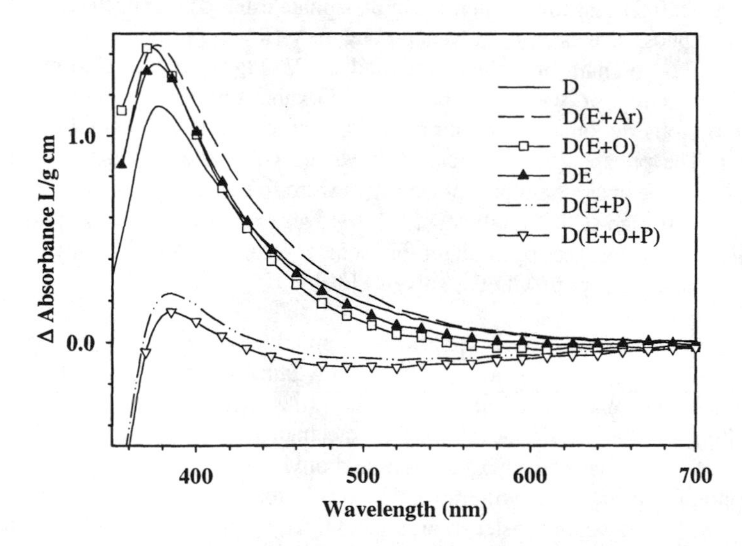

Figure 2. Visible Absorption Difference Spectra for a Series of Residual Lignins.

that this order corresponds to the brightness ceiling results shown in **Figure 1**. The correlation between brightness ceiling values and absorption spectra at the alkaline extraction stage indicates that lignin structural features, such as quinones, may be carried through from a previous bleaching stage and directly impact the final brightness value.

Trimethyl Phosphite Chemistry

***Ortho*-Quinone Derivatization.** Both *ortho-* and *para*-quinones are known to form adducts with trimethyl-phosphite [34-36]. The reaction of trimethyl phosphite with the *ortho*-quinone 3,5-di-*tert*-butyl-1,2-benzoquinone (I) is shown in **Figure 3**. Attack of the trimethyl phosphite phosphorus at the carbonyl is thought to initially give a zwitterionic structure (II). Cyclization of II then leads to a benzo-dioxaphospholene structure (III) [34,36-38]. The phosphorus chemical shift value, determined in this study (-45.3 ppm, DMSO-d_6 solvent), for the benzo-dioxaphospholene is similar to previously reported values: -46.5 ppm (CD_2Cl_2 solvent) [37] and -46.9 ppm ($CDCl_3$ solvent) [36].

The benzo-dioxaphospholene adduct can be hydrolyzed to give a cyclic phosphate ester (IV, **Figure 3**) [34,38]. The phosphorus chemical shift value determined for cyclic phosphate ester was found to be 13.6 ppm (DMSO-d_6 solvent) and this value is similar to that reported by Medvecz (13.4 ppm, CD_2Cl_2 solvent) [37]. During the present investigation it was found that the cyclic phosphate ester (IV) is unstable and further hydrolysis leads to a structure with a chemical shift similar to the open-chain phosphate ester. The expected structure of the cyclic phosphate ester hydrolysis product is V (R=H) [39].

An open-chain phosphate ester adduct (V, **Figure 3**) can also result from the direct action of water on the benzo-dioxaphospholene adduct [37]. Two possible isomeric open-chain phosphate esters products (V, R=CH_3) may be formed. The phosphorus chemical shift values (-2.2 and -2.4 ppm, DMSO-d_6 solvent) for the open-chain phosphate esters were found to be similar to a reported value of -4.0 ppm ($CDCl_3$ solvent) [36]. Medvecz reported a similar phosphorus chemical shift for the open-chain phosphate ester adduct of 3-methoxy-1,2-benzoquinone (-2.3 ppm, CD_2Cl_2 solvent) [37].

***Para*-Quinone Derivatization.** Ramierez *et al.* demonstrated that trimethyl phosphite can form an adduct with *para*-quinones [34,35]. **Figure 4** illustrates the reaction of trimethyl phosphite with the *para*-quinone 2,6-dimethoxy-1,4-benzoquinone (VI). The mechanism is thought to proceed by attack of trimethyl phosphite on the carbonyl oxygen leading initially to a phosphonium-phenoxide zwitterion (VII). Rapid methyl group translocation gives the open-chain phosphate ester in high yield (VIII) [34,35]. Two isomeric adducts may be formed depending upon which quinone carbonyl group is initially attacked. The phosphorus chemical shift value for VIII was found to be -1.4 ppm (DMSO-d_6 solvent). In a related study, Medvecz reported the chemical shift

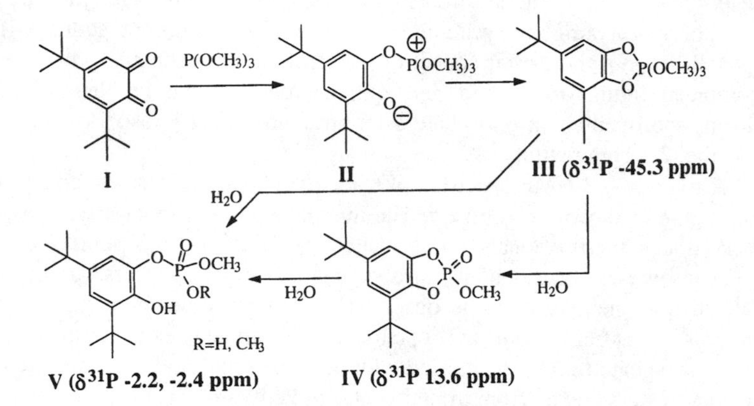

Figure 3. Reaction of Trimethyl Phosphite with 3,5-Di-*tert*-butyl-1,2-benzoquinone ([31]P Chemical Shifts were Determined in this Study).

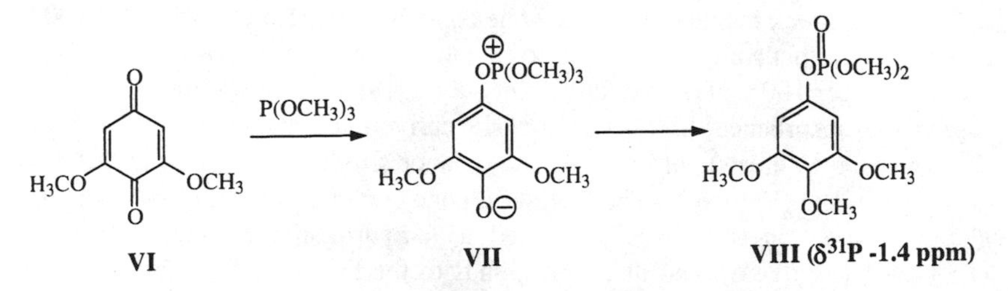

Figure 4. Reaction of Trimethyl Phosphite with 2,6-Dimethoxy-1,4-benzoquinone ([31]P Chemical Shifts were Determined in this Study).

values for the trimethyl phosphite/2-methoxy-1,4-benzoquinone isomeric adducts as -2.6 and -3.15 ppm (CD_2Cl_2 solvent) [37].

Lignin Derivatization. Using trimethyl phosphite derivatization, Lebo and others have developed a solid-state ^{31}P-NMR spectroscopic method for the detection *ortho*-quinones [40-44]. Lebo *et al.* [40-42] and Argyropoulos *et al.* [43,44] both used the cyclic phosphate ester adduct (III, **Figure 3**) as diagnostic for the presence of *ortho*-quinones in trimethyl phosphite derivatized mechanical pulp. For this study, the literature procedure [40-44] was modified and applied to the determination of quinone structures in isolated lignins. The modification consisted of hydrolyzing the cyclic phosphate ester, *ortho*-quinone adduct (III), to the open-chain phosphate ester adduct (V). Therefore, after the addition of water, the combined lignin *ortho*- and *para*-quinone content can be determined by monitoring the open-chain phosphate ester structures with phosphorus chemical shifts in the -2.5 ppm region.

A solution ^{31}P-NMR spectrum of trimethyl phosphite derivatized D(E+Ar) residual lignin is shown in **Figure 5**. The internal standard, tri-*m*-tolyl-phosphate is observed as a sharp resonance with a chemical shift of -16.3 ppm. The broad Gaussian resonance corresponding to open-chain phosphate esters, arising from derivatized quinone structures, is observed with a peak centered at -2.5 ppm. Resonances downfield from the open-chain phosphate ester correspond to trimethyl-phosphate (3.5 ppm, verified with pure material) and an expected series of phosphate esters arising from trimethyl phosphite hydrolysis [45,46].

Lignin Quinone Content. The combined *ortho*- and *para*-quinone content data (after subtraction of the softwood brownstock residual lignin quinone content) for the D_0 and alkaline extraction stage residual lignins is given in **Figure 6**. The brownstock residual lignin quinone content value, 1.6 quinones per 100 C_9, determined in this study was similar to literature values for softwood kraft lignin: 3 quinones per 100 C_9 (*via* reductive acetylation) [8] and 3-4 quinones per 100 C_9 (*via* visible absorbance) [47]. The ^{31}P-NMR derived quinone content data (**Figure 6**) was found to cluster into groups in a manner similar to the visible difference absorbance data (**Figure 2**). The lowest quinone contents were observed when the alkaline extraction stage was reinforced with hydrogen peroxide, D(E+P) and D(E+O+P) (**Figure 6**), and this corresponds to the highest achievable brightness ceilings (**Figure 1**).

Chlorine Dioxide Stage. The application of chlorine dioxide (D_0) was found to cause a dramatic increase in the quinone content relative to the brownstock residual lignin value (**Figure 6**). The D_0 residual lignin contained 0.135 mmol/g lignin (2.5 quinones per 100 C_9) more quinone structures than the brownstock residual lignin. These results are consistent with literature accounts which indicate that both phenolic and non-phenolic lignin structures can react with chlorine dioxide to give *ortho*- and *para*-quinones [12-18]. The ^{31}P-NMR

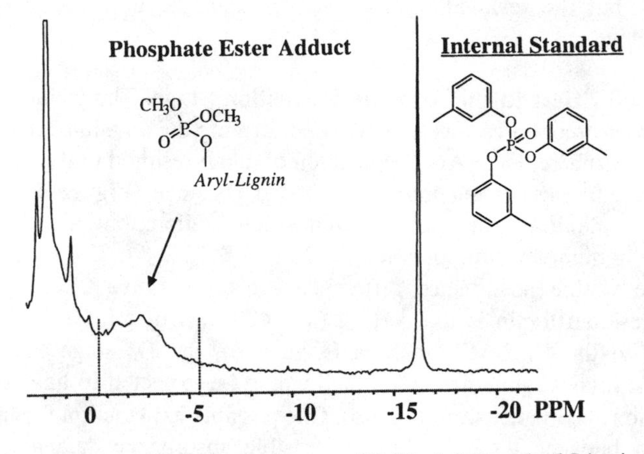

Figure 5. Phosphorus-NMR Spectrum of D(E+Ar) Residual Lignin Treated by Trimethyl-Phosphite.

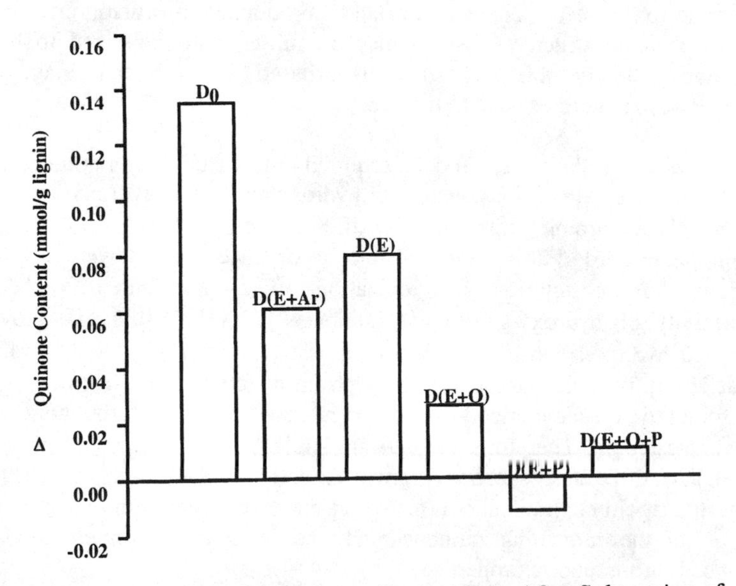

Figure 6. Residual Lignin Quinone Content Data After Subtraction of Brownstock Residual Lignin Quinone Content Value.

derived quinone content data is also consistent with the visible absorption difference spectra which demonstrates that D_0 residual lignin is darker than brownstock residual lignin. Chlorine dioxide is an effective delignification agent (**Table II**), but the residual lignin is darker than the unbleached brownstock residual lignin.

Alkali Effect in the Alkaline Extraction Stage. The influence of alkali on quinone chromophores was studied by performing the alkaline extraction under an argon atmosphere, D(E+Ar). Application of alkali resulted in the destruction of 55% of the quinone content introduced at the D_0 stage (**Figure 6**). Clearly the influence of alkali is not merely lignin solubilization, but also involves the destruction of quinone chromophores.

The visible absorbance difference spectra (**Figure 2**) reveal that the D(E+Ar) residual lignin is the darkest (λ = 430 nm) of all the studied lignins. According to the ^{31}P-NMR analysis (**Figure 6**), the D_0 stage residual lignin contains the highest quinone content and would be expected to have the greatest visible region absorbance (λ = 430 nm). One possible explanation for the apparent discrepancy between the ^{31}P-NMR and visible absorbance data may be that a portion of the quinone structures in the D(E+Ar) residual lignin are hydroxy substituted. Mechanical pulp [48] and model compound [24] studies have both suggested that hydroxy-quinone structures may contribute to the "alkali-darkening" phenomena. The action of alkali on quinone precursors may give rise to poly-phenolic structures, which may be subsequently oxidized to hydroxy-quinones. Clearly, this and other alkali-based reactions that may generate chromophoric structures need to be investigated.

Table III gives spectral data, acquired in this study, for a number of *para*-quinone models with various degrees of hydroxylation. The hydroxyl auxochrome causes a bathochromic shift and intensifies the π-π* transition in the quinone chromophore [33]. Furthermore, visible absorbance spectra were acquired in aqueous dioxane solution and ionization effects also intensify the molar absorptivity of hydroxy-substituted quinones [24,33]. **Table III** shows that increased hydroxyl substitution results in greater visible region (λ = 430 nm) molar absorptivity. In particular, the spectrum of tetrahydroxy-1,4-benzoquinone was found to be characterized by a broad intense absorption throughout much of the visible region. Therefore, the data in **Table III** combined with both visible absorbance difference spectra (**Figure 2**) and ^{31}P-NMR analysis (**Figure 6**) suggests that although alkali is effective at removing lignin-quinone structures, a portion of the remaining quinones may be hydroxy-substituted and display enhanced chromophoric properties.

Oxygen Effect in the Alkaline Extraction Stage. Interestingly, incorporating air in the alkaline extraction stage, D(E), results in a greater quinone content than if only alkali is applied, D(E+Ar) (**Figure 6**). The higher quinone content of D(E) relative to D(E+Ar) can tentatively be assigned to the contribution of quinone formation during alkaline oxygen bleaching [23]. **Table II** reveals that a major benefit of incorporating air in the alkaline extraction stage is greater delignification relative to the effect of alkali alone.

Table III. Visible Region Spectral Parameters for Selected Quinones.

Quinone	log $\varepsilon_{500\,nm}$ [a]	Log $\varepsilon_{430\,nm}$ [a]
1,4-Benzoquinone	0.56	1.35
2,5-Dihydroxy-1,4-benzoquinone	1.35	2.33
Tetrahydroxy-1,4-benzoquinone	2.14	2.04

[a] 90% 1,4-dioxane/10% water (v/v) solvent

When alkaline extraction is reinforced with pressurized oxygen, D(E+O), 81% of the quinone content introduced at the D_0 stage is removed (**Figure 6**). The D(E+O) residual lignin contains 0.026 mmol/g lignin (0.48 quinones per 100 C_9) more quinone structures than the brownstock residual lignin. The application of pressurized oxygen, D(E+O), *versus* air, D(E), suggests that the mechanism of quinone removal by oxygen is dependant on the oxygen concentration (pressure). The general mechanism of oxygen bleaching is known to be influenced by the concentration of oxygen [23,49].

Peroxide Effect in the Alkaline Extraction Stage. The dramatic influence of hydrogen peroxide on quinone destruction is observed in **Figure 6**. Application of hydrogen peroxide in the alkaline extraction stage, D(E+P), results in the removal of more quinone structures than were introduced at the D_0 stage. The result in **Figure 6** is consistent with the known reactivity of hydroperoxide anion towards conjugated carbonyl structures [19-21]. Although application of both hydrogen peroxide and pressurized oxygen in the alkaline extraction stage, D(E+O+P), gives a higher quinone content than hydrogen peroxide alone, D(E+P), greater delignification is a benefit of the concurrent application of both hydrogen peroxide and pressurized oxygen (**Table II**).

Conclusions

The presence of chromophores such as *ortho*- and *para*-quinones may be important contributors to brightness ceiling development during chemical pulp bleaching. This investigation further suggests that brightness ceiling values may be dependent upon the chromophore content established in earlier bleaching stages. Although quinones are only one of a number of potential chromophoric structures in kraft lignin, analyzing residual lignin quinone contents may be of valuable in understanding the origin of bleachability differences between chemical pulps.

The utility of trimethyl phosphite derivatization for investigating quinone chromophores in isolated kraft lignins was demonstrated for the first time. The results of this investigation are consistent with many of the suspected reactions of quinone structures in lignin. For example, the ability of chlorine dioxide to introduce lignin quinone structures and hydrogen peroxide to remove them was clearly observed. Further work is in progress applying ^{31}P-NMR spectroscopy towards understanding the introduction and removal of quinone chromophores in multistage bleaching sequences.

Acknowledgments

The authors wish to thank Drs. McDonough, Dimmel, and Lucia for guidance. Financial support from the Institute of Paper Science and Technology (IPST) and its member companies is gratefully acknowledged. Portions of this work were used by M.Z. and T.R. as partial fulfillment of the requirements for the Ph.D. degree at IPST.

Literature Cited

[1] Glasser, W.; Hon, D. *Poly.-Plastics Tech. Eng.* **1979**, *12*, 159-179.
[2] Imsgard, F.; Falkehag, S.; Kringstad, K. *Tappi J.* **1971**, *54*, 1680-1684.
[3] Polcin, J.; Rapson, W. *Pulp Paper Mag. Can.* **1971**, *72*, 80-91.
[4] Lonsky, L.; Lonsky, W.; Kratzl, K. *Monat. Chem.* **1976**, *107*, 685-695.
[5] Pedneault, C.; Robert, S.; Pellerin, C. *Pulp Paper Can.* **1997**, *98*, 51-55.
[6] Furman, G. S. Ph. D. Thesis, Institute of Paper Chemistry, 1986.
[7] Furman, G.; Lonsky, F. *J. Wood Chem. Tech.* **1988**, *8*, 191-208.
[8] Furman, G.; Lonsky, F. *J. Wood Chem. Tech.* **1988**, *8*, 165-189.
[9] Spittler, T.; Dence, C. *Sven. Pap.* **1977**, *9*, 275-284.
[10] Polcin, J.; Rapson, W. *Pulp Paper Mag. Can.* **1971**, *72*, 69-80.
[11] Dence, C. In *Pulp Bleaching - Principles and Practice*; Dence, C., Reeve, D., Eds.; TAPPI Press: Atlanta, GA, 1996, pp 125-159.
[12] Brage, C.; Eriksson, T.; Gierer, J. *Holzforschung* **1991**, *45*, 23-30.
[13] Brage, C.; Eriksson, T.; Gierer, J. *Holzforschung* **1991**, *45*, 147-152.
[14] Lindgren, B. *Sven. Pap.* **1971**, *3*, 57-63.
[15] McKague, A. B.; Kang, G. J.; Reeve, D. W. *Holzforschung* **1993**, *47*, 497-500.
[16] McKague, A. B.; Kang, G. J.; Reeve, D. W. *Nordic Pulp Paper Res. J.* **1994**, *9*, 84-87.
[17] McKague, A. B.; Reeve, D. W.; Xi, F. *Nordic Pulp Paper Res. J.* **1995**, *10*, 114-118.
[18] Ni, Y.; Shen, X.; van Heiningen, A. *J. Wood Chem. Tech.* **1994**, *14*, 243-262.
[19] Gierer, J. *Wood Sci. Tech.* **1986**, *20*, 1-33.
[20] Gierer, J. *Proceedings CTAPPI 7th International Symposium on Wood and Pulping Chemistry,* Beijing, 1993; Vol. 1, pp 301-307.
[21] Gellerstedt, G.; Pettersson, I.; Sundin, S. E. *Proceedings International Symposium on Wood and Pulping Chemistry,* Stockholm, 1981; Vol. 2, pp 120-124.

[22] Bailey, C.; Dence, C. *Tappi J.* **1969**, *52*, 491-500.
[23] Gierer, J. *Holzforschung* **1997**, *51*, 34-46.
[24] Gellerstedt, G.; Hardell, H.-L.; Lindfors, E.-L. *Acta Chem. Scan.* **1980**, *34B*, 669-673.
[25] Gellerstedt, G.; Lindfors, E. *Proceedings International Pulp Bleaching Conference*, Stockholm, 1991; Vol. 1, pp 73-88.
[26] Pepper, J.; Baylis, P.; Alder, E. *Can. J. Chem.* **1959**, *37*, 1241.
[27] Gellerstedt, G.; Pranda, J.; Lindfors, E.-L. *J. Wood Chem. Tech.* **1994**, *14*, 467-482.
[28] Froass, P.; Ragauskas, A.; McDonough, T.; Jiang, J. *J. Wood Chem. Tech.* **1996**, *16*, 347-365.
[29] *100 and More Basic NMR Experiments: A Practical Course;* Braun, S.; Kalinowski, H.-O.; Berger, S.; VCH: New York, NY, 1996, pp 125-128.
[30] McDonough, T. *Proceedings TAPPI Pulping Conference,* 1996; pp 200-206.
[31] McDonough, T.; Rawat, N.; Turner, M. *51st APPITA Annual General Conference*, 1997; Vol. 1, pp 255-260.
[32] Senior, D.; Hamilton, J.; Ragauskas, A.; Froass, P.; Sealey, J. *International Pulp Bleaching Conference*, Atlanta, 1996; pp 261-273.
[33] Berger, S.; Rieker, A. In *The Chemistry of Quinonoid Compounds*; Patai, S., Ed.; John Wiley and Sons: New York, NY, 1974; Vol. 1, pp 163-229.
[34] Ramirez, F. *Pure Appl. Chem.* **1964**, *9*, 337-369.
[35] Ramirez, F.; Bhatia, S.; Patwardhan, A.; Chen, E.; Smith, C. *J. Org. Chem.* **1968**, *33*, 20-24.
[36] Abdou, W.; Mahran, M.; Hafez, T.; Sidky, M. *Phosph. Sulfur* **1986**, *27*, 345-353.
[37] Medvecz, P. M.Sc. Thesis, Institute of Paper Chemistry, 1987.
[38] Ramirez, F.; Patwardhan, A.; Desai, N.; Heller, S. *J. Am. Chem. Soc.* **1965**, *87*, 549-553.
[39] Schwetlick, K.; Habicher, W. *Adv. Chem. Ser.* **1996**, *249*, 349-358.
[40] Lebo, S.; Lonsky, W.; McDonough, T. *International Pulp Bleaching Conference*; TAPPI Press: 1988, pp 247-251.
[41] Lebo, S. Ph.D. Thesis, Institute of Paper Chemistry, 1988.
[42] Lebo, S.; Lonsky, W.; McDonough, T.; Medvecz, P.; Dimmel, D. *J. Pulp Pap. Sci.* **1990**, *16*, J139-J143.
[43] Argyropoulos, D.; Heitner, C.; Morin, F. *Holzforschung* **1992**, *46*, 211-218.
[44] Argyropoulos, D.; Heitner, C.; Schmidt, J. *Res. Chem. Int.* **1995**, *9*, 263-274.
[45] Scheirs, J.; Pospisil, J.; O'Connor, M.; Bigger, S. *Adv. Chem. Ser.* **1996**, *249*, 359-374.
[46] Aksnes, G.; Aksnes, D. *Acta Chem. Scan.* **1964**, *18*, 1623-1628.
[47] Iiyama, K.; Nakano, J. *Japan Tappi* **1973**, *27*, 530-542.
[48] Leary, G.; Giampaolo, D. *9th International Symposium on Wood and Pulping Chemistry,* Montreal, 1997; pp D4.1-D4.6.
[49] Yokoyama, T.; Matsumoto, Y.; Yasumoto, M.; Meshitsuka, G. *J. Pulp Paper Sci.* **1996**, *22*, J151-J154.

Chapter 27

Chemical Characteristics of Lignins Extracted from Softwood TMP after O_3 and ClO_2 Treatment

D. R. Robert[1], M. Szadeczki[2], and D. Lachenal[2]

[1]CERMAV/CNRS Domaine Universitaire, BP 53, 38041 Grenoble cedex 9, France
[2]EFPG Domaine Universitaire, BP 65, 38402 St. Martin d'Hères cedex, France

This study investigates the role of ozonation in improving the physical properties of softwood thermomechanical pulp (TMP). Lignin fractions were extracted from pulp before and after ozonation and characterized by IR, ^{13}C NMR and GPC. Results show the influence of the extraction method on the lignin structure, with enzymatic hydrolysis of the pulp giving a lignin fraction richer in etherified units but more contaminated with polysaccharides than acidolysis. The number of carboxyl groups formed in lignin was also studied, since this parameter has been correlated with pulp mechanical properties. ClO_2 treatment of TMP, compared to ozonation, resulted in an increase of COOR(H) groups and the formation of muconic structures clearly identified on NMR spectra.

For economical and ecological reasons there has been a spectacular increase in the past few years in the use of recycled fibers. Consequently, upgrading recycled fibers has become necessary. Recycled pulps usually consist of a mixture of chemical and mechanical pulps. It has been shown that the overall quality of recycled pulp is governed by the properties of its lignin-rich fibers (mechanical pulp fraction). Any change in these properties by lignin modification (oxidation) or lignin removal has a direct impact on the quality of the entire pulp (*1*). Attempts to upgrade mechanical pulp by ozonation have been performed in the late 70's, which included pilot scale operations (*2*). No commercial applications are currently practiced, however, mainly because of economics. The mechanism by which mechanical properties are improved by ozonation is believed to be due to carboxyl groups formation on lignin and thereby an increase in hydrophilicity (*3*). Since it is known that ozone reacts 100 to 1000 times more rapidly with lignin than with cellulose (*4*), it is thought that ozonation of a recycled pulp will primarily react with the lignin in the mechanical pulp fraction and thus improve the overall strength properties.

The purpose of this study was to investigate the structural changes in the lignin of a mechanical pulp pretreated by ozonation, and to correlate these changes to the strength improvement of the pulp. Results were compared with another bleaching chemical, chlorine dioxide. Lignins extracted from softwood thermomechanical pulps, before and after O_3 and ClO_2 treatments, were characterized by SEC (Size Exclusion Chromatography), IR and ^{13}C NMR. Since lignin must be extracted from the pulp before analysis, the possible effects of the extraction procedure upon lignin structure was a major concern.

Effect of O_3 and ClO_2 treatment on mechanical strength of TMP

Ozonation and ClO_2 treatment of pulp. Ozonation was performed at high consistency (35-40%) at room temperature in a rotating spherical glass reactor. Ozone was produced from oxygen in a G21 type Ozonia generator. The O_3 concentration in oxygen was 90 mg/L NTP with a gas flow of 0.8 L/min. The residual ozone concentration was determined by iodometric titration.

ClO_2, containing no molecular chlorine, was produced by reaction of a concentrated solution of sodium chlorite with sulfuric acid. Treatment of pulp with ClO_2 was performed at 5% consistency and 70°C in polyethylene bags. The pulp samples were preheated in a water bath before introducing the ClO_2 solution.

Improvement of TMP mechanical properties after ozonation and ClO_2 treatment. As expected, an improvement in the breaking length and burst index of TMP pulp was observed, with increasing levels of ozonation and chlorine dioxide giving better strength properties (Table I).

Table I. TMP mechanical properties after O_3 and ClO_2 treatments

% O_3	0	0.73	2.5	4.27	5.0
break. l. , km	3.41	4.05	4.74	5.20	5.12
burst ind., kPa m^2/g	2.10	2.20	2.75	2.80	3.30
% ClO_2	0	2.0	4.0	7.0	10.0
break. l, km	3.41	3.92	3.78	4.32	4.84
burst ind. kPa m^2/g	2.10	2.20	2.30	2.60	2.80

In the case of ozonation (Figure 1), an improvement in strength was observed when the ozonated thermomechanical pulp was mixed with a chemical pulp, provided that the mechanical pulp represented a significant fraction of the mixture (at least 10%) (*1*).

Influence of extraction methods upon lignin structure

The overall molecular and macromolecular structure of the native wood cell wall, and of the corresponding thermomechanical pulp, should be quite similar. In both cases the lignin macromolecule is partly covalently bonded to the carbohydrate matrix. Release of this lignin from the pulp involves the splitting of these bonds, which should also introduce chemical changes in the lignin. The chemical structure of the lignin recovered will thus reflect both its original structure in the pulp and the extraction-derived changes. The choice of an extraction method which minimizes structural changes in lignin, or at least gives known changes, is important. Among the most efficient methods two were selected: enzymatic hydrolysis of carbohydrates and acidolysis.

Extraction by enzymatic hydrolysis of carbohydrates. In brief, the cellulase-hemicellulase mixture totally hydrolyses the cellulose-hemicellulose matrix. Following the method proposed by Pan et al. (*5*), 5g of industrial softwood TMP, some characteristics of which are given in Table 1, were ultraground at room temperature for 2 days and put in a 100 ml acetate buffer solution, 0.5M and pH 4.6, with the enzyme, 5mg per buffer ml. The enzyme was a commercial cellulases-hemicellulases mixture, ONOZUKA R-10, highly contaminated by carbohydrates moieties (43.3%). After hydrolyzing for 72 hours at 37.5°C in darkness the residue was centrifuged and hydrolyzed for another 48 hours, and then water washed and extracted with a 9/1 dioxane/water mixture. The solution was concentrated under vacuum and the lignin precipitated in water with 2% sodium sulfate at 4°C. An 70% extraction yield was achieved (after correction for proteins and carbohydrates) for both lignin extracted from untreated TMP, designated as TMPe, and lignin extracted from ozone bleached TMP, designated as TMPOe.

Extraction by acidolysis. According to the method proposed by Gellersted et al. (*6*), 50 g of TMP were refluxed for 2 hours in 1.5 l of dioxane/water solution, 82/18, containing 0.1 mole of HCl. Dilution with water and concentration at 40°C under vacuum was conducted until all dioxane was eliminated. The precipitated lignin was washed and vacuum dried. Extraction yield was 30% for lignin extracted from untreated pulp, designated as TMPa, 50% for lignin extracted from ozone treated pulp, designated as TMPOa, and 80% for lignin extracted from pulp treated with ClO_2, designated as TMPCla. Yields are based on the lignin content of the pulp, estimated by the Klason method.

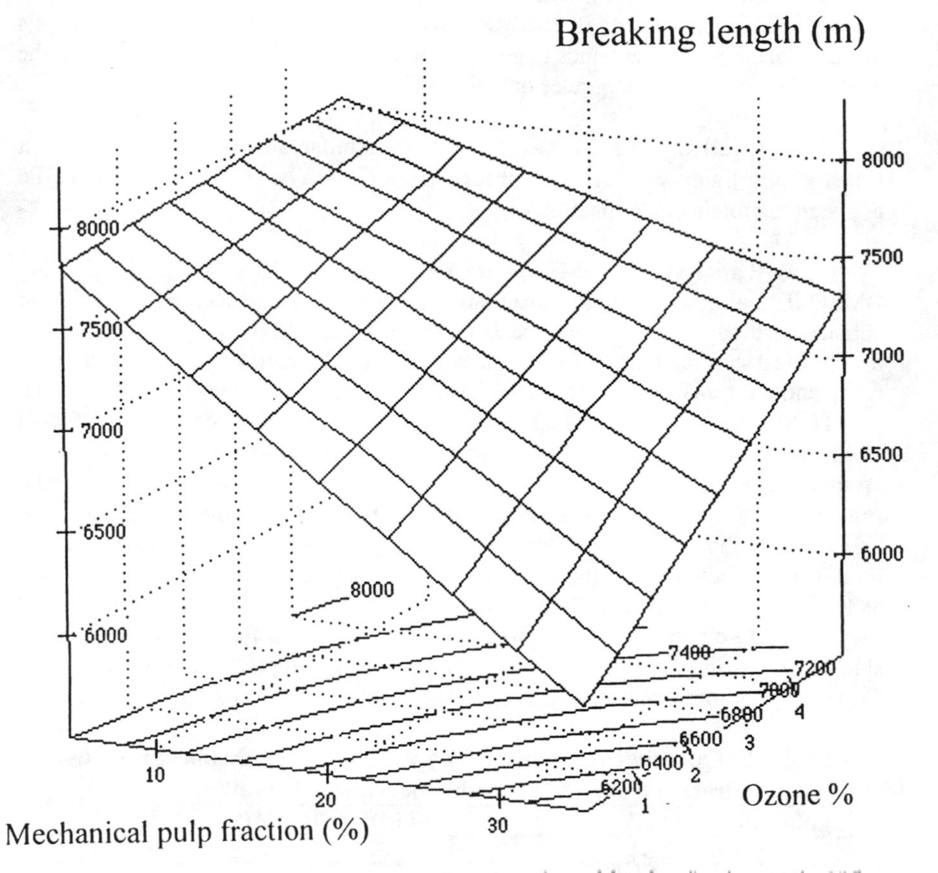

Figure 1. Change in breaking length of a blend of chemical and thermomechanical pulps after ozonation of the TMP fraction.

Structural differences between TMPe and TMPa lignins. Structural differences were investigated by SEC, ^{13}C NMR and IR spectroscopies. Sugars analysis showed a significantly higher sugar content in TMPe, 6.9%, compared to TMPa, 0.1%.

Size Exclusion Chromatography. SEC was carried out on two PL Gel Mixed D columns in series with a 200 to 400000 polystyrene molecular weight distribution. A solution of 10 mg of lignin in dimethylformamide containing 0.01M $NaNO_3$ was injected. Detection was by UV at 280 nm. SEC curves reveal that TMPe lignin has a smaller elution volume than TMPa, which is consistent with an acidolysis mechanism cleaving ether bonds to give partial depolymerization (*7*). Conversely, the enzyme treatment likely degrades only the cellulose-hemicellulose matrix.

IR spectra. The IR spectra are very similar and present the classical fingerprint of lignin structure, except for a more intense band at 1655cm^{-1} in TMPe assigned to protein contamination (*8*).

NMR analysis. ^{13}C NMR spectra were recorded at 75.48 MHz, 323 K, using DMSO-d_6, and qualitative and quantitative NMR analysis performed according to classical methods (*9*). The central peak of the solvent signal at 39.6 ppm was used as the chemical shift reference. An inverse gate sequence with 10 s pulse delay was used for quantitative analysis. For the DEPT sequence, which gives selectively CH, CH_2 or CH_3 signals, a 1/2J = 0.0033 s was chosen as the refocusing delay. Signal intensities were estimated by comparison of their integral to the integral of the six aromatic carbons, assuming one neglects the contribution of vinylic carbons. Some results from the quantitative analysis are given in Table II. Spectra in Figure 2 show, as expected, that the carbon skeleton of both lignins exhibit far less structural modifications than lignins extracted in the same conditions from chemical pulps, such as kraft pulp lignin (*10,11*). In particular, C6-C3-O oxygenated side chains, found mainly in β-O-4 structures, are well preserved as seen from the relative intensity of signals 17, 18, and 21 in the range 60 to 90 ppm, assigned predominantly to the Cβ, Cα and Cγ carbons, respectively, in β-O-4 units.

Table II. Quantitative estimation by C-13 NMR of chemical functions in lignins extracted from TMP (number per aromatic unit [a]).

lignin ext.	TMPe	TMPOe	TMPa	TMPOa	TMPCla
C-O aliph. 60-90 ppm	2.64	2.80	1.80	1.85	2.80
-OCH_3	0.98	0.95	0.92	0.88	0.76
COOR(H)	0.16	0.28	0	0.25	0.60

[a] precision : +/- 5%

There are two main structural differences between TMPe and TMPa lignins. The first one is the presence in TMPe of a broad signal centered at 170 ppm, signal 3, assigned

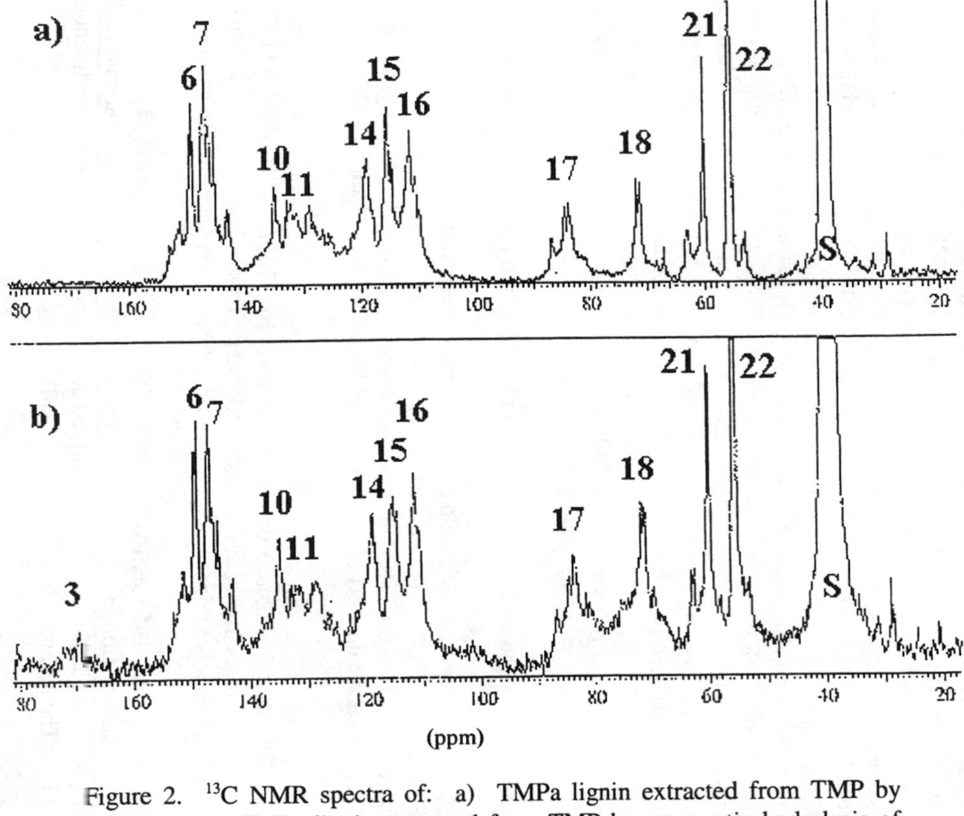

Figure 2. ^{13}C NMR spectra of: a) TMPa lignin extracted from TMP by acidolysis; b) TMPe lignin extracted from TMP by enzymatic hydrolysis of carbohydrates. S, solvent = DMSO-d_6.

to carboxyl functions and representing 0.16 COOR(H) groups per aromatic unit. As there was no chemical pretreatment of the pulp the presence of signal 3 in TMPe, and the absence of peak 3 in TMPa, can only originate from the extraction procedure, the importance of which is thus clearly evidenced. These carboxyl groups are due to contamination from the pulp and/or from the enzyme, which contains carboxyl functions and is itself highly contaminated by carbohydrates. The other difference is of quantitative nature and concerns the higher relative intensity of the group of signals 17, 18 and 21 assigned to C-O aliphatic carbons, estimated as 2.6 carbons per aromatic unit in TMPe and 1.8 carbons per aromatic unit in TMPa. The six cellulose-hemicellulose carbons signals which range between 60 and 100 ppm and indicate carbohydrate contamination, contribute to only part of the C-O aliphatic signals between 60 and 90 ppm in TMPe (the sugar content in this lignin is limited to 6.9%). Thus, the higher quantity of aliphatic C-O carbons in TMPe compared to TMPa is probably due to a substantially higher amount of β-O-4 structures in TMPe, which supports the observation made on the GPC curves. Additional confirmation is found in the aromatic region of the spectra: the relative intensity of signal 6 (C3 in etherified guaiacyl) compared to signal 7 (C4 in etherified and non-etherified guaiacyl units and C3 in non-etherified guaiacyl units), and the relative intensity of signal 10 (C1 in etherified β-O-4) compared to 11 (C1 in β-O-4 non-etherified), confirms that there are more β-O-4 linkages in TMPe than in TMPa. Again, this indicates that cleavage of β-O-4 linkages occured during acidolysis.

Effects of ozonation and ClO2 treatment of TMP upon lignin structure

TMPOe was obtained in 80% yield by the enzymatic method from ozonated pulp, and TMPOa and TMPCla were obtained by acidolysis respectively in 65% yield from ozonated pulp and in 80% yield from pulp treated with chlorine dioxide. Again, lignin yield represents the quantity of extracted lignin based on the Klason lignin content of the pulp.

GPC curves. Figure 3 shows similar curves for TMPa and TMPOa. TMPCla curve is shifted to the right of the two previous ones. This indicates that ozonation of pulp does not depolymerize the lignin significantly whereas ClO_2 treatment does.

IR spectra. TMPOe, TMPOa and TMPCla give a larger band at 1713cm^{-1} than is found in lignins from untreated pulp. This band can be assigned to muconic acids and quinones which are known to be formed during ozonation and ClO_2 treatment of pulp (*12*). The strongest band is found in TMPCla lignin.

^{13}C NMR analysis. Compared to spectra in Figure 2, NMR spectra in Figure 4 show that ozonation does not significantly alter the overall lignin structure, the main change being the creation of a substantial amount of carboxyl groups. Surprisingly,

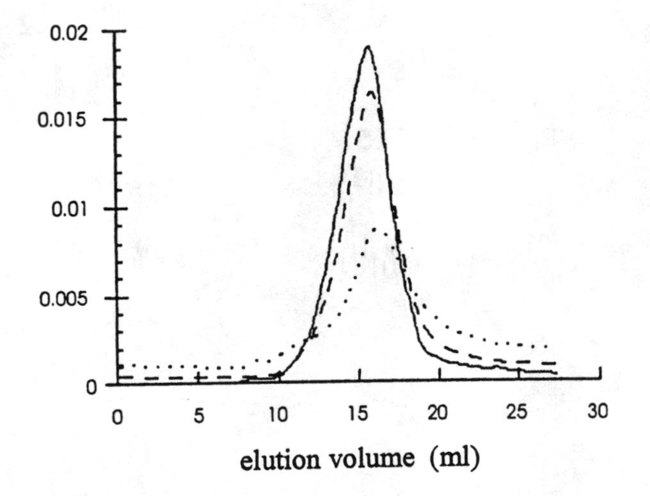

Figure 3. GPC curves of: TMPa (—); TMPOa (---); and TMPACla (...) lignins.

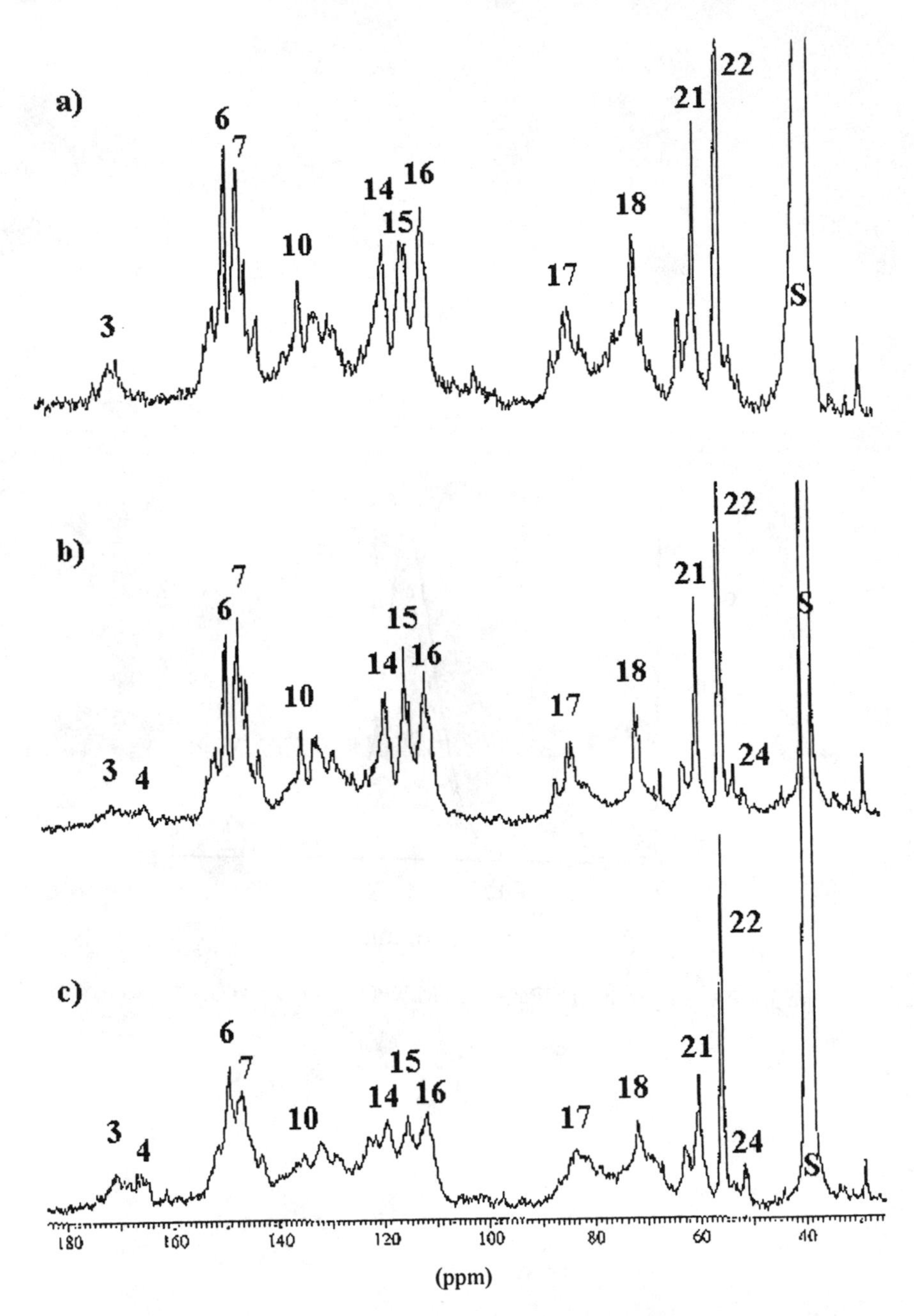

Figure 4. ^{13}C NMR spectra of: a) TMPOe lignin extracted by enzyme hydrolysis from ozonation TMP; b) TMPOa lignin extracted by acidolysis from ozonated TMP; and c) TMPACla lignin extracted by acidolysis from TMP treated with ClO_2. S, solvent = DMSO-d_6

there is only one COOR(H) group, signal 3 centered at 171 ppm in TMPOe, and two carboxyl groups, signals 3 and 4 centered at 171 and 166 ppm in TMPOa. Since the same ozonation sequence was used this difference can only originate from the extraction method, the importance of which is thus again clearly evidenced. Carboxyl groups between 169 and 174 ppm belong to saturated aliphatic chains and between 164 and 168 ppm to conjugated carboxyl groups (*13*). The carboxyl signal at 171 ppm in TMPOe is likely due to both enzyme and carbohydrates contamination and to oxidation by ozone of the lignin saturated side chain. The signal centered at 166 pm is tentitavely assigned to carboxyl groups both in benzoic structures and in muconic acid or ester structures. These structures are confirmed by the presence of signal 24 at 51.6 ppm, identified with the DEPT sequence as a OCH_3 signal, such as muconic methyl ester (*14*). A tentative explanation for the absence of a carboxyl signal of muconic origin in TMPOe might be that these structures belong to low molecular weight lignin fractions and thus are soluble and lost in the water filtrates during enzymatic hydrolysis, while during acidolysis no lignin fraction is lost. In TMPCla, lignin extracted by acidolysis from TMP treated with ClO_2, the two same carboxyl signals 3 and 4 as in TMPOa are present but at greater intensities, 0.60 carboxyl groups per aromatic unit instead of 0.25 in TMPOa. This difference is confirmed by the relative intensities of the corresponding methoxyl signal assigned to muconic structures which are significantly larger in TMPCla than TMPOa. The larger amount of muconic carboxyl groups in TMPCla compared to TMPOa might be explained by the destruction of muconic structures occuring by ozonation of double bonds, whereas ClO_2 does not react with vinylic bonds.

We found 7.65% sugar moieties in TMPOe, and smaller amounts in LTMPOa, (0.52%) and LTMPCla (1%). Since carbohydrates signals fall in the 93 to 97 ppm range, this may explain to some extent the higher amount of aliphatic C-O carbons found in TMPOe lignin (signals 17, 18, and 21), around 2.8 per aromatic unit, compared to 1.8 in TMPOa. Likewise, the untreated pulp lignins extracted by the enzymatic method are richer in C-O aliphatic structures than lignins extracted by acidolysis. It is interesting to observe that TMPCla lignin also has a higher amount of etherified stuctures, 2.8 C-O aliphatic carbons in the 60-90 ppm range vs. 1.80 in TMPOa, revealing less splittings of the lignin side chain and β-O-4 bonds during pretreatment of the pulp by ClO_2. The line widths at mid-height are broader for NMR signals of TMPCla lignin, which might indicate a more reticulated network such a higher number of condensed units.

Origin of pulp strength improvement. As discussed previously it is generally believed that ozonation improves pulp quality by introducing carboxyl groups in lignin (*3*). Similar improvements in pulp properties are observed (Table III) after O_3 and ClO_2 treatment compared to untreated pulp, although the number of carboxyl groups is much higher in TMPCla than in TMPOa lignins.

Table III. Effects of chemical treatments of softwood TMP on pulp breaking length and lignin carboxyl content.

lignin samples	untreated TMP	O_3 (5%) treated	ClO_2 (10%) treated
lignin content[a], %	29	24.2	23.0
breaking length, km	3.86	5.12	4.84
carboxyl groups in lignin[b]	0.00	0.25	0.60

[a] Determined by Klason method ; [b] Detemined by NMR and given per aromatic unit.

Thus, one can conclude that there is no direct relation between strength improvement in the pulp and an increase in COOR(H) groups on lignin for this pulp. Between untreated TMP and TMPOa treatment the breaking length increases with increasing carboxyl groups number, but between TMPOa and TMPOCla, with nearly the same breaking length values, TMPOCla has twice as many COOR(H) groups. Also, the strength improvement is not due to depolymerization of the lignin by O_3 or ClO_2 treatments, since the GPC curves of the lignins are very similar before and after these treatments.

This study also suggests that the pulp lignin content might play a more important role than the number of carboxyl groups on fiber flexibility (and on fiber binding ability). Indeed, after O_3 and ClO_2 treatments a decrease of the lignin content occurred to about the same level, 24.2% vs 23.0%, and an increase of the breaking length to about the same level (5.12 vs. 4.84), although the number of COOR(H) groups is more than twice as high after ClO_2 treatment compared to ozonation. However, the use of other chemicals capable of removing lignin by other reactions would be necessary to confirm this hypothesis.

Literature Cited

1. Szadeczki, M.; Lachenal, D.; Costes, C.; Thieblin, E.; Bailli, A. *Progress in Paper Recycling* **1997**, *6(3)*:22-28.
2. Ericksen, J.T.; Loräs, V.; Soteland, N. *Paper.* **1977**. 31-32 and 70.
3. Vasudevan, B.; Panchapakesan, B.; Gratzl, J.S.; Holmbom, B. *Tappi Pulping Conference.* **1987**. *Vol. 3,* 517-523.
4. Nompex, Ph.; Doré, M.; De Laat, M.; Legerbe, B. *Ozone Sci. Eng.* **1991**, *13*, 265-286.
5. Pan, X.; Lachenal, D.; Lapierre, C.; Monties, B.; Neirinck, V.; Robert, D. *Holzforschung* **1994**, *48(5)*:429-435.
6. Gellerstedt, G.; Pranda, J.; Lindfors, E-L. *J. Wood Chem. Technol.* **1994**. *14(4)*:467-482.

7. Lundquist, K. in *Methods in Lignin Chemistry*, Lin, S.Y., Dence, C.W., Ed., Springer Series in Wood Science, Springer Verlag: Berlin, Heidelberg. **1992**. pp 289-300.
8. Hortling, B.; Turunen, E.; Sundquist, J. *Nordic Pulp and Paper Res. J.* **1992**. *3*:144-151.
9. Robert, D. in *Methods in Lignin Chemistry*; Lin, S.Y., Dence, C.W., Ed.; Springer Series in Wood Science; Springer Verlag: Berlin, Heidelberg. **1992**. pp 250-273.
10. Chirat, C.; Lachenal, D. *Holzforschung*. **1994**. *48*:133-139.
11. Robert, D.; Neirinck, V.; Chirat, C.; Lachenal.; Heuts, L.; Gellerstedt, G. *3rd European Workshop on Lignocellulosics and Pulp*; Stockholm, SE. **1994**. pp 89-92.
12. Gierer, J. *Holzforschung*. **1990**. *44(5)*:387-394.
13. Chen, C.L.; Chua, M.G.S.; Evans J.; Chang, H-m., *Holzforschung* **1982**. *36*:239-247.
14. Evtuguine, D.V.; Robert D. *Wood Sci. Technol*. **1997**. *31*:423-431.

Author Index

Subject Index

B

C

D

E

H

I

J

K

L

M

N

O

P

R

S

U

V

W

Y

Z